Beyond Collapse

Surviving and Rebuilding
Civilization From Scratch

By T. Joseph Miller Jr.

To Sharon - my helpmate, best critic and 6' tall kitten.

Also, with gratitude to James Rawles (survivalblog.com),
Mac Slavo (shtfplan.com),
the folks at shtfblog.com,
and various other places…
Thank you for letting me both ask and answer
a whole lot of "odd questions".

Copyright ©2012, T Joseph Miller Jr.

Please see page ix for Copyright Notice, and
for all rights and restrictions.

Cover Photo: Small debris from Japan, as found on a northern Oregon beach in early 2012.
(© 2012 T. Joseph Miller Jr)

Table Of Contents

Preface   **i**
    Why Did You Get This For Free?   **v**
    (electronic version only)
    Why Bother?   **vi**
    Wait - You're Serious, Aren't You?   **viii**
    A Note About "The Rapture"   **xiii**
    Dumb Questions, Hard Answers   **xiv**
    So What If Nothing Happens?   **xv**

Chapter 1 | Things To Do Beforehand
    Who Are You?   **2**
    Where Do You Live?   **13**
    Who Are You With?   **26**
    Stay Put Or Bug Out?   **31**
    Refugee Preparation 101   **36**
    How Will you Keep It Quiet?   **44**
    What Do You Need?   **49**
    What Do You Know?   **70**
    Who Do You Know?   **76**
    How To Stock Up   **86**
    Things To Do While Waiting   **89**

Chapter 2 | Bracing For Impact
    When (And How) Does It Actually Happen?   **92**
    Do We Stay Or Do We Go?   **98**
    Batten Down The Hatches!   **100**
    Run To The Hills!   **110**

Chapter 3 | Immediately Afterward
    Load Up While You Can!   **126**
    Getting To Know Your Neighbors, Again   **128**
    Establishing Initial Trust   **135**
    Scavenging And Procurement   **136**
    Securing Home And Community   **143**
    Health And Healing   **152**
    Zombies And Refugees   **161**
    So, What's For Dinner?   **166**
    Anatomy Of Post-Collapse Criminals   **170**
    When The Government Comes Knocking   **178**
    Civil Duties And Tips   **185**
    Internal Strife And Troubleshooting   **188**
    Obtaining And Filtering News   **194**
    Going It Alone   **196**
    Conclusion   **198**

Chapter 4 | In The Short Term
    So You Made It This Far...   **201**
    Around The House   **202**
    Food, Food, And More Food   **216**
    Safety And Fire Concerns   **226**
    Scavenging For More   **229**
    Electricity   **236**
    Barter And Markets   **238**
    The Psychology Of Post-Collapse Living   **247**
    Gender, Romance, And Babies   **252**
    Civil War   **265**
    Making It A Permanent Thing   **261**
    Thinking Ahead   **276**

Chapter 5 | In The Long Term
    You're Almost There?   **286**
    Cleaning House   **288**
    Governance And Leadership   **293**
    Defense And Use Of Force   **301**
    Money And Economics   **305**
    Diplomacy   **316**
    Civil Engineering   **325**
    Entertainment And Diversions   **343**
    Education For Young And Old   **348**
    Expansion And Confederation   **354**

Chapter 6 | The Distant Future
    ...Now What?   **359**

Chapter 7 | Recipes
    Household Chemistry   **364**
    Medicines   **377**
    Consumables And Vices   **380**
    Weaponry   **386**
    Government   **389**

Appendix A | Shopping Lists   **408**
Appendix B | Scrounge Lists   **424**
Appendix C | Your Library   **429**

Disclaimer

There are some portions of this book which will discuss things that may be construed as illegal or negligent if performed during peaceful, civilized times. Note that you should only perform such acts or attempt such recipes only under fully legal circumstances (with permission and proper permits/licenses if they apply). Otherwise, you should only wait to do them at a time when law enforcement no longer exists. Anything you do which comes from this book you do at your own risk. The author and publisher will assume no liability, warrantee, of guarantee from any part of this book. If you hurt yourself, hurt others, damage property, get arrested/prosecuted, or in general, if anything negative happens to you at all when performing or attempting anything in this book? You are on your own, and the results will be assumed to be your fault alone.

Long story short? Don't be stupid, and don't do anything stupid.

Preface

"We must prepare for this many-sided inconvenience."
    -Pope Paul VI

I've leafed through a lot of survival books (both good and bad), 'post-apocalypse' preparation books and various other tomes that proclaim to one and all that their particular writings will give you everything you need to survive. I've also read voraciously through numerous novels that deal with similar themes during the course of my research.

To be honest, few of them will make sense to the average human being. Fewer still are realistic. Why, you ask? Well, not everyone has the money, time and effort to be a martial arts master, become proficient with a variety of weapons, gather a force of battle-hardened veterans and, oh yeah – learn trauma medicine, engineering, animal husbandry, agriculture and chemistry. Most, at best, only help you survive for long enough until the initial chaos is past, then dumps you off to your own devices from there. Worse still, a disturbing number of the guides out there will have you looking and acting like a wild-eyed conspiracy theorist, a commando wannabe, or adopting some other personality that you definitely do not want or need. Far too many assume that you have (or are somehow looking to acquire) the advanced skills and tactics training that simply do not exist in the average person.

Let's face it, you and I are ordinary people. We have jobs, kids, a spouse, hobbies, commitments, perhaps church and all the other things called life - which in turn eats into our time and money. That said, if you intend to survive chaos, collapse, or worse, you're going to have to have *something* to start with, or else you're as good as dead and/or exploited (though not necessarily in that order). To that end (that is, avoiding death and exploitation), I'd like to present a more common-sense guide for the ordinary person. I want to help you prepare for upheavals and chaos, protect your family and perhaps even live relatively cozy, in an age where the modern world may well cease to be. This won't require spending hours on end at the gun range, building a bunker, crawling around in a camouflage suit, or blowing tens of thousands of dollars at your local army surplus store. There is also no need to cast aside your entire way of life and go live in a cabin "off-grid", either. Oh and this time you will only use tin-foil for cooking, not head-gear. I also want to keep the lingo and acronyms to an absolute minimum. Spewing such things as

"TEOTWAWKI", "BOL" and "SHTF" is a great shorthand for folks who are really into preparing for collapse, but they're a mouthful when you're trying to teach ordinary people how to live through (and overcome) the unthinkable.

I will get this out right now: I am not an expert outside of my particular professional engineering field, nor will I claim to be one. I don't know anyone in the CIA, NSA, Special Forces, or any survival instructor. I only have experience gained in my years of preparing, hunting, watching global trends, camping, standard survival training courtesy of the US Air Force, engineering, psychology skills that rubbed off from a family of them, a strong sense of rigor and logic developed in Catholic School and a knack for heavy research. I spent my childhood and adult life traveling, living in urban, suburban and rural settings. During that time, I've studied a whole lot of different people and cultures, as well as concepts related to them. To that end, I want to take everything I know and can conceive of and bring it to you in a concise but useful form.

Something else you should know: Anyone out there who claims to be an expert in any sort of global post-collapse/post-apocalyptic survival is either a liar or they secretly own a working time machine. One can certainly gain expertise in living "off-grid" (that is, not being hooked up to municipal or rural power, water, sewer, etc), or living in a pre-industrial fashion. There are precious few English-speaking people who have lived through wartime conditions that would very closely mimic a collapse of civilization (albeit on a localized or regional level) such as Argentina, Bosnia-Herzegovina, Somalia... places in recent memory that have been plunged into a living hell in their own right. However, the modern "off-grid" expert still needs to get parts and supplies from town to build and keep his various gizmos going. The pre-industrial guy (even if Amish) still relies to an extent on modern technology to provide tools, printed books, a police force, advanced medical care and other bits that a post-collapse society simply will not have. The folks who lived through hellholes like 1980's Argentina or 1990's Bosnia knew full well that there was still a functioning world out beyond their borders and many of those survivors still relay their gratitude for international relief efforts that came from foreign (to them) countries.

Quite frankly, what I'm attempting here is to help you survive a possibility that no sane human has ever attempted to study with enough authority to claim an expertise - myself included. Those who have made the biggest inroads into this realm are mostly people who do it as part of a military goal (e.g. SERE/SEARS training in the US Military), which is a bit biased towards, well, military goals and usually assumes that there is always help waiting if you can just get to it. Others have unearthed and distributed some rather valuable information, but mix it in with tinfoil conspiracies or even racial/ethnic hatreds. Then, there are folks who are actually looking forward to societal collapse, so that they can somehow arise as self-appointed leaders in some new pseudo-utopian (or worse, "Mad Max" style) society that they've imagined for themselves.

Fortunately, there are enough sane people out there who have given this matter serious thought and put some serious work into it. Some even host excellent websites, some of which you should visit, that you can go online to and happily print off the very useful information to be found there. Problem is, few have ever really attempted to correlate or collate this mass of useful information into a book that you can use. Some sites even recommend that you keep it on an eBook reader (the site owner is assuming that the eBook and tethered computer will hold up for decades on end, I guess. As an engineer, I can tell you 99.9% of them won't last beyond the 5-7 year mark, even under perfect global conditions.)

The idea of this book on the other hand is simple: Get you up to speed on some basic bits to stay alive and rebuild a working society. I want to get you started in gathering needed provisions (and references) to survive a collapse and its immediate aftermath, but then give you a series of resources that you can refer to, in order to help you along after the dust settles. After all, you probably won't need or be able to use it all immediately. Things in here may be a bit intellectual or crazy at times and I apologize in advance if it sounds that way. On the other hand, I want to drive the point home; and I want to be as precise and as factual as I can. Most of all, I want this book to do more than just help the individual survive. It takes a community to rebuild a civilization and that community is going to need help and at least some practical guidance.

Unlike most survival/preparation resources, I'm going to assume here that civilization will break down completely and will likely take at least a century or more to return to the level it is now. In such a situation, that electrical generator, which too many experts recommend, will end up being a glorified lawn ornament (unless, of course, you feel like making/storing an ocean of fuel and storing a mountain of spare parts...) The idea of having a small backyard garden is a good start, but it's not going to feed you and your family all by itself. Expensive water filters are a good convenience while things are all chaotic, but eventually the filter elements are going to wear out - and you're stuck with having to make your own clean water from then on. Solar panels can work for up to and beyond three decades, but the inverter, batteries and electronics you plug into it probably won't. Anything that relies on batteries (even rechargeable ones) are guaranteed to be worthless once the battery chemicals wear out and there are no more to be had. Guns? A good idea for the more chaotic times; but eventually, the bullets will run out, leaving you (and everyone else) with, at best, a bunch of clumsy clubs made from steel and wood. Any advice that relies on buying and storing something should only be counted as either temporary or the (sometimes literal) seeds of renewable items, be it tools, weapons, or wheat. Some items will be vital in spite of their being temporary, but some will help you make a permanent means of living long-term after it all settles down.

We're going to break it down into some chapters, based on...

- Why you should be reading this book (and more importantly, doing something about it),
- Things to do beforehand,
- Things to do and expect when the balloon goes up,
- Things to do and expect immediately afterward *(for up to six months after civilization crashes),*
- Things to do short-term *(for the first two years),*
- Things to do long-term *(everything beyond that second year),*
- The distant future *(where we get all misty-eyed – or, well, how to really build a new civilization),*
* Recipes for critical items *(for all the vital bits that you will want to use),*
- Appendices *(Resources: shopping lists, post-collapse scrounging lists, websites to visit and print off, etc).*

The good news is, you don't have to read the whole book right now (and given its size, who would really want to?) You can safely read and absorb the first three or four chapters and save the rest for later, when you really need it. Definitely read and put into practice the first one, *Things To Do Beforehand*. The second two, *Bracing For Impact* and *Immediately Afterward*, you will want to be familiar enough with so that when things do go 'splat' on a global scale, you'll at least have some idea of what to do. Also, make up the scrounging lists. The others are a bit of far-in-advance ideas and suggestions, which are perfectly suitable for reading by the fireplace and hammering out with the leadership of whatever new community you build afterwards.

We're not going to go into ridiculous detail on everything, but I do want to give you at least enough information to have a reasonable chance of survival and even doing that will be quite a bit in and of itself. I want to give you things to think about and things to consider, so that you can prepare in advance and have what you need. More importantly, you won't have to become an expert in all of these fields, but will be started in the right direction and when needed, you will be pointed to resources which will help you stock up your library. The library is going to be rather important because later once the turmoil has somewhat subsided, you can focus on learning in-depth those things you need to know by then. A lot of what you read here will look like simple common sense, but that's part of the idea… with the distractions that the world would provide in such a situation, having all of this written down for later reference and as gentle reminders is something I believe to be useful.

Finally, I want to dispense with any fear-mongering. Most folks who sell their guides use and amplify a creeping fear of the unknown (or the partially known or whatever) to push you towards preparing yourself, but mostly they use it to sell books, CDs, eBooks and website memberships. You don't really need to spend anything up front; because odds are good that, if this book is in electronic form, you're reading this sentence (like all of the other sentences in here) for free and with my blessings to do so.

My initial goal is to get you and yours to start thinking ahead, no matter how far along in these pages you get. To start making preparations that will not only save your butt when things get ugly, but can help improve your mind, body and ironically enough, give you a bit of confidence and improvement in how you live today in peaceful times. Once you're at least somewhat prepared, you can face the world with less trepidation, less fear, know what to do and be able to breathe a little easier.

My ultimate goal is to get you and your fellow surviving neighbors together, so that you can rebuild a working community and ultimately, civilization itself. Even if at least one of our existing governments (local, state, federal) manage to recover in some working form, a lot is going to be lost, or possibly disfigured beyond recognition. History has shown us all too many instances where wonderful cultures and knowledge was lost to the ages, taking millennia to be re-discovered if at all. We are still recovering hints and fragments of numerous lost civilizations today. I would prefer that our own civilization and technology not be a complete puzzle to future archaeologists - I think you would agree, no?

So Why Did You Get This For Free?

(Applies only to electronic copies. By the way, the copyright notice is in here too.)

Because I'm a nice guy? Okay, that can't be it. Certainly I'm no saint, but I do believe that you're going to have more than enough stuff on which to spend your money on. So, maybe you should get a break on this part of it all. The idea is to pass this little tome around for free in electronic form, to any and all who want it. I'm not doing this for the money, since doing so would require me to promise you many things that I simply cannot and remain honest. I cannot promise that you'll survive massive turmoil, disaster, or the end of civilization… nor can anyone else, otherwise they are lying to you. I prefer to keep it on the level. The contents in here can give you a leg-up, but they cannot guarantee anything.

I did this for a couple of reasons: I always write things out so that I can better retain and coalesce the information I need. I will be doing something that most authors out there may or may not do - I will be using this book as my own guide. Once I got enough of it written, I realized that the more people prepare with this same guide, the less worrying I'll have to do about how provided-for my neighbors are. I also happen to feel that sharing useful knowledge is simply the right thing to do.

Mind you, this isn't (and won't be) an end-all or be-all guide… only an idiot or a liar would say that and trusting either type is not a safe thing to do. On the other hand, this book should be enough to get you started and definitely enough to get you thinking. It should also cover a lot of things that no one else has up to now: Dealing with yourself, dealing with people and forming the one thing that *can* help you survive - a working community. What makes this thing into something complete and usable? The books you should buy for your little library, the time you spend improving yourself and your community, the skills you gain overall and the materials that you gather for the times ahead.

Copyright Notice

I retain full copyright, but you are free to make as many **electronic-only** copies of this book as you like and pass them around to others for free. I actually want you to print a physical copy for your own personal use as well – seriously, go right ahead. Give your printer a workout or go to the nearest copier/office store and have them do it for you, with my blessings and encouragement. **That said, I do not and will not allow for any commercial redistribution of this book, in whole or in part, under any circumstance, for any reason - alone or as part of a package, or hidden behind a toll or pay-wall of any kind.** I have that one restriction, because I refuse to see others make a profit off of work that they simply did not do.

There's a lot of pages in here and printing it all will likely eat at least two-three times the cost of this book in printer ink alone, will consume most of a ream of paper and you'll need something to bind it all together with. If you have to, you can get by with just printing off the Shopping Lists and building the Scrounge Lists – I suggest at least three to five copies of each, preferably more. When it comes to the whole book or sections of it, if you want something a bit more portable, sturdier and easier to carry around, then buying this book in printed form is probably a very good idea. I have it up on amazon.com (just look for the title and/or my name). I'd really appreciate the purchases and I can certainly put the proceeds to good use. End of advertisement.

So, we've wasted enough time with all the legalese... shall we proceed?

Why Bother?

The short version

Non semper erit aestas ("*It will not always be summer.*") - *common Roman saying*

History proves repeatedly that civilizations rise and fall in cycles and the math keeps piling up against our own global community. If/when things collapse, there are too many people to support for our ecology to bear once the technology and infrastructure grinds to a halt. Unless you feel like dying along with the majority of the planet, you may want to make a few preparations...

The long version

No civilization is permanent and few civilizations ever fall or change gracefully.

The Mayans were taken out by a drought, ending in bloodshed from countless human sacrifices and killings through the inevitable civil war that arose. More Aztecs were killed by imported disease than by Spanish steel. The Romans in the West came apart by political strife, opening themselves up to invasion, constant civil war and what eventually became The Dark Ages. The Romans in the East were slowly choked down and eventually done in by invading Muslim armies (a defending guard in the besieged Constantinople forgot to lock down one of the hundreds of gates - go figure). Samarkand went from independent, to Persian, to Sultanate, to finally being (literally) plowed under and turned into a wasteland by Genghis Khan. The Austro-Hungarian Empire died after essentially causing World War I – years of an inbred and increasingly incompetent royalty led to a strong underground rebellion, which assassinated a minor duke in Sarajevo - which in turn became a flashpoint for various European nations to automatically point fingers of blame, follow their treaties and line up for war. The Hapsburg family (which ran the empire) fell to nothing as a result. The old Russian Empire became the Soviet Union under similar circumstances and at roughly the same time. Numerous native nations throughout the 'New World' were taken out by smallpox and cholera, inadvertently (and in one case intentionally) brought to them by missionaries, conquistadors, soldiers and fur traders.

So... what does all that have to do with us? After all, civilization right now is relatively stable and global. Science and technology expands our collective knowledge almost daily. We have a world-wide network of commerce, communication and cultural trade that is resilient, powerful and rapid. Most nations resort to talk before resorting to guns. The biggest threat to human survival, nuclear warfare, turns out to become a very remote possibility. So we're all good, right?

Well, once you consider that economies are all interdependent to the point where failure in one means failure elsewhere, that rapid global human transport also means rapid global pathogen transport and that as technology progresses, larger and larger populations are demanding resources that are relatively finite? Suddenly things aren't so rosy, are they? However, it isn't the obvious things that should get you thinking... many of these bits have been seen to, for the most part. No, it's the not-so-obvious facets that should really get you to look ahead and wonder. Let me bring back Rome for a moment:

"*The decline of Rome was the natural and inevitable effect of immoderate greatness. Prosperity ripened the principle of decay; the cause of the destruction multiplied with the extent of conquest; and, as soon as time or accident had removed the artificial supports, the stupendous fabric yielded to the pressure of its own weight. The story of the ruin is simple and obvious; and instead of inquiring why the Roman Empire was destroyed, we should rather be surprised that it has subsisted for so long."*

- Gibbon, *Decline and Fall of the Roman Empire*, 2nd ed., vol. 4, ed. by J. B. Bury (London, 1909), pp.173–174.

Now think about that quote for a moment. You have a massive infrastructure of relative prosperity that relies heavily on many, many moving parts. All of those parts have to work flawlessly, must be redundant, or must be able to recover very quickly if they do falter. Otherwise, like the cops arriving at a teen kegger, the whole party comes to a crashing and unorganized halt, as everyone scrambles for the exits to save themselves.

Now, Gibbon was talking about Rome, but look at our own civilization: We have a global network of "just in time" shipments (to keep inventory expenses low). We have a population that relies very heavily on technology and finite resources just to keep itself fed. Most economic policies aren't much more than legalized gambling, with little more than guesswork and politics behind them. Our physical infrastructure has aged quite a bit, with signs of neglect becoming more apparent and maintenance/replacement costs soaring. Many economies are seriously over-extended in credit, to the point where becoming debt-free (be it individual or a nation) is merely an illusion. To top all of that off, there are numerous accidents, politics and mishaps that can, if ever ignored (or unable to fix due to other concerns), remove the "artificial supports" for long enough to take down the massive infrastructure of civilization.

The most important question however is this: how long until it all comes crashing down? The answer could be anytime - months, years, or even centuries. For all we know, it could by some miracle be avoided entirely. The Roman Empire after all took centuries before it became too brittle to hold itself up any longer and only when parts of it fell over did the rest come crashing down.

Can it be avoided? Certainly, if something big, new and disruptive comes along to change the dynamics of society or technology. So far, many things have done just this – industrialization, powered flight, container shipping, the solid-state transistor and Norman Borlaug's "Green Revolution" in agriculture all stand out as solid examples. However, each new development has to have greater and greater impacts in order to keep civilization going. Each advance was able to kick the can further down the road (so to speak). Nowadays, in our global society, things have gotten to the point where the next big disruptive change will need to be something really big – space colonization; the discovery of practical, cheap and commercial fusion power, a peaceful revolution that changes global governments... we're talking very big developments that will be required here.

Personally, I thought the Internet would be one of these changes – and for awhile, it was. While there have been some very impressive changes brought about by it, more and more we see corporations and governments slowly gaining tighter control over the Internet. Where once you could literally post anything in public with little to no fear of repercussion, now potential employers will often demand to see your Facebook page and will use Google (and other search engines) to determine whether or not you should be hired. Law enforcement can, even in the United States, demand and get passwords to your computer hard drives and encrypted files, where otherwise you are jailed for contempt of court (indefinitely if you don't comply). Every political entity that can is devising means of taxation for Internet use. Thanks to media companies, sharing a song or two online with others can get you sued for sums of money higher than you will ever make in your lifetime. Attempts by governments to curtail what you can and cannot do online are becoming more and more effective – both in the US and abroad. Internet service providers spend each year raising prices while providing less (or at best, providing incremental improvements at higher costs). Attempts at network neutrality laws or preserving freedoms online are either a joke or are becoming effectively neutered. Nearly everything you do online has become data that is bought and sold among advertisers.

Because of all this, I am firmly convinced that the Internet isn't going to save civilization – government and corporate control will see that it is eventually neutered and/or locked-down, mostly for personal or political gain. However, until then please use the thing - its storehouse of knowledge can help save your bacon before everything comes crashing down. How to use it? Well, you can take the useful information you find, print it off, bind it into notebooks and save it for the ultimate rainy day.

In the physical realm, we are reaching a point where there are too many people in an economic system that is becoming too complex to handle, on a planet whose resources cannot sustain the existing (let alone projected) population all by itself – at least not without all those complex systems running in perfect order and experiencing constant and substantial improvements.

Overall, we're headed for something that, which viewed through the lens of human history, is normal. Yes, I said normal. In the long history (and even pre-history) of humanity, upheaval and the collapse of societies is more the normal state of affairs than stability. Even civilizations you would consider to be ultra-stable (China, India, Egypt) have had massive upheavals, famines and collapses throughout their histories; the only things that remained continuous throughout are their cultural identities, though even they were shifted rather harshly over time, as well. In fact, China, often pointed at as one of the oldest continuous civilizations, has seen hundreds of millions of fatalities in just the 20th Century (the end of the Imperial throne, the Communist upheavals, Japanese invasion before and during WWII, etc). So you may want to keep that in mind.

Just like any other time that a civilization presses the big 'reset' button (especially now, with our global society), there's going to be some really large upheavals and a whole lot of adjustments to make if you want to stay alive. Unfortunately, come the next big reset, those adjustments are going to be fatal for a whole lot of people globally. There is literally nowhere you can hide from it anymore. My goal is to help you make sure that you and yours: (1) aren't among the fatalities and (2) manage to rebuild civilization. Just know that you're going to have to be willing to put some work into it.

Wait – You're Serious, Aren't You?

"The average age of the world's greatest civilizations has been 200 years."
    -Alexis de Tocqueville

Yes, I firmly believe that it all stands a good chance of crashing. So why is it that this is a problem for you? Well, let's say that everything came to a grinding halt tomorrow. We can all go back to farming for a living and riding horses to get everywhere, right? Live simply, go "green", dance naked and we'll all just get along and sing in drum circles …right?

Well, wrong. Logic and statistics prove otherwise.

Let's get a few facts out in the open about the United States, shall we? After all the United States of America is among the most richly blessed among nations for natural resources, yes?

It is, but consider: Currently, the United States has a population of around 308 million souls (source: 2010 US Census). The amount of arable land is currently 922 million acres (source: USDA). This works out to an extremely optimistic count of just under three acres maximum to feed each human being... plus, whatever comes in from commercial fishing and imports (for instance, most fruits and vegetables that you find in the grocery store come February are actually imported from Mexico, South America, etc). Most estimates allow for modern farming to only require 1.6 - two acres of farmland and pasture per person for a healthy diet; so with modern techniques, we can feed up to two times as many people as we do now... and we do just that, as one of the world's largest agricultural exporters.

Now – what happens when the imports stop coming, the tractors run out of gas and the trains and trucks stop bringing in food (and what few still roll get hijacked at first opportunity)? What happens when your neighborhood is forced to subsist on whatever can be grown locally? What happens when the farmers can't get fertilizer or run their tractors? Well, at pre-industrial farming techniques (no commercial fertilizers, machinery, pesticides, irrigation, veterinary medicine, vaccines, growth hormones, transportation, irrigation pumps, etc), you would need a rough estimate of around 14 acres to support each person. Seems a high figure, but here's why: this includes pastureland as well as cropland, for fishing and aquaculture and includes spoilage and other losses due to lack of pesticides and herbicides. It also accounts for the fact that in a slow collapse, imports will first be curtailed, then stopped. Overall, this shrinks the viable population of the US down to around a somewhat optimistic 62 million (or, incidentally, the US population in 1880). This leaves around 246 million people with nothing to eat. Note that this is also assuming that every current acre of farmland will still be farmed, that fertility is a constant, that food distribution will still exist and further assuming that there is enough skilled farm labor to pick up the slack. Odds are perfect that there won't be (the vast majority of people live in cities); so we can safely cut the assumption down to around 40-50 million people getting enough to eat. As for everyone else? Well, they're going to quite literally starve to death.

Now this is fairly worse-case and the truth lies between this and the most optimistic projection that anyone can make based on facts, as stated. The reason why is because existing supplies/stockpiles will still be around (though in most areas these will dwindle rapidly) and hunting will become very popular for awhile (until the vast majority of local wildlife and domestic animals get wiped out). Before you think all is lost though, consider this: food distribution will be very uneven, as will arable land. Some communities will have plenty, while many others will quickly run out or have none.

We haven't touched on water, but we should. In most of the US, fresh water is going to be hard to come by if the pumps and water purification plants ever slow down, much less stop. Regions like The Great Basin (Utah, Nevada, Southern Idaho), most of Texas and especially the Southwest will return to their natural arid climates, with drinking water becoming more valuable than human life. Even the Great Plains, the very grain belts of our nation, relies on heavy irrigation from a vast underground aquifer called the Ogallala... and it's not going to be very reachable if the pumps ever stop. Top all that off with the fact that most areas rely on aquifers and electric pumps to get their municipal water.

Overall, this means that there are going to be a lot of empty bellies and dry lips around.

Meanwhile, as things get short, society itself will change at the same speed. When things begin to go south, one of the first things to feel the strain is law enforcement. Love them or hate them, it doesn't matter: In the civilized world, the police are your first (and in too many areas only) line of defense against people who want your things worse than you do and they are especially your main line of defense against people who are willing to take these things from you.

During peaceful/plentiful times, crime is low primarily because of two things: almost everyone is usually satisfied with what they get by legal means (since they end up with a lot) and police forces are sufficiently funded and staffed to provide coverage to the largest number of people. In times of crisis and hardship, people begin losing a lot of the things that keep them happy and satisfied. Many folks (especially among the lower income levels) begin to see desperation and even shortage and deprivation of basic goods (food, shelter, etc). Police forces find themselves busier and busier, while at the same time the hardships mean that there is less funding to even keep existing staff on hand, let alone expand to meet the larger amounts of crime. When things get truly desperate and reach crisis proportions, policemen (and policewomen) will begin to stop going to work, if only to stay home and protect their own families. This in turn means that crime (and the audacity of criminals) will skyrocket and the military will have to get involved just to keep order.

As events escalate, those criminals which do get caught will find themselves crowded more and more tightly into prisons and jails, which will be increasingly unable to feed and house them. If things degenerate to the point where no one can keep up and government finally collapses, then new criminals may well find themselves executed on the spot for even the most minor of crimes, such as theft. Those who are locked-up may well be left to die or be released upon society - neither of which is a pretty option.

At some point between the beginning of crisis stage and governmental collapse (say, sometime before the cops pretty much give up), ordinary citizens will begin forming their own protection groups. This may be a group of neighbors forming a vigilante committee, or an existing street gang that decides to start protecting their neighborhood and neighbors from rival gangs and criminals coming in from other neighborhoods. It starts with groups that help the police, such as neighborhood watches and increased general awareness by neighbors. Then as police coverage becomes thinner? Unarmed, and then armed groups begin to appear to perform citizen arrests. Then, they become outright vigilante groups. Eventually? You become your own law. Unfortunately, the same maxim will hold true for everyone else in your vicinity.

While being with such a group will sound like a relief of sorts, these vigilante groups may well begin to make up their own laws as they go along, as due process will be whatever the whim of the lynch mob thinks it should be.

There will be exceptions to this rule outside of the large cities and suburbs. Smaller, more socially conservative populations will try to retain some of the existing structure of law enforcement. On the other hand, actual justice and due process (courts, detention, trial by peers, etc) may be curtailed or simply shortened to execution for most crimes – either real or perceived.

In the initial stages of collapse and for quite awhile thereafter, there is going to be a lot of mistrust, isolation and outright hostility to go around. There will be a lot of mistrust even among family. Any schisms and suspicions that existed during peaceful/prosperous times will suddenly expand a thousand-fold. Any racial or ethnic tensions will literally explode into barely-contained warfare.

For example, the creepy single guy living down the street that everyone thinks is some sort of pervert? He's going to be expected (by everyone else) to rape and kill every child within eye-shot of his house. What about that family down the street from the Mideast? Nice folks; but post-collapse, they will be fully expected (by everyone else) to strap bombs to their bellies and force you to pray facing Mecca at the point of a sword. The fundamentalist Christian family next door? Oh, we'll all expect them to start dancing in the front yard, wearing nothing but sheets in expectation of the impending Rapture – or perhaps wear those sheets while burning crosses in others' yards and taking rifle-shots at anyone who isn't sufficiently Caucasian and Protestant.

Everyone will fully expect to be screwed over and then killed/raped/exploited by everyone else. Those who still have resources will constantly be on guard against those who lack them, and those who lack them will be seeking to exploit those who have - and at the same time will be constantly worried that the first group will take what little they can claim as their own.

Now certainly, there will always be nut-cases, criminals, and folks who will do their level best to screw over their fellow man. We have those folks now, even in peaceful times. As needed resources become extremely scarce however, people who would ordinarily never consider harming anyone will suddenly contemplate horrendous acts, up to and including something like, say, killing and barbecuing someone else's infant child. Why? Because it will mean keeping their own children fed. The sheer creativity of the thefts and takings will astound you, as desperation always breeds creativity. The funny part is, none of this is new news (or at least it shouldn't be) – even the cannibalism. Open the nearest Bible and look up 2 Kings, chapter 6, verse 28 (NIV). There, you will find distrust and deception under severe conditions, even among close neighbors...

Then he asked her, "What's the matter?" She answered, "This woman said to me, 'Give up your son so we may eat him today and tomorrow we'll eat my son.. So we cooked my son and ate him. The next day I said to her, ' Give up your son so we may eat him,' but she had hidden him.
—The Holy Bible, 2 Kings 6:28 (NIV Translation)

The story is thousands of years old, but is still just as true today: when everyone is desperate, you will find that horrendous acts, deception and theft become rampant. You may also notice that, under these conditions, nothing at all is sacred and nothing is off-limits. Until that overall period of desperation passes and sufficient resources become available to the survivors, trust is going to be something hard-earned and checked constantly. That is, if there is any trust to be had at all. Under such a state of existence, any functional society will be chancy and temporary at best, as everyone hedges their bets and only tentatively allow others to watch their backs. When that period of chaos passes? This is easy to state but hard to measure: Society and civilization will eventually return, but only once there are sufficient basic resources (food, water, shelter) for the surviving population.

Once life in general becomes less of a desperate state of day-to-day survival, things will stabilize to a point where a given group of people can begin to trust each other and relax together. Eventually still, a time will be reached when groups will begin to trust other groups.

With a little luck, some areas and population groups will be able to trust each other throughout. **These will, incidentally, become the best places for civilization to eventually recover.**

Once things collapse initially, there will be a whole lot of things that will cease to be, though maybe not all at once: electricity, modern medical care (and medicine!), advanced education, manufacturing, rapid transportation, you name it. People who rely on modern technology to remain alive long-term will die off as the medicines and devices they rely on begin to fail or become unavailable. People in general will find that, without modern farming techniques, food will become impossible to come by and most people will actually starve to death or turn to cannibalism to survive. Without a means to extract and purify water on a large scale, water-borne disease and death by dehydration will be widespread. Without a constant supply of gasoline, kerosene, asphalt and plastics? Rapid transportation will pretty much grind to a halt, eventually becoming a commodity that only the military or the remaining wealthy/powerful among us will enjoy - though for how long is anyone's guess.

Initially, everything you know and expect will still be there. There are very few scenarios that would cause the whole works to grind to a complete stop instantly and most of those scenarios will probably kill the vast majority of humanity instantly. In most cases, most modern conveniences will still be around. It is only when collapse is well and truly underway that you will find that things will stop working, first one by one, then in groups and then in torrents. Creative television or radio entertainment will start to peter out, replaced by nothing but "newscasts" and thinly-disguised propaganda. The Internet will begin to become less and less global, partly by governmental action; but mostly because connecting routers will cease to remain powered up. Electricity will become rationed to a few hours each day (if you can still afford it), then restricted to vital services (hospitals, fire stations, etc). Water and trash collection will likely fail at the same time, or very soon afterwards. Mechanical and electrical devices will begin to break down and not get repaired and eventually spare parts will begin to run out.

Without modern technology, a lot of things that we take for granted are simply not going to exist. However, you will find that a sharp mind for engineering and a creative knack for improvised technology will be valuable and welcome to a community. If you can come up with running water, hot water, efficient (or even useful) home heating, water filtration, electricity, irrigation, improvised medicine, warning systems and a whole host of other things, you can easily carve out a niche for yourself in any community. You know? While we're talking about all of this, now would be a very good time, while things are peaceful, to either find a way to adapt your existing skills to a post-apocalyptic world or to gain the skills that will become necessary. There's no need to be an expert in everything needed (it's impossible anyway), but you can gain an expertise in some skill that others will not only find useful but with which you can barter for. The good news is, this book will in many ways help you achieve that.

The downside of it all is that the knowledge which remains among the survivors will be based on whatever the survivors in your immediate community know or can learn from surviving texts. Because not every surviving group of people contains a full complement of PhDs in every known subject, in general, you can count on technological knowledge often being inadequate, incomplete and in many cases even flat-out wrong. However, that's what you're going to be stuck with, so make the best of it …and maybe even prepare to fill in the gaps a little.

As time passes, you may well eventually become the local source of knowledge for your community in a given skill and it may fall on you (or someone suitably skilled) to become the impromptu civil engineer, local doctor, teacher, or whatever. Eventually, years ahead, those among the newly-born survivors who are adept and technically-minded will, with luck and enough information, become the new scientists and engineers to help rediscover by-then lost knowledge and rebuild a new world from the ruins of the old.

A word or two should be said about that dirty, nasty, root of all evil: money. We all work towards getting at least enough to satisfy our needs or at least, know and love plenty of people who do. It is the main medium of trade and without it, getting what we want and need is pretty rough. Well, as civilization goes, so goes its money. So what happens when things come crashing down? Paper money will become very worthless, very quickly. Since most folks keep theirs locked up in a bank computer, it will become very inaccessible once collapse begins going full-speed.

Odds are pretty good that money is probably going to be the main reason that our current global civilization comes to a grinding halt. As economies go sour, people have less money to spend. Goods become more expensive as less of it is being made or delivered. As prices go up, more businesses (and thus more jobs) will disappear, meaning less goods or services available for sale. It quickly becomes a vicious cycle and one that rarely is pulled out of in time.

In many cases, as things go really wrong, money will cease to be trusted. That or prices will skyrocket to the point where you could leave a wheelbarrow full of it on the street and thieves will steal the wheelbarrow... but leave the money in it behind. It's happened before. Germany suffered this between World War I and World War II, where postage stamps were printed that once literally cost a million Deutschmarks (DM), whereas 10 years prior you could get a book of 100 stamps for just one DM. In other cases, the existing currency (in our case paper money) will be discarded in favor of another form. This could be gold, silver and precious metals, which have throughout history been readily accepted. Barter, being a most basic and the oldest form of commerce, is not going to go away. You may even find (or create) locally-invented money, backed by precious metals.

The take-away from this section is that you really shouldn't count on your credit cards, debit cards, credit rating, home equity, or any such instrument to get you by when things get really bad. You could have a billion dollars in the bank; but when things truly go south, you won't be able to touch so much as a penny of it. You may not even be able to count on precious metals to get by (after all, one cannot eat a gold coin). Your best bet is to have a few extra supplies to barter with, some precious metals (just don't go overboard) and perhaps a few extra bucks locked up somewhere in the house for any last-minute spending before things truly go dark.

Overall? Like it or lump it, what are the odds of civilization going through a big reset, or at least, an extremely turbulent transition period that will pretty much ruin everyone's lives? Disturbingly better than they should be. Most sane indications show that these odds are getting closer to even with each passing year. Barring a major advancement in global technology, things cannot stay at steady state; it will either rise overall, or fall overall. As I said at the beginning – it could happen next century, or it might happen next month. We may already be going through it even now (at least the earliest stages of it). The trick is to actually do something about it now, while it is still possible (and far easier!) to do so and learn what it will take to survive if and when it all goes down the tubes.

A Note About "The Rapture"

There are likely many reading this book who, perhaps before getting this far, will put the book down, make a '*tsk tsk*' noise and quietly dream of being taken into Heaven minus having to deal with the crash and bang of a falling civilization. Meanwhile, all us dirty old sinners will just have to suffer for our Heavenly reward, though we've likely been condemned anyway, right?

To those among you who think you'll get to avoid such a calamity, well, I have news. Bad news. There is absolutely no guarantee that the alleged Rapture will take you away. Without getting into any deep religious debates about the validity of your beliefs (heck, let's assume you're right for a moment) here's why I think this will fail in theory:

- There have been many civilizations that have fallen since 33 AD and none of those populations got raptured either. Even the newly-Christianized Roman Empire was flattened, all without a single mortal soul being taken up directly to Heaven rapture-style.
- You don't know when it's going to happen. Even Jesus Himself says you won't.
- What makes you so certain that you'll be among these lucky folk? We're pretty much all sinners (this means you, too). There's a reason why the Bible calls them the Chosen *Few*, no?
- Even if you were among those going, everything that you assume about the theory behind it were true and you're otherwise good to go, what about your family? Your wayward kids? Your parents? Your friends? Your neighbors? Do you intend to leave them all stuck down here with nothing? If you say 'yes', then you're likely not among those who gets taken up (and the first one to claim that everyone they know gets to go is either deluded or a liar).

Hopefully this should be enough to convince the most fundamentalist among us to at least consider equipping themselves and their families for the crash.

That said, if you're still fully convinced, then please do the rest of us a favor: Even if you're 100% certain of being raptured and have no living family, pray faithfully, etc. – at least get up a stockpile of sorts, a few good books just in case, then leave this book and a post-it note on the whole pile which says "a parting gift from a Heaven-bound soul" or such. That way, if you are indeed right and you get to avoid the muck of surviving collapse, at least one of us sinning types will sincerely thank you (or your departed spirit) when we find the goodies; and it'll likely get you a few bonus points when you do meet your Maker.

As for all of us sinners? We're still going to have to get on with the business of surviving. So…

Dumb Questions, Hard Answers

Q: *"We can all go back to living simple lives like our forefathers did, right?"*

A: We answered that one earlier, but the answer is actually "yes"… once the population drops enough. You'll have to stay fed and hydrated before that point in time, though. So unless you're currently Amish, I wouldn't start buying horses and piling hay up in the garage just yet.

Q: *"I can just go to Uncle Bob's Farm in Idaho - I'll be alright… right?"*

A: Nope - and you have some big problems with this plan. First, you have to actually get there and unless you're very lucky or very psychic (enough to get there before things crash), getting to Bob's place will be a problem that will grow exponentially fatal with every mile that you have to travel. Second, assuming that Uncle Bob knows you're coming and will actually let you stay, he likely won't have enough resources to support you. Resources? You know, things like food, a bed, clothing, etc. So unless you drag all of that along with you (making you quite the juicy target along the way), you're only going to put your family (and Bob's family!) in jeopardy. Third, if you or your family have any special dietary requirements or medical things going on, Bob is not going to have the stuff to help you deal with it. Finally, Bob's neighbors aren't going to appreciate your existence, either. You're just another mouth or four in a county full of people that want to be fed. Besides, they don't know you well enough and you likely won't have the rural skills needed to be useful there.

Q: *"I don't have to worry - I'm coming to your house, buddy!"*

A: You might get a small bucket of food as charity, you may get told to go away, you might get a warning shot, or you just might get shot. Or, me and my stuff may be long gone, leaving you with nothing to ask for. It all depends on how things are in the neighborhood at the time. Don't depend on my stuff and my shelter for your survival, because it probably won't be there for you. It was all I could do to stock up from my own meager budget (while you were buying that new boat and kitchen remodel, incidentally). I'm not sure how much of it will last in any event. By the way, this all goes double if you have a family - I simply can't budget for that many mouths to feed. But hey - maybe two years or so after the collapse is done? Assuming you're still alive and not a criminal by then, you can come on over any time; I'll see about getting you set up in your own place or something, depending on what's around.

Q: *"I don't need all that "prepping" stuff. I have guns, and I can get anything I need!"*

A: You'll be dead within a few days (weeks if you're lucky) and will end up providing a nice firearm or two for someone else. Be sure to carry extra ammunition when you do try, as the recipient may not have your particular caliber handy. Most people who take the time to seriously prepare also take the time to arm themselves very well and to maximize their homes (and even neighborhoods!) for defense. You might get lucky and surprise someone, or you might take from an unarmed family or two, but your luck starts getting worse with each passing day. As time ticks along, more resources are consumed, desperation increases all around and people learn to start playing for keeps (oh and you'll start running out of ammo). In a very short time, the only ones with supplies will be armed and very willing to use those arms to defend what they have.

So What If Nothing Happens?

Then nothing happens. At best, world peace breaks out and an immediate era of Star-Trek style global utopia becomes the norm. If you want to drop by with an 'I told you so', I'll be happy to hear it. However, I'll likely be out exploring the solar system or might be busy over at one of the 24/7 Gameatorums that are sure to pop up under such a circumstance.

However, nothing is lost in the process!

Unlike the big Y2K scare, you're not going to be stuck with a big generator or a .50-cal machine gun that you'll have to unload at fire-sale prices. Unlike the Cold-War 'duck-and-cover' era, you won't have some big, ugly, ill-equipped bomb-shelter moldering under your backyard. This book is structured so that you don't end up spewing odd politics like the stereotypical conspiracy nut. There is no need for living in a remote cabin, wearing a tinfoil hat, eating cold beans and spewing about the world being run by a conspiracy of Illuminati/Aliens/(Some Ethnic Group)/Bankers/Microsoft/Lucifer/Hottentots, or any other such nonsense.

The whole idea of this book is to get you - the ordinary person - sufficiently prepared without having to sacrifice your life, your intellect, your religion (or lack thereof), or your credibility. For physical preparations, the idea is to stock up on things that you can use anyway – the rest is up to you. Canned food can still be eaten at any time that you decide to not bother preparing anymore or it can be donated to the local food bank, which would be grateful to have it. Garden seeds can be put to use in the garden (any garden) at any time. Books can be donated, or kept around for your kids to learn from whenever they stay home sick from school. Tools can be kept for ordinary household chores or given away. If you decide to keep and become proficient in firearms (highly recommended where not otherwise illegal), then they can still come in handy during camping trips (as protection from predatory animals and for signaling if lost), hunting trips, or for home defense. Many of these items (books, tools, precious metals, etc) can be passed on from parent to child. To top all that off, if/when natural disaster strikes, you will have everything you need to get on just fine, with perhaps enough to share with a few of your neighbors.

Preparatory skills such as foraging, blacksmithing, or bushcraft? They become fun hobbies to to enjoy at your leisure. From the mental preparations, you gain a sense of being more alive, of being assertive without being aggressive and of thinking through any given crisis, large or small. These are skills that are just as useful in the boardroom as they will be in any post-apocalyptic wasteland.

The best part is this: you're going to be keeping largely quiet about the whole thing anyway (and you will want to for a lot of very good reasons – none of which will have anything to do with your reputation). At the worst, you have some extra stuff that you can put to use anyway. At middling, you can be at least somewhat prepared for many lesser, just-as-ugly situations: hurricanes, tornadoes, earthquakes, ice storms, you name it. At the absolute worst, you have a leg-up on staying alive for long enough for civilization to recover and/or be reborn. Either way, it's always good to be prepared, no?

It is assumed that the majority of the population won't read this book or even care about collapse and upheaval. Fair enough. I hope that at least some of them have some sort of contingency plan in place. The idea is that hopefully some folks learn enough to at least get the idea. My ultimate goal is that long-term, our civilization may possibly (and even probably) die, but our knowledge (and perhaps the best parts of our culture) won't have to die with it.

Chapter 1

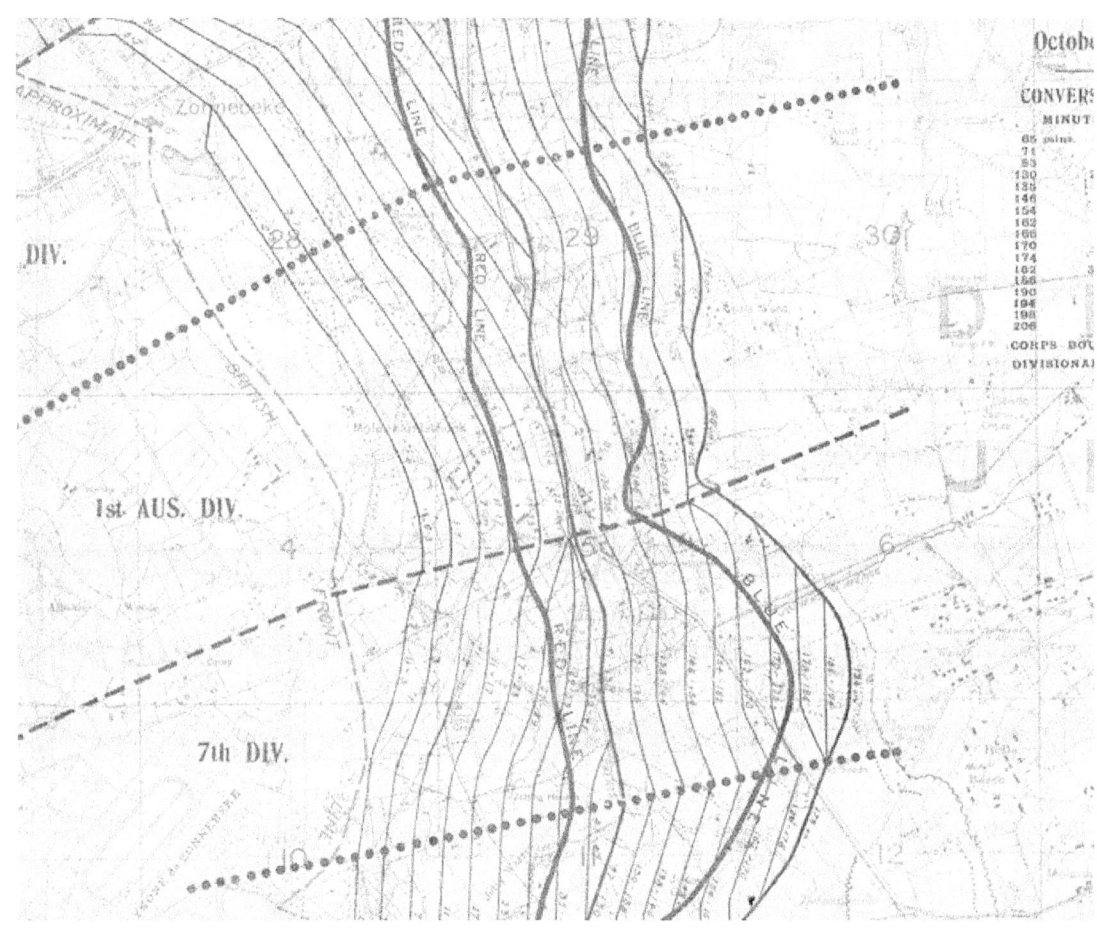

Things To Do Beforehand

Who Are You?

Before we can prepare for calamity, I need you to know a little more about yourself. Yes, you likely know yourself very well. However, you need to view yourself in a more critical light. First, you and I must dig down a little, to find out if you are mentally equipped to make it and if you are physically capable and able to survive, especially in a world without modern technology and equipment.

This isn't to scare you off or anything, but unless you are sufficiently able-bodied, mentally stable and have a personality that will help you survive a crisis, the odds are definitely not going to be in your favor. This certainly does not mean that you should simply give up, but it does mean that anything which hinders you must be honestly assessed, overcome if possible and accounted for in your preparations. Through each step in this chapter, you seriously need and want to question yourself: *"Can I really do this?"* *"Can I get my mind into a state where this becomes second nature?"* *"Am I able to do this?"* *"Does this violate my moral or religious beliefs?"* (generally this chapter won't violate that last one, but some parts might give you pause.)

Keep those questions in mind as we move through the chapter. For now, let's start with the first thing that should happen when you become aware that something is way out of order. Specifically, let's start with your...

Reaction and Attitude

First and most important is your reaction to things. This one factor will often make or break your odds of survival. If a major source of stress arises, how do you deal with it? Some of the things you definitely do not want to do are the following:

<u>Bad reactions when faced with stress</u>
- Fly off the handle with anger
- Withdraw into yourself and become unresponsive (otherwise known as going catatonic)
- Take charge and do an action without thinking, or try to assert leadership in a group even if you have no idea as to what should be done
- Become incapable of making any sort of decision, but moving or speaking randomly
- Immediately look for something or someone to blame
- Automatically expect someone or something else to resolve the issue

None of these reactions are useful and most of them will get you killed in the crises that we're talking about. Most people will fall into one of these categories to some degree, even if it is an initial impulse that is quickly smothered. The idea is to reach a stage where you can do just that with bad reactions – immediately swallow the impulse and unless the situation absolutely demands a snapshot decision or action, *focus on the situation, not the people in it.* (including yourself, by the way).

There are mountains of psychology books out there which tell you to always consider a stressful situation as it arises, then act in some idealized and reasonable manner. To be honest, neither you or I will ever reach that utopian ideal. We are human after all and we will react in one or many ways to a stressful situation. We all have faults in this regard. For instance, when faced with a heavily stressful situation, I start talking. A lot. Not mumble, but actually speak. My brain decides to use the spoken word as a temporary place to store thoughts while I consider multiple avenues towards finding a solution to a stressful situation. This obviously helps no one, since the words are only partial bits of thoughts that only I can follow (not because I'm any sort of super-genius, but only because no one else can read my mind well enough to get the context of what I'm saying and why).

My solution for this is to bite my tongue (sometimes literally), stop and think of nothing at all for a split second, to allow my thoughts to recall the situation on their own. At this point, one of two things happen: Either training kicks in and I have a course of action to decide on (the ideal), or I force myself to mentally break down the situation into bite-sized chunks. If the situation requires absolute split-second decision-making, then I only observe and break down enough of the situation to get me (and anyone else with me) out of (or away from) the worst of it, until I can safely think through the rest of the situation and come up with a full decision.

<u>Good reactions when faced with stress:</u>
- If time does not permit, have the mental training to immediately take a direction that will maximize your chances and act on it without contemplation.
- If time permits, quickly observe as much as you can of the situation.
- If time really permits, inform yourself of all pertinent facts as much as possible before deciding.
- Take a mental inventory of what resources you have on hand to solve the situation.
- Put your emotions away. You can sort them out later.
- Blame is a luxury you don't have right now, so don't.
- If the stress is prolonged, force yourself to get some sleep regularly, even if it means using (a little!) alcohol or sleeping pills to do it.
- Always be ready to change your mind when circumstance or facts change.

Every situation is different, but I promise you, they all involve two things: observation and thought. If you don't have the luxury of time, try to come up with a decision that gives you that time. For instance, if a rush of flood water is bearing down on you, most folks are able, in fractions of a second, to have some idea as to how fast it is moving, how big it is and what general direction it is moving in. You should also be able to mentally take in potential things to climb up on, or if you see it is moving slower than you can run, be able to know which way to run. If it is too small to overtake the small hill or item that you're standing on, you have the luxury of a few more seconds to take in what's coming, to consider potential escape routes and things like that. If you see that there is higher ground close by you can get to quickly, do that. Only then can you give yourself the time to think of what to do next, even if your legs are still running while you're thinking.

Nobody is asking you to become the lord and master of snap decisions and to be honest, unless the situation is immediate and life-or-death, snap decisions are the worst kinds of decisions to make. However, that's not the idea. The idea is to force yourself to think and do the best action in that moment of realization. You should first and foremost observe with full attention and if there is time, think. Fortunately, even in a post-collapse situation, you will more often than not have time to react properly to a stressful situation.

For those situations where time is not something you have, you can train yourself to react in a way that gives you the best chance of survival, depending on the situation. These situations can fortunately be classified somewhat. We have:

- Fight or Flight
- Impending Disaster
- Injury or Sudden Medical Issues
- Fleeting Opportunity

There are of course others that do not fit into these, but this is the vast majority and will serve to help you out a little, especially when it comes to mental preparation. We'll go from worst (and hardest) to best (and easier) to train for.

Fight or Flight

While this basic human instinct covers a wide variety of things and is often uncontrollable, we're using it as a term to talk about what happens when you find yourself faced with people (or animals) that want what you have and are trying to take it. This is the toughest reaction to train for, because it is the hardest to overcome. As a civilized person, you've rarely (if ever) had to contemplate this before and if you had, it is likely that you usually thought it through enough to map out a plan of action that involves dialing 911 or meeting at the front yard until the firemen come. The problem is, in a post-collapse situation, you won't have the luxury of a professional police or fire-fighting force to protect you and if one still exists, they will likely be stretched too thin to be of any help. The only one who will protect you is, well, you. You might have help in the form of fellow survivors, surviving family, or your neighbors, but they may not be close at hand and even if they are, you must always do your fair share of the defending.

The variety of attacker comes down to two basic types, each under two situations:

Thugs: These people are strangers and are after something you have: food, medicine, gear, your daughter, etc. They may not even be after your stuff, but simply want to harm you for the fun of it. In either case, they should mean nothing to you and you certainly mean nothing to them.

- If you are away from home when this happens, get the hell out if you can and if you have a firearm, get it out immediately and start using it if needed while you retreat. Always move towards cover and towards friends if you have any nearby. If you do not have a firearm, get out any weapon you have. If you see a handy weapon laying in your path, pick it up, but don't spend too much time weighing the options. Only stand your ground if you have a sufficient number of friends and firepower, but the idea is to get away.
- If you are home or at a trusted friend's house? Either way, the answer is unconscionable to most, yet must be done: Kill the invaders without hesitation. Do not try to scare, do not try to injure, do not try to warn. Just kill them by any means you can, period. Anything less means they will only come back with more attackers, more weaponry, or both. If they manage to get away alive, they will likely be injured, which may just scare them into staying away – but don't count on it. Train yourself mentally to go for the throat, eyes, stomach, or inner legs if you're using an edged weapon. If you're bare-handed, kick and punch for the head, groin and torso, but keep thinking about the nearest weapon you can pick up. Go for the head, arms, or legs if using a blunt weapon. Aim for the torso (preferably the chest) if using a

firearm. Killing should be your first reaction, period. Even if you realize that you're unable to overpower them, you can always continue fighting while walking backwards and with your weapon at hand, they're not going to be in any hurry to tackle you. If you fend them off and they begin to run, keep trying to kill them. Only stop when they can no longer be seen from your property line. If there are any survivors among the attackers, you will probably want them dead, but use your own judgment. Just keep in mind that if they leave the premises, they may come back later on.

Note that none of this is to make you into a 'badass'. To be honest, it will likely be the hardest thing you will ever do and many of you will not be up to it. It will scare you senseless. Many of you will freeze up and fail at it. Even if you win? Afterwards you will puke, wet/crap your pants, be covered in bruises and cuts and/or shake uncontrollably - but the important part is, you will still be alive. I found myself in a similar situation and puked my guts out after they left, in spite of having a pistol. Post-collapse, the alternative to fighting is to end up dead anyway, since you will either be killed (and all of your possessions stripped, your family exploited in the worst ways, etc), or you may escape but end up dying of exposure, starvation, dehydration, etc. Given all that, you may as well put up a fight either way and don't stop fighting until either the intruders end up dead, or you do. If they want to steal or harm, make them earn it.

I apologize if this sounds harsh, but a post-collapse world isn't going to be pretty, it won't be glamorous and definitely will not leave anyone unscathed or unscarred. The only ones who survive are those who are willing to do whatever it takes to insure their continued survival. If you are not up to that, then I suggest putting down this book and conditioning yourself with one goal: once the collapse becomes obvious, plan to commit suicide by whatever means you feel comfortable with (recommendation? A bottle of compressed nitrogen with an oxygen mask hooked to it is fairly quick, mostly quiet and completely painless). Just don't take anyone else out with you.

Rebellious Allies: These are people you know and maybe even love, but they've decided to attack you for some reason or other. Perhaps it is a former neighbor who has become desperate and is coming for what you have. Perhaps it is a desperate young person (kid, teenager, etc) who can be subdued, then reasoned with or co-opted into seeing things your way.

- If you are away from home, fend them off, but get deadly if they refuse to stop and cannot be subdued, then leave as soon as it is possible to safely do so.
- If you are home, the same rules apply: Kill. However, odds are good in this case that the attacker would be more likely to break things off and run, or try to reason with you. If they do reach that state, try to disarm them before allowing them to speak any further, confine them and then you have time to decide what to do about it, which means it leaves the scope of this discussion.
- Kids were mentioned and something needs to be clear – don't think you can simply subdue one. Take any and all steps necessary to stop them. Children can and will kill you if they are desperate enough. They aren't mature enough to have a full conscience, so don't count on them having one.

Impending Disaster

This one seems like a no-brainer, doesn't it? A fire, an impending flood, a tornado, an earthquake... these are covered wide and well and what to do about many of them has likely been taught to you since childhood. On the other hand, the one thing few books and/or courses cover is how to tune your initial reaction. Far too many people mentally vapor-lock, standing stock still while the doom keeps coming. What follows is not a comprehensive list of what to do, but exercises that teaches you to react quickly and to have your initial reactions to occur correctly. There is a little trick you can use to help break that initial 'deer in the headlights' reaction and it consists of just one word: "*MOVE!*" Repeatedly train yourself to hear that word in your head at every unexpected big thing. Have your spouse, friends and (older!) kids sneak up on you at random times and shout that there's a tornado, an earthquake, a flood, or such (...but not fire – don't want any false alarms on that one). Do the fire drills, the earthquake drills, the (depending on area) tsunami drills, whatever.

If you're alone or want to do this without others knowing, no problem. Watch the news alone – cable news if you got it. Every time the announcer mentions a story about a fire, a tornado, a flood, or any other disaster, immediately get up and pretend to do something about it, then look back and see if you did and thought the right thing. Do this often enough and you start reacting by "muscle memory" – that is, you'll begin to react automatically depending on what the claimed emergency is that reaches your ears or eyes. Eventually, if/when the real thing hits, you're immediately on your feet, evaluating goals and in all cases moving towards doing what needs to be done... which in turn puts you ahead of 99% of the population.

Injury or Sudden Medical Issues

A bit complex, but out of the situations so far, this one may well give you the most time to think. On the other hand, you cannot spend too long trying to figure things out. The big three to keep in mind are Airway, Breathing and Circulation, or basically insuring that the person injured is breathing, has a pulse and isn't bleeding out. The rest can be covered later on (and in other more complete books), but your first and initial reaction should always be to check those three things and just keep in mind that there will likely be no "911" to dial.

Fleeting Opportunity

This one seems the easiest and it is – if you know what to do. This can involve a barter at the last minute for something you desperately need (or want), a chance to join a group of survivors, or perhaps a deer shows up in the road on your way home. Each answer is wildly different, but the first impulse should always be the same: *What do I win and what do I lose?* Some of these will be no-brainers (like the deer in the road), where you immediately go for it and figure out the consequences later. On the other hand, some of them will require a lot of thought in a short period of time. However, if you keep asking yourself that one question and can do a quick inventory of what is won and lost by making the decision, you stand a good chance of coming out ahead. If you can stall for time, then do so, but only long enough to make a good decision and not so long as to let the opportunity slip. Be sure to ask questions then and there if the situation demands it. Sometimes you won't have that luxury, but always approach a fleeting opportunity with a positive but tentative answer. Even if the answer is "yes, but..." You will have gained a bit more time. That said, since most human beings are wired this way, it should be fairly easy to do, no?

Attitude And Reaction

Now, let's chat a bit about your attitude. Is it aggressive? Is it passive? Do you have a balance between the two? We're not talking about your initial reaction (since we've already covered that), but, as you can probably tell by the first part of this chapter, we're looking for a balanced approach. The trick is to be aggressive immediately *when you have to be*, but stop and think in any other situation.

Sometimes the passive approach wins out. As Calvin Coolidge once said: "*When you see ten troubles rolling down the road, if you don't do anything, nine of them will roll into the ditch before they get to you.*" The trick is to pick out which emerging problems will be the ones you have to do something about and figure out which ones you can safely ignore. If a gang of criminals are looting stores a mile or two away and taking big-screen televisions, this is not going to be a problem you have to deal with (well, yet – you'll know why later). If they start breaking into houses in the neighborhood, then you have a problem you have to contend with - hopefully alongside your fellow neighbors. Being somewhat passive about the first problem (the big-screen TV looting) means you don't draw too much attention to yourself, which is a good thing.

Let's give another, more complex post-collapse example: If your teenage daughter had her heart broken by a boyfriend down the street, no big deal – it isn't something you have to worry about. If that boyfriend tried to rape your teenaged daughter, then you have a situation that definitely needs to be dealt with. Say the boy got stupid, but his dad is a vital part of the group helping to defend your neighborhood. Finding the boy and beating him to within an inch of his life (or worse) is obviously not the solution (in spite of your impulse to do it), since his father will instantly want revenge on you for doing that. On the other hand, leaving the situation alone means the kid may try again. A balanced approach is to find the father of the boy and calmly but firmly inform him that this needs to be taken care of immediately. Odds are good that the father will feel shame for his kid's actions (at least if he is sane he will), fear that the neighborhood will find out and take action as a group (this is your greatest leverage) and will worry about how to deal with the kid (after all, he still loves the boy). The two of you will likely work out some scenario that, while not perfectly satisfactory, will at least stand a good chance of preventing another incident. However, if the father mouths a bunch of empty words then does nothing about it, or he gets arrogant, then bring all the other neighbors into it, where he and the boy can then explain themselves to everyone. Odds are good that as a group, the neighbors are going to see to it that it never happens again, or else the boy and/or his father will be injured if not killed. What may happen is that, at the very least, his whole family is forcibly ejected from the neighborhood with (maybe) enough goods to travel on for a few days. Or, if you were contemplating leaving yourself, you might just skip all of the confrontation and leave anyway, with as much as you can carry (later on, we'll be giving you the tools to help you decide how and when you should leave anyway… but really? Unless you were already planning to leave, don't. Deal with the situation first in either event.)

The idea of that scenario is to show you that there are times where you have to take a balanced approach to situations, be flexible and even adapt. Most importantly, the lesson is this: Suppress your own emotions for long enough to get the facts and get the situation stable (in your mind), before you come up with a solution to the problem.

Overall, your mind is the greatest weapon, tool and device you could ever hope to have. But, like any other tool, if you don't use it and tune it, you will find yourself not knowing how to use it well when it is needed the most. Take the time to tune your mind, your emotions and your reactions. Before long, you will reach a point where these skills will not only save your life, but may well help you get ahead in the civilized world too.

Personality

One of the most important things about you is your personality and especially how you use that personality to deal with other people. A lot of this depends on how you introduce yourself to others, how you act (not react, but act) in a group and how much tolerance you have for other human beings.

In a post-collapse situation, the best personality to have isn't exactly what you think it might be. Let's look at the situation overall. You have to understand that everyone and I mean everyone, will be frightened to the core - they will be afraid of you and you will be scared stiff of them. Nobody will know what to expect. We will all misinterpret any aggressive words, movement and gestures as a warning of attack. Anything that irritates us will irritate us with 1000 times the intensity.

Let's face it – in a post-collapse world, for quite awhile you will find that people in groups stop being people – they become what human beings minus civilization are: dumb, panicky animals. They will also be dumb, panicky animals with guns, clubs, torches and knives. Needless to say, this isn't going to be a good time to go around scaring, irritating, or angering everyone you meet.

The best personality to have is one that is calm, quietly alert, assertive, a little cautious, confident, a little bit tolerant and at least somewhat friendly. You know, all of those things which will be nearly impossible to be or do when the world is blowing up all around you. Even worse, you're going to need to have somewhat of an independent streak. Why? Because you may well find yourself having to go it alone, at least for awhile. It also helps to not blindly follow anyone claiming to be any sort of authority, as doing so will likely get you killed.

However, you can start working on at least some of that right now. How about we try a few exercises, shall we?

- If you find yourself in line somewhere, or waiting at a bus stop, or similar, strike up a conversation with the nearest friendly person and see how long you can keep going with it (just don't keep pushing if the other person does not want to talk, or it gets obvious that they're done with talking for awhile). Make it a goal to get that person to smile (doing that in a dentist's waiting room will be your toughest challenge, by the way). Another goal is to fall silent for a moment after starting the conversation and see if they strike up the conversation again. While you're doing all of this, pay attention to your facial expressions (I know, but you can still guess by the way your face feels). Pay attention to your tone of voice, how loud you speak and what you're talking about. This will give you the opportunity to learn how to read other people and in this modern yet isolated age, it will begin to give you the ability to put people at ease. <u>Bonus exercise</u>: while talking, see if you can, using only peripheral vision, keep tabs on everything in the room.
- The next time someone does something minor that irritates the crap out of you, try to let it go. Say you get cut off on the freeway, or someone nearly causes an accident. Say your spouse begins to pick his/her teeth, or does some other thing that bugs you to no end. Smile at the person, wave to them and push your brain to let it go. If you're in the same room, quietly sit there and ignore the thing that irritates – see how long you can hold out. You would be surprised at the number of times the irritating activity wears itself out

before you do. This will begin to teach you tolerance, which is going to be a vital skill if you expect to work and live with other people (especially complete strangers) in a stressful situation.
- If you find that friendly ignoring the irritant isn't working, then smile and quietly tell the person who is irritating you that the activity is irritating and to please stop doing it. Be sure to wait for awhile afterward, as again, the irritating activity will likely wear itself out before you do. If it doesn't, then go to them and ask them again, this time you can be a little bit firmer about it (just not too much). See how long it takes escalation-wise. This teaches you to start acting calmly, but assertively.
- See if you can make a game of it – have you and a friend (or spouse, or even kids) find out something about each other that really irritates, then do that action to (or in front of) each other. See who can hold out the longest (odds are good you'll both break down laughing, but try to see how long you can hold out anyway. Smiling is okay, but laughing or getting angry before the other person does means that you lose). This actually does two things – one, you learn a little something about the other person and two, you actually start building up a tolerance for whatever it is that bugs you so much.
- If you see a news item that drives you nuts (especially politics), force yourself (no, I mean it – *force yourself*) to come up with a valid point of view in support of the person or statement that got your blood pressure up. It's okay if you don't quite manage it, or if it even makes sense. The idea is to at least try and to always try. It doesn't even have to be a correct point of view, just something that will begin to teach you to listen to the other person in spite of not wanting to. Now certainly you can dismiss it if someone is spewing actual racist garbage or obvious things like that, but the goal here is to do it for opposing political viewpoints, opposing views on fashion or opinion, or even opposing religious views. This will teach you two very important things: how to read others' motivations (which is why you need to be serious about it) and how to tolerate other people who may not look or think as you do but are still your allies. The more technical term for it is called building empathy, which can work to your advantage in reading people, among other things. It also forces you to be creative, which is always an excellent skill to have in a world gone sour, no?
- Politely but assertively question authority if something doesn't sound right, even if you only do so mentally. If the answers make sense (minus any emotional response on your part) and have a basis in reason or logic, then you can accept the premise. Otherwise, reject it if you can. If you can, demand answers and look for signs of 'weasel words', or signs that the speaker is avoiding a direct answer to your question (after restating the question to clear up any misunderstandings, if possible). Know that you do not have any obligation to blindly and without question follow an authority figure except under very specific circumstances (e.g. you're in prison, you're an under-aged kid and your parent or teacher is speaking, etc). If the request or demand is reasonable and logical, you are certain that it presents no potential for physical harm to you and there is no imminent danger, you should be good, but take the time to weigh it in your mind before responding.
- Take the time to research and know your rights under existing law. This helps you become forearmed with the knowledge of what can be asked of you by an authority figure, what is clearly unreasonable and what would otherwise make no sense.

There are of course some things and activities which deserve no empathy, tolerance, contemplation, or even mercy. However, if you were to draw up an honest list of them, you would be amazed at how few those things are. Your chance of surviving long-term lies in how well you get along with other people (that is, outside of combat or attack situations). If you get along, you stand a better chance of living longer, simply because you don't end up standing out, or wind up aggravating the neighbors. It also helps give you a good reputation, which in turn means you are more likely to receive help when you do need it and it is available.

I know full well that this stands as a direct contradiction to what you have read earlier and honestly, it is supposed to. You see, you treat friends and allies (as well as potential friends and allies) one way, which is what this bit about personality covers. You treat attackers, criminals and suchlike a totally different way - both are needed in order to survive.

There is one other aspect of your personality you may want to tend to and it boils down to one question: How badly do you really want to live? Without a positive attitude and outlook towards the future, you'll probably end up dead anyway. The reason why is that the mind is the biggest factor in deciding whether you will survive or not. It can help carry you through times when food is scarce, or when clean water is hard to come by. It can help distract you from pain, boredom, fear and deprivation. It can help push you on when you start feeling hypothermia and it can help you find a creative way to carry the day. Learn to gain a fierce love of life. Learn to treasure each breath as if it were your very first and your very last. Find a reason to live each and every day until it becomes a habit. Even if that reason makes sense only to you, that's okay - but always find something new. A positive, upbeat (but not naïve), tolerant (but not a doormat) personality is likely the absolute best way to not only survive, but in the very long-term, thrive. Shoot for having both.

Physical Capability

You know the funniest statistic? It is not the massive weightlifter, nor the marathon runner, or the fat guy, who best survives adversity and disaster. It is usually the ordinary person who manages to make it more often than not. Here's why...

In a post-collapse situation, surviving has nothing to do with how much weight you can bench-press and has zero to do with how far or how fast you can run. In fact, those types of people usually have a harder time than most. Why? Because extraordinary physical ability often means turning one's body into a finely-tuned (and thus fragile) instrument, or in allowing one aspect to be hyper-focused on, while wasting away all the others. The weight-lifter trades in his stamina in exchange for the ability to move massive amounts of weight with just his body. The long-distance runner trades in her immediate strength in exchange for the ability to perform an action that lasts for extraordinary distances or times. A severely overweight person can go a very long time without food, by virtue of having prodigious amounts of stored fat. On the other hand, the massive fat store removes his ability to do any useful work for himself, or for others.

The best body shape to survive long-term is going to be a mixture of all three... or in other words the average person with maybe a touch of love handles or a slight (note that I said "slight") paunch on board. This leaves you with some fat to survive the leans times for awhile, enough strength to move, lift, or shove something when needed and enough stamina to do something (walking, running, etc) for reasonably long periods of time if needed.

I should mention something at this point and it isn't going to be very politically correct. If you have physical or medical disabilities, they will impact your ability to survive. If you are bound to a wheelchair, or are blind, deaf, or similar, you're going to have a lot more work to do in order to prepare and protect yourself. You will also need friends, who will have to carry a bit of an extra load in order to accommodate you. In return, you really should learn a few critical skills in order to make yourself highly valuable to the group, so that you're not the first one ' kicked off the island ' if things get rough. Medical skills, extreme impromptu engineering skills, electrical or mechanical skills, blacksmithing... skills you can actually do and do well enough to be considered an expert and most importantly, considered to be *needed*. A paraplegic doctor is someone that another person in your community wouldn't mind literally carrying on his back if it comes to that.

Of course, if you are perfectly healthy, it pays to have the skills too, but for those who are disabled, it is doubly important. This is because when supplies run short, there will be a lot of talk about rationing and empty bellies tend to be the most eager to ' prioritize ' who gets how much. Anyone seen as pulling less than their fair share will be the ones most likely to go without first. On the other hand, if you have skills that are considered vital, you end up pulling more than your fair share, enough so that others will never consider you as a waste of food, water, defense, or heat.

There is one important you need to do, especially for such a simple part of this chapter. You need to sit down and honestly take stock of your abilities. Can you lift your own body weight if you had to? Could you drag your spouse or kids out of a burning building (ladies, this means you too)? Can you walk at least two miles continuously, while carrying 40-50 lbs of stuff on your back? How much do you weigh, anyway? If the answers come up short, there's no need to panic, but you may want to get a bit of exercise in, at least to the point where you can do these things, even uncomfortably. Even if you're confined to a wheelchair, see how far you can go on the thing and how much you can carry on it. Otherwise, you're going to have to set your plans for something that doesn't require going anywhere and a lot of praying that you get left alone by people with bad intent.

And yes, this is becoming a prelude to...

Health and Fitness

True confession here: I'm the biggest hypocrite on the planet in this department. I'm saying this because as I type this sentence, I just polished off a fast food combo meal and I have a lit cigarette sitting in a nearby ashtray. I'm willing to bet that a lot of you out there are similarly situated. On the other hand? I know that my immune system is still well above average, that my last physical checkup was as close to perfect as is possible (the only difference being a cholesterol level just two notches above average). I can run two miles and walk twenty-five without stopping. I walk or bike to work daily. I make it a point to get up and walk around at least once each hour-and-a-half during the workday and I watch my posture. I eat a balanced diet throughout most of the week and I don't drink anything stronger than the occasional glass of red wine or dark draft beer (as in, perhaps a couple times a month at most).

Thing is, you really should too. No one here is going to demand that you buy a treadmill, go on a crash diet, or start eating twigs and granola. What you do want to do however, while easier, will take a bit of doing nonetheless:

- Start walking more. It's good exercise and helps you gain more stamina. Maybe carry a backpack with a bit of weight in it once you're comfortable with walking for long distances.
- Try to eat a more balanced and healthy diet. No one is saying to give up on the tasty stuff, but keep it in moderation.
- Keep the booze down to a minimum, and avoid it if you have a history of alcoholism.. This doesn't mean cast it aside entirely, but keep it in moderation – and never, ever, drink to get drunk.

- If you smoke, try to find ways to quit - if not, cut it down to as low as you can get it – a pack every 2-3 days if you can. If you insist on continuing to smoke, know full well that tobacco will run out quickly when civilization goes splat and that it will affect your attitude and disposition during the first 2-3 weeks that you're forced to go without (so keep nicotine patches or gum on hand).
- Try to avoid having to take any kind of drug on a regular basis - even aspirin.

Note that drugs (legal or not), will radically affect your chances of survival long-term. This is because once civilization collapses, that supply of drugs will (by rot or consumption) run out. This means your body is going to be in for a nasty shock. For those who rely on beneficial drugs, this can range from a return to chronic pain, to slow, lingering death. Diabetics who are insulin-dependent will be the most acutely aware of the impending kidney failure, coma and death that comes from blood sugar overload. Heart patients who run out of various cardiotropic drugs suddenly become very susceptible to a heart attack. Folks who suffer from chronic pain will see that pain come back a hundred-fold. If this is you, then you will definitely want to stock up on medications that you need – at the very least three months' worth. See if there are alternatives that are easier to come up with. For instance, people who suffer from chronic pain can find natural alternatives, such as marijuana, in those states which allow it. For type 2 diabetics, perhaps a large supply of Metformin can help bridge the gap between the insulin tapering off and full recovery after enough weight is lost for the pancreas to come back to full operation (for type 1 diabetics, I'm afraid it's going to either be insulin or death and the insulin will probably run out).

For those who rely on illegal drugs and alcohol (for recreation or addiction), I strongly suggest getting sober as soon as you humanly can. Otherwise? When the goodies run out, you will probably die a rather horrific death and place your loved ones in danger of getting killed along the way. The only thing that will save you (for awhile) is the fact that illicit drugs will likely remain around for longer than the legal ones, but even home-grown stuff like Crystal Meth needs base chemicals that will quickly run out. Meanwhile, getting drugs usually requires being in contact with people who will rip you off and happily plunder your goods without the slightest bit of scruple or morality. If you have a family or live with one, you place them in danger as well. Except for marijuana (which grows under nearly any circumstance), everything else *will* eventually run out. When it does, the withdrawals will at best incapacitate you (leaving you vulnerable to anyone who wants to take your stuff), or as is the case with some drugs, the withdrawal itself may very likely kill you. During the initial period of that withdrawal, you will also become a hazard to yourself and to everyone around you. It's better if you try and avoid that now, when there are medical facilities and therapists around who can help you get sober safely.

A note to people who have family members who are on illicit drugs: At the first sign of collapse, cut them off and throw them out, no questions asked and no conditions given. It doesn't matter how much you love that person - they will place you and yours in mortal danger just to get that next hit. They will quickly steal your supplies and let you starve, just to trade those supplies to their dealer for the next high. I also strongly recommend that you never let that person know in advance that you're preparing for anything, so that they have no idea as to what you have after you throw them out.

For folks who need certain types of instruments to survive? Unless you can somehow stock up extra parts, or fashion reasonable parts from local materials, there isn't going to be much hope. This isn't just things like dialysis machines, pacemakers, or oxygen bottles, but things as innocuous as C-PAP machines and the like. If you can live without them, try to. If you know of someone that cannot, then your best bet is to find a way to either make do, find a working alternative, or make that person as comfortable as you can on their way to the hereafter.

The good news is, many of these items can have alternate means of accomplishing the same thing. For example, if you snore or have sleep apnea, you can replace a full C-PAP machine with a mouth guard designed to keep the airway open. Do you need oxygen, but live in, say, Denver or Salt Lake City? Try living at a lower elevation… you'll see an increase of ambient oxygen by 200% or more and may even be able to dispense with the need for an oxygen concentrator or bottle. A manual wheelchair can come in handy instead of electric scooters and things like catheters can be boiled to re-sterilize them. Glasses are simpler and sturdier than contact lenses and Lasik surgery can obviate the need for either one.

For most of us, this will not be a big thing to deal with. Just keep healthy, eat right, do a bit of exercise and avoid becoming dependent on medications. That way, when things around you come crashing down, you stand a reasonable chance of not having the loss of modern medicine kill you off.

Where Do You Live?

Almost as important as what you are and what you're capable of, is of course where you live. This will strongly affect what you must do and what you are able to do – both in the short and long-term. In this section, We're going to do a couple of things... one, we're going to let you in on what to expect, depending on your location. Then, we'll quickly explore options for both short-term and long-term survival. This is because survival depends not on just you and your family, but on your community.

The Big City (Urban areas)

If you live in what is more commonly known as a "Metro" area, or in a completely urban area, things are going to be, well …interesting.

Pluses:

1) Any and all governmental assistance and shipments will get to you first.
2) Military protection, if it occurs, will be most likely be here.
3) You will likely be the last to lose electrical, water and sewage facilities
4) Commerce will last the longest in the city and will be among the last to fall
5) Technology, medical and specialty items will be the easiest to obtain here – both before and during the collapse.
6) Long-distance communications (Internet, telephone) will remain up and running here the longest
7) Weapons will be the easiest to find here (well, initially).
8) Barter and trade will become somewhat quickly available here

Minuses:
1) Unless you get that governmental assistance, supplies will run out in very short order.
2) Once supplies are generally exhausted, nearly everything critical and useful will be nearly impossible to come by.
3) Violence will be dramatically higher and criminals far more numerous.

4) Evacuation will be dramatically harder to do, since everyone else will either be busy packing to get out of town, or will be already out there clogging the roads by the time that you decide to.
5) If martial law is declared, you will have the hardest time doing anything about it (like evacuating town, or staying put if ordered to leave, etc).

Living in an urban area is a tough call to make. On the one hand, if the collapse isn't complete, you stand the best chance of riding things out in an urban area. On the other hand, if the collapse comes all the way down, you're going to be in for the worst ride and will only survive if you become the meanest, baddest, or cleverest individual still breathing.

Cities do have distinct advantages, however. During times of major crisis, the military, state and national governments will do their utmost to protect you first and since there are so many of you and your neighbors in one spot, the supplies will be directed your way as a priority. You also have neighbors who can help you far more easily. Finally, you have closer access to more things which you can convert into machinery or weapons as needed.

Not all cities are equal, however. Let me lay out some differences that you may want to keep in mind:

- Defense: If you live in a city where there are strict gun-control laws, the criminals will already be well-armed and you (along with your law-abiding neighbors) will not be. On the other hand, most towns where gun laws are more relaxed will see you and your neighbors on a more even footing with the criminal element and the criminals in turn will be less likely to attack you directly.
- Evacuation: If you live in a city next to a major river or coastline, you might be able to evacuate by boat, which will likely be less clogged than the freeways and highways. Cities with rail lines will also be (somewhat) easier to evacuate as long as you follow the railway out on foot, motorcycle, or even bicycle (just stay off the tracks directly).
- Supplies: People living in cities with large industrial areas may find it easier to procure specialty, rare and bulk supplies. Just know that criminal gangs will likely be wanting to hold these areas as well.

Next item to consider? Well, not all abodes in a city are equal…

- Living in a high-rise apartment provides the absolute best defense you can get (easily-covered hallways, one entry/exit or two at the most counting fire escapes), but will prove to be the hardest to keep supplied.
- Living in an apartment complex, or on the first three floors of a building makes resupply a little easier, but you lose a lot of the advantages that height would give you.
- Living in an actual house will be the hardest to defend, but the easiest to access and in spite of the openness, can even provide the best defense if the place is intelligently fortified.
- An industrial area provides the absolute best resources at hand, especially if there is a warehouse full of needed supplies in it. The problem is, if known about, it will be a very popular destination for both military and criminal groups.
- Business districts are going to be mixed, but likely will be stripped in very short order. Same with malls and shopping centers of any size.

Short term, a city will have plenty of basic resources initially, but these will run out quickly. Let's assume you intend to stay in the city, post-collapse. As long as all of the basics continue to come in and the power stays on, you will actually find it relatively comfortable. However, if anything stops (power, water, garbage collection, food shipments), things will get very ugly, very quickly. Your best bet here is to stockpile as much of everything that you can, but most importantly, form a strong bond with your neighbors, set up a defense perimeter, enforce it and fortify both your home and neighborhood against intruders. Be prepared to fight for what you have and always be on the lookout for things you need. Form small teams that can go out and scavenge (we'll show you how later on).

Long-term, living downtown is going to be very hard to do. You can set up small industries of sorts and use the rare and specialized goods (and skills!) to trade for basic goods from the countryside. Once the population comes down enough, your (now smaller) community can even start to establish small farms (or enough gardens) from parks and open spaces to care for your basic needs and small factories from remaining industrial areas.

The only real problem you're going to have here is crime and criminals. Short-term, gangs and individual thugs will rule the streets. Eventually, if you want a thriving long-term community, you are going to have to hunt down and either convert, co-opt, or kill off the remaining criminals. The chances of pulling that off will be pessimistic to say the least, but it will be required. Expect large parts of the city to be simply abandoned, but on the fringes, expect a lot of surviving entrepreneurs to set up shop for various needed skills, services and goods. Scavenging and rebuilding will be a huge part of the skills that the city can provide.

The Suburbs (both suburban and "bedroom communities")

I'm including "bedroom communities" in this category as well, since while most of these towns were once independent in their own right, their complete independence has withered away in favor of the residents who now work in the nearby big city. In many cases, these towns have been absorbed into the larger city wholesale, with no difference or marker between the two (aside from the obligatory road signs and tax records).

If you, like most people, live in a suburban area, you've actually got a somewhat easier time of it than the folks downtown due to the larger spaces on which to grow food and neighborhoods that are often coincidentally designed to be cordoned off into somewhat defensible areas. However, you still have a lot of neighbors - they're just not as close by. If the collapse is not complete, you still stand some chance of receiving governmental and military aid. You also, unlike the urban folk, have some options that they simply do not.

You will also find neighbors who are more willing to help out and a wider variety of survival skills that one wouldn't normally find in a densely-packed urban area. Whereas downtown you would likely find lawyers, doctors, architects and business types, out in the suburbs you begin to find mechanics, carpenters and folks with more blue-collar skills that can become highly-prized in a post-collapse society. These are also areas where you will most likely find hospitals, college campuses, larger warehouses and light industry.

Like cities, not all suburbs are equal...
- Defense: Small suburban communities are somewhat easier to defend, as streets can often lead to cul-de-sacs and similar limited-access route plans in and out of the neighborhood. If the terrain is naturally rough anyway (mountains, hills, etc), defense becomes even easier in many neighborhoods (especially if your neighborhood is

along or on the top of a steep hill). More often than not, there are other natural (but light) defenses built-in, such as fences, canals, streams and small rivers, etc.
- Evacuation: Evacuation is not as tough as the strictly urban areas, but can be if your neighborhood and surrounding areas include a lot of winding or maze-like access roads. The odds are not usually in favor of there being a navigable water route nearby, either. There are however more open areas (small farms, parks, etc) that can supply needed escape routes if absolutely necessary. Evacuation from an apartment complex will be the toughest of all, since nearly everyone is going to try for the small number of (usually narrow) driveways leading in or out of the complex.
- Supplies: There is a good mix of supplies here initially – shopping centers, clinics/pharmacies, hospitals, parks and even schools. These can be used as depots to gain raw and manufactured supplies (covered later), land for farming and sources of water (ponds, streams, etc). Just note that they will empty very quickly.

Within suburbs, it is more often than not that you will find exactly three types of homes: Single-family dwellings, apartment complexes and duplex/triplex/condo homes.

- Apartment complexes are the easiest to defend if you live on an upper floor, due to restrictions in access – both in roads/driveways and with hallways. There is more often than not a single walk-in entrance to each apartment, with a second entrance leading to a balcony or patio. The downside is in the second entrance – it is often a sliding-glass door, which may have to be barricaded on second-floor apartments and must be barricaded on first-floor ones.
- Duplex-triplex (condo-style) homes have an advantage of being covered on one (or more) side(s) by an adjacent home. This can possibly be a disadvantage if your next-door neighbor wants to break in and get your stuff (he can dig through the wall), but you'll likely hear it happening before the neighbor gets in. On the opposite of that, if you trust your neighbor completely, you could, together, create doors or passageways between your homes for mutual support and defense.
- Single-family dwellings will be the most common and have the widest variety of results. These homes are usually made of wood framing, with sheet-rock (gypsum dry-wall panels) inside and a thin layer of siding (vinyl, wood, or aluminum) or brick outside. Homes with a brick or (real) stone facade are going to be easier to defend and homes made with at least an outer shell of brick, stone, or cinder blocks are going to be the strongest that you can typically find available. With more property surrounding the home, you can also begin to consider defense in-depth and plan for things such as gardens, an outhouse and (possibly) a water-well. You also have the greatest room for modifications, such as cleaning and re-routing gutters to catch rain-water, out-buildings, reinforcements and similar changes that an apartment landlord or HOA won't allow. This type of dwelling also gives you the greatest amount of storage.

Short-term, you as a suburbanite will find many of the same conditions that the city folk will find. However, with a lower population density and homes equipped with amenities that help with independence, a lack of power actually becomes tolerable and lack of water can, in many areas, be fixed by simply digging a well, or by purifying water from nearby creeks and streams. A lack of food and other critical items will however, still need to be addressed. This means you can set up priorities as to what to stockpile. A good suggestion of top items would be food, weapons/ammunition, a means of purifying water and medical supplies. If you live in an area that doesn't have a nearby large forest, you will also want to consider stockpiling firewood (unless you feel like raiding empty houses for furniture and wall studs just to keep warm.)

Organizing the neighborhood for mutual defense will also be a very high priority, as gangs and criminals will be scavenging very heavily and will eventually come out and attack anyone they see as having supplies.

Long-term, your community can more easily establish farms (in parks and playgrounds), gardens (any large backyard) and will have more than enough of scavengable materials for local needs. Skills will be more generalized, but sufficient to thrive on. Most of the markets that exchange city goods for country ones will likely be established in the suburbs. Eventually, larger industry may establish itself here.

Like the big city, criminals and gangs will recognize that there are plenty of things to take here and will, like in the city, establish bases here. Unlike the city, these criminals will focus their raids on shopping centers, hospitals and any place where there are lots of supplies in one spot. This might spare you for awhile, but you will likely need many of these supplies as well and defending your own neighborhood is still a top priority. Eventually, like in the city, you will have to band together and hunt them down, or they will start raiding you directly once they consume their own existing supplies.

Small Rural Towns

Many of the more popular (at least recently popular) post-apocalyptic shows feature characters who live in a small town (we'll explain why in a moment). These are usually communities of up to around 3,000-4,000 residents. The number is a rough estimate, because it encompasses a ratio of acreage to people, as well as isolation from the larger towns and cities. In general, any town that is more than 30 minutes' drive (at freeway speeds) from any neighboring town, has less than 5,000 permanent residents, is isolated from its neighboring towns by a large number of farms, large national parks, or large tracts of wilderness, but is still large enough to have its own basic amenities (post office, city hall, library, police force, etc) qualifies.

Most of these towns gain their income from one of a few industries: farming, mining, logging, or tourism. That main industry will usually determine the makeup of the town' general character, supplies that can be had in bulk by the community and various other factors.

Let's explain why living in the sticks is such a romantic idea: Nearly everyone in film, TV and books has some sort of apocalyptic fantasy of living either out in a small town (a'la the television series *Jericho*), or living in the country and roughing it in a cabin somewhere. Why? Mainly because movie producers find it cheaper to film in a smaller town, but there are other factors as well: small/rural towns usually means a ton of old technology lying around, a lot of horses and a lot of people who know how things were done 'in the old days' and very little crime aside from the one family of inbred villans that inevitably gets portrayed. As a bonus, the sense of "isolation" in such flicks leaves open a lot of plot possibilities. Problem is, very little of this is true in rural areas today. Nowadays, most rural residents drive pickups, four-wheeler ATVs and cars - horses are a somewhat rare (and costly!) hobby nowadays, even in many (if not most) rural areas. Most rural folk who grew up to adulthood minus electricity and with primitive living skills? They're currently in nursing homes, dying by the hundreds of thousands each month from old age (rural electrification was largely complete in the 1940's). Rural crime is pretty high in a lot of areas throughout the nation, especially when you consider that Methamphetamine is just as common and available in rural areas as it is in the largest urban centers.

While such things as isolation (thus relative safety) and more abundant natural resources may be appealing, riding the storm out in a small town isn't always the best idea. The reason why centers around the fact that small towns usually mean tight-knit cliques, a large distrust of anyone who has no family history there and being the last place to expect any sort of official relief, such as governmental/military supplies, security and technology. Even in peaceful times, one thing that will be in short supply in most small towns? Jobs. Unless you live in a company town and work for the predominant industry, or own a farm and are able to make a living at it, you will have a very hard time pre-positioning yourself there in advance of a collapse and an impossible time afterwards. Meanwhile, if you already live in a small town, well, here's the situation…

Not all small towns are equal...
- Defense: Many small towns are impossibly hard to defend if assaulted. In prairie and Midwest/Great Plains areas, roads are often laid out in a straightforward grid pattern and everything is likely visible from a great distance. In mountain areas, defense is far easier to accomplish, as roads will likely be limited to whatever the terrain offers. In heavy forests, you get a mixture of both. More often than not, your neighborhood will consist of the entire small town, or at least a sizable fraction of it.
- Evacuation: In an isolated small town, evacuation is rather simple and easy to accomplish, no matter what part of the town you happen to be in. On the other hand, odds are nearly perfect that, post-collapse, you will not be getting any help from anybody if you do have to evacuate.
- Supplies: Small towns that are centered around the tourist trade will likely not have too much of anything aside from basic infrastructure (to insure top-end amenities to tourists) and will get nearly all of their supplies via just-in-time delivery. Small towns centered around harvesting natural resources (farming, logging, mining) will likely have more than sufficient resources centered around the particular product in question (wood, a particular type of grain or livestock, a certain mineral, etc), but not too much of anything else, which will rapidly become hard to come by.

Within these small towns, you likely will not find large apartment complexes, and definitely not any large apartment buildings. You will however find plenty of these...

- Single-family dwellings: Unlike the typical suburban home, these dwellings will have been built with a lot more attention to sturdiness and detail (excepting pre-manufactured homes). Otherwise, these are covered in the previous section and before and the same rules apply here - just with far larger zones of defense around the home.
- Pre-manufactured/trailer homes. While these can also be found in suburbs (and even some urban areas), you will find them predominately out here, in the small towns and farms. To be honest, unless you really reinforce one, or desperately need shelter, these are usually not the kind of homes to live in long-term and for a number of reasons. Among these are the need for specialized chemicals to maintain them (e.g. roofing sealants, anti-corrosives), the poorly insulated construction, the thin walls (usually aluminum), ease of corrosion (usually of the steel undercarriage) and the typical window size/placement that makes keeping watch almost impossible from indoors.
- Above-store Apartments: These are usually the second (and occasionally third) story of the small businesses and shops in the middle of town. You can often find them in larger towns as well, but they are more predominant here. These are going to be older structures and will need more than the usual maintenance to remain useful. They are similar in scope and detail to a typical 2nd/3rd level apartment in a downtown/urban

building. Finally, they are not always occupied (and haven't been for decades), so you may have a whole lot of work to do in order to make such a space livable.

Short-term, you're likely going to have most of the basics covered, as most homes in rural towns are (out of necessity) equipped surprisingly well for self-sufficient living. A new water source will likely have to be established and power can be done without for most things. Community defense will have to involve the whole town and perhaps a couple of outlying areas. You will have a harder time obtaining anything specialized (medicines stand out as a big example), but these can possibly be bartered for.

Long-term, small towns will be very likely to thrive - assuming they ride out the worst of the storm. A working relationship with farmers, combined with an increased population from manual laborers moving in to work the farms (okay, escape the city), will cause a lot of rural towns to swell in numbers and experience the most growth. Eventually, this is where prosperity will likely return first (so long as the town not only survives, but works intelligently towards that end).

Your biggest problem will be the isolation. With huge distances between towns, roads and pathways will be chock full of hiding places for raiders - both those who attack (or steal from) the town itself and those who lay in ambush on the roads outside of town. The former will either attack openly, or will more commonly try to sneak in and out. The latter will lay in wait on the roads, looking to steal from refugees at first and eventually from those traveling with supplies to barter or bring home. You will require a bit of reconnaissance and intelligence to survive such journeys once collapse is in full swing and eventually these bands of raiders will have to be either converted to living a more civilized life, or will need to be hunted down and killed.

Isolated Areas: Backcountry, Islands, Coastlines, Boating It

For those folks who live alone way out in the sticks, you already know what you have to deal with. Living out there in the farmlands or wilderness means that you already have to be at least somewhat self-sufficient and that you definitely have to pay attention to security and safety. You also have to insure that your home is built to better withstand the elements, which tend to be a bit less friendly in these areas.

Just like anywhere else, not all isolated areas are equal and not what you think:
- The nearest help you'll get is likely miles away. The more isolated you are, the farther away that help will be. This means that defense, medical help, supplies and pretty much most of what you need will either need to be on hand already, or will require a lot of transportation. This in turn exposes you to revealing where it is you live if anyone cares to follow you home.
- As demonstrated by the collapse of Argentina's economy during the 1980's, the folks who had the hardest go of it were those who lived in isolated farms. Gangs would often form large numbers and quickly overwhelm a farm. They would then steal/rape/pillage whatever they wanted. Whether the residents lived through it or not depended on the mercy of the gang at that time.
- Isolated areas in mountainous regions are going to be tough-to-impossible to grow food on, but easy to defend with smaller numbers. Isolated areas on a coast or riverbank will provide better transportation (and a very solid escape route if planned intelligently), but will likewise be more open to view (and thus attacks) from the water. Isolated regions in the desert will require irrigation to grow any food at all, but have natural defenses in the form of the surrounding desert. Islands are the ultimate in

easily-defensible places, but the amount of land (and the number of people it can support) will be very limited. Boats are the penultimate in mobility, but are limited in the supplies it can carry. Isolated areas in the middle of large forests offer a large amount of concealment, but in turn it can limit what you yourself can observe around you.

Nearly all of the home types out here will be single-family homes, or at smallest will be large vehicles (e.g. trailers, large boats, etc). There are some exceptions and variations (e.g. homes partially underground, sailboats versus trawlers, large RVs versus single or double-wide mobile homes, etc), but by and large you will have one of those two types. See above sections for the applicable type, but we'll cover boats quickly here:

- Boats can vary between sail, engine, or a hybrid of both. The only way to survive in one for any long period of time is to have it large enough to carry a lot of supplies - especially fresh, potable water. The disadvantage of engine-only boats, in addition to fresh water, is fuel - you will need a whole lot of it. In either case you will have to go back for supplies and eventually to settle. The disadvantage to sail is in slow speed, but with a proper boat and rig one can literally circle the globe in it.

  You can however procure one thing with a boat that most folks on land cannot – seafood. You can use seafood as a plentiful barter material, depending on local conditions, the amount of storage you can set aside for it and resource availability.

Short-term and with sufficient supplies, you might think that you will find things rather easy and (relatively) the most comfortable. Because of the sheer isolation, most criminals and raiding parties may well find it unprofitable (and in some cases impossible) to reach you, but this should never be an assumption to rely on. You must not be complacent. You may be a target for local gangs or individual thugs because you have less manpower for defense and if you fail, there is no one around to stop them from taking their time in stripping your homestead empty. Also, as long as gasoline and operating vehicles are available, you might even find yourself a desirable target for those criminals who have local knowledge of not only the area, but of your location. You would be partially correct in that it is tough to reach you and as long as no one knows you have a large stockpile of goods, you will appear to be too much effort for the results. On the other hand, some may find your location to be a prime place where they can move in to themselves (of course that would be after they kill you off). Your priorities here are going to be a large stockpile of everything, but especially on home defense, defense of your own perimeter and perhaps the laying of traps and warning systems. A number of large and loyal dogs would also be helpful.

Long-term, you're still pretty much on your own though. If criminals and gangs are indeed driven out of the cities and towns, they will have only one place left to go... out where you are. You may get lucky, but the odds are not good. Assuming you can remain free of detection and thus criminal incursions, your life will remain isolated and largely dependent on trading for whatever you can in order to get the consumable things you do need. Contrary to popular belief, a 'mountain man' (or anyone isolated) is still dependent on commerce for many of the consumable basics and at the very least for ammunition. If you're on a boat, long term you're going to have to keep up maintenance on a vessel of which spare parts will be hard to find, then impossible to find. If you're isolated on an island (preferably with a group), you will have to work out some sort of self-government and utilize the local resources of that island as best you can. In all cases, you will have to integrate with the nearest community and even provide some services and goods to it.

By Region...

While we won't get too comprehensive, we do want to spend a little bit of time on some differences depending on the climate, since this will affect your priorities and situation.

- Northern Temperate regions (across most of the Northeast US) will have warm summers, cold winters, deep snowdrifts and overall will see moderate temperature extremes. With a shorter growing season, you will need fast-growing plants. The biggest concerns will be the temperature extremes (most particularly the cold) and population, since this is a densely-populated region – for good or ill. Weather is what most would consider normal, though with heavier-than-usual snow in the winter (especially along the Great Lakes areas). Your big priorities from this region will be a warm shelter (with few/insulated windows) and a means to house livestock in their own indoor situation during extreme winter weather.

- Southern Temperate regions are marked by relatively mild winters, hot and humid summers and long growing seasons. Population density will be somewhat lower, but still numerous (and may become more numerous as refugees in general flock to warmer climates). Summer will usually bring severe electrical storms and the occasional tornado (or even hurricane) can occur. Winters are relatively mild when compared with Northern regions and are usually accompanied by ice more than snow (snowfall/ice gets rarer the further south you get). The priorities here should be stacked towards defense, since this will be a somewhat desired area to live for survivors.

- Plains and Prairies are among the best for farming and ranching (especially for grains), but winters will generally be fierce and the summers boiling hot. You will find yourself out in the open and far more exposed, especially in the flattest areas of this region. Firewood will likely be hard to come by, as trees are generally short and/or clustered into copses and pocket forests. Natural resources will abound, but you will have to know how to harvest them (and note that many of them may end up in short supply.) Water will become tougher to find, but can be had with sufficient local knowledge. A common weather pattern in the summer involves severe electrical storms, with a good chance of tornadoes, at any time of day or night. In the winter, there is a danger of severe blizzards, where the problem isn't so much snowfall amounts (though these can drift to dangerously high levels), but the winds. Winter winds can become hazardous and will cool down even a moderately insulated house in a short period of time. Priorities here include securing a steady supply of firewood (or knowing how to burn things like dried cow dung), a good source of clean water and at least some means of concealment. A concealed ("bug out") emergency shelter a good distance away from your main shelter would be a very good idea out in this country.

- Deserts are hard country. Lack of water, massive temperature extremes (even within a given 24-hour period) and lack of resources in general (unless you really know how to find them) will pretty much kill the uninitiated. This is compounded by those cities in the desert and near-desert areas (Phoenix, Salt Lake City, Las Vegas, Los Angeles, San Diego, etc) which will be full of people (up to 7,000/sq. mile in LA), but very short on vital resources (especially food and water). Unless you have knowledge that others do not, resources will be practically impossible to get hold of and unless you're lucky enough to be among the first wave of refugees leaving (or have knowledge of a lot of unused back roads), transportation will be almost impossible as well (due to a combination of large distances and an environment that can be very hostile to human life itself). You have two big priorities here: Water and Shelter. You will have to know where to find year-round springs and/or wells (most sources of water out here are seasonal at best). Shelter, because the naked sun can literally bake you to death in some areas of the desert. After that, a constant and renewable source of food will become pretty large on your list of things to secure. One other big thing to consider is electrolytes, especially salt. In the desert, you will sweat a lot and the dry air will wick it away in very short order. This also means that salts and electrolytes will leach out of your body quickly and you need to replace them.

- Coastal Areas have one big advantage that the others do not – the potential for seafood (fishing, shrimping, oysters/clams, crabs, you-name-it). However, in many areas they have one huge drawback – large populations. The coastal regions of the United States (excluding Alaska) contain 53% of the total US population (source: NOAA). This means you're going to have a lot of neighbors here. Coastal climates are generally like those of the inland regions they are near, but are generally less extreme and on ocean coasts, temperatures are generally far milder by comparison. Coastal regions around The Great Lakes are prone to higher snowfall in the winter ("Lake Effect") and coasts in general are prone to stronger storms, due to having that big open body of water next door that isn't blocking the weather. All coasts will have to contend with occasional high winds. Your big priorities here should include finding some means of living off the sea (or lake), since you may be somewhat limited in the varieties of food you can grow (mostly due to salt air, salty and sandy soil close to the shore, etc). Fresh water may also become a big priority, though wells dug about a quarter-mile inland are likely the most productive in these regions (you're using the soil as a big filter). Weather can be variable, depending on what other regions are directly inland, though as stated earlier, temperatures are usually much milder on the coasts. Winter (and in some cases summer) storms can be fierce however, meaning that your shelter will have to be tough enough to hold up to it. If you are very close to the ocean (close enough to smell the salt air), know that metal objects (e.g. tools) will rust much faster unless scrupulously cared for.

- Mountainous regions (both eastern and western) are going be a mixed bag, depending on which mountains we're talking about. Along the Appalachians and Ozarks, the effects aren't as pronounced as they would be in the Rockies. In either situation, temperatures will be a bit more extreme than most other climates, but defense will be far easier. Resources will be somewhat harder to come by, though this will depend largely on latitude, proximity to larger bodies of water and altitude. Altitude? Certainly! In many mountain regions above 5,000' (mostly in the Rockies), the higher altitude means less air to go around (about 80% of sea level). Unless you have spent a long time in such areas, you will find yourself winded (and sunburned) much faster and until you acclimate, you will find yourself unable to do quite as much as you may be used to doing. The largest priorities will include finding land that can be cultivated (mountain soils are usually thin, rocky and tough to farm), shelter from the occasional violent storms (and heavy snowfall) and because of the higher altitude, more extreme temperatures.

- Semi-Tropical regions are going to rarely see cold, but you will see heat and lots of it. While not the life-threatening sun that you would normally associate with a desert, the heat will still dehydrate you, so fresh water will become something you will want plentiful access to. You will have to keep a sharper eye out for certain disease types, but otherwise if you hate cold, this is likely among the warmest places to live (at least temperature-wise). Foliage is usually thick and in many places can conceal your home (but also conceal any potential raiders). Weather is often fast and furious. Aside from hurricanes, summer storms can occur almost daily, but are usually short-lived. Your priorities here will involve shelter, though not necessarily for temperature reasons. Mosquitoes can carry some rather ugly diseases, bugs of nearly every description will want to get in and eat your food, so keeping them all out will be pretty big on your list of priorities too. In this region (as in the desert), electrolytes (and sources of them) will become valuable, because you are guaranteed to sweat them out.

- The Pacific Northwest sees relatively mild extremes in temperature, with occasional snow in the winter and the occasional heat wave in the summer (though both are somewhat rare in the depths of the area). Water is not a problem in this region, since rainfall is among the highest and most frequent in The United States and Canada. Note however that the rain usually confines itself to the winter months. In some areas, this rainwater can be more than sufficient to live off of exclusively. This region is often mountainous, coastal, or both. There are areas where the ground will support farming on a large scale, but usually the ground will be moderately rocky. The wilderness areas can be treacherous, though not necessarily due to climate – the largest danger lies in the hidden cliffs, crags and other falls. Foliage will hide practically everything. One thing unique to this region is a high incidence of molds and fungi, which include varieties that can cause chronic diseases. The coastal areas of this region involve very cold ocean water (even in summer), as well as winter storms and winds that can often exceed hurricane force (that is, greater than 62mph or 100kph). A semi-active earthquake zone, this region can potentially trigger tsunamis along the coastline and even a rare volcano eruption inland (Mt. St. Helens, which is still active). Your priorities here will be to try and take advantage of what features you can, stock up on foods (or pills) rich in Vitamin D (due lack of sunlight in winter) and to find areas where farming is possible.

- Arctic/Sub-Arctic regions (Alaska, Northern Canada including Hudson Bay, etc) are quite extreme and not recommended for any but the absolute hardiest (or absolute most well-stocked) souls. Summers are very short, often measured in mere weeks, so you can forget most kinds of agriculture (except perhaps along certain southern-end coastlines). Winters are long, very cold and in the Northern portions of this region, very dark. Weather is going to be mostly cold, with swings into extreme cold. Water is usually not too much of a problem here, though you will occasionally have to dig it out of the permafrost or melt it. If you are south of the treeline, you will have quite a bit of wood, as most parts of this region are largely untouched. North of the treeline and you're simply going to have to do without much, if any wood at all. An advantage to this region, out of very few, is the very low population density and fairly plentiful wildlife. People will be very far and few up here. Your priorities? Assuming that you are crazy enough to try at it, you will have to find a good, strong (against deep snow), very well-insulated shelter and a means to heat it. Solid, well-insulating clothing is not optional up here and you will need a lot of it, in layers. Also, a way to defend you and your stuff from wildlife is a must. Food and growing vegetables/grains will be quickly sniffed out and eaten by any animal that comes anywhere close to it and many of these animals will kill you quite easily if you disturb them – this includes animals such as the Elk, Moose, Mule Deer, Bear, Wolves, Cougar, etc.

- Swamplands and Wetlands, while not very common, do exist in enough numbers (and size) to be worth mentioning. Most of these will be in the Deep South, but you can find a few large swamps as far north as Virginia (The Great Dismal Swamp) and wetlands/marshes as far north as New Jersey. Swamps are generally areas where traditional living is going to be a bit tough. Digging a hole will result in a small quickly-filled pond (which makes burying anything nearly impossible), transportation will be a hardship through most areas due to extremely soft and/or waterlogged ground and growing any kind of food will require finding a rare-but-treasured patch of dry land to do it on. On the plus side, the soil will generally be very fertile. Also, it will be generally hard for the casual raider to attack your home without you knowing it and without them getting lost in the process. Among your priorities here will be a lot of local knowledge, both to keep from getting lost and to know how and where you can and cannot go. A means of keeping snakes, bugs and worse (in Florida and South Georgia, this means alligators) out of your home? That will be a big thing on your to-do list. You may also, depending on just how swampy things are, want to keep and maintain more than a couple of shallow-draft boats around. Be sure that as many of them as possible have a means of manual propulsion (that is, oars and/or paddles).

"Oh, crap – I do not want to live here! I'm moving!"

Okay - so you look around and all things considered, discover that maybe you might not be able to make a go of surviving societal collapse in the place where you are right now. Let's further say that you get the idea of moving. Well, this is not going to be as easy as it sounds. In fact, the following reasons will make relocation to an idea place quite a tough thing to do. Leaving aside family (especially families that thrive by being close-knit), you're going to have some obstacles to overcome:

- Jobs. Odds are rather good that the job you're working, or the one you want or need for your career, is simply not going to be available at your desired destination. This is especially true if you're looking to move to a lower-population area, since the industry and technology (and businesses that thrive on such things) will be less readily available.

- Wages, Cost of Living. Even if you find a job in your dream spot, wages are likely to be radically different than what you may currently enjoy. The cost of living may change wildly depending on the items you need and buy the most, even if the average remains the same. For instance, out in the country pre-packaged foods will cost quite a bit more than they would in the city, while fresh goods may become dirt-cheap. Same goes for gasoline; with longer distances to drive outside of town, you'll likely burn more of it in a given month. Conversely, in the city rent will likely be dramatically higher, fresh produce will likely cost more, etc. Long story short, whatever the area produces the most of will be cheaper than those things which it must import. Wages will vary wildly as well (usually tied somewhat closely to cost of living), so an offer of a really high wage in San Francisco may make you worse off due to cost of living than getting ½ your current salary in Mississippi, where cost of living is almost dirt cheap.

- <u>Culture, Religion, Race</u>. Not all places may be readily accepting of your particular religion, cultural lifestyle, ideology, or habits. For instance, if you happen to be homosexual, Muslim, Catholic (disclosure – I am one), Jewish, Hispanic, or the like, you're going to have to choose a little more carefully where it is you want to move to (to varying degrees) and I daresay that you may well want to continue living where you do. This is because there are many places which simply have little tolerance for people who are not like themselves (or are otherwise deemed harmless) and communities of these folks are going to be even less tolerant. By the way, this can happen to folks who are white and Protestant as well, so if you are, don't skip this point over so easily. If you find yourself desiring to live in what turns out to be a predominantly Native American or Hispanic area, you're liable to be heavily discriminated against, post-collapse. If you're not Mormon, living in many parts of Utah may put you in a bad situation (both pre- and post-collapse). Also notice that ideology fits into the picture as well. If you happen to be strongly 'conservative' in outlook, you may have a hard time fitting in with a community of people whose outlook is strongly 'liberal' and vice-versa. Now please do not take this as a signal to only seek out what you consider to be "your own", but do keep it pretty high up in your mind while you're looking.

- <u>Outsider Syndrome</u>. The smaller (and older) the community is, the stronger this phenomenon is likely to be. In many small communities, if you weren't born there and in some cases if your parents and grandparents weren't also born there, you may likely be treated with suspicion (and sometimes even veiled contempt) until such a time as the community decides that you can fit in. This can be partially alleviated if you have a highly valuable skill (to the community), but note that this amount of adjustment time can vary from a few weeks, up to generations in the smallest communities. Time, a home purchase, active participation in the community and a cheerful, helpful personality tends to make the acceptance happen sooner. However, if the collapse happens within a year or two of your moving in, odds are good that you may not exactly be high enough on the community's priority list of people to count on or to help. On the other hand, if you're a doctor, police officer, combat veteran, or the like? Your experience will likely be radically different, as your skills will be desperately needed too much for the community to keep you at a distance. However, unless you have a critically needed skill (most folks don't), you may consider making a move soon, so you can maximize the time you have available to fit in.

Of course, there are also the investments you have already made in your own present community. Not everyone can sell their house and suddenly move. The kid(s) will be acclimated and integrated into the local school(s) and you may well be all settled in to your local church, community groups, etc. You may know your neighbors intimately and have known them for years, which is a bigger advantage post-collapse than being in a perfect setting but with total strangers for neighbors.

As I mentioned at the beginning, this is how you can survive a collapse of civilization but at the same time *not immediately or radically change your lifestyle*. If you're still nervous about where you are, then make small changes, not large ones. If you're still nervous about your neighborhood, look for another in a neighboring county before talking to a real-estate agent in South Dakota, eh?

Conclusion

By and large, you can make it in most settings and regions, but some will be better than others. All of them have at least some advantages, but each will present a unique challenge to continued survival.

Who Are You With?

Are you alone? Do you have a spouse or girl/boyfriend? Are you married and have kids? What about your parents? What are your neighbors like? Who you live with affects what you're going to do, what you can do and how much you have to put together.

Before we start, we want to briefly bring a few concepts to the table. The following may at first make you seem paranoid and secretive, but honestly, these concepts will help you no matter what bad situation comes your way. In the case of a collapse of civilization, they will keep you from being pestered and perhaps even assaulted by your neighbors.

The first concept is security. In the United States Air Force, this is broken down into two pieces: OPSEC and COMSEC. OPSEC (OPerational SECurity) is the concept of preparing quietly, in a way that doesn't tip off the neighbors (or even family members) to what you're doing. You will find this a lot in other books and it is quite valid. However, it is often misidentified or confused with COMSEC, or "COMmunications SECurity) – the practice of not letting others know what you're up to by your speech, emails, Facebook postings, etc. They apply here, because of a simple fact: The only true way to keep a secret is to keep it to yourself. However, if you live with others, this is going to be impossible. Therefore, if/when you start letting others in your household know what you are up to, it is vital that they keep quiet about it. The summarized fact is this: the less people who know that you have a massive stash of food and supplies, the less likely you will be sought out, attacked, vandalized, assaulted, or looted when things get ugly. Note that this is the last time you will see those acronyms, because we did promise to not use those.

The second concept is ability vs. disability. If you have healthy, mentally normal family members, you're quite good to go. However, if you have an disabled parent, disabled kids, or a disabled spouse/partner, things are going to get tricky, depending on severity. You're going to have to figure down what these lesser-abled family members will need (on top of the usual food/water/shelter/etc) to live from day-to-day and from month-to-month. If there are medical needs involved, you will have a big obstacle in making sure those needs are met during a disaster and even post-collapse. If these needs cannot be met, then you're going to have to either find an alternative means of helping them out, or of helping them to be as comfortable as possible when the inevitable happens (no, this doesn't mean abandonment, either). The problems generally break down three ways:

- Chronic (pathological) disabilities are going to range from the easiest to the hardest of all to deal with. Most of them involve medications and without those medications, things can go from bad to worse in a hurry. In short-term disaster situations, you can often have enough medication on hand (even insulin) to bridge any drastic gaps in medical care. However, in long-term situations and collapse, things are going to range from permanent and chronic discomfort (e.g. Psoriasis, Crohn's Disease, Fibromyalgia), to death (Type 1 Diabetes).

- Physical disabilities are also something that will present an obstacle. While not necessarily life-threatening, a handicapped family member will present many obstacles that you will have to account for. These include finding a means of long-distance transportation that doesn't involve gasoline (in case you have to evacuate), maintenance/repair of any apparatus or devices (wheelchairs, prosthetic limbs, crutches, braces, etc) and even someone to help that person along when necessary (e.g. in the case of those who are blind, deaf, mute, etc).

- Mental disabilities, ranging from psychoses to mental retardation, but also includes degenerative conditions such as Alzheimer's. These are obviously a different matter altogether. The condition may or may not be treated (or even treatable) with drugs and may or may not require more than the average care. The family members with conditions of which there is no ongoing drug treatment, will survive just fine and their care will deviate little from what you're doing now. However, those members with mental conditions that require a daily (or even weekly) regimen of psychotropic drugs? This will become a great big question mark once the drugs stop – not only from the original condition, but from the withdrawal symptoms and from any drug interactions as one drug dries up while the others are still on hand. One further thing to note – these particular family members may well need to be kept in the dark about your preparations and plans, since it is possible they will be the most likely to broadcast it to strangers and others (not like it can be helped, but they may do so just the same).

In all cases, take the time to determine if and how often professional medical help is necessary during normal times, what is performed during those times, how it is performed and take an active role in trying to perform some of them yourself (within reason!), even if only for practice against the day when you will be the only one doing it.

The third concept is the skill-set. If your spouse is a doctor and/or you are an engineer or an infantry combat veteran? Your family's value in a community just went up immeasurably and you will likely be actively defended by your neighbors once things settle down. If you and your spouse are Intellectual Property lawyers who have no manual skills to speak of and you can't even make ice-water without a recipe? You're screwed and it won't matter how big the bank account or how pretty your Porsche in the garage is. Take the time to figure out what your family is capable of doing, both together and individually. As you go through the book, take the time to figure out what it would take for you to acquire some basic but needed skills that will make you more valuable – not only to yourself and your family, but to your community. Many of these skills can even become a fun hobby that you and the family can enjoy now, while things are still hunky-dory. We will cover quite a few of them later as we progress through the chapters.

(True story: I once worked with a lady who had a PhD in mathematics, was an excellent software project manager, spoke three languages and did blacksmithing for a hobby at home. She had her own forge and everything. I'm guessing that if civilization indeed goes down the toilet, her hobby will help her eat a lot more regularly than any of the other skills will. And just in case she's reading this book: Emily, you can skip this section – you're good to go. :))

Now, about your neighbors...

Let's step outside the house for a moment. Everything you just read up there applies to the neighbors as well, except you don't have to worry and/or prepare as much for them - or do you? A little charity does go a long way, especially during hard times (and it may even save your butt if it's done right).

At a start, define your neighborhood. How big is it? A good measure would be to check the maps and figure out what parts of the neighborhood are actually defensible. If you live in a neighborhood with lots of roads leading in and out, you'll have to radiate outwards in a circle of sorts from your house. If your neighborhood is in a large cul-de-sac or on a dead-end road, the road's main entrance would be a good first place for a boundary. The idea is to look around, find a chunk of houses that make up a smallish but cohesive group of homes and make friends out to there. If it comes to too many neighbors, you're going to have to scale back a bit. The idea is to be on good speaking terms with at least 10-20 of your neighbors. Sounds like a lot, but really, it isn't, especially when you consider that most of them may not be there post-collapse (not necessarily due to dying off, either – evacuation at the last minute, or simply moving away before it hits also cause the numbers to drop).

Before we get into preparations to consider for them, let's find out who they actually are. And by the way, how many of your neighbors do you actually *know and have met*, anyway? Maybe now (or very soon) is a good time to get out there and introduce yourself. But, while you do, introduce yourself one neighbor at a time. While you're visiting, here's a couple of things to (*discreetly!*) keep mental notes on (you can write them down later at home)...

- What skills do they have?
- What kind of provisions do they keep around? (Note – don't ask and don't sniff around. Just keep an eye out and guess a little).
- Tools (non-electric and non-gas ones preferably)?
- Abilities/Disabilities?
- Political (or rather, ideological) leanings?
- Religion?
- Means of warmth, if applicable? (if you see a fireplace or wood stove, cool. Otherwise, they may have a hard time keeping warm when it gets cold).

Now is a great time to get to know these folks, because you're not going to have much time to get to know total strangers when they get hungry, cold, scared and frantic.

Of course, maintaining secrecy as to why you met them in the first place and not blabbing about what you're up to preparations-wise, is an extremely good idea. The fun part is, you can discover all of the above things and more about your neighbors without really having to ask. You just have to keep your eyes open and talk about whatever (and since conversations usually drift to many of the other topics anyway...) People generally like to talk about themselves... a lot. This means you can ask questions that go a bit deeper (but not too deep - use your judgment) and mentally stock up on knowledge that you can put to use later on.

As for talking about yourself to the neighbors (you will have to do that, you know), just be honest, but don't even think of describing any preparations you have made for disaster unless directly asked. If asked, don't say anything other than something vague, like "well, we keep a few extra cans in the pantry and the grill full of gas just in case of a tornado/hurricane/earthquake/etc" and "Oh, I've always found it a good thing to have the kids think ahead and be prepared, in case we're camping and something goes wrong..." One exception though, is to brag a little on any critical post-collapse skills you have, but not in that context. For instance, if you're a rather good shot, you can brag about it as a hunter (but not anything stupid like talking about how you would shoot looters and such). If you work in the medical profession, that's a neat idea to talk about too. Just be damned good and certain that you can perform these skills that you're bragging on (especially any specifics), because a disaster situation is a real bad time to have your neighbors find out you were just full of hot air.

All that said? Honestly and meeting and getting to know your neighbors is generally a good thing and can benefit you even if civilization holds up for the next ten thousand years. You gain friends, opportunities you would have never otherwise had and in general the whole neighborhood becomes a happier place.

One thing to keep in mind is that knowing your neighbors and being friendly with them is an ongoing process. Folks move out, folks move in, things change and keeping in contact with everyone makes for a tighter-knit community, which is always regarded as an excellent thing.

Making Friends, Expanding Your Community

Now that you're (working?) on good terms with many of your neighbors, it's time to figure out which friends can be trusted and which ones you only keep as acquaintances. As you get to know people, you can start performing a few harmless tests on them... simple things that make a big difference.

Qualities you are looking for are things like these – will the person help you if you asked? If they ask you for help, are they willing to work together with you and put some effort into insuring the success of what they're asking for? Do they reciprocate on favors and assistance if you help them first?

For those who have made it past that initial testing, see if you can take it to the next level. Are any of your new friends willing to split the cost and go to a sporting event you both enjoy, camp-out/hunt together, or participate in some other shared common interest or event with you? Have a BBQ or other backyard cook-out... are they willing to bring supplies (food, charcoal, extra grills, etc – according to their ability and means of course) and more importantly, how many will (without being asked) stay around and help clean up afterward?

For friends that are of the same religion, how many would be willing to really go to church with you, instead of simply sitting in the same building? Now mind, prayer is often a personal thing and don't expect to be chatting together during, say, a Catholic mass or a Friday evening meeting at the local mosque or synagogue. However, take a quick and discreet peek at how fervent they are (or are not). What you want to do is to get to really know the person and have them get to really know you (outside of the prepping thing, of course).

This will require a bit of stretching on your part, too. Here's a way to take it to the next level and an explanation of why you should do it. No matter if you are complete atheist or a fervent Evangelical Christian, ask to attend a service with someone of a religion that you are not already practicing or familiar with. You don't have to pray they way your friend does, or violate your own religious principles, but you do have to remain respectfully seated (or standing?) and quietly take in the service. Point out the positives on your way home and ask questions about anything you're not sure of. Whatever you do, avoid any sort of debate or negative commentary, at all. This shows that you are (and you should be at the very least) interested in others' beliefs and that you respect them for who they are. It also comes in handy post-collapse, so that you're not handing out canned hams to a Muslim or Jewish friend, or if you're scavenging an abandoned Catholic church, you may learn to consider the wafers or wine in that little gold-trimmed box (it's usually under a red candle near the altar) as something that you emphatically do not pass around as casual snacks. This way, you don't accidentally start angering folks by doing something that, while appearing perfectly innocent, turns out to be highly offensive. After all, you're going to need all the help you can get to survive in a world gone rotten and you don't want to do it by having to apologize a lot.

Building A Group

Once you have gained a fair number of good friends, find out which ones are the best to be the closest ones. These are people you know you can trust and have had that trust proven multiple times over. These are the people who, in a crisis, are the most likely to come together and help each other. These are also the people you want to (gently!) bring up the subject of survival to when you're alone with each of them (probably best to not do it all together in a group or anything). Mention that you are worried about how things are going in general and that you have been somewhat preparing just in case.

Above all else - don't mention any politics, or go off onto any rants – those are not going to be necessary. It is sufficient to mention that you are preparing in case of natural disaster, deep economic troubles and the possibility of worse. You don't need to start rattling off your inventory and even with close friends, don't give anything beyond a vague impression. If they are curious, answer any and all questions as honestly as you can, without giving away any critical information. Discuss how the neighborhood at large would survive together. Explore topics that you find throughout this book. If your friend gains an interest, feel free to give them this book. If it is an electronic copy (how to find it can be found at the beginning of this book, or via your closest search engine), copy the book onto a thumb drive, a memory card, or email it to them. Let them work on how to prepare their own families – that is no (close) business of yours.

As time goes by, if you have at least one friend or many who get into it, start meeting up on a casual basis. Help each other out. Use the time not to discuss how the world will end or when, but how to best prepare themselves and their families. Start gardening together, or help each other garden. Share harvests as well – for example, if you have a grape vine, each year you may be swamped in grapes, so give them away freely to the group. Another friend may reciprocate when his crop of tomatoes come in and another may end up almost drowning you in squash.

You'll find that, far from being a bunch of tinfoil-wearing doom-and-gloomers, you're working together to build each other up and sharing both knowledge and produce. Start taking walks together. Start hiking together. If someone has a family member that is handicapped or otherwise going to have a rougher time of it than most post-collapse, put your heads together and find ways to make things easier for that family and especially for the handicapped/disabled person. You may well be surprised by the solutions you come up with.

Even if you never have a need to put these skills to use, with a bit of luck you will gain friends and friendships that may well last a lifetime. You also gain a lot of very useful skills around the home and neighborhood; you will find that people will come to rely on you as you come to rely on them. Overall, it makes life much more enjoyable for both you and them. By the way, in the career/executive world they call it "networking" and that is the most important survival skill of all for a species as gregarious and social as humans, even in a perfectly peaceful world.

Stay Put or "Bug Out"?

Before we go any further, we need to have you stop and think for a moment, because your decisions in this chapter will affect how you prepare and what priorities you will place on your preparations.

So... are you going to stay where you are when it all collapses, or do you have a better place in mind (and perhaps own property or have family there), but for some reason cannot live there now? Are you convinced that it's just better to pack up and bail, that anywhere is better than where you are now? In some cases (especially if you live in places like Los Angeles County), you may well find a better life somewhere, anywhere, other than where you are, even if you have nowhere to go within a reasonable distance (that is, the distance your car can go on ½ tank of gasoline)

This breaks down into three categories: Staying Put (pretty obvious option), Bugging Out (that is, having somewhere you know you can go to), or simply becoming a Refugee (not knowing specifically where to go, but knowing that it has to be safer/better than where you are now).

Staying Put

The advantages:
- You can stock up a lot more supplies.
- If you have good neighbors, you have a potential community to help you rebuild civilization.
- You know where everything is in the locale (or at least you should know).
- You have a better claim to property, distributed goods and rights to live there should the government (be it existing or new) start sorting things out.
- If you have good neighbors, you already know each other, making team efforts easier and smoother to accomplish.
- If you live in a heavily-populated area, what's left of the government will try to get aid and relief supplies to you first.
- No worries about false-alarms: If it happens, you're already where you need to be.
- Most everyone else is likely trying to run away from where you are.

The disadvantages:
- If things get too dangerous to stay, you may be less prepared (or even able) to leave with all of your supplies, especially since it would be impossible to simply pack it all into a car.
- If it ever gets to the point where you need to leave, transportation and infrastructure may be bad enough that you'll have to leave many critical supplies behind, in addition to the comfort items.
- If you eventually do have to flee, anywhere you go will already be full of other refugees who got there first, leaving you with less options and little-to-no chance of being welcomed in.
- Your neighbors, if unprepared and desperate, may turn against you.
- If you live in a heavily populated area, you stand a greater chance of governmental interference with your life.

In spite of the disadvantages, staying put has a lot of good advantages that overcome it in most cases, especially if you have a lot of good neighbors and the capability to make it long-term. It is preferred if you can do it, mostly because it gives you existing shelter, less danger (compared to the open road) and less competition among others who are fleeing the area. Because odds are good that a lot of people in your area might also be bugging-out, it gives you the chance to scavenge what they leave behind. You'll already know the area and will be better able to take advantage of it than someone who just arrived but has no clue as to where anything is. You'll know what neighborhoods or areas to avoid and stand a better chance of becoming trusted among your neighbors and among others who move in.

Most of this book is geared specifically towards staying put – either the whole time, or after you find a place to settle after evacuating your original home for some reason or other. Staying put will be the eventual state of being if you expect to survive long-term, with very few exceptions. This is because you will need the relative stability, the community and a permanent shelter if you expect to eventually grow your own (or help grow the community's) food, raise your kids, rebuild society and die of old age – and not just wander around until you drop dead from disaster or violence.

Bugging Out

In this case, we refer to the act of evacuating to a pre-determined haven. This could be land you have purchased and pre-provisioned, a relative's home (where they expect you and you have supplied pre-positioned there), a 'vacation' home or cabin that you have stocked, or etc.

The advantages:
- If you leave soon enough, you have a place to go, with supplies pre-positioned and waiting for you.
- If you have family waiting there, you may have a foot in the destination community's door.
- You can stock supplies there as well as supplies at your usual home
- You will be able to remain quieter about it all if it is remote.
- If your neighbors, well, suck? You don't have to worry about convincing any of them to help, or of trying to help them when things go splat (though perhaps you may want to leave some supplies behind with a note telling them that it is for their use, as a charitable act).
- You have a solid option available to you if things get too bad in your own neighborhood.

The disadvantages:
- You're going to need advance warning in order to beat the traffic jams if that location is more than 20 miles away; anything over 200 miles will need near-psychic powers of perception.
- If you do not have family (or some very trusted friends) waiting there and prepared for your arrival, you may arrive to find your property occupied and your stuff being consumed by one or more (probably armed) refugees.
- If no one is waiting there who will welcome you, you may not be welcomed at all, even if you own property at the destination.
- You may not be able to get there at all (a vital bridge could be down, widespread fighting could be going on between you and your destination, winter storms could block the roads, etc).

Bugging out is usually what most folks think of if you asked them *"What if...?"* and then latched a global or national disaster onto the end of that question. Most will claim that they can travel to a family member living in some remote farm or cabin, or on their own to a far-flung forest, mountain range, desert, or countryside. There is usually zero thought given towards how many other people will also be heading there, how much in the way of supplies (potential or actual) are waiting, or how many people that destination's local ecology can actually support.

The big problems with that instant theory are obviously manifold. This is true even if family is not involved and you happen to own a cabin or small farm that usually isn't occupied – especially if it is out in the boonies.

First, you have to have some reasonable chance of *getting there first*. This needs a bit of explanation: unless there are family or very trusted friends living in/near your remote cabin or doomsday palace, some stranger will have occupied your presumed sanctuary and will probably not want to be kicked out of it. This means not only will you likely have to fight your way out of town and fight your way along the route you're taking, but now you have to fight to get someone out of your new home. Oh and you have to get them out of it without destroying or losing all the stuff you have stored there. Secondly, all your stuff could have been scavenged and your new home left empty and broken - if not by criminals, then by otherwise honest refugees who figured that you weren't able to make it, so why let the supplies go to waste?

Another problem is that you have to actually make it there. If it takes a full tank of gas to get there, what do you think your odds are going to be if the tank is only half full? (Speaking of which, you do keep at least a half a tank of gasoline in the car at all times, right?) If it takes more than ½ a tank of gas to reach the destination, then you had better be psychic, because otherwise you'll never find a reliable way to re-fill that empty gas tank once the horde of refugees really start moving. Even if you have a full tank, most of it will be burned off sitting through all of the stop-and-go traffic (mostly stopped) along the way if your timing isn't absolutely perfect.

Finally, because of all the above reasons, you have to worry about false alarms. If you time it wrong and think that things are collapsing in full speed when in reality they aren't, then you get to take that drive of shame back home, then try and keep (or re-get) your job (assuming you just dropped the trigger and ran without so much as notifying your manager). If you were stupid enough to laugh at any neighbors on your way out, well, guess what they're going to do in return when you sneak back into town? Oh and your wife and kids are liable to be rather surly and somewhat merciless about it.

Don't worry too much though, in the section "Do We Stay or Do We Go?", we will at least try to minimize the chance of false alarms happening...

Going Refugee

In this case, we're exploring the option of not having a specific destination in mind, but instead intend to head for a general direction, some semi-deserted island, a wilderness area, what-have-you. For instance, you may have the idea of taking a fully-stocked RV or boat out and away from everyone else, to wait out the majority of the population and chaos.

The advantages:
- You're not stuck with being in one place, or a target that cannot easily move if needed.

- You can stay or leave at any time as circumstance may demand.
- You don't really have to plan too far ahead in many aspects.
- You're not stuck with going along any particular direction or route and can change plans as the circumstances dictate.
- You will be more able to take advantage of opportunities as they arise.
- If planned and executed properly, you can hang back in some ultra-isolated area and avoid the majority of chaos.

The disadvantages:
- You can't really carry all that much in the way of supplies with you, so organization will be essential. Whatever you have is what you will be able to use, unless scavenged locally later.
- If you're in a vehicle? Even a large RV or ocean-going sailboat will need replenishing (fuel, water, etc).
- You're a ripe moving target for any criminal group that sees and can reach you.
- Shelter will be incredibly hard to come by if you wreck your vehicle.
- Unless you're among the absolute first to get a hint and evacuate, your getaway vehicle will become useless and stuck in traffic, permanently (or stolen, if it's a sailboat).
- You had better be at least somewhat physically fit, because otherwise this option may well likely kill you.
- You may be stuck (and permanently) with a crowd of similarly deprived people.

In spite of all claims of being prepared, even among the serious 'preppers', this category is probably where the vast majority of folks may end up. Even if you are seriously prepared, you may have to spend some time as a refugee at least once in your post-collapse journey. However, while you do want to pay attention, this is more for those who know full well that there is no hope of staying where they are when it all begins to collapse, but they have nowhere specifically that they can run off to.

As an example, let's assume that we live in some rather huge city: Los Angeles, New York City, Washington DC, Miami-Ft. Lauderdale... Seattle-Tacoma, San Francisco... it doesn't matter in particular. Assume further that we live in a neighborhood within one of these cities that isn't exactly all that great for long-term survival. Now, traffic is pretty ugly on most days, even now when things are normal. These areas are also teeming with human beings, so any massive collapse situation will lead to anyone with a vehicle trying to get themselves and their families out of town, so you can forget about trying to drive your way out. This means you're going to need to either walk, or perhaps load down a bicycle with what you can. Or, if you're lucky, you have a large sailboat (or a boat with plenty of fuel and/or a route that takes you downstream). It also means that you're going to have to walk/bike/boat for a very long distance (up to 100 miles or more) before you can get sufficiently away from everyone else to settle down.

As dreary as it may seem, all is not lost. There are two options that you can prepare for in this situation that gives you enough of an edge – enough of one to hopefully find a place to settle down and survive long-term...

The first is to get (or already own) a large RV. Not a trailer, but a full RV camper, with enough room in it to pack with supplies. Yes, the same amount of food, tools, weapons, etc. Used RVs capable of holding that much stuff (and spare parts for the RV itself) are surprisingly inexpensive if you're willing to put some sweat and parts into it. You can pack it all up and have it ready to go, or you can have your supplies 'pre-staged' to be loaded on a moment's notice.

Meanwhile, you (discreetly!) can unload the RV as needed to take camping or hunting trips in, which helps keep the fuel from going stale and lets you get some actual use out of it. Just remember that whole timing thing, because the roads are liable to be very jammed, very quickly.

The second option and one that is little used, is to either own or get a moderate-sized sailboat, or a very large motorboat. You'll need one that is large enough to carry all of the supplies that you would normally have if you stayed home (well, as many supplies as you're able to fit in there.) For a family of three, this would mean at least a 30-40' sailboat. Expect things to get a bit cramped in there and expect to either live very close (within 2-5 miles at most) to a launching ramp, or to have it permanently docked at a close by (within 5-7 miles at most) marina. While a good used sailboat of this size will cost quite a bit (depending on area), if you live near the ocean, a very large (Great-Lakes-Sized) lake, or a large, open river, this could represent your best option for evacuating the city, since the vast majority of people will be trying to get out by road. The only trick is that there are only so many regions where you could conceivably set up such a thing and expect to succeed. On the other hand, it does open up a lot of other options that land-based folk simply do not have – namely, if there are islands or empty stretches of coastline that are isolated, you can conceivably have them all to yourself (and whoever you choose to bring along).

The one thing to keep in mind about either of these options is that you will have to know well in advance when it is time to leave and to act on that knowledge as soon as you can. This doesn't mean keeping an eye on the television's evening news, where the news would be too stale and too late by the time they get around to broadcasting what you need. It means keeping your eyes and ears open on the Internet, with your brain engaged and a close ear on news radio stations. You'll want to leave before the crowd does, but not so soon as to jump-and-run over a false alarm.

Final notes On Your Decision

No matter what option you plan to take, get up your endurance skills. Start walking... a lot. You may well need to use those feet to carry you and a heavy pack a very long ways. Try and get to the point where you can literally walk and be on your feet all day long... and you will be well on your way to being fit enough to survive, even if you never leave the house.

While none of this is a set metric or plan for you (yet...), it can give you some good ideas and the pros and cons to what you may decide to do. The one thing you will definitely want to keep in mind is that no matter what course of action you intend to take, you will need to research it further. If you decide to escape by boat, but the river you live next to has locks and dams blocking it within 20-30 miles in both directions? You're likely to be stuck within just that range before you're forced to start walking it. If you're staying put but you live in a crime-ridden urban neighborhood, expect any criminals that stay behind (hint: they all will initially) to try and become the new dominant force in your neighborhood and try to take everything you have. If you have a bug-out location all dreamed up, picked out and the closing papers are ready to go? Be absolutely certain that you can get there by more than one route, that you can get there in a reasonable amount of time (but that it is far enough away from the crowds) and that it is secure enough to withstand anyone else who may get there and find it before you do.

Whatever you decide, just know that you're going to be somewhat committed to it - in time, preparation and money. You can always change to any other options while civilization holds up, but be prepared to go through with it fully once civilization stops holding up.

Refugee Preparation 101

Even if you decide that you're roughing it out in your home no matter what, you had better be prepared for the potential of having to leave everything behind and travel.

The reason why? Well, what if things blow up suddenly and you're stuck at work, miles away from your home and family? What if you're on vacation or business trip when it all comes down and you're unable to get home? What if, post-collapse, your neighborhood catches fire? What if a large and well-armed gang starts moving in and taking over? What if the military forces you out? Quite simply, there are a large number of reasons why you want to have the means to travel long distances without dying of cold, starvation, disease, or dehydration in the process.

To this end, we're going to pack what is often called a "Bug-Out Bag", a "Get-Home Bag" and similar. Basically, these are small to medium-sized bags (or backpacks) that are full of goodies that you will need to get to where you want or need to be, even if that place is '... anywhere but here!'

How Many Of These Things Do I Need?

Ideally, you should have one backpack for you and one each member of your family over the age of six. You should also have an additional "get-home" bag for each adult member of the family who is gone from the home more than four hours per weekday and has a job or classes more than five miles away from home.

What Kind of Bags Do I Use?

First and foremost – unless you are actually a member of the military or are widely known for wearing military gear at work, school, or wherever, do *not* use military-style backpacks. Avoid them for two reasons: First, they generally spook the hell out of everyone if you walk into work with it and second, they're made for carrying mostly military gear – firearms, ammunition, grenades and etc. They can also be useful for carrying what we're putting in there, but are certainly not optimized.

For those bags ("get-home bags") that you keep handy in your car at work or school, a large well-made laptop backpack is just fine. It's unobtrusive, doesn't stick out, fits with your clothing and can carry a surprising amount of stuff for the size. If you work manual labor or construction, then a backpack of similar roughness in appearance will work perfectly. The idea here is to blend in and not catch attention from criminals, police, or anyone. In a mass evacuation, you want to look like one thing: just another anonymous working schmoe in the crowd trying to get home.

For those bags that you'll use as main refugee bags (or "bug-out bags"), things are a little different. Invest a little money and get some good, solid backpacks from a sporting goods store. These bags are going to be riding your back for up to weeks on end or longer. The more comfortable it is, the less of a beating your back and legs will get from it. Look for backpacks that can carry a lot for their size. A strong suggestion is to choose a backpack that is a muted earth-toned color, or colors that blends in well with the natural colors of your area (an appropriate shade of green for forests, beige or tan for desert areas, etc). Bright orange or yellow backpacks stand out and you really do not want to attract attention while fleeing across the countryside.

Incidentally, if you have a dog that is big enough and decently obedient (and you intend to take it along), look into getting a couple of saddle-bags for him. If he can carry his own food (and maybe a couple of non-essential items), then why should you have to?

Where Do I Keep These Bags?

This all depends on what the bags are for. For the get-home bags that are used to get you from work or school back home? Keep it in the car. You can also keep them under your desk if absolutely necessary, but usually stashing them in the trunk is a good idea. If you take the bus or train to work, then keep it under your desk, in your filing cabinet (just don't lock it while you're at work), in your locker, or at the work site. You carry it to and from work on the train/bus/whatever, just in case. This also holds true if you work in a high-rise or in a building, keep your car parked in an underground garage, carpool, or keep it parked at any distance beyond a single city block from your office.

The main bug-out bags can stay at home, preferably somewhere that is easy to get at, grab and go. If you're driving any distance to visit relatives, go on vacation, etc, stash one or two of your main bug-out bags in the trunk alongside your suitcases and such. Even if world events remain perfectly peaceful during your trip, those bags can be very helpful if your car breaks down in an isolated area and you have no means of calling for help.

In all cases, you want these bags in places where you can grab them quickly and literally run if you have to, not waste time trying to find them, rummage through a pile of useless stuff to get to them, or do anything that will slow your progress down as you leave.

How Much Do I Pack In There?

For get-home bags, the idea is to look at the distance between where you work/study and where you live. If that distance is less than 20 miles, then enough gear (food, water-filter, a small weapon perhaps, a lighter, small flashlight, etc) for a day or two will be fine, but pack for three days, just in case. If the distance is from 20-60 miles, then enough gear for 5-7 days will be needed at a minimum, in addition to some means of a quick shelter, or at least some means of keeping dry. Male of female, you will certainly want to keep a comfortable pair of walking shoes alongside that bag, unless you already wear them to work or school. High heels or dress shoes aren't very waterproof, are lousy for hiking and are guaranteed to tear up your feet. Ladies, if you do wear heels, a pair of comfy socks would be a very good idea as well – just keep 'em stuffed in the shoes.

For bug-out bags, you will want a lot more. The Shopping List in Appendix A is a good start, but you will want to eye each component critically. Keep an especially strong eye on getting the most bang per ounce, because you will be carrying every last ounce. A Bug-out bag should ideally carry about a week's worth of food (preferably two) and more if anyone in your party is under the age of six, since you will be carrying their food too. Each bug-out bag should also have its own water filter (there are many backpacking models out there) and enough goods to keep you fed, warm and hydrated for a week or two, plus some tools that will allow you to catch, procure and cook/boil food as you find it. You shouldn't have to live out of the bug-out bag indefinitely, but you should be able to survive under primitive conditions with it for up to three weeks if absolutely necessary. It should go without saying that you should test out and actually use every item you pack in there at least once. Buy an extra packet or two of the food you pack in there, then cook and eat the stuff under primitive conditions - just to be sure that you can do so, that it tastes decent and that it fills you up a little.

Again, please see the relevant shopping lists in Appendix A. It may cost a little bit, but you don't have to go too crazy – just make sure the bags and their contents are sturdy and that every last ounce or square millimeter is put to good use.

Planning The Trek Home (using a get-home bag)

This is the basic plan: If you're out of the house when things finally get bad enough that you can pretty much quit your job or school and go home (or if disaster strikes and you need to get home *right now*), you try and get home by whatever means you usually use. If it's by car, drive as close as you can get to home before traffic blocks your way. If you take the bus or train, try and ride it as far as it will go towards getting you home. The less distance that you're stuck walking, the better.

Let's get into some details about that route: first off, is your usual commuting route the only way home? If there are alternate routes home, especially those which don't require being stuck on a freeway, highway, or main road, they will be worth looking into. Try to find as many alternate routes as you can without having to go too far out of the way. Secondly, have you really measured how far you drive each way? If your car has an odometer, record it when you start your daily drive and when you get there. Drive directly to work or class and do not stop for any errands along the way. Another method is to go to a search engine mapping page and get "directions" from your job to your home. It should tell you how many miles it takes. If you use Google, it also has a feature that allows you to set up routes and approximate your walking times – just click the little walking-guy icon above the fields where you enter the addresses.

Now that you have some idea of how far off it is and even some idea of how long it takes to walk the route (as well as getting multiple routes), this is going to be the basis for your planning.

Take a good look at that route: How much of it goes through tight urban areas? This is not necessarily a bad thing, as long as the crime rate is low (we'll get to that), because the sight of someone walking there is not going to be all that big of a deal. How much of it is out in the middle of nowhere, or goes through an industrial park? This is a bigger deal because you'll stick out more; you'll definitely want to physically drive/walk that route and find ways of getting through them more discreetly.

Now, take a look at the crime rates of the neighborhoods you'll be passing through (how to track them in detail can be found towards the end of the "Who Do You Know" section). If your route (or most likely, the freeways and highways) pass through some bad parts of town, you will want to immediately re-route your trip. Steer well clear of these areas and for two reasons: If you're stuck walking home (likely in a mass panic or disaster), walking through one of these areas is likely to get a whole lot of unwanted attention, especially after dark. Come to think of it, walking openly through any densely-populated urban area, as everything collapses, after dark? It is a rather bad idea all around.

All that said, if you have no choice but to walk through industrial areas or bad neighborhoods, your best bet is to do it as quietly, quickly and as safely as you can. If you can, stick to roads where there are a lot of people walking – the idea is that criminals, like predators, prefer to pick off the weak, the sick and the **solitary**. Don't be any of those. If you're stuck all alone, then you're going to want to keep a very sharp eye out and walk down from one block to the next in a measured and confident manner. At each corner, quietly (so as not to be obvious) take a good look ahead of you, around you and *behind you*. Once in awhile, quietly and carefully cut over to a parallel street to avoid having anyone track you without you knowing it. Whatever you do, walk on the street-side edges of all sidewalks and slow down a touch as you approach corners or alleyways, making a wide arc around them.

Avoid the temptation to stop at an evacuation center, homeless shelter, or any such similar place. Going inside of these places may get you warm and dry, certainly. However, the management of these places will quickly and happily relieve you of your get-home bag and all of its contents and for the following reasons: First, no weapons allowed in the shelter. Second, any additional food in many of these shelters will probably make you a target for other occupants and the managers would prefer that such attacks not happen. Finally, in the case of impending collapse of civilization? Local, state/provincial, or federal governments may well use those areas as places to round folks up and forcibly evacuate them to refugee camps. The last thing you want is to not only not get home that night, but to wind up hundreds of miles away with no knowledge of your family's condition. Worse, you'll be at the complete mercy of a rapidly decaying government.

If you possibly can, avoid routes with choke-points, but this may not always be possible. Bridges, roads that go through rough country (e.g. cliffs and drops on either side), tunnels and other similar places may present points along the route where you may be stuck fighting crowds, get ambushed by criminals, or may be cut off, even if temporarily. If the route is not completely blocked at such points, get across them as quickly as you can without running (unless everyone is running) and move rapidly away from it on the other side. Under no circumstance do you linger in these places unless you have no other choice.

Know your routes! Practice driving along them. Walk them if you can, or at least parts of them. If there are any choke-points, have at least one alternate route to get around each point. Use some creativity whenever you can, but remember to keep it safe.

So how long will it take if you had to walk the whole thing? Well, if you are in decent physical shape, you can walk 20 miles a day easily under normal-but-leisurely conditions. If you were to walk for 16 hours straight (more or less) at the average human walking speed of 3 miles per hour, it would get you 48 miles in one hard day's walk. Now obviously, no one outside of an endurance race is going keeping that pace constantly... you'll be stopping for bathroom breaks, stopping to eat, stopping to rest on occasion... but 20 miles in one day is quite doable. A healthy and fit person can do 30 miles in the same period of time, but always assume in your planning on 20 miles at most. If you manage to do more, good on you, but do not skimp on your planning, thinking that you'll make a 35 mile journey home in one day. If you are older or have small children, assume 15 miles a day. If you're elderly (over 65 years of age), it goes down to 5-10 miles per day, max. If you're overweight by more than 40 pounds, I strongly suggest exercising and losing weight if you can, because you're going to have a 5-mile absolute limit. Anything more risks a heart attack or worse (yes, there is worse), especially since you're already under stress, you're tense and you're carrying a load (that get-home bag) with you.

Planning The Trek Out of Town

Unlike getting home, getting out of town is another problem entirely. A lot of the same rules apply (avoid bad or crime-ridden neighborhoods, choke-points, etc), but there's one great, big problem in addition to that need to get out of town: You probably have nowhere to go. So, this should be the first thing to tackle. Plan a general route that takes you away from town, but get some sort of destination in mind in that direction - somewhere to head for once you're safely out of town. It can be specific, or it can be general at this point.

What kind of destination? This is a tough one to call, mostly because most places are radically different from each other. In some places (say Miami, Florida), the choice is going to be obvious because you simply don't have many (if any) choices. In other places, (say Des Moines, Iowa), almost any general direction you choose to go can be workable.

However, before the folks in Central Iowa cheer, or the citizens of South Florida hang their heads and cry, let's go through some general guidelines, starting with what to avoid:

- If another city larger than 50,000 is nearby, go around it or go the other way.

- Do not plan to simply live out in the deep woods, the wild backcountry, or the mountains. Unless you're a trained forester, mountaineer, bushmaster, or survival expert? You're more likely to die out there than in a post-collapse city.

- If you have a boat and plan on living out on the open water until things calm down, it had better be one very, very large boat – enough to carry all the food, *fresh water* and fuel that you and your crew will need - for at least six months and perhaps up to a year or two, because if civilization collapses, it will probably take at least that long before everything begins to settle down.

- Small towns are a good idea, but know that the population there may well turn you away unless, as mentioned, you have some very critical skills, or goods that are in great need.

- Is Canada or Mexico nearby? Forget it. If civilization collapses, it will be global. This means the country you're heading for will either have an army busy with turning you away forcefully, or it will be in worse shape than the one you're leaving, making things even more dangerous. You may well have to deal with a huge wave of refugees leaving there anyway, ironically trying to get to where you are.

- Unless it cannot be helped, avoid all major highways and roads outside of town. The crowds will be taking those routes, eagerly stripping the countryside along the way of anything that can be used as firewood, food, shelter, fresh water...

Looks a bit depressing, but let's see what we do have in the way of places we can go:

- If you have relatives that you can trust, living at least 50 miles away but not more than 150-200 miles or so? You may want to ask them to store some supplies for you, prepare together in advance and you can at least try to make it in that direction.

- Check the news a bit (you'll have to dig around for it online), especially county and small news sources... you're looking for small towns that have been depleted of people or are declining due to unemployment and mass business closures. These places will have quite a few empty buildings and the remaining population may just welcome you, because they are less likely to have the needed skills to be a self-sufficient community.

- Look for places that are (depending on the population of your city or town) at least 50 miles away, but not more than 250. If you're in good physical conditions, you could walk 250 miles within 10-14 days - as long as the weather cooperates, the terrain is somewhat level and you encounter no troubles. Anything longer or farther than that and you start courting excessive risks, delays and trouble.

- Start taking drives out to the countryside. Look for the aforementioned near-ghost towns, or towns that are mostly empty. Get to know (in a casual way) the folks who are still there (by frequenting the local businesses there). Odds are good that the majority of them will be older, friendlier and such – if you do stumble upon one of these places and it is within a week or two of walking distance? Treasure it and plan to aim for that town if you have to.

- On those drives in the country, look for a places, even out in the country, that have reasonable road access, but is somewhat remote and out-of-the-way. Places where there are a few abandoned homes you could walk to in a relatively short period of time. Places that have clean flowing water year-round, decent soil (if possible) and enough nearby residents to help form a small community.

The places you do want to look for are going to be a bit out-of-the-way, but are not going to be stuck way out in the howling wilderness. Places you can drive to in a couple of hours if you're lucky enough to be ahead of the curve when it comes to warning, but that you could walk to in a week or two if you've no other choice for transportation.

Something else to consider: Plan for different times of year. If you're forced to be a refugee in the middle of a Midwestern winter, you may want to insure that you have enough clothing to avoid hypothermia and some sort of portable shelter. You'll have to plan on walking at best only half (or even a quarter) as far each day due to snow, weather conditions, etc. This may also alter which direction you can ultimately go.

Overall, it pays to plan ahead, look around and get a good idea of where you want or need to go. This will put you ahead of the vast majority of your fellow refugees and will give you something to shoot for. Most importantly? Having a specific goal, while certainly way less than perfect, at least gives you hope.

Note that even if you intend to stay put at home, its is a very, very good idea to start looking around and scouting locations anyway, just in case. You never know when you'll need to get out of town.

The Best "Bug-out" Vehicle...

This topic has a bit of controversy about it. A lot of the more hardcore types will immediately tell you that you should purchase the biggest, meanest, beefiest off-road diesel vehicle that you can still legally drive on pavement. On the other hand, there are distinct disadvantages to having something resembling an up-armored Hummer. First off, while an off-road vehicle dripping with testosterone can make short work of many obstacles, it certainly won't be able to tackle them all. Also, a big, stout, brush-eating vehicle also has a very bad habit of gulping fuel at rates that would put a locomotive to shame. This in turn means you would have to carry a lot of extra fuel to get the requisite distance from a major urban center, or that you will have to settle for dangerously closer destinations. You can somewhat compensate for it by having a larger fuel tank (or in many pickup truck models, multiple tanks). However, you will then be stuck with filling those tanks and not only paying for the privilege, but running the risk of not having (or getting) enough fuel for the vehicle during a crisis, which tends to become a rather important consideration. Such a vehicle also tends to make you a far juicier target for criminals with enough firepower or smarts to corner you.

Going the other way, while giving you the important advantage of being unassuming, well... that isn't too smart either. A car with a range of 600 miles on one tank can get you far out ahead of the fleeing crowd (assuming you leave early enough), but the cargo capacity is going to be incredibly small and such vehicles require roads that are relatively smooth and free from obstacles (you know, obstacles like stalled and burned-out cars, criminal ambushes, fallen trees, government checkpoints, etc).

So what do you want, then? It's going to be handy to have a vehicle that has at least some bit of capability off-road, though believe it or not, this is only a top consideration if you know you're going to be late getting out of town, or if you actually do live in some rough terrain. While it may seem that the obvious choice would be an SUV at this point, a minivan would be just as useful (if not more so) in gentler terrain due to its larger cargo capacity. A small pickup truck could be useful, but most of them tend to have grossly undersized engines and are incapable of doing much in any serious uphill drive, so be careful about what kind you get.

Your best bet? You'll probably be better off choosing a vehicle that has a moderately large range on a single 15 gallon fuel tank (say, 350-400 miles - better than 25 mpg). Choose something with a decent amount of cargo capacity, so you don't have to lash too much stuff onto the roof (which could get wet, fall off, get removed easily by someone else, etc). Choose something that runs well in the terrain (mountains, plains, desert, etc) that you live in. Most of all, choose something that doesn't stand out in a crowd. This way, you don't attract attention. Note that a camouflage color scheme attracts attention just as much as a jacked-up suspension, a luxury car emblem, or a bright red or yellow color will.

For You Business Travelers

There was once a time when you could simply take your 'get-home' bag and stick it on the airplane, just like it were just another piece of carry-on luggage. Well, the year 2001 changed all of that. Carrying any more than 3 ounces of liquid, knives or weapons and the like? Well, they're strictly verboten now. So... what's a business traveler to do? Well, if you can afford it, be sure to carry at least the following bits in your *checked* luggage:

- 7-14 days' worth of concentrated freeze-dried foods (at least 7 days if you can swing it)
- A national-scale car/road atlas, with at least some detail. This is usually not too bulky and can fit flat on the bottom of your suitcase.
- If you often fly to a specific destination on a frequent basis, more detailed maps from that destination point to home will certainly come in handy.
- A compass – and learn how to use it! You won't be able to recharge any GPS kits, so...
- Warm clothing – at least one set, with a spare coat.
- A hiker's water filter pump or kit (not the liquid kinds, but those with a ceramic filter).
- Two large pocketknives – they're in checked luggage, so it should be okay.
- A fire-starter kit – the flint-and-steel kind and at least a small survival kit.
- A pair of comfortable hiking or walking shoes.

If you find yourself stranded in a different city when it all comes crashing down, immediately try to get hold of your checked luggage, see if you can rent/borrow a vehicle by *any* means necessary and start heading for home. Get as far as you can, because walking hundreds of miles is going to really suck. Keep your luggage in the car next to you while driving (not in the trunk!) and follow the rest of the instructions that you will find in Chapter 2, in the section titled "Run To The Hills!" The only difference is that you're getting home (if your family is 'staying put', that is.)

One other thing to keep in mind – whenever you're out of town, have some sort of plan set up for the family and agree on whether they should stay and wait for you, or try to move on (say, to a bug-out location, to a safer residence with extended family, etc) and make your plans accordingly.

How To Practice?

Practicing is actually not that hard, but it will require patience and a will to improve over time.

A large component of it involves walking for long periods of time, which can be executed and practiced anywhere. A pleasant walk in the park then? Well, not exactly. Let's try hiking – first with just your ordinary street clothing. Once you are comfortable walking a couple of miles at a stretch, try doing it with a small backpack containing about half the weight of your bug-out bag. Walk for as many miles as you can with it, each time walking further if you can. Once you're comfortable with walking that way for at least 5 miles, start walking with a backpack that weighs as much as your bug-out bag does. Again, walk for as long as you can, each time pushing yourself a little further. Once you're comfortable doing that for 5-10 miles straight, then start exercising with a backpack that weighs about 10 pounds more than your bug-out bag. Why do that? Because we want to compensate for tiredness, stress and just in case you add stuff to the bag anyway. Continue with that (same routine, pushing yourself further each time), until you're comfortable walking 10 miles with it in the space of 3 and a half hours. This speed will put you at around 2.8 miles an hour, or just under the normal average walking speed without any weight (3 mph). This in turn will put you ahead of most folks, who will be unused to any weight they carry. Therefore they will take far more frequent breaks and over time, but your better endurance will put you well ahead of them under similar circumstances.

To keep yourself in shape, walk up the stairs at work, walk your dog (if you have one) for longer distances than usual and as long as you don't overdo it, you should do well in keeping up to snuff. Occasionally (say, once every month or two at least) hike with that big backpack for 10 miles or more at a stretch.

The second skill you need to sharpen is your sense of awareness. We've covered a few mental exercises, so this shouldn't be too tough to do. While you're doing all of that walking, try and take in as many details as you can along the way. Write down any non-moving details as you go (no sense in slowing down just to take notes...) Things like trees, trail conditions, particular items of interest, turns, signs, statues, things like that. Take in as many details as you comfortably can without stopping. The next time you walk that particular path or trail, compare your notes with what you see. Then, take another trail or path and do the same thing – write down details, then compare notes the second time around. After that, start doing it in ordinary circumstances.

Once you're used to noticing details that don't change, let's start noticing things that do change. On your walks, start noticing people, and looking for animals. There's no need to stare (in fact, don't even let them know you're looking), but try and figure out the emotions and intent of the people you see as you walk. Remember, try to do this without breaking stride. Do this every time and try and plan all of your exercise walking routes so that you walk by more and more people. Eventually, you'll find it entertaining in a way (people-watching usually is), but the reason why you're doing this is deadly serious: Post-collapse, you want to gauge the mood of a crowd that may become a mob, or determine if a couple of guys up ahead are just minding their own business, or if they are scoping you out as easy pickings.

Being able to read people and notice details is also quite useful in normal life - certainly in your career, sports and even in romance. The more you practice that, the sharper you get and it's an improvement all around.

A third exercise to get you in shape? Look for opportunities while you're walking. Keep an eye out for water sources. Look for places where you could rig a quick shelter for the night. Mentally note places that look like you could defend yourself from within them, but still have a means to escape. As you're driving around, take quick mental notes of places where you could go slightly off-road and around traffic if the pavement were suddenly jammed full of cars.

As a bonus, once in awhile take along a book that describes edible plants and greens. Start to identify those plants and see how many you notice as you go on your walks. Give one or two of them a small taste (and nothing more until you're certain that a given plant won't make you sick!) Supplementing your backpack foods with edible plants is a great way to make the food stretch out for longer than usual and they provide vitamins and minerals that are lacking in most commercially-prepared backpacker foods.

Conclusion

Everyone should think ahead and fully expect that you may be forced to walk long distances as a refugee would. The more preparation and planning you do for that, the less likely that a forced evacuation would mean an automatic death sentence.

How Will You Keep It Quiet?

Security is going to be the biggest aspect that, in the absence of any disaster or collapse, makes this whole prepping thing stand apart from ordinary disaster preparation. No, this does not require the adoption of code-words, nor does it require that you start encrypting your email or any such thing. What it does mean however is that you have to show a little discretion in your preparations. The closer a secret you keep it, the better off you are.

So... Why All The Secrecy?

It seems kind of strange that you are reaching out to your neighbors on the one hand, while on the other you are busy trying to keep the whole preparation thing a secret. Well, as much as we're trying to get you to love your neighbor, we're also wanting to keep you from getting mobbed by them, or worse.

You see, ordinarily, individual people are thoughtful, smart, caring, altruistic and kind. However, in a situation of great stress and compounded by the thought of starvation or worse, people will suddenly become panicky, desperate, thieving, thoughtless, heartless and even murderous. If even one of your neighbors know during the collapse that you have huge stockpiles of food and they don't? They will beg you, cajole you, insult you, assault you and even kill you to feed themselves, their families and especially their kids. However, if they don't know any differently and think that you're just as bad off as they are, you will have ready allies in hunting down, foraging and procuring additional food sources.

I can guess at what you're thinking already. My neighbors are kind, they're harmless and probably wouldn't hurt a fly... right? Wrong.

Put yourself in a pair of their shoes for a moment. Here - I'll help:

It's obvious now - things have really gone to Hell. The government promised to help see everyone through, but it's been weeks since any relief at all had arrived. There's no place that you know of to evacuate to and no camps close by where you will be taken care of. There is no Red Cross or other relief agency coming by to help any longer and what little initial help they did provide has dried up weeks ago. You emptied your pantry four days ago after a long, hard rationing of what was there. You're now living off the last few crumbs and scrapings you could find in the empty bags, boxes and cans. The local stores have been picked clean a long time ago - well, except for that one that a local gang camps out at but they're armed to the teeth. Your kids are hungry and your youngest daughter began feeling faint today. Practically every animal in the neighborhood has been killed and eaten by now by the other families. Last night you heard gunshots and found out that the old guy down the street was killed, his house looted and even his body is missing. Your wife, the strongest person you know, has started crying this afternoon in fits and starts. Your oldest son ran away two days ago, probably out on his own search for food, or maybe to join that gang at the store, just so he can eat. Your own stomach is roaring with hunger...

...but wait – your neighbor down the street has food! He's one of those "prepper" types. He and his family are definitely eating just fine, because you can smell them cooking dinner almost every night! You can even hear them laughing it up while they eat all that good, juicy food! Maybe if you went over there and asked for a bit of help, just until the government gets back out here with a shipment? Oh, wait, you did that two nights ago and got told that no, he only had enough to see his own family through.

But... damnit! Your family is starving! Maybe tonight, when he and his family are asleep, maybe you can go over there. You still have that shotgun you used to go hunting with, your old service pistol, and there's still a box of shells left. 15 years on the police force have taught you well about urban tactics, so you know you can't just bust in there alone. But hey - the Johnsons next door are in even worse shape - their little baby boy died three nights ago, after Mrs. Johnson's milk dried up. Mr. Johnson's got a pistol and a small rifle. As a bonus, he spent two tours of duty in Iraq. The Coopers two doors down have been grumbling about the smell of food wafting from that survivalist's house too and Mr. Cooper carries his shotgun with him everywhere now. He's a crazy old man, but he did a tour as a Marine in 'Nam. Maybe you three can sneak over there, do a quick B-and-E, catch 'em in their sleep and maybe scare them into giving up a meal or two, just so that all of you can eat tomorrow. Maybe take just enough food for yourself to get you and the family out of here and to get by until you can reach that government refugee camp you heard about in the next county. That smug bastard doesn't need all that food anyway! You've seen the rows and rows of canned food in his garage! He'll be fine until the government gets back out here! You and your two neighbors reassure yourselves of this over and over again, as you slowly discuss the details and quietly load your guns...

As your wife and kids go to bed hungry again tonight, you whisper to them that you're going out to get them some food and that tomorrow morning they will eat again...

Do you think I'm playing up the drama? Honestly, I'm not. I actually toned it down. Think this through for a minute - when it's your wife and kids who are starving, or even if it's just you, you're going to do anything you can to get fed. I do mean *anything*. Try to tell me otherwise, I dare you.

Now, on the other hand, if your neighbors don't know about your food supplies and you can successfully keep it to yourself even after the collapse, then they'll likely go pick on someone else who isn't so secretive about it - and not pick on you.

It's a lot easier to keep it quiet now, than it will be to keep the hungry neighbors at bay once things collapse. Then again, you might be able to help them out a bit, if you know how to do it right (and do it anonymously!)

Out of (their) Sight, Out of (their) Mind

This whole preparation thing requires not only keeping quiet about it, but also keeping the goodies out of sight. It makes absolutely zero sense to quietly stock up on provisions, but have them all lined up on the garage shelves and leave the garage door open all day long. Having a delivery truck pull up and unload pallets of food isn't going to do you much good in the ' keeping-quiet' part of things either. You're going to have to gather up your provisions slowly and quietly. For things like food and equipment that you go get, buy them a few at a time and make sure they're kept in shopping bags or boxes when you pack them in the car and as you bring them into the house.

In your home, find secure, cool and dry locations to store them. A basement closet would be perfect for this (assuming you have a relatively dry basement), as would other similar out-of-the-way places. Find places where even the nosiest neighbor wouldn't go snooping around and never open that area when you have company in the house. A better idea still is to find more than one place to stash things, so that even a burglar would have a hard time knowing instantly how much you have and where it's all kept.

Your kids should never, be allowed casual entrance to these areas, under any circumstance, especially if they are younger. Your spouse or partner on the other hand should know about them – but insure that they remain just as tight-lipped about it.

Speaking of family, those adults who live with you need to know not only what you have and have going, but they need to participate in the discretion. That is, they need to know how to keep things on the down-low as well. This means no chatting with the neighbors about it, no bragging on it online and certainly no going around trying to scream to your neighbors about how the sky is falling and how they should prepare as well *right now!* The whole idea is to still be an ordinary person, living an ordinary life... which is a really big goal of this book. Call it a family secret – you know, like having the proverbial crazy aunt living in your basement, but this time without the booze and the screaming jags.

Excuses, Excuses...

So, at some point someone, somewhere, is going to know about you buying some of the more specialty items. You can avoid some of this by buying online, but if there are good deals to be had locally, why not just get them there?

True story: Recently, Wal-Mart has begun selling "disaster ready" food items. They are very reasonably priced, can be stored for 20+ years, come in a wide variety and each can holds something like 20-30 servings per can. Each trip my wife and I make for groceries, we buy a few cans of these to bring home. Usually we have a separate cashier (it's a big store after all), but one of them asked point-blank what we were buying them for. My wife looked at me immediately and fortunately, the cashier was as well. My quick explanation was to bring up the severe snowstorm that we had in our area a few years ago. The cashier quickly picked up the story and described all the problems she had during that time. She then mentioned that it wouldn't be a bad idea if she and her boyfriend stocked up a bit on some goods, just in case. I suggested that she have a grill (gas or charcoal) handy and stocked up on some quick canned foods too, since electric stoves are usually worthless when the power is out.

Notice how that story mentioned nothing at all about the collapse of civilization, or of governmental conspiracies and the like? Notice further that the conversation was easy-going, friendly and explained everything perfectly. Since I only had a couple of the cans, nobody is going to get the idea that I'm some sort of "survivalist", or, translated, someone whose house will become the neighborhood's new post-apocalyptic grocery store. The only thing I'll have to keep in mind from here on out is to be sure I find a different cashier the next time I pick up supplies there (out of the literal dozens that work there, this is certainly not a problem). One other thing you should know: The store in question is a good 45 minutes' drive from my home. Nobody that I know (at least neighbor-wise) goes out there to shop.

As for other, more obvious things? After all, it's going to be a bit hard to explain a rainwater catcher in your backyard, or solar panels on your roof, now isn't it? Fortunately, there's a whole lot of folks out there who have been advocating these things (and more!) for years, all in the name of distinctly non-collapse reasons, such as environmentalism, saving money, etc. No harm at all in helping out the environment and in many areas, it also translates into saving money. ...and notice how there's no granola involved (unless of course you pack it in your food stores).

One final thing to keep in mind is this: Try not to leave any catalogs, literature, survival/prepper-related books, or the like laying around the house. Honestly, half of it is useless anyway and the other half you will want kept in a safe place anyway. Keep such books (including this one!) With your supplies, since you'll more easily find them when you go for your supplies than you will if you keep them in the open.

Keeping it all locked-up

This probably goes without saying, but you really should pay attention to the physical security as well. If you keep it in the basement and in closets with strong (but discreet!) locks, perfect. If not, you will probably want to do something more than just hide the stuff. For instance, if you have a kitchen pantry that's closed by only a curtain, you probably do not want to store anything really vital in there. This goes for any supplies or equipment that you will need to survive.

The only exception I can think of would be a bag or backpack full of critical supplies that you can simply grab and run with (you remember – they're called bug-out bags) and perhaps, once things begin to collapse, only the supplies that you intend to use that day should see any daylight.

Staying Undercover By Supplementing Your Supplies

There is of course one way to help your neighbors stay fed, stretch out your supplies and at the same time provide a plausible means of explaining why you're not starving during the more chaotic times of collapse. That one way is foraging.

You should have at least one guide to edible plants and animals local to your area, with enough info in it to make a solid go at foraging for local edibles. Take the time now to look about your local area and neighborhood. Plan for (and stock supplies against) capturing local animals (pigeons, rabbits, etc) and turning them into livestock. For both seasonal and year-round edibles, take the time to harvest and test-eat as many as you can *positively* identify to be edible. For instance, I have blackberries in my neighborhood. In mid-to-late September of each year, the berries are ripe and ready to harvest, from wild-growing patches that can reach up to an acre in size or greater. The trick is to get to them without scratching your skin to shreds (hint: lay large and thick sheets of cardboard or wood down and walk on that into the shrubbery), to know what to look for and know how to prepare them properly once you pick enough of them.

Once you get good at this, you can do one thing to help your neighbors out and to help them form a solid community: You can teach them the same thing. By helping them learn to forage, they can feed themselves (and you can still collect a few yourself to make it look good.)

Here's the trick - you stop by a few trusted neighbors' homes and let them know you're going foraging and invite them along. When asked why, offer to bring some back and eat the results with them. If anyone comes along, be sure to teach them carefully what to look for and what to avoid. Be certain to be seen preparing and eating the same plants, insects, etc. Build up the trust. After awhile, they're likely to go foraging on their own, or go along with you (for mutual support/protection, etc).

If you do this early enough (when things begin to get bad, but still not quite in a crashed state, civilization-wise)? You can begin building up trust and building up the seeds of a new community. You can use the time (while things are starting to get bad) to discuss mutual defense and to discuss and organize potential scavenging sorties. You can even do it in perfect peacetime conditions, as a way to introduce adventure and at the same time begin the earliest seedlings of becoming a community, even when it isn't desperately needed.

All you need to pull this off is a good, solid field guide to edibles in your area, a couple of classes if your county extension or community events program offers them and some research online while things are hunky-dory. Then, you practice a few meals with just your family, or you can invite the neighbors. Eventually, you could even hold cooking contests to see who can come up with the best foraged meal at a potluck get-together.

Conclusion

You don't really need to launch on this huge cover-up or conspiracy to prevent your friends, neighbors and world from knowing what you're up to. You can even begin subtly teaching them the skills they will need to keep themselves fed (to take pressure off of you). However, you do need to keep things discreet and to avoid making things obvious. Note that we will be touching on security a **lot** as we go through the book, and this is just the beginning (note that tinfoil headgear will never be needed, though).

What Do You Need?

I will not lie to you. For a worst-case scenario, this list is going to be quite long. However, instead of simply writing down a checklist of everything you're going to need, let's stuff the checklists to the back of the book. Meanwhile, we'll cover some of the basics first, then we can get into details. By then, you'll likely have a good idea as to what and how much you will really need and can start making up supplementary lists of your own.

Shelter

This one kind of goes without saying, though all too often, it does need to be said. The idea is that, according to the climate and weather you live in, you will need something to live in that is fairly warm, keeps the sun and rain off of you, keeps the bugs out, can keep intruders at bay and can more or less serve as a place of repose and rest.

Fortunately, in most cases, you already should have that – your home. If you own your home, you're much better off. If you're renting, you may have some other bits to think about. But before we go off on tangents, let's see what it is you want to keep in mind as you prepare your home.

- Something secure. If you cannot keep the bad guys out, then nothing you put in it will stay yours for very long. Securing your house means going over it like you would in keeping out burglars, but with a twist. The difference this time is that there won't be any police or 911 to come along and help, so you'll have to do more than merely discourage or scare off intruders. Strengthening doors, finding a means to add storm shutters (or even steel ones) to larger windows, replacing the sliding glass door with a pair of strong french doors (or at least something that will hold up to a pounding better) and similar. While you're at it, take a good look at your yard (front and back) and see if anyone can hide in there without you knowing it from looking out of the windows. Keep a few concrete blocks in the garage (or maybe make a big sandbox out of it for the kids in the backyard?) in case you need to reinforce and shore something up in a hurry.

- Something warm. If you require a utility delivered by the city to get your house warm, you're going to be screwed. You should, first and foremost, try to find/install some alternate means of heating your home. Even in sub-tropical and desert areas, temperatures can get dangerously low on occasion, so you will definitely want to consider either installing a means of heating your home without electricity/gas/unicorns, or you should make sure you have those means handy. The easiest way is to do this is to have a fireplace. The second-easiest (but better) means is to have a wood stove. Other means (passive solar, etc) depend on what you can afford and on what your local climate happens to offer. You don't necessarily have to install something permanent, but at least a temporary heating source that can become permanent (and have proper ventilation and fireproofing if you're using fire as the source of heat) is highly recommended. You could even build something and keep it in the garage for now, with materials and tools to get it indoors and running once things go splat.

In all honesty though, you really should put in either a wood stove or a large (-ish) fireplace, or insure there is one in any potential home you intend to buy (or even rent, if you can help it). This becomes a requirement if your place is outside of any suburban or urban area.

Also, if you can do it, insulate the place to within an inch of its life. It saves your energy bills now and saves fuel later (and in some climates, can push off or even eliminate the need for heating). As a bonus, the extra insulation in the attic or crawlspace gives you more places to hide stuff.

- Something dry and bug-free. Needless to say, keeping the rain (snow) and sun off of you is a good thing. Here, just keeping your roof in good repair should be sufficient. Keeping bugs out is going to take a bit more work, though. A tight home with screens, no cracks and no always-open entrances is a must in some climates, at least if you want to keep the bugs out of your food, your bed and your hair.

As a helpful tip in this direction? You can keep bugs and dirt to a minimum if you get rid of the carpeting. If you use hard floors (wood, stone, tile, whatever) and area rugs, you end up with a home you can keep clean with a broom and a stick (to beat the rugs outside with). In a post-collapse situation, carpet becomes a huge liability for fire, dirt, bugs and mold/mildew. Without a vacuum cleaner and an electric-powered scrubber/steamer (or chemicals to kill any fleas or lice that get into it), carpet becomes a big problem once the power goes out and stays out. Certainly there are non-electric carpet sweepers and the like, but they don't work as well and you still have all the other hazards.

Other very important considerations for your home is to have supplies and materials on hand that can be used for repairs. Try and set it up so that all non-electrical portions of your home can be repaired without power tools and/or rented equipment. Things to keep on hand may be things like:

- Extra 2"x4" and 2"x6" boards (about 10-20 of the 2x4s and maybe 10 of the 2x6s)
- At least one 4'x8' sheet of 1/2" plywood for each window of your house
- Extra fiberglass and 'fluff' insulation
- Extra roofing shingles and roofing compound (which, if kept sealed, won't go bad for decades)
- Extra utility blankets (e.g. moving blankets and the like)
- A few large boxes of 16d nails and perhaps a large box or two of smaller nails. Spray with a light machine oil once every couple of years to keep rust away.
- A large roll of heavy-gauge sheet plastic.
- A few large tarps.
- Plaster (preferably in powder form so it doesn't go bad)
- A full complement of hand tools (screwdrivers, a couple of hammers, a couple of larger hand saws, a full set of wrenches for both standard and metric sizes, perhaps a socket set, etc)
- spare electrical wire (seriously – even with no electricity, it can come in handy)

- a few bags of concrete mix (be sure to keep it dry)
- See Appendix A (Shopping Lists) for the rest of the list.

The good news is, you can actually use all of this stuff at one time or another if you own the home. Just be sure to immediately replace what you take from your stocks.

Water

Water is going to be a great big concern. Even in places where it rains all the time and it seems like you cannot get rid of the stuff fast enough, clean fresh water is vital to your continued survival. Without water, you can die in as little as two days wandering in the open through a harsh, hot desert. There are countless examples of folks stranded on the ocean who have died of dehydration. Let's face it - water is critical.

Now I know a little of what you're thinking at this point – well, there's a stream or pond nearby and you can just drink that, right? Well, not without a little preparation you won't. You see, natural water contains a lot more than just dirt... it also contains a few nasty things that you really do not want inside of you. Aside from the usually pile of bacteria and viruses, most unfiltered open water often contains traces of man-made chemicals, toxins, pharmaceuticals and pesticides.

The absolute best way to clean up the water before you can drink it is to distill it. This does a lot of good things at once: it kills germs, it separates out the dirt and bugs and the more volatile chemicals are either evaporated long before the water turns to vapor, or it stays behind with the rest of the crud. For the long-term, you can build and maintain a decent high-capacity water filter (it only takes a large container, sand, gravel, charcoal which you can make and some cloth) and follow it with a good boiling.

It may take some time between when the city water stops flowing and the time where you can mass-produce water. However, water is still available to be had, if you know how. Start by turning off the hot water tank infeed line (the pipe that feeds your water heater... there should be a valve that lets you shut that off). Then, shut off the pipe leading out of the hot water tank into the rest of your house. You now have anywhere from 20 to 80 gallons of drinking water that you can use once the contents cool off. Got a waterbed? You can bathe and do laundry with its contents for up to 200 gallons more.

Meanwhile, at the first sign of things going awfully bad, fill up as many extra-large containers with water that you can. This also includes the sinks (the kitchen sink with water to wash in, the bathroom sink for extra drinking water). While you're at it, if you have any spare bathtubs, fill them up as well and cover them with plastic. The sinks will give you up to 15 gallons extra per basin and the tub an additional 40-80 gallons (depending on the tub's size). If you have a working, covered hot tub in the backyard, then bonus! You can use that extra couple hundred gallons in there to wash your face and body with (but for Heaven's sake don't drink it - it's full of chemicals that you do not want to ingest! Also, if it has a chlorine system, using it on clothes should be done only sparingly at most.) Got a swimming pool? You then have enough water to wash with for a very, very long time... just keep it covered.

All this said, you may want to think ahead. As in, way ahead. The absolute best way to do it is to dig a well in an area with a clean water table and make it large enough to get a person in, so that you can get down there and clean it out once in awhile. Failing that, a small pipe well with a simple manual or solar-powered pump will get you what you need. Don't have a well and can't dig or drill one? Well, the next best method is to have a clean stream, pond, or lake nearby and a permanent filter that you can maintain over the long haul. The next best method after that would be (if your climate allows it) a means to catch and keep rainwater and filter that.

Something to keep and mind and scout out while there is plenty of time to do it: What you want is for your future water wells or other sources to be at least 50' from septic tanks, septic leach fields, livestock yards or pens and silos of any kind. You have to keep it at least 100' away from any petroleum storage (like gas stations) and from any fertilizer storage and handling areas. Keep the drinking water sources and/or wells at least 250' away from any manure piles (source, United States CDC) while you're at it. Figure these sites out now, so that you won't have to scramble for them later when things get rotten.

Next to a proper and secure shelter, water is going to be the most important thing you can secure, plan for and get hold of.

Food

Everyone's gotta eat. In spite of the fact that air, shelter and water are the most important things (in that order) to sustain human life, food comes in very close behind them. When it comes to disasters, everyone immediately thinks of four things: Food, food, food and plywood. Without water, people get listless and eventually die within a few days. Without shelter, people get listless, hypothermic (or overheated), way too calm, then end up dead within hours in some cases. However, without food, you can go for up to 4-5 weeks before you end up dead. This also means that if you're an ordinary healthy individual, you will be able to go for a a solid week (perhaps two) in a somewhat energetic state without eating.

Thirst and chill (or sunstroke) don't have nearly the same psychological effect as hunger does. When you're too cold or too overheated, you don't really get all crazy-desperate to find warmth or shade. When you're dying of thirst, you're often too weak to do too much of anything about it after a day or two. However, when you're hungry, you have a whole lot of time (and an astounding amount of energy) to think up all kinds of crazy and desperate means of getting your belly full. This is why you almost never hear of folks doing shocking things to keep warm, rarely hear of folks drinking toxic or odd substances to keep hydrated, but very often hear of things like cannibalism, eating spoiled foods, eating weird and strange creatures, eating dirt or grass, etc etc...

So, yep, you're going to need some food stored away. Question is, how much and for how long? Depending on the situation, a mild and/or regional disaster usually means you will get food again in a couple of days to a week, so a typical full kitchen pantry is more than sufficient. For a full-on collapse however, you're going to have to plan on feeding yourself from now until the day you die, and it will be a year or two before you can grow any. Long story short, food is going to occupy the vast majority of your storage and getting or keeping/preparing food will consume a huge percentage of your time once things collapse completely.

A common misconception is that you only need to run out and buy a couple cases of military MREs (Meals, Ready to Eat) and you're all set. That's a bad idea, for two reasons: One, the stuff really isn't nutritional after eating it nonstop for a year and two, it packs in way more calories per serving than would be healthy for the ordinary human being. This is because MREs are made for combat use and troops in combat burn upwards of 4,000-5,000 calories per day. Sucking down all of those calories when you don't really need to will cause you to get fat, causes intestinal problems in the long run and even runs the risk of giving you a whole bucket-load of health problems that you really do not want. An average human being uses about 2,000-2,500 calories a day with moderate exercise and 3,000 calories a day with heavy exercise (or in colder climates during the winter). You can get by on 1,600-1,800 calories a day for quite a few months, especially if you're moderately overweight to start with. Elderly adults can consume as little as 1,500 calories a day for a few months or so (depending on weight) before any adverse effects are noticed. The only type of person that comes close to easily swallowing an MRE-sized calorie intake is a late adolescent male, who can swallow up to 3,500-4,000 calories a day if he plays hard, but can certainly get by on less if/when necessary.

Another hazard of simply stocking up massive quantities of one type of food (be it canned, dried, freeze-dried, long-term, or whatever) is that you begin to lose a lot of variety and therefore a lot of nutrients.

The idea behind the right foods is to have a wide variety, stocked three ways:

- Short-term foods that you can consume quickly. These are the foods you would normally keep in the freezer, refrigerator, or cupboard. No more than about a week's worth (two at the most) should be kept on-hand. If you happen to have more (e.g. a chest or standalone freezer), then only count on the first week or two. Anything after that will thaw out and then spoil by week #2.
- Moderate-term foods that you would ordinarily eat, but does not require refrigeration. Examples include canned foods, dry pre-packaged foods and similar. These are foods that can last up to a year or more before losing flavor and are relatively easy to prepare. The idea would be to start eating these foods once the refrigerator is either empty, or its contents are no longer safe to eat.
- Long-term foods that can be stored for years on end (usually from 5 to 20+ years). These are the freeze-dried foods and long-term grains, which can be in bulk, though you will likely want at least a few of them in a quick-to-carry form, preferably stashed in your backpacks and bags... just in case you have to leave in a hurry.

Seeds are your fourth type and are like the others, only the most vital of the types. Before you run out to the hardware/garden store and throw a ton of packets in the shopping cart though, you may want to stop and do some research first. Get hold of at least two different (and well-respected!) gardening guide books that have information applicable to your area (you can use the Internet initially, but be extremely certain that you buy the books – the Internet will likely not be around post-collapse). Learn which seeds are good for your particular region and determine how many of them you will need. The idea is to keep the following things in mind, in order to find the best balance:

- The seeds must be hardy and put up with the climate while growing.
- The resulting plants should be low-maintenance, to minimize the work you have to expend on growing them. This also means minimizing the amount of fertilizer required to keep them growing.
- The resulting harvest must be easy to store for long periods of time, or can be dried or canned for long-term storage (say, at least 12-18 months canned, preferably more.)
- You must be able to harvest and save seeds from the vegetables, fruits and grains you grow (for grains, you save a percentage of your harvest for next year).
- There must be a balance of fruits and vegetables wherever possible, so as to provide the widest nutritional value.

- The best and most ideal produce will be those kinds which require the least amount of space to grow. Not only is this because of the limited availability of land, but because these will require the least amount of work.
- The seeds must be organic and not genetically modified or hybrids. Many common seeds and varieties nowadays are the type which do not produce seeds at all. You want those plants to produce seeds so that you can harvest them to plant the next year's crop.
- As a good option and if you can spare the budget: look into plants that can provide some extras – tobacco, chili peppers, herbs, things like that. They can provide income (via barter) and a little bit of welcome flavor - in a world where such things are either gone or in short supply.

The perfect stockpile would be to have moderate-term (canned) goods that will keep everyone in the household fed for about 6-9 months (at 2,000 calories per day or so), Long-term storage (freeze-dried, grains, etc) foods in enough quantity to feed everyone in the home for about 2.5 years if necessary and enough seeds appropriate for your region to grow two years of full crops of vegetables and grains. These seeds will have to produce enough to feed everyone in your home for a year, plus 15-25% more to insure next year's seed crop, as well as to bank against crop loss (the second crop of seeds sit around in case your first crop fails for some reason, gets stolen, etc).

Small livestock (chickens, pigs, etc) can get pretty impractical (and hard to conceal in a suburb or city), but would make a good addition as well, if you can swing it. If not, then perhaps rabbits and other small animals will suffice. Post-collapse, you can make your own 'livestock' out of captured pigeons (they taste like small chickens when done right) and other small animals such as rats, squirrels and etc. Note that cats and dogs are not really good livestock. Dogs are however useful as guards, portable hot water bottles of a sort and as companions and playmates for those times when things are cold outside and everyone is bored.

Incidentally, if you have pets, make sure they're fed as well, for at least a year. Small dogs are easy and relatively cheap to stockpile food for and even large dogs can be provided for if you plan ahead.

Obviously, all of this food is going to need to be stored somewhere, but before you order up a large shed and park it all in the backyard, stop and think for a moment. It is a better idea to keep the food in your house, where it will be the safest, with perhaps some food cached in a hidden location a goodly distance away from your home, just in case you're driven out of it. Set aside some space for it (in either place), but keep that space concealed. As mentioned before, the best idea is to put your stash all around the house (and perhaps a little bit concealed in a tough waterproof bag, buried or hidden in your backyard), so even if you get robbed and live to tell the tale, you will still have some food to access once the robbers leave.

A final bit you want to keep in mind is that all of this food is going to require a little bit of upkeep. The good news is that it won't take much more than what you already do. You go through the short-term stuff as usual and odds are very good that you stock up about that much food already. The moderate-term foods? Just rotate your stock – use the oldest cans and packages first and replace what you use as time goes by. Keep an eye out for expiration dates, as the contents lose flavor from that point on (though with canned foods, as long as the can is not rusted or swelled, can keep for quite awhile beyond that date, up to years beyond in many cases).

The long-term stuff? A little different in how you buy and handle it, but not too rough to do. Buy a sample of a brand before you buy any large batches of it. Make a meal or two out of the sample and give it a good taste test. This will also teach you how to use the stuff and give you a good idea as to how much water you need, preparation time and how it will taste. Once you find what you like, buy as much as you can (within your budget) and pack it away. Once every couple of years, do a taste-test from the long-term storage foods and replace what you use. If it tastes/looks/smells bad or rotten, replace the whole batch with a different brand and throw out anything similar to it, perhaps sampling from other containers in that same batch. Keep an eye on the dates and if anything is close to expiring (as in, within five years of expiration), replace it. If the old stuff is still good, you can use it for hiking, camping, or keep it aside to give to neighbors (anonymously) if collapse happens between then and a year or two after the expiration date. These long-term foods can either come pre-canned, or you can make a long-term storage bucket out of a sturdy plastic bucket+lid, some electrical tape, some duct tape and one of those typical air-activated hand-warming pouches. The pouch removes oxygen from the air - you open and chuck it in right before you seal the bucket. (Don't worry about too much heat - it'll cool as soon as it sucks the little bit of oxygen out of the air in the bucket and will warm up again when you open the bucket, as there's more oxygen introduced into that bucket when you open it.)

Medicines

You may be perfectly healthy. You may be riddled with chronic diseases. You may be perfectly able-bodied, or you may be bound to a wheelchair. You may live perfectly clean and sober, or you may have addictions. Here, we're going to cover all of that and help you prepare no matter what.

Even in perfect health and with no problems, you're going to want to keep at least a few things in your stockpile that you will find useful and even life-saving. Beyond the basic first-aid kit (which you should have different kinds of anyway), there are a lot of things you will definitely want to keep in stock.

Below is what a normal, healthy family should keep on hand:

- At least one large first-aid kit. This should contain: supplies to stitch large open wounds, large bandages for large cuts and abrasions, an eye patch, an arm sling, at least 200-300 standard-sized adhesive bandages, a splint kit for broken bones, 2 large tubes of antibiotic ointment, bottles of pain relievers, each with 100 tablets (one aspirin bottle, one acetaminophen bottle, one ibuprofen bottle). Pain relieving creams (one menthol-based and one capsaicin-based), anti-fungal ointments (athlete foot and nail fungus), Benadryl(R) (for severe allergies), petroleum jelly, temporary tooth/filling repair kit, snakebite kit (if you live in areas where snakes are frequent), three pair of tweezers (of various sizes), single-edged razor blades (a large package – 50 to 100 of them), alcohol pads, rubbing alcohol, hydrogen peroxide, a cheap cigarette lighter, Dermabond(R) (or similar) to 'glue' moderate cuts, at least 50 'butterfly' bandages, at least four 3" wide Ace(R) bandages, surgical tape (at least two large rolls), latex gloves (at least 25 pair), An anti-diarrhetic (e.g. Imodium(R)), Iodine solution, a small quick First-Aid guide, a larger, more comprehensive first-aid guide and a couple of small, clean towels.

- At least two smaller first-aid kits, usually sold as home first-aid kits.

- At least three small ' pocket' first-aid kits, usually sold as first-aid kits for bicyclists or hikers.

Some optional pieces you may find useful? Crutches, 'boot' style foot immobilizers (if you've ever had an accident involving the use of one to heal, just keep it in a clean, plastic bag when you're done using it). If you can get hold of a used manual wheelchair, it may be useful as well, but there's no need to run out and get one.

Your medicine cabinet, post-collapse? For a normal, healthy family, you'll want to stock it with things like the following:

- Laxatives, in pill form. At least 500 pills.
- 4-6 bottles of anti-diarrhetic pills (80-100 pills each). Trust me – you may well need them to remain alive until you figure out your permanent drinking water situation.
- 3 sealed bottles of either Milk of Magnesia, or Bismuth Subsalicylate (the pink stuff). Same reason you're keeping the anti-diarrhetic pills around and extreme stress (like, oh, civilization collapsing all around you) often leads to an upset stomach, so it tends to be a good idea.
- at least 3 large sealed bottles of Acetaminophen, to reduce fevers
- at least 3 tubes of toothache pain gel (the reason should be self-evident)
- four large (500+) boxes of cotton swabs
- four large tubes of muscle pain cream
- four large bottles of Calamine lotion (not just for poison ivy, but also good for bee stings, jellyfish stings and etc).
- A package of 10 earplugs (or more...)
- 2-3 bottles of eye drops.
- at least two disposable enemas (you'd be amazed, but it can be used for more than you think).
- A large bottle of talcum powder
- 2 large bottles of anti-athlete's-foot powder
- 2 large bottles of ipecac syrup (in case you ate a bad mushroom or such)
- 2 large bottles of activated charcoal (see above, but for those poisons the first aid book says you don't want to throw up)
- 3 large bottles of white vinegar
- a douche/enema/hot water bottle kit
- a large jar of petroleum jelly (which can incidentally be used for more than medical purposes)
- two large bottles of iodine solution
- two large bottles of unbuffered contact lens saline solution (for the saline content, not for contact lenses)

That is just what you should have for a normal, healthy family. The idea is that if there are no more drugstores, you pretty much have to become your own. We'll have a more complete list in Appendix A.

Now if you do have medical problems, you're likely used to seeing me harp on how you have to assess what the problems are and how you intend to overcome them. Note that this is not optional... if you are unable to get around and do much, you're liable to end up dead and not for lack of caring by your family members, either.

Depending on your condition, it breaks down this way:

- If you require medications to keep a chronic illness or pain at bay, you're going to have to stock up on as many of them as you can humanly can, or find alternatives. Once they are gone, they are quite simply going to be gone. Therefore, if you can stock up on them up to a year in advance (opiates and prescribed narcotics will be impossible to stockpile that far, but most other medications should be okay), do so. Meanwhile, look into and try (with doctor's advice) as many legal pain or condition-relieving alternatives as can be found in nature, in your locale. If you cannot create it locally, you won't be able to get hold of it after the collapse and will be at the mercy of whoever has what you need. Some conditions (e.g. mild asthma, strong allergies and the like) can be kept watch over and sometimes avoided, even without medication. You will however have to know and recognize these signs and what to do about them minus the medication until the symptoms pass.

- If your medications are to treat mental illnesses or maladies (depression, ADD, ADHD and the like), you will have to learn how to get along without them. If a family member relies on these drugs, you will have to plan for a way to help the person deal with living without them, up to and including physical restraint if necessary. Note that you will also need a plan to wean them off the medication slowly.

- If your condition requires an external device, then buy and keep spare parts and at least two spare devices. This means things like crutches, walking canes, permanent splints, wheelchairs (and take note to get very sturdy non-electric ones) and the like. If you use catheters or ostomy/stoma supplies, then buy as many of them as you can lay hands on... post-collapse, they're likely going to be all that you have.

- Some conditions are going to be tough to provide for. Folks with pacemakers, or people who require oxygen, dialysis and the like are going to have the toughest road of all. Sometimes there are ways of making do – for example, people who require oxygen can buy a portable oxygen concentrator and a solar charging kit and have some hope of continuing to get what they need until the parts wear out. Sometimes, there are ways of doing without, though note that this will increase your risk of something catastrophic. For instance, people who require C_PAP machines when they sleep at night (and cannot use any alternatives), are going to experience sleep apnea all over again once that machine has no more power.

- If you require the services of a home health care nurse, well, you're still going to require that and either you should get a younger family member to take his/her place, or you're simply going to have to try and do without.

If you yourself need certain medications or devices just to remain alive and you have no way of continuing to receive or somehow replace their effects post-collapse, then unfortunately you have a very hard decision to make. It is very easy to say, but very hard to do: Prepare for the worst and prepare your family for the worst. Try to do it in a way that shows grace and dignity and do not endanger yourself or a family member just to procure these things after things get ugly – doing so will only harm you and possibly them. Relying on some shady black-market dealer to keep yourself alive means exposing your family to a rather large danger. The less the wrong people know you exist, the better your family's odds of survival.

No matter what your state of health, one thing you should look into is natural alternatives. For pain, this could well mean looking into growing marijuana (any other illicit drug is going to be out of the question, as it will attract the wrong type of attention, or will be unfeasible), though obviously you do not want to even think of procuring the seeds or start doing the growing until well after the collapse. Fortunately, there are plenty of non-illicit drugs that you can make from local plants and herbs, if you know where to look. A book on how to find and make such mixtures in your area will prove to be invaluable and will be in the list of books to get and keep later on.

Withdrawals are going to be a huge problem with some medications. Check your medicine cabinet or prescriptions to see if you will have to face this issue. If you do, research (and possibly discuss with your doctor) a safe plan to slowly wean yourself off of the medication. Some medications (especially certain heart medicines) you simply cannot stop taking, ever, unless you want to risk severe results (for instance, there are heart medications that, once you begin taking them, you cannot stop taking unless you enjoy having a real risk of cardiac arrest). However, for most drugs, there is usually, if not always a means to slowly withdraw from using them. If you can safely do so, you may want to consider doing that now and switching to medicines that aren't as evil about presenting severe withdrawal symptoms.

Even if you never need these items, it is a good idea to procure and keep as many of them as you can (legally) store before things go wrong. After the collapse, scrounge as many medications as you safely can, because they will come in handy later on – if not for you or your family, then for someone who can barter for them. We'll get you a list of things to keep an eye out for later on in the book.

Keep in mind that most medicines have expiration dates and that you will have to rotate a lot of these items in your medicine cabinet and first aid kits. Be sure to keep a good semi-annual schedule to look into these and replace what you can if it starts to go bad. Keeping most over-the-counter medications past their expiration date usually means a small loss of potency over time, which in turn means you'd have to take more of it to get the same effect. Otherwise, they can last an amazingly long time. However, many prescription medications expiring can often have some very ugly consequences, especially those drugs of a psychotropic nature (that is, drugs used to treat mental illnesses).

Defense and Firepower

I already know how this is going to start, so let's get the politics out of the way first...

There are usually three schools of thought when it comes to arms and armament, especially firearms: The first group is passionately hateful towards the idea and will have nothing to do with firearms, or even many other effective weapons. The second group is just as passionate, but in the other direction – that a wall of guns and a mountain of ammunition will protect them from anything out there. The third group are the ones who will know a little bit about them, perhaps use them for hunting, maybe used them in military training long ago, keep a legitimately-owned rifle or shotgun in the house somewhere, etc.

The first and second groups are the extremists. They are either going to be victims of the first armed criminal to wander by, or they will be found in a pool of their own blood, surrounded by dozens of empty or jammed weapons. Both will likely be exploited and/or killed due to overconfidence, bravado and idiocy.

Let's get this straight, folks: Owning a firearm does not make you a god, nor does it make you a criminal. It is just a tool. A firearm is neither malignant or benign, but firearms are extremely intolerant of ignorance, carelessness, or neglect. However, you will probably need at least three for each adult member of your household and you will definitely need to know how to use and care for them. One other thing – you and everyone who carries one in your household, will have to use them without hesitation should the situation arise: Even if another human being is on the other end of the barrel.

No matter what your stance on gun control and gun violence may be, there is one undeniable fact: If civilization collapses, you're going to need firearms, because it is almost a certainty that most everyone else will have them – a very large number of whom will not hesitate to use theirs on you if they can get at your food and supplies by doing so.

For defense, you need three basic firearms for each adolescent and adult member of your household – a mid-sized rifle for game and long-distance shots (good to at least 100-500 yards distance), a shotgun for closer quarters and easy aiming for up to around 100 yards (if necessary) and a good, solid pistol for combat in areas where swinging around a shotgun or rifle is impractical (such as indoors). My own recommendations for each would be as follows:

- Long-range: A .30-06 rifle with a scope is your best choice here, mostly because the ammunition is cheap and plentiful and the bullets tend to fly straight and flat for up to 500 yards without too much trouble. The scope is used to make things at least a little easier for long-distance shooting and while it makes the rifle a bit more delicate to carry and move around, the benefits are well worth it. Now, for those who are wondering what you need a long-distance rifle for, the reason is two-fold: One, you can go hunting with it – while the opportunity may be rare, it is always a good thing to take advantage of it if you can. Two, there are instances where you will have to shoot someone from a distance – and a rifle makes for the best all-around stand-off weapon, where you can fire at an attacker and keep him from getting too close.

- Medium-Range: A 12-gauge pump-action shotgun with an open choke is your best bet here. The shotgun gives you a wide variety of ammunition types, but your best bet are a good bird-shot (say #2 or #3 shot), #00 buck shot and rifled slugs. Always buy "hunting loads", as they will contain more powder than the typical target loads. The rifled slugs are the least used, but the #00 shot will kill both deer and human alike and can be used in tight quarters. Bird shot will injure and incapacitate people, but kill birds. Just note that using the lighter birdshot is not what you really want for defense. If you only injure an intruder, you'd better be prepared to follow-up and kill that person – you can't afford to use up your medical supplies and an injured intruder that gets away will likely come back with friends.

- Short-Range: This is a toughie, but only in that you can choose between revolvers (simple, reliable mechanisms, but limited in capacity), or automatics (higher ammunition capacities, but a bit more complex in mechanism). If you decide to go with a revolver, be sure to have "speed-loaders" on hand to quickly reload if needed and make sure you're using a caliber of .38 or higher. Also insure that the barrel is at least 4" in length, else the accuracy will be worse than worthless. A .357 Magnum with a 6" barrel is (just in my opinion) the best setup for a revolver, as it provides plenty of power and enough accuracy for short-range hunting, should the opportunity present itself. As a bonus, a .357 revolver will just as happily use .38 caliber ammunition. If you decide to go with an automatic, then you will want to insure that the caliber is large enough to do the job (9mm would be the extreme low-end here) and that the mechanisms are easy to keep clean and serviceable. The best automatic in my opinion would be a Colt M1911-style .45 auto. While the typical .45 magazine only gives you seven rounds, an extended magazine can give you ten. Also, an assailant that would withstand multiple 9mm shots to the body will fall much more quickly from one or two .45 caliber slugs (this is because the .45 bullet has far more mass, which in turn does a lot more damage in a lot faster period of time). One thing to note about automatic pistols – the movies are flat-out wrong. You cannot simply "double-tap" a person or animal and expect both shots to hit the target. Doing so is a waste of ammunition and a great way to empty the magazine without hitting anything. This is because the first shot's recoil will throw the muzzle (the barrel tip) upwards and the next shot's recoil will throw it up further still. Take your shots one at a time.

The decision of whether or not to use a firearm is sometimes obvious and often not. Earlier in this book, we went through some situations and mental exercises that you will likely have to face. Use those as your guide and once your mind is set on a course of action, do not hesitate to carry it out.

Defense is more than just whipping out a gun and firing away. You have a lot of other non-firearm tools to help you defend your home or land and in many cases, it would be far preferable to use those instead. Options include knives, swords (yes, a real sword, or at least something that can be used as one), bows/arrows, crossbows, a long steel rod or pole, axes (the wood-chopping kind, not the Medieval Viking variety) and even a length of moderately heavy chain.

The obvious primitive weapons, bows and crossbows, have a nice dual-purpose use. Either against animals or against people, they work equally well. You can always, with a little skill, make more ammunition for them, or just re-use and/or repair what you do have (note that you cannot safely make wooden arrows for compound bows, however). They're whisper-quiet, which means you don't have to initially give away your position (at least until the target discovers where the arrow came from...). They're excellent for hunting. The technology for either one has been around for thousands of years. A basic recurve bow is somewhat easy to make from native, local materials. They are effective for ranges of up to 50-60 yards for a compound bow and up to 200 yards (with enough skill) for a crossbow. Most commercial ones come in colors and markings that camouflage easily. Finally, you can buy as many bows and arrows as you want without the US Bureau of Alcohol, Tobacco and Firearms getting all nosy about it. Crossbows are however somewhat restricted by state and local laws. A crossbow is the easiest of the two to use (you aim it like a rifle), though it does take a moderate bit of strength to load one.

A compound bow and arrow gives you all the killing power of a pistol or shotgun, but takes a decent amount of practice to master and requires a moderate amount of strength to use. Crossbows, as mentioned before, do require some strength to pull the string back into the trigger mechanism, though there are devices that can ease this. Both are also very hard to conceal if you're walking around town with one, due to the larger size. A crossbow could possibly be used indoors if necessary, but a bow would be an iffy proposition (although you could draw one in a room, it would be iffy indeed).

Knives and swords are great multi-purpose weapons and can even be handy tools to have around. Knives can be used as pretty much anything and in a more primitive post-civilization world, will likely become a good all-purpose tool to keep on your person at all times. It's a great last-resort weapon if you're stuck fighting up-close and personal without a firearm and can in extreme circumstances be thrown if the situation requires it. A good knife to have would be a fixed-blade one (not a pocketknife), with a blade at least 4" long. Pocketknives are okay tools if you have nothing better, but they're notoriously hard to get to, open and put to use – and the blades are generally too short for combat, unless you have no other choice.

A "sword" (notice quotation marks) doesn't necessarily have to be something you carry around in a scabbard and whip out to slay the occasional dragon. You can just as easily carry around a Machete and it would make a great dual-purpose tool. On the one hand, you can use a Machete to clear brush, hack a small tree limb, break up a large hunting kill into quarters, or other various outdoor uses. On the other hand, if you have no other weapon, it makes a great impromptu slashing sword, as long as you have some skill with using edged weapons. As far as carrying a real sword, it's not entirely a bad idea. An authentic solid and sharp Japanese Katana (also known as a Samurai sword) can perform some of the (non-chopping) tasks of a machete, but makes an excellent close-quarters (less than 15') fighting weapon if you know how to use it. Another great alternative would be the Gentleman's Cane Sword – easily concealable, useful for non-combat situations and in a pinch can be drawn and put to use in pretty short order. I would suggest at least learning how to use one anyway, both as a hobby that gives great exercise benefits and because you never know if anyone will ever get around to making gunpowder again after civilization collapses.

Other weapons begin to get more primitive and pretty much any length of chain or sturdy stick can be used as a weapon if you have to and it's available. This can include poles, long walking sticks, chain, most large/long hand tools and the like.

Overall, it is a very good idea to train with the firearms you have (perhaps at least once every three months you go out and practice), any bows or crossbows you have (you can practice this in the backyard with a bale of hay or a commercial target foam block, just try to do it at least once a month or so) and definitely learn how to use a more primitive weapon, such as a knife or pole (this can be done in the privacy of your own home, though if you're practicing with a pole or a sword, make sure you have enough room, or do it in your garage).

The idea is to keep yourself in some sort of physical shape if possible and at the same time to become proficient in the weapons that you do keep around. Speaking of physical shape, a few Martial Arts classes wouldn't hurt you either. You can go to a classroom, or just practice with a DVD at home if you don't feel like wearing the funny uniforms. No matter how you do it, do it with a trusted partner, or in a classroom with other people. You need to know how another human being moves, how they react, how their bodies react and it is good practice for those times when you may be stuck with facing down someone in a real hand-to-hand fight.

The most important weapon of all, unmentioned up to now, is your mind. You have to train your mind. Hopefully, you've been doing this throughout the book, but in this case, you need to start doing it for real. This does not mean running out and kicking someone's ass, but it does mean that you need to spar with someone on occasion (keep it friendly) and build what is called "muscle memory", so that certain styles of kicks, punches and grapples become automatic to you. This in turn means you don't have to stop and think through every single movement – you can just select what you need when the opportunity presents itself in a fight. It also a good way to get over that huge hurdle that most civilized people have – the indecision during that moment of ' fight or flight' and the momentary paralysis brought on by fear.

Criminals and thugs rely on paralysis and indecision on your part. They rely on it in order to throw you off-balance, which gives them the advantage. You have to become able to react without contemplation, so that the attacker doesn't have that luxury. This in turn puts you at the advantage: most of them rely on you to be compliant and indecisive and if you're not, they likely have no clue as to how to react, which allows you to place them into that position of fear and indecision. Even if they don't, that quick reaction could well mean the difference between you and your family living and all of you getting killed. The post-collapse world will be rather rough for awhile until the population dies off enough to balance against local resources and ecology, so unless you want to be a part of the population that has to die off, you're going to have to train your mind.

I began with personal weapons and personal training, because there is one universal fact of defense: It all begins and ends with you. Most experts preach "Defense in Depth", which is a very good thing, but they always seem to start with the outside and work their way in, sometimes leaving you as almost an afterthought. But, since you and your your mind are the ultimate means of defense, we needed to start with that and work our way out.

So let's move out a bit and look around the inside of your home. How many floors are there? Is there a place where you can lock everyone in to keep folks out if necessary (and maybe still be able to fight back from)? A good, strong, bulletproof room that has its own food and water (and weaponry and means of fighting outwards from) can keep the bad guys out until they either get bored and leave, die from the firepower coming out of your mini-stronghold, or they die of starvation, exposure, dehydration, etc.

Do you have more than one way in or out of your home without resorting to ladders? You'll need that, since even natural disasters (fire stands out as a good example) can leave you otherwise trapped. How many windows do you have and what kind are they? While nobody is recommending that you run out and buy bulletproof glass, a means of barricading those windows from the inside is a very good idea. As an alternate, installing hurricane shutters with slits cut into them (for firing outwards) isn't a bad idea. Converting that big ol' sliding-glass door into a pair of strong-but-elegant French Doors will save you a lot of grief and barricading should the neighborhood get lawless.

How are your stores spread out in there? Having multiple secure locations for your food, water and supplies is an extremely good idea. You don't necessarily have to do it now, but reinforcing certain closets and other small places around the house (then spreading your stores out to these places later when things begin to get ugly) helps bring the odds up a little if someone breaks in and cleans out one of the storage depots you have spread around the house.

Something to consider: set aside one semi-obvious storage area and place fake food and fake goods in it. The idea is that even if you're taken by gunpoint and forced to give the intruder food or supplies, you give them the fake stuff instead, leaving your real supplies still hidden and intact. Instead of flour or powdered milk, fill the containers in there with Plaster of Paris or the like, mixed in with just enough of the real thing to give it a taste. Your "Powdered Sugar" is actually Borax (it's cheap and great for keeping ants out of your house). Replace liquid medicines in that fake storage spot with pesticide or antifreeze. Things like Milk of Magnesia? Used House Paint. Salt? Granulated white fertilizer or white beach sand, mixed 70-30 with real salt. Cake mix? Ground-up Cat Litter. Speaking of kitty litter, bulk buckets can be mostly filled with it, then a false bottom put in with maybe a layer of the good stuff on top, laced with Boric Acid or a similar pesticide.

Yes, it's unethical, but ethics aren't exactly going to be observed much by criminals post-collapse. If you feel that badly about it, you still have alternatives - use food coloring and castor oil and the like for the liquids and as the real stored goods expire, move them to the fake storage area once you've replaced them. While you're at it, buy a few non-working guns that look like they should work and have those hanging on the wall, or sitting in the fake storage area. Remember those TV commercials that sell the "gold clad" collector coins for $5 each? Buy up a handful of those and put them in a bag in the fake storage area. Let your imagination run wild, but make it look/smell (and even in some cases, taste) real. Just be sure to have a couple of containers open and when things begin to get ugly, make the whole thing look like you've actually been using the fake supplies.

Post-collapse, consider moving the family into the basement for their sleeping areas, but have at least one of you standing guard upstairs in shifts. This way, you have better protection against anything that goes boom in the night. Do you have a dog? You really should get one, even if only a smaller one. A breed like the Chihuahua, Jack Russell Terrier, a trained Miniature Poodle, or Dachshund eats very little, can travel along with you reasonably well, but barks at damned near anything that comes even close to the door.

Moving on to the outside of your home... what is the siding made of? Vinyl or Aluminum isn't much for stopping bullets, but having some empty sandbags stored away with a few large bags of sand will help reinforce the inside areas of the house around where you and the family sleeps (got kids? Build them a sandbox to store the sand!)You might want to consider or get a home with brick siding. While not perfect, it does a better job of stopping the smaller caliber bullets and still looks nice from the outside.

What does your yard look like? Chain-link or wood fencing? What would it take to reinforce portions of it? Are there decorations or areas you can hide/duck behind in a hurry? How easily can you get to any out-buildings (sheds, garage, etc)? How much open space is around your home? The more the better, as it gives intruders less to hide behind (and if you live in any kind of wooded area, it helps keep forest fires at bay).

Looking even further out to the edge of the property, how many ways can you get in and out of your property without being immediately noticed? If the answer is less than two, you really need to fix that. Most suburban and other lot homes usually have only one way in or off the property, usually by way of the front yard. You should put in some sort of quickly removable (from the inside) portion of your backyard fencing if you can, so that you can quickly cut through a neighbor's yard if necessary without having to climb a fence (which would slow you and your family down). You cannot always be certain that you can stand your ground and a means of grabbing a few things quickly and leaving the home without having to fight your way out is a very good idea.

Now, it's time to take a look at the neighborhood...

How many roads and paths lead in and out of the neighborhood? Could you and your neighbors barricade some of the extra roads/paths off if it comes to that? How are your neighbors' homes situated? Can any of them be used as points of defense from an attacking group, or strong enough to use as a shelter for multiple families? How far away are your neighbors from each other? Too far off and help could be slow to come. Too close by and attackers can use the confined spaces for cover and as a way to sneak in or out.

How many of your neighbors are able and willing to fight to help defend you and how willing would you be to help defend them? If the answer is less than an immediate affirmative for most of them, then you may want to reconsider where it is you live and find a different neighborhood to live in.

If you live in a small town, how is it situated? Do you have the high ground, a source of clean fresh water within town limits and a means to defend the town? If you live in a suburb or larger town, you will most likely want to forget about defending anything larger than a block or two. Anything larger and you risk spreading your defenses too thin and it rapidly becomes a logistical nightmare as the size grows beyond a dozen families. We will be covering this bit in more detail later on.

Be it a small town or a large neighborhood, how easily could you and your neighbors set up a patrol that covers the perimeter? At this point, maybe a few cheap walkie-talkies are a good idea to have stashed, even if you could only use them for a few weeks until the batteries are all gone. Failing that, a system of firearm shots could also be used to raise the alarm.

Something else to think about – is there a way for the entire neighborhood to escape as a group if it had to? Many of these things are at this point going to require cooperation between you and your neighbors, but there's no need to run around and corral everyone while everything is still peaceful. As mentioned before, we'll cover this one a bit later on.

In conclusion and summary: inspect things in your mind and train it to be more aware of its surroundings and to face the unthinkable. Get your body into shape. Procure and get some skills in a few good, solid weapons – whatever you can legally purchase and keep around the house. Inspect your home with an eye towards defense. Go over your property and improve what you can to your advantage. Think about your neighbors (if you haven't already gotten to know them, you really should). Take a good, long walk around your neighborhood (or town if it is small enough) and look over maps of it... what can be done and what cannot?

Transportation

In these relatively peaceful times, getting from Point A to Point B is pretty easy to do: Get in your car and go there. Transportation when civilization collapses will be just a little trickier. If the collapse is sudden, roads will be jammed tight with cars trying to flee the cities, suburbs and towns. If the collapse is slow enough, then the roads will be relatively clear, but may be full of police or military checkpoints, or they may be unusable for automotive traffic due to a lack of gasoline. It all depends on how it all goes down. However, no matter how things come crashing, you will want to take a serious look at how you intend to get yourself and/or your stuff from one place to another.

But wait! You decided to stay put! Why do we need to know about transportation? Why not just hang around the house, grow our own food and scoff at the hordes of people that are certain to be wandering around? Well, fact is, there may come a time or circumstance where you will have to be a bit more mobile than you wanted to. The biggest reason is that your home may end up becoming a place that is too dangerous to remain living at any longer (and there are lots of sources for that reason). Another reason is to get to what is commonly known as a "Bug Out Location", or a secondary place where you are among friends and family, or where you are likely to better survive long-term if your primary home is in danger or destroyed. Another perfectly valid reason is, eventually, commerce. You may have a good or service that is rare and you stand to profit greatly if you take it on the road to other locations that don't have it. Or, a rare good or service is only available if you get yourself to where that good or service is. Any or all of these things may crop up and this means transportation.

With post-collapse transportation, you will have a few considerations to keep in mind, especially if you want to get to your destination alive, healthy and unmolested:

- Ambush: You stand a huge chance of being ambushed and harmed (or at least robbed) on the open roads. This could be done to you as easily by common criminals, as it could be done to you by a family desperate to keep themselves fed and warm.

These will usually happen by a blockaded road that forces you to stop, by 'chaser' vehicles that can run you down, or by deception (e.g. an "injured" young woman that lures you in, only to be overcome by a group of her companions). Another method is to incapacitate your vehicle by using something to puncture your tires. The general idea behind an ambush is to get you to stop so that they can rob you of your stuff and possibly to kidnap any of your family that they deem useful or profitable.

- Checkpoints: This is a mixed bag. Police or military checkpoints (especially frequent ones) can be a means to keep ambushes at bay, but as society breaks down further, they can also become a means to relieve you of your possessions. They can also present obstacles that prevent you from getting to your destination and odds are good that even if it isn't a uniformed robbery, it stands a good chance of being an operation that will "confiscate" any "contraband" goods (excess food, medicines, firearms and the like). If the collapse involves any outbreak of disease, it will likely serve as a means to enforce a quarantine – either to keep you in a quarantine zone, or keep you out of one. If quarantine is the case, expect armed officers and/or soldiers to be a bit more forceful than you might expect.

- Infrastructure Rot: Washed-out or collapsed bridges, rock-slides, landslides, fallen trees, massive car wrecks... these and more will begin to collect on the roads as chaos grows. Many of these obstacles will also serve as ambush points, so either approach them carefully, or consider turning around and taking an alternate route immediately, if you can.

- Lack of Fuel: Let's face it, unless civilization collapses literally overnight and you're one of the few survivors in the aftermath, getting fuel is going to be close to impossible and even harder still as time goes by. Even if by some miracle there's a gas station that has fuel on-site, odds are nearly perfect that the fuel is locked up tight in underground tanks, with no real means of pumping it out. If you're going to travel any kind of distance, you had better make certain that you have enough fuel in the tank to get there, with enough extra to account for any detours and hold-ups you may have to endure. A good rule of thumb for one-way trips is to have enough fuel to get there and have half a tank left. For round trips, you should have enough to get there and back with at least 1/3 of a tank of fuel left.

- The Elements: If you're not going by an automotive vehicle, you're going to be at the mercy of the weather. This means in addition to carrying whatever provisions you can bring along, you will have to carry sufficient clothing, shelter and/or water (depending on temperature and climate). If you are going by automobile, you're going to have to put up with roads that won't be plowed when it snows, nowhere to really duck out of a hard storm, ice on the roads that won't be salted and all the joys that come with it.

- Time: Even by car, getting there is going to take a lot more time, mostly due to either avoiding the above, or putting up with the above, or perhaps both. If the highways and freeways are clogged or too dangerous, taking the back roads will consume a **lot** of time, often two to three times as long as you would normally expect.

Getting from here to there (and possibly back again) in a post-collapse society is going to take a bit more effort than simply loading up the car and driving over. Some of the things you will have to contend with in your journey immediately post-collapse will be the following:

- Why are you going? Unless there is some immediate and/or desperate reason for traveling, you're going to want to really think this one through.

- Who is coming along? If it is just you, the options and logistics are going to be a lot simpler to decide on than if your entire neighborhood is traveling with you. A small group is less able to defend itself, but is more easily concealed. A well-armed large group can fend off most ambush attempts, but will move slower and require more stuff along the way.

- What will you use to get there? This doesn't always mean doing it in a car or truck. Sometimes, it is faster and safer to take a boat if there are waterways that will accommodate that boat from start to finish. If you live in a heavily agricultural area, there may be some literal horsepower (as in, horses) available to get you and your things to where you need to go. Maybe everyone is stuck with traveling on foot. No matter what means you decide to take, each will require its own calculation of time, supplies (gasoline, hay, bicycle tire repair kits) and tactics (can you move only at night on foot, or can you move openly during daylight with a large armed force?

- What's the route like? You're going to have to know what route is the safest and most possible to get from one place to another. It will help to have some very good maps on hand (hint: buy them now) that show not only the main routes, but back-roads, trails, streams and rivers along the way. It also means knowing as much as you can about people and activities along the way. No, we're not talking tourist attractions, we're talking about criminal activity (e.g. a certain stretch of road may be prone to constant ambushes), natural events (landslides, rockslides, etc), military activity (checkpoints, convoys, etc) and similar bad news.

- How isolated are you along each point of the route? Heavily populated areas present different dangers than lightly-populated ones and each place will have its own priority as to what it is they want from you the most.

- How welcome will you be once you arrive? If you're intending to go somewhere but have no property, trusted friends, a market, or family awaiting you at your destination, you may well be screwed and probably do not want to attempt it. This is because in a collapse and post-collapse situation, any coherent communities left will be struggling to keep themselves going as best they can and unless you have some critical skill or service they need but do not have, they will very likely turn you away. If a community only has so much food or fresh water available to them, you're going to be one more mouth to feed that they can ill afford to provide for. Even if you bring all of your supplies with you, a new community may well confiscate them all and then drive you out anyway, because sometimes, people are like that – just plain evil. Now if you have property at the destination, or have family or a lot of friends waiting for you there, then you stand a better chance of being welcomed into that community.

With all of this in mind, how do we apply that to preparation? Well, first off, take a look at where you are now, during peaceful times. Are you intending to stay here and ride things out if it all goes sour? If not, then you will want to make it a large priority (right after deciding where), to find the safest route to take between where you are and where you want to be.

The first thing you want to do during this planning is to answer all of those questions up there but with one other thing to keep in mind: Where is everyone else going to go? In a sudden catastrophe, everyone in the suburbs and cities are going to have one thing in mind: Get the hell out of Dodge. They're all going to pack their cars with whatever they can and then they're going to get immediately out of town. This is going to clog up pretty much every major road in and around a suburban or urban area. It will also be a playground for thieves and gangs, as they prey on the stranded motorists and their families. The idea is that you do not want to be in and among them - before, during, or after it hits.

During your planning, get some very good maps. Plan a way from start to finish by taking nothing but back roads and side-streets.

Next, drive the most obvious routes now, while everything is peaceful. Even better, have a spouse and/or trusted teammate do the driving while you take notes. Take very careful note of everything along the way.

population centers, overall travel time, fuel consumption (then multiply that consumption by 150% as a safety margin), the kind of neighborhoods you drive through (both good and bad – especially bad) and any stores or shops along the way. Break the route up into segments beforehand and take note of how long it takes to cover each segment. Check the odometer before and after the trip to see how many miles are being covered for the journey in each direction. Note the road conditions (smooth, potholes, narrow vs. wide, etc). How many forests, wooded areas and other natural hiding places are there along the way? As you go along, write all of this down!

When you get back, take all of your notes and set them aside for a moment. Now, drive that route again, but this time at night, preferably sometime after 10pm. Take notes of the same things along the way that you did before and especially take note of any and all differences. Also write down which stores are closed and which are still open.

When you get home from these two trips, ask yourself a few questions... What parts of the route were the most heavily populated? Moderate-sized and "bedroom" communities along the way can quickly become traffic snarls in an emergency. Were there any traffic lights to contend with from the time you left your town to the time you reached your destination? If so, those can become choke-points, especially if the power to those lights goes missing. How narrow were the roads and where were they too narrow? If they were two-lane roads (most country roads are), you may want to go back and identify the portions where things are even more restricted (such as deep ditches along the roadsides, embankments, rock walls, etc), or places where you couldn't easily drive off the road in a loaded-down vehicle, in order to to go around a potential wreck or obstacle. How busy were the roads (both during the day and during the night)? If these roads were typically busy in peaceful times, you can count on them to be far too busy during a mass evacuation or disaster. How many bridges were there along the way? Hopefully as few as possible, as bridges are natural choke-points for either clogged traffic or for ambushes. How big or plentiful were the state patrol or highway patrol? If there were plenty, count on there being even more in a disaster, but this time with checkpoints. How many farms or woods were along the way and how big were they? A lot of large farms means plenty of places with lots of hiding places if it ever comes to that. How much gas was left in the tank when you reached your destination? If you had less than 1/3 of a tank, you will definitely want to include a gas can full of fuel that you can chuck into the vehicle on short notice (the good news is, that gas can likely be used elsewhere around the house in mowers and such, or you can simply dump it into the car's gas tank once every few months and put fresh gas in it). How many military or large police installations were along the way, or close to your route. We have nothing but love and respect for our armed forces and uniformed officers, but in a large enough disaster, those places (and more importantly, the roads around them) are going to be locked down tight and you're going to have to find a detour around them, if not another route entirely. How many large tourist attractions are along your route? The fewer, the better, so you don't have to worry about traffic (again). Is any part of this route a favorite of truckers and other large vehicles? This will be hard to determine, but fortunately, most large trucks and the like stick to the freeways and major highways. The idea is to not be where they are, since they add to the traffic snarl. Same goes for any mines or large factories along your route.

Take a second look at your map with the route marked on it. Take the time to mark places where there may be heavy traffic or problems and look for detours around them if you can. Take the time to mark alternate routes entirely if possible, just in case there's a solid reason why you cannot take your original intended route.

Most folks expect that they will have a large enough motorized vehicle to pack their stuff in and evacuate with, that they will be among the lucky few who see it coming and get out in enough time and etc. However, the reality is that this is simply not going to happen. Unless you're willing to dedicate a tough vehicle for just this task and you have your eyes and ears peeled for the first sign of trouble, it simply isn't going to be feasible. There are also a lot of other 'what-if's that can put a serious crimp in your evacuation plans, if you're crazy enough to believe you can do it this way. What if your gas tank is almost empty when the sudden need to bug out comes? What if your car is broken-down, or otherwise out of order? What if the roads are unavailable? What if you can't even get the car out of your own neighborhood?

This is where a bit of planning comes in. A second vehicle, a bicycle, a boat or canoe (if there's a good clear waterway between start and finish) and similar are things you can plan for and take into account. Just note that if you're going to plan on alternative vehicles, then you're going to have to insure that these are also in top condition, ready to go and in the case of boats and the like, can be launched quickly and easily, in spite of traffic and hostile surroundings.

If you're stuck with walking it on foot, then make certain that you take the added time into account and that you (and your family!) are physically capable of walking the 15-20 miles per day needed to get there before the food and supplies that you carry on your back(s) run out. Speaking of which, make sure you and your family can actually carry those supplies for that daily distance as well. Also know that hiking it is going to severely limit what you can carry.

Another means of getting out of Dodge, or to transport yourself from point A to point B, is to possibly go by boat. You will have to (carefully!) check and see if you can get from here to there by water, because any interruptions will mean unloading the boat and carrying your gear from that point the rest of the way. If there are locks or dams along the way, forget it – you cannot count on locks being operable along the route. If part of your water-route involves the ocean or the Great Lakes, be sure your vessel can handle a typical storm out there. If the destination is upstream of your starting point, make absolutely certain that you can get there on whatever fuel you have on board (don't guess, *know*!) Finally, make certain that you can get that vessel into the water in the first place (or if it is berthed at a marina, make certain that you can get to the marina with all of your supplies in very short order).

For those folks who happen to have a horse farm and/or a couple of horses very close by, make sure that your animals can carry the weight of you and your stuff and that they are in top physical health. Most of the maintenance that goes into a horse is perfect for keeping one ready to use as transportation, with one exception – if your horse is only used to light riding, you're going to have to train it to work with heavier loads. You will also have to procure things like saddle packs, in order to pack goods onto the animal and set your route so that plenty of grass and fresh water are naturally available along the way.

Bicycles and motorcycles can be a good means of transportation when nothing better exists (just remember that motorcycles eat gas...) Both have a far more limited amount of cargo and have weight limits. A Bicycle can usually carry up to 200 lbs of rider and gear, before you need to worry about the wheel rims warping, so if you weigh 180 lbs, you only get to carry 20 lbs of stuff on a 200-lb limit bike. Larger, sturdier bikes mean more stuff, but know that you're the one pedaling all of this stuff along and those uphill climbs are going to be murder if you're not used to it. Motorcycles are sturdier, but you're still limited as to how much you can carry on one. Too much and the frame will probably still hold up fine, but you'll end up dumping the bike over on the first sharp turn that you make at any speed over 20 mph.

For the really crazy among us, there are of course exotic vehicles – airplanes stand out as an example here. Odds are good that if the world is collapsing, a good, clear runway will be hard to find – either on your way out, or at your destination. However, it is still somewhat feasible, if your aircraft is built for grass or rough runways. You will also have to be certain that you can reach the aircraft, load it with your gear, take off and manage to get there and land - under any known weather condition that can be found here, there and along the way. There is also that little bit about you possibly having to do this without any kind of ATC help, so you're going to have to keep your eyes peeled even harder for other aircraft, potential near-misses/collisions and anything else that may be airborne (including military patrols). Get too exotic and it rapidly becomes impossible. You're not going to find a hovercraft that can carry your whole family, your provisions and go a couple hundred miles, all on a single tank of gas (the military has such vehicles, but odds are perfect that you do not). Anything crazier than that is only good for comedy value, not for saving your life or as viable transportation.

Money

Strange that this subject would come up, considering that after civilization collapses, paper money will become pretty much worthless. This doesn't mean that all forms of money will disappear, however and until that point in time, money can still be a rather useful too. The trick is, you do not need to have it all in the bank, or have it all in paper money.

The first and smartest thing you can do is to keep around an alternate form of money. This more often than not means precious metals. Gold and silver are prominent metals here and for good reason: These forms of metal have been used as money for nearly all of human history. This does not mean immediately emptying your bank account and burying a ton of gold in your backyard. What it does mean is, you may want to take a few thousand dollars and purchase at least the following:

- A few ounces of gold, preferably in small coins (if you can afford it).
- 2x that total weight in pure silver coins (much easier to afford)
- 3x that total weight of the silver in what is called "junk silver", or pre-1962 US Dimes and Quarters (both of which are 97% silver by weight).

The reason you buy the third variety is to best avoid counterfeiting and fakes (most counterfeiters shoot for faking gold coins). Buy all of these only from reputable dealers and do not buy them in "shares" or any such nonsense. Buy only the real thing that you can take home. Local coin shops will be the best means of finding these dealers.

When it comes to where to keep your money, put your paper money in your local credit union (they're cheaper than banks, have less fees and often have the best interest rates). Keep a bit of cash at home, kept safe and hidden. Keep most of your precious metals at home as though, perhaps storing some extra in a safe-deposit box.

Storing the metals at home is a tricky thing. The idea is that you do not want to keep it all in one place, lest you get robbed. Bury some of it next to your home in your backyard, in the crawlspace, or in your basement. Have a small, relatively cheap safe to store just a little bit (and any fake gold/silver if you can find any) – that's what you leave in a semi-obvious place for any thieves to take. Keep any passports and/or marriage licenses, insurance paperwork, etc in that small safe as well – it'll make things more believable. Have a strong safe in a well-hidden place to store the majority of your money, your birth certificates, etc.

Tools

Note that you probably won't use any of these immediately, but you will most assuredly need them as time goes by, chaos dies down and you have to get on with the business of living for the long-haul. Since these tools are likely going to be something that you hand down for at least two or three generations, you had better select them carefully and buy the best you can on the budget you have. If treated with care and not abused, nearly all of these tools will likely last well over a century if you let them.

This isn't going to be as easy as trotting out to the hardware store and picking up a handyman's kit. Instead, you're going to want to do a little research. Find tools with excellent reputations (ask any experienced factory maintenance department, experienced construction worker/contractor, a long-time professional auto mechanic and the like – odds are good you'll have a list very quickly). Then and only then, go and get those particular brands. Most will have a no-questions-asked lifetime warranty. Most will cost quite a bit, but will be worth it. Also, avoid the 'gimmick' tools – they often rely on moving parts, are often more delicate than you want them to be and likely won't hold up under continued usage. What you want are solid, well-built, simple tools that can be used by hand, without electricity.

A good listing of basic tools, broken down by type and use, will be found in our shopping list at the end of this book. It is suggested that you take advantage of it.

Summary

To summarize, there is a lot of stuff that you're going to need if you want to survive a collapse of civilization. However, nearly everything listed is actually useful, even if civilization goes on for the next two lifetimes. It also doesn't require you to start wearing tinfoil, or to walk into an Army/Navy surplus store with a high-limit credit card. You just have to think it through a little, prepare a little and look ahead...

What Do You Know?

Knowledge is usually one of the single greatest things that gets ignored in most books about survival and post-collapse living. However, if you are going to survive for the long haul, you're going to have to know a lot of things and be able to pass those things on to your children (or on to the community's children). It seems a waste that the end of civilization would have to mean that everything the human race has ever discovered and known would have to go out the window as well. In fact, it seems wasteful.

Even if our own present civilization turns to dust, a new one can always rise. It would help those future generations if they didn't have to re-discover everything from scratch. This is why you're going to want to keep some books around. Relying on human memory and oral tradition tends to break down after awhile, going from complex concepts and fact, to myth and fairy tale. Relying on skills that have gone unused due to time and chaos tends to introduce errors that future generations will discard, along with the original useful concepts and theories. The idea is to keep it in a form that can be read as-is by your descendents, thus the printed book.

Skills You Will Need

There are a few skills that your great grandparents once took as ordinary chores, but will be vital for you to learn or become re-acquainted with. The skills you want to get good at now, before the collapse, include the following:

- Gardening and agriculture: If you don't have a green thumb, get one. Your ability to feed yourself and your family will very likely depend on your ability to grow food, so the sooner you practice and gain this ability, the better. Start with growing more than just houseplants – experiment with growing a little bit of everything, but do it without any commercial fertilizers, potting soils, or the like. Use what you have and use only things you have at hand, such as compost, hand tools and such. Try and find ways of increasing the yield per cubic foot of garden. Note that I said "cubic" foot – you can use that vertical space above and below the soil to increase the yields of vine crops such as beans, tomatoes, or root crops like potatoes and carrots.

- Weapons and combat skills: Odds are very good that you and/or your community will have to fight to keep what you have and have built. If you don't fight for it, it will be taken away from you. Spend the time to learn how to use the weapons you choose – firearms, archery, clubs, sticks, edged weapons... learn to be comfortable with as many as you can and learn how to handle them safely and efficiently. Also learn how to maintain them. This obviously doesn't require that you go all-out, but it does require that you get at least exercise, practice with a sparring partner, practice with the weapons and even take a few classes here and there.

- Heating With Wood: You have to keep warm somehow. Unless you live in a sub-tropical region, this is most likely going to mean burning wood. It takes more than just chucking a pile of wood together and throwing a burning piece of paper at it. You'll want to lean how to start fires without matches or lighters, learn different types of burning material (tinder, kindling, fuel) and different kinds of wood (hardwoods last/burn longer than softwoods, softwoods burn hotter but occasionally 'spit' as resin flash-boils, etc). A book which covers this should definitely be in your library, but you will certainly want to practice it once in awhile.

- Old-Time Cooking Methods: This means cooking without blenders, mixers, choppers, the microwave oven, the stove (unless it's a wood-stove). In the initial chaos, it likely means cooking in a fireplace. You could use a grill, but most commercially-made grills aren't all that good with wood and gas will run out quickly. Also, cooking on a grill often more easily pin-points the smell of cooking food to your house, something you may not want to advertise in a time of crisis. Cooking in a fireplace will dissipate the odors a bit more and ejects those odors higher up (out of the chimney), making it at least a little harder to detect and track. This skill also means knowing how to make foods tasty without a bunch of MSG or exotic spices. Note that this will also mean getting a few pots and pans that likely aren't in your cupboard – these books should definitely be in your library (and are listed), but it can also be fun to practice these methods now.

- Old-Time Food Storage Methods: Your refrigerator and chest/upright freezer are not going to run for very long after the power goes out and unless you live where winter is permanent (or you have access to permafrost), refrigeration is going to be pretty much a non-starter. This means storing food by drying, curing, smoking, canning, root cellars and the like. If you can learn and practice these things now it's a benefit and can save you money that would otherwise be wasted in food spoilage. It is a lot easier to get good at this when you don't have to, than to try and hurriedly learn it and risk screw-ups (thus starvation) when you absolutely need to do it.

- First Aid and Trauma Skills: If someone gets injured, severely sick, or worse, it will likely be up to you to take care of the person and his or her very life may well depend on how well you are able to stabilize the disease or injury. Any first aid book that includes the phrase "dial 9-1-1" in it is a book you do not want – get something geared towards military trauma and first aid and a good resource will also be college-level textbooks written for nurses and EMT. First-aid and trauma classes for EMTs and entry-level Paramedics are fairly inexpensive and are well worth looking into and taking.

- Tracking, Botanical and Hunting Skills: While honestly, anything that moves, crawls, or flies will be shot and eaten in pretty short order as chaos deepens, it never hurts to gain these skills, in case opportunity presents itself, or in case you find yourself in a place where the local wildlife may still be plentiful. The botanical skills come into play when you learn to (correctly!) identify and seek out edible plants in your region.

Skills you Will Want To Gain

Eventually, there are other skills that you will likely want to pick up on as the chaos dies down and as communities re-form. These skills include...

- Sewing and Textile Skills: Your clothing isn't going to repair or replace itself. You're going to have to start looking first into means and methods of repairing/patching existing clothing and of eventually making new cloth and clothing from plants (flax, cotton, hemp, etc) and animals (wool, leather, etc). The first people who can do this well, will become quite well-off in their eventual community.

- Blacksmithing: Sounds strange, but the ability to take existing steel or raw iron and make it into useful tools and implements will become rather highly prized. With basic blacksmithing skills, you can make, weld, or cast nails, tools, weapons, chains, knives, scythes, plows and a whole lot of other things that will in turn make life a lot easier to get on with in a community.

- Engineering and Fabrication Skills: This is a bit general, but it encompasses a basic set of skills in building things from basically scrap parts and in taking things apart to get at the parts you need. It requires a little creativity and a bit of looking at a problem, then imagining a solution to it using the things you have. This is likely not as great of a necessity as the others, but if you are good at it, you can parlay the skills into something you can barter or trade for and create things around the home that can make a hard life at least somewhat easier. You can also use it to keep electricity flowing (at least somewhat) and to repair or even build some basic luxuries that people may well gladly pay and trade you for.

- Woodworking Skills: If you're able to turn raw logs into finely-crafted furniture, tools, or lumber without the use of electricity or power tools, then you may well be able to find a niche that can keep you in both food and goods from the community.

- Shipbuilding/Boatwright Skills: If you live near any moderate-to-large body of fresh water or ocean, then fish may be plentiful (with no commercial fishing rigs around, fish populations can become rather massive in just a few years). Getting to those fish means having boats handy that do not require motors or fuel, since spare parts and fuel will quickly run out. If you're able to build sturdy, usable boats (or even small ships as a group) that can fish these waters, you become a valuable member of a coastal or lakeside community and even to other coastal communities nearby. Boats, barges and small ships are also useful for transferring cargo from one community to another, or as ferries. It may not even be necessary to build from scratch, but to convert existing hulls into something that runs well under sail.

Transferable Skills

So, you're not a woodworker, a blacksmith and don't feel like you have the time to learn post-collapse skills if such a thing comes, if ever. No need to fret – just be certain that the skills you do have are useful in an age without electricity, gasoline, or other modern technological advantages.

Skills that an eventual community needs up-front aren't that much different from the skills that are needed now, just that they're a bit, well, not as complex. A community post-collapse will still need some sort of security/police force, a militia to repel criminals and invaders, doctors, teachers, farmers, ranchers, blacksmiths/woodworkers, some form of textile making, diplomats of a sort to communicate with other communities, a preacher or priest to help tend to the spiritual needs of at least most of the community, possibly fishermen (if the community is along a coast, or next to a very large lake or river) and things of that nature. A leadership of sorts to help it run smoothly is also a good idea and housewives are going to be absolutely critical.

Things that, while not critical, will be useful to a post-collapse community? An accountant or two to insure that supplies and inventory is kept in good order, engineers to provide and design clean water and efficient sewage systems, engineers who can build primitive-but-useful machinery, electronics techs who can repair and even build remaining lower-level devices, someone who can procure or produce minor luxuries, perhaps a chemist, someone who can brew safe and consistent alcohol (for drinking, sterilization/disinfection, etc), maybe someone who can brew biofuel to feed any remaining engines, veterinarians (who can double as doctors if necessary), perhaps even prostitutes (laugh all you want, but they've always been around and they aren't going away anytime soon).

However, the things a post-collapse community will have little-to-no need for? Lawyers, middle-managers, computer programmers, car salesmen, marketers/MBAs, fast-food restaurant workers or management, bureaucrats, dog groomers, program managers, wedding planners and basically any occupation that cannot exist without a complex technological civilization around it.

If your job description or skill-set happens to not fall into a 'useful' or 'critical' category, it will be a very good idea to gain some skills that will be useful after a societal collapse. Actually, let me amend that – it's not just a very good idea, it is a vital idea. Otherwise, you'll likely end up being a manual laborer, doing what's needed and largely doing what you're told. Of course, you could always hope that civilization holds up just fine until you're dead of old age, but you wouldn't be reading this book if you honestly believed in that, now would you?

The good news is, there are many, many other skills that are useful post-collapse which I haven't listed, but also make a decent hobby pre-collapse: pottery (plates/bowls, jugs, etc), knitting, stone-masonry (sculpting gives you a good start here), machining skills and the like. Use your imagination, then take an honest, hard look at whether or not a family would be willing to trade food or other goods in exchange for it. Something else to consider – even if you have a vital skill, it always pays to learn a second one. It never hurts to have your spouse pick up a craft. The more skills you can pool together, the more important you become to a future community.

You can even look back into earlier portions of your career for ideas. For instance, your humble author makes his living by working in and around computers and networks. However, I am an engineer by training and education, so I keep those skills sharp, for eventual application in a community that has no further use, need, or ability for anything concerning computers and networks. I can (and have) brew and distill alcohol. I can build nearly anything I need to, given enough time and materials (e.g. a printing press, which while initially not critical, will become so in the long-run). As a former associate professor, teaching is certainly within my grasp as a skill. You can do this too – or, if you'd like, find a skill that you can enjoy and you think may be useful, then practice it. It's fun, doesn't require too much time and effort and you actually gain not only the skills, but an appreciation for the craft. Even if you only want to make bricks and tile? Those are useful skills – *go for it!*

Note that if your skills and/or useful hobbies require any specialized tools or materials, that you have a local long-term source of these materials and that you have enough spare tools (or the means to make more) to last you a lifetime. This cannot be emphasized enough, since you likely won't get a chance to stock up on needed materials after civilization has blinked out.

Stocking Your Library

A good library is worth a thousand pounds of food in the long run. With knowledge, you can do almost anything in the face of bad odds. This is why we have a large-but useful checklist at the end of this book of items you will want to fill your library.

Note that running out and buying all of these books brand-new is sheer idiocy. Unlike most any other critical item you need to get, books do not spoil (unless you store them in a humid place), they do not expire and buying used makes them no less useful than buying new. A good place to start hunting them down is used book stores, local library book sales (where they dump off excess books and gets a bit of fund-raising out of the deal), the local thrift stores and places like that. It takes a bit more looking, but is well wroth it as you find and collect these books for pennies on the dollar.

In addition to the books listed in our checklist at the end of this book, you will seriously want to pick up and keep around some collegiate-level history books, perhaps a few collegiate-level introductory books to science and engineering, as well as college-level books on anatomy, physiology, moderate and advanced mathematics and other subjects that are important to a functioning civilization. To top it off, there are certainly some cultural books that you would likely want future generations to enjoy, to give them a look into what we were like back when everything still ran. Perhaps a bit of sheet music and a basic book on how to read it would be a nice touch as well. Even if you yourself never crack these books open, I'm certain that future generations will be grateful that you set them aside.

So where to keep them all? Store them in a place that is cool, dry and not exposed to direct sunlight. Set aside a room just for the books and maybe put a comfortable chair in there. Make it a retreat of sorts, where you can study these books in those rare moments of peace and quiet. Be sure you keep them under reasonable security, because there will come a time when those books are liable to be valuable to others (either as information sources, or as toilet paper - you pick.)

By the way, when things start to get ugly out there, start keeping a diary. A daily one if you can, catching the reader up on those days that you may have missed. It will provide a study that will help future leaders understand what went wrong, why it went wrong and more importantly, what to do to avoid having all of that happen again. It will also give your kids, grandchildren and other future generations a good hard look into what you had to put up with in order to make sure they live more comfortably than you did during those times.

Now vs. Later

The whole idea behind all this accumulated knowledge is to learn now what you need to know in preparation, but to have the tools to learn later those things which you will need to learn later. So far, we've given you a good idea of how to sort those out and where to start. The rest falls into place... those things which you do not need to know for immediate survival you can safely keep in mind, or keep stored in the books. Complex skills will need to be acquired and practiced in order to keep them sharp. You will need to be working on them soon, as you should be up to speed before things get ugly. Fortunately, most of the complex skills also make good hobbies.

One thing to keep in mind: Whatever you decide to pick up on, try and get as good at it as you can. The work of a master is far more valuable (and the products worth far more) than the work of an amateur. If you can, for instance, turn out finely-crafted dinnerware from clay, your products are going to be in greater demand than a shoddy half-skilled bowl hammered out from scrap sheet metal. If you're able to provide a better and sturdier repair than someone who is winging it, the community will start coming to you to get their things fixed. If your apples are larger and healthier than those of someone who doesn't know how to naturally keep pests away from his or her trees, you can command a higher trade in bartered goods.

You may at this point be wondering why not try to learn it all now? First, that's impossible to do and even if you devoted your entire waking days to mastering these skills (which will leave you no time for anything else), you'll likely forget most of it in the stress of societal collapse. Second, I'm here to tell you that, assuming you make it past the first couple of months, you will likely find enough spare time to catch up on the not-so-critical parts, especially during the winter. This is because unless things on a perfectly lucky scale, you will likely have to wait before you can plant any crops (though you may be able to harvest any existing ones if you live in the right areas at the right times). This in turn means that there may be long periods of doing a whole lot of nothing. These are good times to start reading up, start reaching out to surviving neighbors and to start the tentative steps towards forming a community. And that brings us to...

Pooling Your Knowledge

It is impossible to know and do everything. On the other hand, as you begin to make the tentative steps towards building a community, there's a lot of information that can be shared around. Once you find a niche that you really enjoy (and that provides for your family), why not find a trusted neighbor who could use some skills of his or her own and treat them to a few of those books showing other skills that you keep stashed (they can borrow them, you can trade them, or whatever is mutually agreeable)? In return, they can provide these skills - skills that may well be needed by the community.

As years pass, many of those books can become textbooks from which to teach future generations. At first, teaching could be something done part-time, but if there are enough children, a full-time teacher can be provided for and the kids can receive somewhat of a decent education, at least equivalent to what kids received before the age of computers.

One other thing you can consider is to write down everything you've learned in skills, especially as you rediscover them. If you can, write them down in a separate diary or other set of blank pages in a book-like format. As you gain mastery over these skills, your notes can be handed down to your children, or to whomever rises to eventually replace you as you cease to perform them yourself.

Conclusion

What you know can often make the difference between living and dying. The idea is to know what's important beforehand and learn it before things get ugly, because the middle of a crisis is no time to try and teach yourself the necessary skills for those moments. On the other hand, there are things you can pick up later as time permits. The trick is knowing which is which, but it isn't too hard to figure out. Whatever you choose to gain skills in, try and get as good at it as you can, because there may be times where it will make the difference between life or death and later on, the difference between relative comfort or desperate poverty.

Who Do You Know?

There is no such thing as a long-term lone survivor, period. Even the most badassed human alive can be taken down by a determined and organized group of opponents, if he doesn't go insane first. You have to eat sometime. You have to sleep sometime. You have to defecate sometime. To stay awake and alert 24/7 for every day of the rest of your life is beyond any human being's ability. To make and maintain, by yourself, everything you need to survive? It will eventually reduce you to the state of a Cro-Magnon cave-dweller. Families who do not reproduce outside of their own gene pool eventually die off due to inbreeding-induced disfigurement and miscarriages.

Long story short, you're going to need a community. You're going to need folks who will trust you and folks that you can trust. You will need skills you simply do not have and others will need skills from you that they do not have. Any man who serves only himself has an idiot for a master – sometimes you desperately need someone to tell you that a contemplated action is a really bad idea to carry through. You will always need someone else to prove a sanity check and even sanity.

While it is still imperative to prepare quietly for your continued survival in advance, it is just as imperative that you reach out to others once things begin to calm down and perhaps even while it's all still in a state of chaos. Having extra supplies procured and set aside for potential friends is a must. Having a tolerant mind for your neighbor and a kind heart for the innocent is a must. Having the ability to lead men and women is an excellent trait to have and practice, but only if it is done for the good of those being led.

Reconnaissance

If you haven't already, it's time to take a good, long look around your neighborhood. This means literally walking around. If you have a dog, you have a perfect excuse to do it. The first time, just walk around, take a good look at the neighborhood and make mental notes. When you get home, compare what you've seen to a map of your neighborhood. If you don't have a high-resolution map of your neighborhood, get one, or print one off of a search-engine map. This will give you an idea of the terrain, where all the roads go, what those roads look like, where the open areas are, where the nearest fresh water creek or stream can be found (if there is one) and what kind of houses are around yours.

Take a walk again the next day, this time with that map in your mind. The third time you take that walk (yes, we're going to make this a habit – it's good exercise, after all), start taking quiet mental notes of the houses as you go. Which ones have big yards? Which ones belong to folks who are obviously trying to ' keep up with the Joneses' (so to speak? Those are easy to spot – just look for all the big boats, RVs and other ' toys'). Which ones have yards and homes that are unkempt? How many of these homes are empty? How many are for sale, or have been foreclosed on? Again, take notes along the way.

After awhile, your daily walk (if you can do it) will bring you into contact with folks. Never pass up a chance to engage in friendly chit-chat and always go out of your way to be kind. Be sure to be as tolerant as humanly possible and take a friendly interest in any differences in culture, ethnicities and the like. This also gives you a chance to later meet folks for dinner, or to have a barbeque or other activities with them. If you have kids, you can get a chance to meet with the parents of their friends.

The whole idea, in case you haven't noted it, is to help promote community. This word, community, is not to be confused with neighborhood. A community is a neighborhood that actually operates together, helps each other out to varying degrees and looks after each other. Just remember that you're not out to convert anyone just yet, nor are you out to spread wild tales of the End of Times or any such nonsense. You're just getting to know your neighbors and getting to let them know you a little bit. Be honest about everything (save for information about your preparations – if pressed and on-the-spot, be very vague and under-estimate it severely) and just be your open, friendly self. As long as civilization is still around, enjoy the luxury of being civilized. You may find it to be a very good thing.

Gaining Allies

The best way to win friends and gain allies is to know who your neighbors are while things are peaceful and to keep mental tabs on them as things progress. You end up making a few rather good friends, you begin to learn who (and what) to avoid and overall, when things collapse, you will learn much faster who you can trust and who to watch out for.

Here's how you do it – remember back in the section "Now, about your Neighbors...", how we talked about going out there and discovering who your neighbors are? While it would be crazy (and potentially problematic) to write any of this down, you take what you've learned about your neighbors and start figuring out who you can trust and who you cannot. This will be a bit like dating – you don't want to give everything away, but you are going to have to trust somebody.

The best way to go about it is to identify those folks who are also preparing, or has that kind of mindset. Start with subtle 'what if' scenarios, working them as smoothly as you can into into private conversations. You don't have to give away what you have at all. If you're nervous about any answers you hear, you can be all nice and vague about it, with phrases like: "*we have a couple of weeks' worth of stuff stashed away, but if it gets real ugly, we'll go to my Uncle Steve's place in Wyoming*" and the like. If their reply is along the lines of "*well I'll just come along with you!*", then they're liable to try and do just that and you should immediately work to encourage them to stock up themselves for at least a few weeks and inform them that "Uncle Steve" is a crazed gent who is armed to the teeth, etc. Or, you can use it as an opening to say something like: "*Well, since it'd probably be dangerous to go there by the time things get that bad, why don't you and I start preparing together here? We can both stock up on some things and help defend each other if it ever came to that.*" Pay careful attention to the reply and any subsequent activities, because it will tell you right off if the neighbor is serious, or if they decide to blow it off. If the neighbor is seriously preparing, do not, under any circumstances, inquire as to how much, or of what... the both of you can keep things generalized, yet somewhat specific (when needed) about the kind of things that you should both stock up on. Above all, not only do you keep the exact nature of your supplies secret, but both of you need to remain secret about it from the rest of the neighborhood.

If you're lucky enough to find neighbors willing to prepare as well, then together and secretly, you should start being specific about one thing: the resources and skills that you will pool together to assist those neighbors you know of with critical skills, yet are not themselves preparing. For example, if there is a doctor, medical student, nurse, EMT, or the like in your neighborhood, prepare to take care of that person and his or her family... their skills are going to be more than worth the effort and resources that you spent in preparing on their behalf.

All this said, post-collapse, your community might just consist of no one until things begin to calm down a bit. For awhile, the only way to stay alive may well be to trust no one and to look out for yourself and your family. However, if you can organize everyone, things will go a lot smoother and will do so a lot faster. As population pressure eases (a nice way of saying '*when enough folks die off...*') you will want to start reaching out to and trusting nearby, similarly kind souls and start banding together. The sooner you can successfully do this, the better your chances in the long run.

Something else you should consider, even if you're all alone in the neighborhood preparing: Buy some extra goods as charity. For neighbors who are in the military, have medical knowledge, are engineers (especially mechanical and electrical engineers), mechanics and the like, you can buy a cheap-but-usable bucket of of emergency meals for them that will contain about a month of food for three people (@ 3 meals per day) per bucket. Some nearby big-box stores for example sells such a bucket for $100.00 USD or so. You can make a home-made month's worth of nutritious meals for two, fit it into two 5-gallon buckets and it will cost you less than $70.00. If you think any of that is expensive, consider this: How much would you be willing to pay to have a medical-trained person help heal your sick or injured spouse or child? How much would you be willing to spend to get a broken-down car running again so you can evacuate if you have to? How much would you spend to have someone help defend you from a desperately hungry gang of thugs who find out that you have food?

We'll cover the wheres, whys and hows later on, but for now, take the time to think about this: If you simply dealt with all of your post-collapse neighbor problems with gunfire, how long do you think you can survive? How trusting would they be of you (or each other) if the only social interaction in your neighborhood involves flying lead? It's time to consider more than one means of communication with the folks who may well become the core of your new community.

Watching With Clear, Jaundiced Eyes

While we're getting all chummy with the neighbors, somewhere in the back of your mind, you need to do some sorting-out. Some things to run through your mind (later, when you're alone) would be things like:

- Which neighbors would handle themselves well in a crisis?
- Which ones would the most likely to simply pack up their things into their shiny new SUV and go clog up the freeways?
- How many would run out of supplies in very short order?
- Who among your neighbors could handle themselves in a fight (no, not necessarily in case they come after you, but in case you need defensive help from them)?
- Which of your neighbors have special needs (medical or physical)? Maybe you can help them out a bit in advance.
- Do any of your neighbors have family members or residents who are addicted to drugs, alcohol, etc?
- Which of your neighbors may have special, critical skills? Doctors, Cops, Mechanics, Nurses, EMTs and folks like this are going to be well worth doing favors for.

Note that these questions aren't automatic judgment calls, nor should it mean you suddenly shun those who may not 'measure up'. Well, I spoke too soon there. You see, one exception to this would be anyone who is addicted to illicit drugs or alcohol. Keep a respectful distance from them pre-collapse and reveal nothing to them, at all - no matter how friendly they may be. When things come crashing down, they're going to attract the absolute wrong kind of friends, will likely cause problems (burglaries, etc) and will become huge problems in and of themselves once their personal supply of illicit substances run out.

For those families with members who have special needs (handicap, chronic illness, etc), they're going to be hard-pressed to prepare and it wouldn't hurt to set aside a little extra for them. You can send it over to them anonymously once things get ugly.

Charity and Kindness

Your basic priority is going to be preparing for you and yours, first and foremost. However, I want you to spare a few thoughts (and more importantly, some supplies) for those who are honest, kind, hard-working, but less-fortunate. Families who have handicapped members, the elderly, honest-but-poor neighbors, working single mothers and folks like this. They're not going to have many (if any) means to prepare, if they can even spare a thought or a dime towards doing so.

Even if you aren't of a Judeo-Christian religion (and if you are, your eternal soul depends on your being charitable, so...) look at it this way: helping others almost always pays back by helping you. Sparing a few supplies and sending them to a starving family nearby will probably prevent them from attacking you for your supplies. Taking in an orphan child will likely pay you back by having someone who helps care for you in your old age. Taking in an honest refugee may blossom into a friendship that last for life. Many hands make light work and the more of your neighbors who survive in good health, the stronger your eventual community becomes.

Preparing to be a charitable soul isn't all that hard, though it will take a bit more in the way of time and money. However, this, like any other preparation can be spread out a bit. Every time you go grocery shopping, you buy a couple of extra cans of food for your pantry anyway, right? Why not set aside a couple of cans for the 'charity bin'? Say, a couple of bucks' worth of canned food each trip. You would be amazed at how quickly your set-aside for the neighbors will grow.

The trick to being charitable post-collapse is to do so anonymously, since you do not want to tip your hand quite yet, lest your act of charity become an act that endangers your family. It will also help you out greatly if you are able to identify those folks that you want to be charitable to and those folks who, in spite of being commanded to love them, you will need to avoid.

People whom you should keep a charitable eye towards are folks are going to be generally honest, friendly, kind, are willing to stand and fight for what they believe in, have skills that are useful to the (future) community, have a bit of an independent streak to them and are not easily panic-prone. Finding these qualities are going to be a bit hard to do when things are cozy, but when things get rough they will shine out fairly quickly. However, you can get a feeling for it even now and keeping an eye out towards these personalities is a very good thing.

The people to avoid are going to be a bit easier to figure out...

- Any neighboring home with a resident who is addicted to alcohol or drugs. Help them if you can during peaceful times, but unless they eject the addict from their family, avoid them at all costs once things begin to collapse. If the addict cleans up and goes sober, give anonymously, but withhold any trust until that sobriety lasts non-stop for at least three continuous years.

- Any neighbor who is listed on a sex offender registry, who has been convicted of assaulting a minor, of rape, or of dealing with illicit materials in connection to a minor. This is not necessarily out of any automatic reaction, but because of the mental illness that may well put children and women in danger if they are welcomed into the community.

- Any neighbor who is overly aggressive, constantly rude, bullying, or overly brash. These are going to be folks who will try to seize leadership, but may not necessarily know how to lead much of anything. In fact, these are neighbors who will become a danger to the rest of the neighborhood once things collapse if they hang around... because they're liable to be the ones most likely to start attacking other neighbors (and even you) for supplies that they want or need. Letting them stay, post-collapse, is a very bad idea, no matter what critical skills they may have.

- Any neighbor convicted of multiple felonies within 10 years of the day you're reading this. A single conviction, or a distant checkered past always leaves room for repentance and forgiveness, especially if the person is kind and non-aggressive. However, people who have a demonstrable history of repeated criminal behavior are going to react quite differently in a collapse than ordinary folk -that is, they're going to be among the most liable to start trouble, or to start committing crimes to keep themselves fed and warm.

- Any neighbors who are constantly shouting or fighting. In the case of an abusive spouse or parent, offer to give the victim(s) shelter and protection if you can spare it. In a collapse situation, be certain that you and your neighbors escort any abusive individuals (not families - individuals) out of your neighborhood. Be well-armed when you do it and do not hesitate to insure that the abuser never returns – by any means possible. It is hoped that you will have turned such individuals in to the police by now, if only to break up such a relationship and save a few lives.

Overall, this is a good start as to what kinds of neighbors to look out for, both for good and bad. Those neighbors with solid skills, are honest and kind, or are ordinary folks who hold up well in bad situations are going to be your ideal compatriots in a new society. Folks who fall in between are going to require a bit of a judgment call, both on your own part and on the part of any neighbors who are actively working alongside you in preparation.

Refugee Sources and Migration Patterns

Whether you live in an isolated log cabin in the Rocky Mountains, or live in downtown Manhattan, one thing you will have to study a bit is the refugee - specifically, very large numbers of refugees. When civilization goes down, there will be a lot of them - more than most you can possibly conceive of overall. When things crash in the United States, you have a potential for hundreds of millions of people packing whatever they can amd moving away from areas of high population density, crime, chaos, or possibly whatever area was struck the hardest by disaster or war. This flood of humanity can represent the greatest threat to your own survival.

While we just spent a whole lot of ink on how and why to be charitable, there are instances when the demand for charity will overwhelm you and leave you in sincere danger of starving or getting killed. Refugees are that one instance. Think of civilization collapsing as the very real story of the RMS Titanic, which struck an iceberg and sank just over 100 years ago. I'll save the story and just spit out the quick synopsis of why you yourself need to be careful: Not enough lifeboats, way too many people that needed saving. The same problem applies here. You cannot possibly hope to help everyone - not even the US government can do that. Trying it will quite literally kill you and your family and you'll end up becoming a refugee yourself in your final days, as you yourself desperately cast about from place to place for food, water and shelter.

Just like sitting in a lifeboat filled to capacity (that is, your home, with just enough stores to keep you and your family alive), you're simply not going to be able to pull everyone you see out of the water without swamping the lifeboat you're sitting in. If there is space in your boat for one or two (that is, having enough excess resources to do it), then pull in one or two. Otherwise, you're going to have the sad task of keeping them out, even as they fight to get in. The worst part is, all those folks really, really do not want to die and will desperately try anything to stay alive. They will do so with every degree of violent and deceptive attempt that you can imagine and the vast majority won't really care what happens to you, as long as they get to eat.

Mind you, this does mean hating them. Just like a swarm of locusts, they're not doing it to spite you, or to hate you. They do it because they simply need to eat and there are a lot more of them than there are resources to support them. Once they use up all the reachable resources in a given area, they move on to the next area and then the next and then the next…

A large migration of refugees will eat, drink, burn and use everything in their path - just like locusts. The only real differences is that we're talking about human beings (which means they're more creative about it) and that they may have brought a few things along with them, lessening their impact only by a tiny amount. As a mental exercise, think of what you would do as a refugee and what you would use. You would burn wood to stay warm. You would forage and scrounge for anything edible along your path. You would seek out and consume any clean water you find (and use wood to boil it). You would take anything useful that you find and can carry. As you pass, the path behind you would be less useful for anyone who passes the same way. Now - multiply your activities by a factor of thousands. Imagine thousands and even hundreds of thousands of people on the move, in streams. Imagine what would happen to every natural and scroungeable resource in their path. By the time they (finally) pass, the vicinity behind them would be stripped of wood, stripped of anything remotely edible, the water sources (and the path they took) would be stomped into mud and likely contaminated with urine and feces. A crowd of one hundred thousand would stomp out a path that is literally miles wide, rendering it useless to anyone else for months, if not up to a year or more (depending on the fragility of the local environment).

Your job, if you expect to remain alive for very long, is to hopefully keep the hordes at bay, but avoid being in the path of where the large hordes of people will eventually be going. Failing that, you and your community will have to divert them away from you, deceive them into thinking you have nothing until they pass you by, or try and hold them off, waiting them out as long as you can. Failing that, even worse may be required.

I hate to say it, but you may have to try and turn away a whole lot of desperate, dying, but otherwise kind and loving people. It will suck. You will hate yourself. However, if you're not up to this task, then stop reading, put this book down once and for all and when things begin to collapse? Set your whole pantry out on the front porch with a big sign that reads "free for the taking". Then go back inside, have your whole family lay down and you can all wait to die. The choice is yours. Did I mention that this is going to really suck? With a little luck, it may never come to fending off starving mobs and civilization will be alive and well for the rest of your life. On the other hand, if it does all go down, what exactly do you intend to do about it? This is what we're all here in this book for and there are going to be no easy answers.

However, let's first see just how much of a problem is may be for you. We begin by anticipating where large masses of potential refugee-type people live and what routes they might try to take once the food runs out in their homes and the relief trucks stop coming.

Get out a map of your county, city (if applicable) and state map. If you live out in the sticks, get out a county map, a state map and a maps that shows all cities within 100 miles of your home. You should have at least two copies of each type of map anyway and the local ones should have as much detail as you can find. Take a look at the town you're in first. If you live in the countryside or backcountry, you can 'zoom out' a bit and skip getting into any fine details (just have a general map showing the closest towns bigger than 10,000 souls in it). On the other hand, those who live in any town with a population of over 10,000 should start paying attention here.

Which parts of town get most of the bad news on the evening local TV broadcasts? You know, where all the shootings, robberies and other violent crimes occur? A pattern will form pretty quickly over the next month if you pay attention. Also check with local sheriff office and police crime reports – you can get these online, or at the local police/sheriff offices. Every time the evening news comes on, mark the location(s) they mention with a purple dot. (For those living outside of town, just mark the general corner of the town)

Once you get a very good idea of that, then we start checking to see who is most reliant on governmental assistance. Yes, this is likely to be highly discriminatory. No, this has nothing to do with race, creed, or any other demographic. Now, start checking to see where in town that the most governmental assistance goes; to do this, begin looking for government housing project locations if you can find them (the Internet is your best tool for this job). Note that this means local, state *and* federal government assistance. You can find these out by checking in with your local city, county, state and federal Health and Human Services offices and/or directories, with a few discreet inquiries. Find out which neighborhood grocery stores in the larger towns take the most Food Stamp or EBT purchases. Pretend to be a college student researching a sociology paper or such. Remember - you may find it easier though to look around for this information online, too. In either case, once you start accumulating physical locations, begin marking those spots on the map with an red dot.

After you know where the crime and the governmental assistance happens, start looking up the locations of the following businesses: check cashing and "payday loan" stores, "rent-to-own" stores, Planned Parenthood clinics, discount or "dollar" stores, homeless shelters, bars and the like – these are establishments that more often than not cater to the less fortunate and are often placed in areas where there is more than the usual poverty. Mark these areas with an orange dot.

Again, we're not necessarily picking on the poor here – we're just figuring down where it is that things will likely come apart first. I know full well that it may even look bigoted or worse to do this, but I assure you, we're not looking at individuals, we are looking at general states of preparation by area. Folks who are on government assistance (e.g. welfare, EBT/food stamps, WIC, housing and the like) are statistically going to have the least amount of food, water and critical supplies stored in their homes. As things break down, they're going to get hit first and hit the hardest. They are also going to reach the breaking point first. These are going to be areas where any military curfew or enforcement will be focused first and foremost. Most importantly, these are the areas where you will find the most desperate individuals taking to violence first, sometimes long before anywhere else. Are there areas where protests occur more often than not? For every protest of any kind, involving over 50 people, mark those spots (and routes) in red. Odds are good that as things degrade, there will be a lot of protests and they will likely get rather violent as desperation sets in.

Once you start getting a large number of dots (say, at least a dozen or so of each color dot), you will begin to see a pattern emerge, though the three different colors may not necessarily all merge together. The more information you have, the more accurate your planning will be. So… what are we planning? It will tell you two things: Areas you will want to avoid like the plague if you can as things get worse and source areas where hordes of completely unprepared folk will wander outward. Keep this map with you and use it to mark and update progress (or in our case, collapse) as events do get worse.

Marking possible refugee routes are going to be a bit tricky, but rather important; here's how we do it. On your map, mark every major highway in red. These are routes that you want to steer clear of and where you yourself will not want to go, even during evacuation. Unless you own a boat and are taking the waterways, this also means marking every large bridge (especially if they are in city limits) in purple and not expect to cross them in a vehicle, for any reason. You avoid them for two reasons: First, because as natural choke-points, bridges will likely be crammed with stuck and backed-up vehicles and second, because post-collapse they will be heavily targeted by scavengers and/or criminals laying in ambush (depending on how navigable and/or crammed with abandoned stuff the roads are by then). Mark these routes as unusable until they reach about 20-30 miles out of any town limit. Mark any roads and thoroughfares larger than four lanes wide (larger than 2 lanes per direction) in orange. These will likely be clogged-up as well and likely used as travel routes for folks evacuating, but not as likely as all the previous routes are. Note which roads are at the edge of town and into the suburbs (and beyond).

For those of you living out in the smaller towns, countryside, or backcountry, you get to do a bit more work. Take out any maps that show land (and roads) between the nearest larger towns/cities and you. Any interstates, toll roads, or state highways that head in your direction you can mark in red. These are the main routes by which the refugee groups will be taking. Paved county roads with two lanes or more you can mark in purple or orange, especially if they branch off of the main freeways and highways – these will be secondary routes.

Doing this should give you a good general idea of where the refugees are all going to be coming from. This comes in handy both when you are figuring out how soon you need to brace for any incoming hordes (distance), roughly how bad it will get (size) and where they'll be coming from (location). It will also help you figure out any escape/refugee routes that you yourself may want to take, as the best idea is to avoid the general routes that everyone else will take (again, because they will have likely stripped the countryside along the way).

Planning Your Migration As a Potential Refugee

Up until now, we've been viewing this whole thing from the eyes of some small-town hayseed anticipating inbound hordes. However, what if *you* are going to end up being one of these refugees? Well, the processes are the same - really. However, instead of marking the space between the sources and where you live, you get to study the spaces and roads between you and where you want to go. This is important because as someone who will be viewed as just another refugee, the absolute last thing you want is to be in that crowd (or any large migratory crowd for that matter.)

The reasons you want to avoid being part of the refugee hordes is both obvious and subtle. The obvious reasons? Avoiding the crowds means less fighting for resources, less chance of being intercepted and/or rounded up by remaining government forces (for 'resettlement' or other reasons) and way less chance of being robbed by fellow travelers for whatever it is you're carrying. Even if your fellow refugees don't rob you, bands of criminals will happily travel along the periphery - picking off the stragglers, the isolated and the weak. The subtler reasons for going your own way? Moving in small groups or alone means you can change plans as needed. A smaller group has a greater chance of finding and gaining acceptance into a small, isolated community. A large moving crowd doesn't have too many options - once they start moving in a general direction, they must keep moving.

When I say that you should travel away from the crowds, I mean traveling far, far, far away from them. As in, miles away. We will be covering more about how to travel as a refugee as you read along, but suffice it to say at this point that you will definitely want to start planning routes that few others will want to take.

Anticipating Criminal Activity

As things begin to deteriorate, you will notice that crime rates will shoot way up (the only exception being a large disaster that strikes suddenly). As criminal activity goes up, you yourself will have to step up your observation efforts. Keep a close eye not necessarily on the evening television newscasts, but more on the radio, on local "police blotter" sections of the paper (if you have one) and on local news websites. Keep a partial eye on state/province and national news, but mostly on the local crime reports. You should be doing this anyway to keep the map up as we discussed earlier. The important part is to keep a watch over how often these crimes happen and what time of day they occur. Pay special attention to crimes involving theft of property, robberies, burglaries and the like and not so much on those with motives that don't involve procurement of goods or money.

This is why we look at crime activity – it is usually one of the best indicators of how things are going and at any scale – local, state, or national. As things break down, crime goes up. You can accurately gauge how civilization is holding up at large by way of national reports, as well as reports from other industrialized nations (you can usually find these online – you only need to search for national crime statistics, which are usually annual). Note that on any scale, there will always be places that are going to be somewhat crime-ridden. However, when you notice that crimes begin happening more frequently outside of areas that traditionally get hit by it, then you can see it as an indicator of things going bad all around. This spread means that either criminals are willing to risk the increased police patrols (and alarm systems, etc) common to the more crime-free areas, or that the residents of those areas are starting to commit more crimes than would be expected.

As part of your monitoring, pay attention to the unemployment rates, in all regions (local, state/provincial, national and even international). As they go up, things will get worse. Look into state budget numbers for welfare and other social services. Forget the political rhetoric and perform a simple calculation: how many people are being helped by state assistance? These figures should be available on the department websites, or in public reports (which you can demand and get) on either a monthly, quarterly, or annual basis. If the numbers are rising sharply, then this is an indication that trouble is looming on the horizon, as more and more people seek governmental assistance and as things in general are starting to get worse. Go back and compare numbers from previous months (or years), which should give you a good idea overall at the rate of change. If your state or local departments don't list numbers of people, look for the money – how much are they spending? It's harder to tell that way, but it should still give you a rough idea that you can still use.

Conclusion

As you might have gathered, people tend to be the biggest factor in your preparations for long-term survival and for both good and ill. The best way to improve your long-term odds is to figure out who can be trusted, who is generally a good person but cannot be, who to help in times of need if you can, who to avoid and who to drive off at gunpoint if necessary. It will also help greatly to know where all of these people are going to come from and where they will be moving once the cities can no longer sustain them. The knowledge you gain from studying mass movement will help you better plan what to do yourself and in many cases, where to go.

How To Stock Up

It is hoped by now that you have some idea as to what your general plan might be, what you will need and to have a good idea of how well (or badly) you are positioned to prepare for it. If not, try and work towards doing that. Just as hopefully, you've begun to exercise. It will also do you well if you've begun training your mind to be calmer, more rational and more alert - to the point where it is almost automatic now. Notice how none of it involves a tinfoil helmet? This is a good thing and everything you've done to this point will have perfectly valid uses without civilization actually collapsing, which is an even better thing.

Meanwhile, it's time to start putting some of that planning into action. First and foremost, look at your monthly budget. What do you spend it all on? We're not going to ask you to demand your paycheck in gold or silver, nor do we want you to live like a hermit. It is perfectly valid to continue living as you do now, with only a few exceptions and a small shifting of priorities. That said, let's look at that budget... if you're blowing $5/day on exotic coffees at the local espresso chain, you could instead make your own at home,and use the difference to buy extra canned foods and seeds. If you're spending an extra $50-$100 on premium cable television channels that you rarely if ever watch, well, maybe trimming back on that a bit will free up a few extra bucks to stock up on water filtering supplies. Going out to exotic vacations around the globe every single year will give you fond memories, but those memories won't keep you warm in January with the electricity and gas gone.

Timing Is Everything, So Start Now

Long story short, it's time to put preparations a bit higher on the priority list than the obvious luxuries. Of course, once you have a solid stock of food and supplies, you can throttle back a bit and go into a maintenance mode (replacing expired stock or consuming it before the expiration, etc). Until you reach that point of relative comfort, you may want to catch up on the business of stocking up. The sooner you're stocked up, the better off you'll be when things do begin to come crashing down. Doing it sooner rather than later means that supplies of the items are plentiful, cheap (or relatively so) and accessible. If you wait until things begin to get ugly before stocking up, the prices will have begun to rise and this means you can get less of what you need, putting you in a bigger bind. Waiting also means that you'll be furiously trying to stock up when everybody else is.

Our first perfect example of this is aimed squarely at the folks who live in the Gulf and Southeast costs of the United States. Every single time a hurricane is predicted to arrive, everyone in the area suddenly gets this big epiphany, runs out to the store and buys as much plywood, drinking water, propane, batteries, radios and flashlights as they can. Items that were cheap and plentiful any other time of year suddenly go way up in price and supplies of them practically vanish. Oh and before you folks up north begin to snicker, consider how you behave around snow shovels, tire chains, rock salt, ice scrapers and the like come the first snowstorm …not so funny anymore, is it? In a typical Northern summer, you can get these things dirt cheap and in large numbers. Come the first big snowstorm, good luck finding one and the few you do find are going to cost a mint.

It's the same with prepping, really – except this time, the consequences are deadly serious and the shortages and price hikes are liable to be far more painful (and permanent). Trust me, it's better to get what you need now - while you can still afford it and find it.

But wait! This does not mean running out and dropping a huge wad of cash on supplies! Doing that is a fundamental mistake that leads to even bigger problems down the road. Let's approach this with at least a little intelligence and planning, shall we?

How To Eat An Elephant

Given this rather massive pile of stuff to buy and collect, it may seem a bit overwhelming at first, especially if you're on a tight budget. That said, do not lose hope, because it really isn't as bad as it seems. There's an old Indian (as in India) joke that has a uppity tourist asking a wise old Guru the following question: *"How do you eat an elephant?"* The Guru's answer: *"One bite at a time!"* The humor? Yeah, kind of lame, but the wisdom in that answer is clear – you can do big things if you tackle them in small increments, accomplishing your goal by way of accumulation.

Start small, but continue building up - constantly and often. Buy a few extra cans of food, or a large bag of grain (rice, flour, sugar, salt, etc) each time you go shopping. Every other shopping trip, pick up some of the common medical items on the list. Once every couple of paychecks, save up for or buy a higher-priced item (water filters, home supplies, etc) and if you own a firearm, pick up some ammunition for it. Perhaps on that same time period, pick up some items for the bug-out and get-home bags. Once a month or so, pick up a book or two for your library. Once every couple of months, pick up a really big item (say a month's worth of long-term storage food) and store it. This would be a schedule that will get you almost there in around six months or so.

If you're really on a budget, then do what you can without breaking your budget. If you're hurting financially, you can still scrounge up a month's worth of food for two people for around $70. Here's how you do it:
- 15 one-pound bags of dried peas, beans, rice, lentils, or etc
- 1 64-oz box of powdered dried milk
- Four cans of tomato paste
- 1 26-oz can of salt
- 1 16-oz jar of honey (real honey - check the label)
- 1 5-lb bag of self-rising flour
- 1 small can of baking powder
- 1 large box of cheap tea bags (100 bag size)
- 4 16-oz cans of ham or other cheap canned meats
- 1 60- or 100-capsule bottle of daily multivitamins
- 1 small can of ground pepper
- 1 4-lb bag of sugar
- (Optional) - a couple cans of cheap spices (seasoned salt, garlic powder, etc)

And now to make it storable:
- 1 small box of sandwich-sized zipper bags
- 2 sturdy buckets with tight-fitting lids
- 2 chemical hand-warmer packets
- Electrical tape
- Duct tape

First, get the buckets and toss in the bags of beans, rice, lentils, etc. The reason we use one-pound bags is for proportioning and convenience. Next, get out the small zipper bags and punch a couple of pin-sized holes (the smaller the holes, the better) in them; you will use them to hold your powdered milk, tea-bags, flour, etc. Next, use the bags to hold your flour, sugar, tea bags, powdered milk and salt - be sure to label each bag with permanent marker so you know what's in them. Be sure to tear off the directions for the powdered milk and put that in one of the buckets. Place all of your goods except for the hand warmers into the two buckets.

Just before sealing each bucket, open a hand warmer and put it in the top, then immediately close and seal the bucket with electrical tape, then with duct tape. So why the hand warmers? Well, hand warmers usually operate by rapid oxidation of the ingredients in the pouch. The side effect is that it consumes oxygen as it does so. This means that all of the oxygen in your sealed bucket will be consumed by the hand warmer - at first the warmer will be hot, but will quickly cool back down as the oxygen is absorbed in your bucket. The lack of oxygen means your food won't spoil as quickly and the warmer will always be present to consume any oxygen that leaks in through the plastic bucket walls via osmosis.

Properly sealed, this can keep your goods from spoiling for up to 20 years. Be sure to label each pair of buckets with a name or number that identifies them as being a unique pair, so that you only open those two (and none of the others) as you need them. As a bonus, you can wrap each pair in multiple layers of trash bags and keep them in the attic, crawlspace, or you can bury it after placing it in a suitably-sized (and seal-able) plastic trash can or drum. The buckets are small enough to keep a couple stashed into the trunk of a car, in a garden shed, or even literally hidden under couch cushions (seriously - many couches can hold a couple of them or more underneath the springs if you modify the couch a bit). Incidentally, smaller buckets (using four small ones instead of two larger ones) can be hidden in more places. At this point, you're only limited by your creativity.

Now, if you're independently wealthy? Well, get all your stuff as soon as you can safely sneak it in to your home without anyone noticing it, but take the time to get a good all-round diet before you do.

Priorities, Priorities

The priorities are going to be pretty cut-and-dried and are as follows: bug-out bags, then get-home bags, then at least one firearm per adult family member (if you are legally able to, otherwise a basic set of weaponry capable of killing people and/or animals). Next up is food, water filtering and seeds. After that, defensive supplies (other weapons, extra ammunition, windows boarding, etc), then everything else.

The bug-out bags can be used as soon as you get them filled and ready and are familiar with what is in them. The get-home bags, same deal. At that point, you have barely enough to get to your home, grab the bug-out bags and get out if you need to.

The weaponry is to help you defend yourself on the road, or at home as you build up your stockpile, should things come apart while you're stocking up.

The food and water-filtering supplies are next, along with the seeds. This helps to establish a means of both short and long-term survival where you are. You will likely already have an initial supply of food in the refrigerator, freezer (if you have one) and pantry, sufficient to keep you fed for at least two weeks. However, that is only going to be temporary and will empty out fast, so once you have your evacuation gear set up, then it's time to start stocking up for real, in addition to the usual household stuff.

Defensive supplies, other than the initial weapons, help bolster security in and around the home post-collapse and helps you establish a permanent presence (well, as much of one as you can hope for). You should also take the time to give your property a good looking-over and spend the time and resources to renovate things towards making things more secure.

Only when you have these basics in place do you want to start piling up with everything else in the shopping lists. While you're doing that, any spare change or small windfalls you come across should go towards increasing your supplies of all the vitals, especially food and ammunition.

If you think you can skip some steps, please don't – your very life depends on getting the basics right and in preparing as much as you can. Why? Because there's a common, but stupid (and frighteningly sincere) answer that many people give when confronted with the question of preparing for permanent disaster: *"Oh, I only need to stock up on guns – then I'll come over to your house when I need to eat"*. Don't be that guy, because that guy is likely to end up dead as soon as literally everyone begins playing for keeps.

Too Much Of a Good Thing?

Surprisingly, in many aspects the answer is "yes". Unless you fully expect civilization to end within a year, your food supplies will go bad over various lengths of time and will be a waste of time and money if you don't use them. If you buy enough 'survival' food to feed your family for 20 years, but the stated shelf life of that 'survival' food is only 5 or 10, guess what you just wasted? If you buy enough ammunition to run a small army, but can't store it all in a cool, dry place, a lot of it will be rendered useless by humidity and heat, but you won't know which ones are duds. If you build an obvious bunker and stock it to the rafters, you become a massive target by every starving and speculating refugee who can rub two neurons together and pick up a gun.

The idea is to get just enough to get you started and maybe just a little more beyond that. To supply you with enough of what you need, when you need it. To give you an edge in a situation where the vast majority of your fellow human beings will die off because they did not prepare at all.

We're not just talking about having too much stuff here. There is certainly such a thing as over-preparing your mind, your family, your home and the like. If you spend all day every day preparing for the end of the world and reading nothing but survival literature/websites/etc, you're going to end up a frothing, card-carrying member of the Tinfoil Haberdashery Society. If you spend all day every day trying to force your spouse and kids to do the same, you'll shortly be signing divorce papers and paying child support, as your new ex-spouse broadcasts your inventory all over town. The simple answer is, don't do that. Spend the majority of your time having some fun, going to work, advancing your career, watching birds and smelling roses, getting out and in general enjoying life. Enjoy the fruits of civilization, already! That's what it's all there for! The good news is, everything I've outlined so far for mental and physical exercises can all be done while doing just that and won't take up too much time or effort. So enjoy this world to the fullest and improve yourself, doing so in ways that are not only useful to you, but useful to society while it's still around.

Things To Do While Waiting

There's not too much that really needs done once you're reasonably up to speed and have what you need for both mind and body. However, if you recall the scrounge lists we mentioned earlier, well, this is when you will want to get them filled out and set aside, with perhaps a little maintenance on them as time goes by. Instructions are included there, with examples and rationales.

After that? Relax a little. On the other hand, it doesn't mean sitting on your porch with a gun in your lap waiting for The Big One, either.

For the physical gear and supplies, keep them in good storage shape, check them once in awhile, practice with them occasionally (you know, like take the guns out to the range and actually shoot them?) Stores with any sort of expiration date need to have somewhat of an eye kept on them and when it comes to food, eat (and replace!) the short and moderate-term storage supplies.

Pursue and master those hobbies today which may become tomorrow's critical skills. Keep your mind sharp and improve it as you get chances to. Get to know your neighbors as they move in and any alliances and trusted friendships you gain with fellows of the same mindset should be kept alive, but kept on the down-low (that is, keep it quiet).

Do not go overboard! Do not seek out conspiracies. Keep an eye out for the big changes, but do not use a mental microscope to find them. Collapse not happening is a good thing, after all. At most, keep a loose watch for it - if you're sufficiently prepared, you will most likely get enough warning to set your plans in motion. Keep up on world, national, regional and local events. Always seek valid reasons for the opinions you form on a given topic, be prepared to change or modify them as new facts come to light, and treat all political speeches with cynicism. For politicians: Never ascribe to malice what is best explained by incompetence.

Encourage and foster a friendly community around you. Don't be the family that gossips or starts drama. As long as things are still reasonably civilized, be a polite (but assertive) citizen, even when you don't want to be (otherwise, the toes you step on today may end up sheathed in a pair of boots that you'll be forced to lick tomorrow). *It's time to bring the people skills back to life*. Blend in, and never appear to be any sort of rabble-rouser. Besides, the last thing you want to do is end up on anyone's watch-list – be it personal or governmental.

Otherwise? Get out there and have some fun while you still can! Kiss that babe or dude that you have a crush on. Go home at lunch hour and ravish your wife once in awhile. Don't just teach your kids how to fish and hunt (but seriously, make sure they know how) - teach them how to whistle, how to laugh and how to do fun stuff. Take them out fishing for crawfish, but then use the creek as a swimming hole and picnic spot while you're out there. Take a camera with you while out hunting. Don't just teach your family how to spot wild edibles while you're out camping - tell them cheesy ghost stories and burn the crap out of some marshmallows. Lie your ass off to your friends about your last fishing or hunting expedition. Take in a basketball game once in awhile. Grow some flowers in that backyard garden. On a hot day, get out the garden hose, sneak up on your spouse and kids with it and soak the crap out of them with the water.

A massive reason why we're wanting to survive the collapse of civilization in the first place is so that future generations can someday enjoy all the little things in life. As long as you enjoy them as well, you won't forget what it was like when it's all gone. Prepare ahead to remain alive, but never devote your entire life to preparing.

Be an actual human being, you know?

Chapter 2

Bracing For Impact

When (And How) Does It Actually Happen?

Good question and a very tough one to answer.

The bad news: There is no magical formula, since each collapse in history is almost as unique as the civilization that was wiped out by it. It could be a rapid crash, taking only a few months. It could be a long, slow decline - taking a few years, or even decades. There isn't even a formula for predicting how it will all come down. Will it come down quickly and hard, will it slowly glide to the dirt over generations, or will it start declining slowly until critical portions of it break down, causing a massive and sudden fall? To be honest, there's no credible way of telling *what* will be the events that cause it.

Anyone that claims to know, or makes too-confident predictions, are all liars – except for the one guy who happens to guess it correctly and he'll be ignored, discounted, or nowhere to be found. (*True story: Did you know that Adolf Hitler actually predicted the correct location and week for the Allied D-Day landings at Normandy, in early June of 1944? His generals thought he was so deranged and loopy by then that they actually persuaded him out of the idea, for fear that troop concentrations would be in the wrong place at the wrong time. Moral of the story? Even the most unlikely candidate of all can predict big things accurately at least once in their lives.*)

The good news: We actually have some mental math on our side for this, even if we can't pin down the specifics or the timing. We also know in a general sense how each part of society reacts to changes in the other parts. We know how people tend to react in large groups. We have an idea as to how nations react to large events no matter what their ideology (the only differences are in how rapidly --and violently-- they do react). We can even get a grasp on how resources are distributed and therefore which parts of the globe, nation, region and even town will get hit the soonest, the hardest, or both.

With all of this in mind, let's go through some different types of collapses and give you an idea as to what to do about them and when to act on them.

The Sudden Crash

Possible Causes: Asteroid impact, rapid and global pandemic (with mortality rates of at least 60% or greater), thermonuclear warfare, sudden (as in less than a year) climate change, certain types of massive and global economic crashes, "supervolcanoes" (see also sudden climate change).

Likelihood: Low to (statistically) none, though history has seen instances of some of these.

Societal Indicators: Nothing beforehand, but as it strikes, you will see overall shock, mass evacuations, mass panic, probable rioting, overall confusion. Massive initial fatality counts that may continue to rise for some time, with no let-up. There will be an initial coming-together of people in general under most of these scenarios, until the supplies begin to run short (excepting the global pandemic scenario, in which everyone will stay as far away from everyone else as possible).

Governmental Indicators: Disarray, confusion, possible forewarning but little that can be done to stop it. Some initial massive aid efforts may be mounted, but these will quickly run out of supplies, personnel and other elements.

How you will know: You find out when everyone else does, though under some circumstances you may have some forewarning. It will be obvious beyond belief and no government or agency will be able to credibly deny or cover it up.

What to do: Stay out of the way if you can, hunker down. Otherwise try to predict what places and communities are best able to survive long-term, then try to get there by any means possible. If one of them happens to be yours, then sit tight and work to make the best of it.

The Slow Spiral Down

Possible Causes: Societal degradation, chronic governmental corruption, societal unrest in spots, a series of progressively worse economic failures, loss or gradual reduction of a critical resource (e.g. fossil fuels, agriculture, etc), slow but negative climate change.

Likelihood: Quite Possible. Well over half of the civilizations in human history bit the dust this way.

Societal Indicators: Overall malaise, or a shifted focus from actual critical societal issues to increasingly obscure and inconsequential ones. Fracturing and distancing of human interaction outside of family groups, increased crime as things begin to get worse, lack of goods or prices on existing goods skyrocketing, large inflation growth that is not correlated to mere spikes, etc. Political discourse degrades from common solution-seeking to partisan bickering, even to the point where the result is inaction towards facing upcoming crises.

Governmental Indicators: Increasing complexity in laws and legislation, to the point where ordinary citizens can easily break the law without even trying to. A slow by unmistakable shift of power from individuals to the wealthy and merchant classes, increasingly blatant bribery and corruption and increased spending directly to the population for a short-term distraction (e.g. "bread and circuses"). A police presence and reaction that can either decrease to ineffectual, or increase to the point of fascism.

How you will know: This one is the absolute trickiest to know for certain, as all civilizations have experienced peaks and valleys of decline and recovery. However, the trick here is to know when the ' valley' will no longer rise and become a 'peak', but only keeps deepening. Generally, once the point of no return is reached, it will be easy to figure out - well... at least you'll know for certain shortly after that point has passed. Some indications that can be picked up on (based on past collapses) include a government that over-reacts in an uncoordinated and misfiring manner. You will see increased lawlessness, especially in isolated and thinly-populated areas and wild economic swings.

What to do: This is actually easier than you'd think, because you have a whole lot of time in which you can prepare, right up until the last days. Get out of debt and for those debts you cannot avoid, keep them under tight control. Avoid impulse buys and extravagance. Keep a low semi-anonymous profile. Live in (or move to) areas where you can best survive long-term after your civilization collapses. Encourage community and friendships on a local scale. Get to know your neighbors. Begin to look out for each other. Encourage local economies over national or global ones.

The Stall Pattern

Possible Causes: Almost any, but usually an initial Long Spiral Down, followed by brief revival/recovery of sorts, but the force of recovery loses steam with no fallback, so it all comes crashing down.

Likelihood: Probable, as most other types of collapse can easily morph into this one.

Societal Indicators: Initial shock, rioting, a rapid rise of crime and other general collapse behaviors, but are controlled and tamped-down by the government. Optimism returns as things appear to recover, but then cracks begin to appear as the recovery falters. Jobs begin to not appear as quickly, followed by layoffs and a rapid decline.

Governmental Indicators: An initial crack-down of crime and decline, but the recovery plan is often based on an unsustainable policy or resource. As the underlying weak basis for recovery falls apart, so does the recovery and then the civilization.

How you will know: Due to the deceptive nature presented by the pre-crash recovery, this is the hardest to predict. Most other collapse types can appear to be a Stall Pattern if there is any general optimism and signs of recovery, but that recovery will have to last for at least a year, yet show signs of being temporary, before it meets this criteria.

What to do: The same preparations that you would make as described in this book – stock up, build up your community as much as you can, live where you think you will best survive, etc. However, you can use the period of recovery to stock up even more, replenish any lost or consumed stock and to prepare for the possible final crash.

The Failed Revolution

Possible Causes: Rapid change of government type or overall organization, change of basic economic policy, change of societal structure, or reaction to an increasingly repressive government.

Likelihood: Sometimes inevitable, but usually not global. This means that the odds of crashing civilization on a global scale from this are slim at best, unless multiple governments or a select few superpowers are the ones in revolt.

Societal Indicators: Unrest, revolution and all of its attendant indicators: factions, heightened ideological differences that often split communities and form new ones and an overall restlessness in society. Propaganda will fly in both directions within every level and type of mass communication.

Governmental Indicators: There are usually two or more factions that arise which come into conflict. At least one of these factions will be reactionary, or preferring to keep things as they are. At least one other faction will demand and try to introduce radical changes. The few not involved will demand and work for "reforms" and compromise, but with no power or ability to get anywhere. Note that we're not talking about the usual give-and-take or heated debates, but full-blown charges of "treason" (or similar), or a demand for changes to the fundamental foundations of how the government is formed and run.

How you will know: It will become rather obvious over time if you are looking for signs of growing unrest/revolt and are rarely spontaneous. The propaganda will slowly become more obvious, the calls for change more strident, protests may become more violent and violence may begin to be called for by the protagonists for change. As society begins to come apart, no one faction gains the majority. Guerrilla warfare begins to make itself known and gains traction. Terrorist acts arise in multiples, with factions claiming responsibility and with no letup in attacks.

What to do: Get and/or stay out of the way and quietly do your best to insure that your neighbors do the same. Rid yourself of all debt and governmental contact if possible. Own your home rather than rent, but try not to own too much property or other large stakes in the economy (stocks, bonds, etc), as they may be lost. Use your supplies only when you absolutely have to and utilize whatever aid comes along. However, do your level best to not subscribe to, or participate in, any particular ideology or political faction and distance yourself from anyone/everyone who would try to recruit you to a given cause. Avoid volunteering for (or participating in) any political party, rally, or demonstration. Avoid showing sympathies for any faction or ideology that is violent, or that condones or calls for violence. Keep your politics to yourself and don't become a target. Quietly stock up while you can, preparing for the time when it all comes crashing down - one way or the other.

The Lock-Down

Possible Causes: Actually, this won't be a cause per se… it'll be a reaction. It is the polar opposite of 'The Failed Revolution' - and here's how it plays out: Things begin to go downhill in general (either due to cyclical economic measures, a crisis, etc) and the government panics right along with the people. Government officials, keenly interested in keeping things peaceful (and themselves in power) begin to clamp down on the populace.

Likelihood: Plenty of historical precedent: The first half of the USSR's post-czarist history, 1936 Germany and 2001 Argentina stand out…the only real effect on civilization depends on how big the victim country is. If it occurs to a smaller country, civilization as a whole will withstand it rather easily. If it happens to a superpower, the odds of pulling civilization down with the falling country is very good.

Societal Indicators: Initial acceptance of the situation, but then a growing fear of the government as the increased power begins to saturate the entire structure of leadership. At first only a couple of small 'niche' actions and rights are curtailed, but as things get worse (or stay steady-state), more and more rights begin to disappear, until such things as interstate travel is restricted and ownership of certain items becomes forbidden (especially weapons). If a crisis of shortage arises, citizens and businesses may be accused of "hoarding" critical goods. Black markets will begin to spring up for common/ordinary goods and individuals will find and exploit every loophole they can possibly find.

Government Indicators: The leadership enacts measures which are distinctly against original liberties and freedoms, but these will begin subtly and each step is sold to the public as a temporary measure designed to correct things to a more peaceful and stable state. The problems arise when those "temporary" powers are not relinquished or returned to the people and even more rights are weakened, to the point where subtle and temporary measures become naked power grabs. You will see gentle propaganda and subtle misapplications of the truth at first, but eventually the media becomes saturated with propaganda and outright lies.

How You Will Know: When government first suggests and then demands that you inform on any neighbors who act in ways opposite that of the government's wishes? That's a good sign of this type of crash happening. When you see less news and more propaganda on the evening news - especially when you see a distinct lack of criticism towards the government. When most important and basic economic statistics turn out to be falsified or hidden. When the local police become armed with more powerful weapons on a regular basis and are held less accountable for bad behaviors among them. When neighbors disappear under police custody and no news is broadcast or can be found as to why.

What To Do: Keep your head down. Don't stand out, or give even the slightest reason for any official to look at you twice. Quietly prepare as much as you humanly can without getting caught doing it. Security will be paramount in this particular case. Not only will you have to worry about unprepared neighbors, but you will now have to worry about neighbors becoming informants and criminals targeting you before things finally come crashing down. You also have a new worry: government (national/provincial/local) punishment from searches and sweeps. Find (or be) somewhere quiet to live if you can and stay there. Hide everything that even smells like it doesn't belong in a government-approved household. Don't buy anything new if you can help it. Look and live humbly. Appear cheerful (or at least hopeful) in front of government officials.

I know there are some reading this, who will begin screaming epithets at me right about now, followed by demands to "*rise up and fight!*" Well, here's the deal. A gent named Marvin Kitman once said: "*A coward is a hero with a wife, kids and a mortgage.*" Well, mortgage aside, the wife and kids are another matter altogether and if you have the latter, you need to spend your time quietly educating them once they're old enough to figure out how to keep their mouths shut in public. If you're young and single, hey - go for it… but use your brains and do it quietly. For a good primer on how to resist, I heartily suggest a science fiction book by a gent named Robert Heinlein. The title? *The Moon Is A Harsh Mistress*. Besides being a highly entertaining read, it is an excellent primer on how to resist an overbearing government.

On balance, history has shown (repeatedly) that most fascist and dictatorial government have burned themselves out in fairly short order and if civilization is collapsing anyway, then it makes a lot more sense to ride the storm out until it crashes once and for all. In this particular scenario, if a superpower goes fascist in a desperate response to the world crashing around it, then odds are nearly perfect that the dictatorial types will be among the first to die.

If things get too ugly, there is another suggestion that bears consideration: Emigrate if you can possibly do so. There are going to be circumstances where you will seriously want to get out of the country. Consider a Jewish family living in 1936 Germany… their best possible chance of survival was to simply get the hell out of there. The alternative involved hiding from the local police, somehow fooling the authorities into thinking they weren't targets, or trying to survive being locked up in a *Konzentrazionslager*, such as Dachau or Auschwitz. The last bit wasn't smart, as the *KZ* usually killed you.

Long story short, in this particular type of crash, you may seriously want to simply get yourself as far away from the government as humanly possible, or at least find a political 'hole', jump in it and pull that hole in after you.

Common Patterns and Signs

In spite of all the differences, there are things with are common to them all once civilization finally gives out.

These include most (and not necessarily all) of the following:
- Increased violence
- Increasingly optimistic proclamations that do not appear realistic, or even possible
- A decrease in police presence
- An increase in crime
- An increase in overall economic troubles above historical and/or cyclical norms (be it unemployment, inflation, interest rates, etc)
- An increase in the number of people on governmental assistance (welfare, food stamps, etc)
- A decrease in the number and quantity of out-of-season foodstuffs
- An overall (and rapid) increase in prices for basic goods, above and beyond the rate of inflation.
- A collapse of multiple and unrelated economic markets.

Knowing When It All Comes Down

Aside from the big and obvious civilization-enders (asteroids, massive global pandemics, etc), it's going to be hard to tell for certain when to cut off all ties with the infrastructure and start living on your own. Fortunately, if you planned things right and according to the book, this isn't as big of a problem as you think. You can alternately live on and off of the moderate-term food stores and supplies for as long as you can buy replacements for them. You can swap between barter and monetary purchase as opportunities arise and use whichever gives the best "price". You can happily continue going to work each day up until things begin to get (oreven appear) too rough during the commute, at which time you can try and get your employer to allow telecommuting (which they will likely do at that point, since the management has to get to work the same way you do). In any case, continue working there until your employer ceases to pay you sufficiently, or getting there becomes too dangerous.

The one thing to keep in mind is that unless there are mass evacuations going on, you can continue playing along, harvesting as much as you can and as things deteriorate, increasing the pace of your preparations and stockpiling. The only real hazard to this is if government (at any level) outlaws hoarding, in which case you'll just have to be more discreet about it.

If it ever gets to the level of rationing, curfews, or martial law, or to the point where travel outside of a populated area is literally dangerous to life and limb, then you know it's time to hunker down (or run to the hills if that's your plan).

Until then, carry on as usual, but with an eye towards increased stockpiling, coming together as a community and increasing communal neighborhood patrols.

The False Alarm

If you do this right, there won't be any, but it is quite possible that you may get the timing off by a week or more. However, if you miscalculate, let your emotions take the place of reason and logic, or if you are just that much of a 'nervous nellie', then a couple of things will save you from exposure, embarrassment and worse:

- First, always use an automobile in your evacuation plans wherever possible. This way, if you run for the hills, the worst that can happen is that you drive a lot before

you realize that it hasn't all blown up yet. It's also easier to get home that way and the radio makes for a good means of knowing either way.
- Second, make it a habit to use the radio you threw into the bug-out bag as you travel (it's listed in the Shopping Lists... you do have one, right?) That, or keep the one in your car tuned to a reliable news station. That way, when you hear no news of mass evacuations, you can turn around and (tentatively) head back in a day or two.
- Third, before you run (or if you prefer, while using the cell phone while you're running), tell a plausible lie: Call in sick to work, or claim that a parent, in-law, or sibling has died and that you have to rush out to attend/plan the funeral. This way, if things are still running civilization-wise after a week or so, you can come back with less chance of being fired.
- Make it a habit to lock-up tight and set all alarms when you leave the house (unless it's on fire, obviously). That way you don't come back to find your stuff gone from burglary.
- If anyone asks on your way out the door, tell them you're going to a funeral of a family member, taking stuff to the in-laws, camping (if it's summer), or etc. Ideally, you'd be leaving at night anyway (more on that later), so you won't have to explain all the packed supplies.

Long story short? Unless it's a screamingly obvious reason (e.g. asteroid impact) always stop and think before you make the decision to run for the hills, head for your bug-out location, or etc.

The good news is, by the time you've reached the point where you're packing the car, events will likely be plentiful (and local!) enough to support your decision.

Do We Stay or Do We Go?

This one decision will demand to be answered on the spot - and yes, in spite of already making it before doing any preparation work at all. This is due to the fact that between then and now, a few variables may have changed. Your neighborhood may have deteriorated while trusted families slowly move out, to be replaced by untrustworthy ones. Population patterns may have shifted, putting your once semi-rural home smack in the middle of of high-density suburban homes. A disaster may render your home area at higher risk of flooding, fire, or worse than it otherwise would have been. Your neighborhood may be at risk of being part of a quarantine zone. Insurgent forces may decide to use your neighborhood as a free-fire zone, or as a base of their operations. Nuclear war may be looming and your house is within a couple miles of a major military base.

For whatever reason, if things start to get ugly out there and civilization is in crisis to the point of threatening to come apart, your life will depend on how you decide this one small but powerful question. Do you stay, or do you leave?

Factors To Consider

The most important factors to consider before deciding once and for all, rests around three things: Risk, Prospects and Community.

Risk is the most complex to evaluate, but only because there are different kinds. You can however boil it down to risk from people, risk from government/politics and risk from nature. If any of these, now or later, become large and imminent, then running for the hills may be your best option. However, unlike the others, these may be temporary, so you will have to think about your chances of survival until they either pass or die off on their own. Risks from nature and government/political actions can be weathered and/or prepared against (with a few obvious exceptions), but risk from people are the most likely to make things untenable. On the other hand, risks from people are things that you can possibly do the most about. If your community is tight, sufficiently armed and together, you can fend off an equally-sized (or even larger) gang of raiders. If you're sufficiently defended and clever, you can scare off or discourage (or even disguise yourself from) a large horde of refugees passing through. Therefore, when looking at the risks, look at what you can do about them along with your community (just make sure you have one).

Prospects are going to make or break a decision to stay, but will require a bit of knowing what you have and what is around you. Questions to ask yourself (and you should already know or at least somewhat guess) are things such as *"...is there a supply of fresh water around?" "is there enough nearby open land to plant crops sufficient for the community?"* and *"how soon will things become realistically peaceful enough around here to allow for long-term survival?"* If the answers to these questions are are definitive "no" or "never", you'll want to consider leaving.

Community will be (or rather, should be) a known factor and the answer to these questions should come quickly: *"how many trustworthy neighbors do I have left around here?", "can I trust my neighbors?", "are they reasonably prepared or at least self-sufficient?", "have the majority of the trustworthy ones bailed out or are about to leave?", "can we even hope to form a coherent team?"* These answers to these will definitely determine whether to stay or go. If the majority of your neighbors have run for the hills, if they're completely unprepared and without any sense of self-sufficiency at all and if they are unable, even in a time of crisis, to form any sort of team? You may want to consider leaving. If more than 2/3 of the trustworthy neighbors in your area have left, you will probably need to leave.

Something you may want to think about if you are sufficiently able-bodied: If there's not a mass evacuation going on (or is about to), wait a day or two and see what happens. If things are crashing slowly enough and there's not a volcano with hot lava rolling towards your house, you can hedge your bets a little. Take a day to see if the madness stops, or at least calms down a bit. You may find that things are safer and better long-term where you are, than on the road or at whatever remote destination you may have had in mind.

Making The Decision

The decision to stay or run will start to form up before you even finish considering. However, go through them all, then make a tentative decision. Then, take a quick peek around, make sure you have all the facts right and only then, only then, do you make the decision. Be sure to not make it alone. Involve all adult members of your family and definitely involve any trusted fellow preppers you may have been working with. Make the decision together and make sure the decision is unanimous if you can.

Take the time to listen to all sides of the argument, if you can spare the time (and odds are good that in most cases you can still spare a little, even if it is just ten minutes).

Once you make that decision, immediately start putting it to use. If you decide to stay, keep the option of bugging out in mind, but start digging in. If you decide to run, start packing and leave (together) the first night that you can, before the roads become impassable or too dangerous. Either way, keep an eye out for any developments that may change your decision, but stick to it for as long as the factors weigh in favor of what you have decided.

If for some reason you're still not certain?

- If you had previously decided to stay put back when you began preparing and intend to remain? Keep the bug-out bags close at hand, but know that you may well have to walk the whole way by the time (if?) it does come to leave.

- If you had made a previous decision to have a "bug-out" location and have stocked and prepared it, get out and go as soon as that very night. The sooner you can get to your sanctuary, the easier it will be to be welcomed by any existing neighbors there, the faster you can get set up and the less likely that someone else has snuck in first and taken up residence there.

- If you had made a previous decision to get out as soon as things got bad, then get out as soon as possible.

Points Of No Return

The window of opportunity to leave will close rather rapidly in most cases and when it does close, traveling with be very hazardous and will be so for quite a few years. You will have a little time to think it through (unless it takes you totally by surprise – then you have none), but you won't have a lot of time, so use only as much of it as you need. On the other edge of the sword, we have the fact that if you leave too soon and somehow need to come back, there is no guarantee at all that your home will remain stocked and unoccupied (or that you would make it back at all). Once you leave in a true collapse situation, you cannot conceivably return. Your home will either be occupied by a refugee, stripped for firewood, or any usable supplies and materials left in it will be put to use by someone else. Or it will burn down.

In the face of all of that, there is still the matter of having to stick with your decision. Once you've made it and left, you're going to need to stay gone, possibly for good. Once you've decided to stay, you're in it for the long haul, because once the refugees begin to flow, going out there will be dangerous (if not fatal) by orders of magnitude and you won't want to do it unless your odds of survival by staying put drop to practically nil.

Whatever you do, make your decision, make it in a short amount of time while the opportunity is still open, but make it with all of the known facts at hand. Once you make it, stick with the plan, because once you put it in motion, you will not easily be able to change it.

So let's go through the two options, one at a time. We'll start with staying put.

Batten Down The Hatches!

If you've made the decision to stay and ride it out where you are, don't think you have it easier than the soon-to-be refugee. You have some work to do and it will be just as hard. It will also mean having less options, but you will certainly have more security... well, relatively so, anyway, as long as you have a good community and solid prospects.

Stop, Think!

Whatever happens or how you came about this decision, your first job is to stop and think. To get all the facts. To take a good, hard look around your neighborhood. To figure out which neighbors are staying for certain and which ones will cut and run at the first hint of trouble. To figure out who already left. To take stock of what is around you now and be as certain as you can that you can indeed stick it out here.

Something else that will be different now is that you're not going to be doing this alone any more. It is time to reach out to all of the neighbors you can trust (at least those who have remained behind). It will be time to meet them, even if you have to do it covertly. You will have to formulate some sort of plan and begin to band together. But, before you do that, you have a few things to get in order at home.

First Things First

Before you do anything, stock up as much as you humanly can on everything you can. Don't go too far away from home, but get to any stores left open and spend as much cash as you have to get as many critical supplies as they have for sale. Be sure to go armed, just in case (because by then things will have gotten bad enough to pretty much require it.) Buy whatever seems useful, take it home, then come back for another trip (never leave anything sitting around in your car unattended) and continue doing it until you're out of cash. Forget the grocery stores at this point – go for the hardware stores, local drugstores, electronics stores (for things like radio parts, batteries, etc), home and garden stores (seeds – get seeds. And pesticides and hand tools. If you have the space and means to haul it, boards and plywood are useful. Forget anything else). Also seek out small clothing stores, thrift stores and places where people generally don't go in a crisis. Most people in a panic head for exactly two types of store to the exclusion of all others: the supermarket and the gas station. Avoid both unless you have no other choice.

While one of your adult family members are shopping (*with the neighbors? We'll explain later*) someone should be at home and stocking up on water. Start storing clean water into clean containers, everywhere you possibly can, for as long as the water supply continues to flow. Get out and prepare water filters. Clean your bathtub out as much as you can and rinse it out as thoroughly as you can. Place a clean trash bag over your bathtub drain and tape it down, then duct tape some plastic over the upper vent (it's that silver disc under the faucet, or the holes along the top). Fill the bathtub. Tape together enough clean trash bags to completely cover the top of the tub and lay it directly on the water's surface, letting the excess drape over. Tape any portions of plastic that touch the back wall or the tub's rim. Leave a corner un-taped to draw the water out with a straw or a large, clean cup. This will give you anywhere from 50-100 extra gallons of fresh drinking water. Shut off the inlet and outlet valves of your hot water heater and turn the gas/power off to it. Doing this will mean an extra 20-50 gallons of drinking water (depending on size). Leave one side of your kitchen sink open, but rig up the other side as you did the bathtub and fill it up. That's 5-10 gallons more. Shut off the water valve to every toilet tank. That means 3.5 to 5 gallons of water per tank. Put four to six layers of clean trash bags together, one inside the other and set it in a laundry hamper or spare (clean!) trash bin, then fill it up. That gives you another 20-50 gallons each time (depending on bag and can sizes). As long as the water continues to run, use it – draw that water from the un-stocked side of the sink and leave all the other water supplies alone. Use the garden hose and put cans in the garage if you have to.

When the water starts to sputter out, shut off the main water valve into your home and leave it sitting there until you need it - this leaves 5-10 gallons of water in the house pipes that you can drain from the lowest faucet in the home.

Get out the weapons, start wearing them and keeping them close at hand and meanwhile bring in any critical items that you have stored in any sheds, detached garages, or any other storage area that requires you to step outside of your house in order to reach it.

Finally, if there is time, get as much wood as you can lay hands on. Stack it close to your home (a good idea would be to fill the garage with it if there's room.) As long as it is fairly thick and it burns, bring it in. Hopefully, you had already stored and kept at least enough wood to keep your fireplace or wood stove burning for an entire winter of daily use. In northern climates, shoot for two winters' worth.

Even if you're staying put for the long-haul, get out your bug-out bags and quickly check their contents. Replace anything broken, expired, or spoiled immediately if possible (from your household stores). Bring in any get-home bags that aren't going to be used and see how much of it you can stuff into the bug-out bags. Keep these bags near the beds.

Heating Preparations

Next up is to take a good, hard look around your home. You're going to want to put together (or have) a means of heating your home when the electricity and gas finally go out. If you don't already have a fireplace or wood stove, it's time to buy one (if you can still do so), build one (if that is still possible), or at least locate and/or collect enough materials together to make one later, once collapse is obvious.

If you decided to wait a bit before building one, your shopping list and a bit of ingenuity can save you here. Start by planning and locating your materials. If you have metal ductwork, you can put that to use in making a chimney flue. A bit of cement-board (commonly used in bathroom floors) the size of a nearby window with a duct-sized hole cut into it will allow the ducting to pass outdoors, where you would run the ducting up the side of the home and cap it with something that keeps the rain out, but lets the smoke escape. On the inside, your metal stove hood can be put to use as a fireplace flue and hood - you would of course remove the motor and all filters. For the hearth, you would want to remove all carpeting and replace it with brickwork or cinder-blocks (or both), packed tightly with sand and/or mortar, sited directly underneath the hood.

An alternative is to have a couple of 55-gallon steel drums and ducting pre-purchased (and partially pre-assembled and pre-cut), then you install it once you are certain that the electricity and/or gas is never coming back on.

Medical Preparations

Once you're done with that final round of shopping and gathering-in, take stock of all of your medical supplies. Know what you have and how much of it you have. Set up a protocol and a space to treat any injuries and/or disease. Sort out which room(s) you want to use for quarantine and treatment. Set up that room with the warmest blankets, spare linens and first aid supplies (not the portable ones, mind you – use the home kits).

If you have a trusted neighbor in the medical profession and happens to have the skills necessary to save lives and heal, then take a portion of your home first aid gear to him or her, in exchange for assistance when you need it. Before you do, be sure that the person is here to stay, will not run off at the first sign of trouble and is trusted. You'll have to tell a white lie or two as to how much you haven't given him/her (after all, nobody needs to know what you really have unless it is a life-or-death situation). Just make sure you can get there almost instantly and that he or she doesn't live more than a few houses away, or is within ½ mile of your home if you live in the country.

The reason we're getting the medical situation squared away right now is because it will be the one most needed when things go wrong. You can never go wrong with having what you need right there and handy.

Stop, Love, And Pray

I've done my level best to avoid any religious connotations up until now. If you are an atheist, take it as you will and skip the spiritual bits. However, if you are religious at all, take it as something you better do at this point. In either case, take a few moments to gather your family together, take turns telling everyone else, one at a time, of your love for them. Take a little time to explain what you love about each person in your household. Ask for forgiveness for any wrongs that you may have done. Encourage each other as best you can. Get on your knees together and pray for forgiveness, mercy, forbearance and rescue. Pray for those about to die and pray that you are able to rebuild a new society on the ashes of the old. Pray that the new society will be one that will keep Him in mind and in heart. Pray that His will be done and that you recognize and act on that will wherever possible. Take the time to recite selected readings from whichever books you call holy.

Most importantly, take this time, a few hours if you can spare it, to be together and to quietly discuss what could be coming and to strengthen each other in both mind and heart.

Ground Rules At Home

Once you have the medical situation squared away and your family at peace with your Maker, then it's time to gently but firmly establish some ground rules for you and your family. Most of these will come naturally to you, but some may not, so let's cover them generally. The situation may require a few modifications, additions and even subtractions, but in general, this is the routine you should set up. All that said, these are going to be the basic rules and should be explained and agreed upon by every household member who is able to speak and understand them.

Ground Rules should include:

- Nobody says anything to anyone about what and how much your family has stored. This goes especially for the kids. Give the younger children the impression that you're rationing what little you have and that if anyone else were to find out what you did have at all, you would be in danger of being hurt or killed. To be honest, you're not lying to your kids here. If your neighbors, trusted or not, discover that you have way more in supplies than they do, they may decide to redistribute the wealth a little – by taking it from you (by means fair or foul) and spreading it around.

- Children under the age of 17 are not allowed to be out of earshot or view of at least one parent. They are to stay close to home at all times, possibly not even leaving the front or back yard unless accompanied by a parent or close relative. This is to prevent kids from violating the previous ground rule and to prevent them from wandering off into a neighborhood that has just become infinitely more dangerous.

- Anyone that enters the home uninvited after dark is to be shot at and/or attacked with any available weapon, without question or hesitation. If someone wants in, they can either knock on the door or wait until they are invited.

- There will be no leftover food after meals. If you take it out of the pot or bowl, you eat it, or find someone else at the dinner table who will. Wasting food is to be considered a no-no.

- Everyone will wash themselves at least twice a week whenever possible. Cleanliness may well make the difference between survival and dying of disease.

- If you make a mess, you clean it up as soon as possible.

- Everyone takes turns pitching in on the cleaning duties – laundry, sweeping, the whole lot. The home will be cleaned top to bottom (picked-up, swept, laundry) at least once a week if time permits. Unless seriously soiled, clothes will be worn at least three times before being washed (to save soap usage), except in summer, where only two times is sufficient to consider it washing-ready.

- All waste of all kinds are to go outside.

- An outhouse will be dug at the side of the backyard and some sort of privacy tent or building is to be put up around it. The hole will be at least 4-5 feet deep.

- If you can do it, keep a night watch, in shifts. All household members over the age of 10-12 will take turns keeping watch while the others sleep. Those who keep watch at night will be allowed to sleep in for up to 3-4 hours after daybreak. If there are only two adult members of the household, then one sleeps from 4 hours before sunset to midnight and the other keeps watch, then switch for the hours between midnight and four hours past sunrise. If you're all alone at home, then try to take turns with your next door neighbors, watching over both homes in shifts. The idea is that you do not want to give criminals an opportunity to sneak in while everyone is asleep.

- At the first sign of an intrusion or attack, an alarm will be raised loud enough for the neighborhood to hear. Bells, whistles, gunshots, whatever it takes.

- If your neighbor is in trouble, you will defend him (unless your neighbor is the one attacking, in which case you simply treat him like any other intruder).

- Once a week, if you are religious, hold an in-home service dedicated to that religion. Depending on religion, this can certainly fall on a Saturday or Sunday and on any holy days as your religion dictates.

Pretty obvious stuff, but they all need to be stressed. Setting up a routine and sticking to it is not only good for your sanity, but it gives everyone something to do, especially during those times when it seems that the whole world got boring and scary all at the same time.

Planning Food And Water Consumption

Certainly, you have a mountain of food stocked up (right?) Perhaps you also have untold gallons of water as well. However, it is all too easy to chow it all down, waste it profligately and be without either or both in a matter of weeks to months, with nothing to eat or drink for months ahead. Never assume that you have enough. Some of it you will have to use right away, but most of it can be preserved, or left in storage until you need it.

Planning for consumption of food is easy to do. Take a good, solid inventory of everything you have in your home that is edible. This includes your storage, but also includes the refrigerator, the freezer (if you have one) and everything in every cabinet, pantry and so on.

Once you know what you have, start working immediately to dehydrate and preserve everything in the refrigerator and freezer. Salt meats and begin to cure them. Dehydrate vegetables if you can, using a towel in the back of a car with the windows cracked open, or a commercial dehydration rack (are you minus electricity already? Just put the racks in a hot windowsill or near the fireplace and the results are the same.)

Once you are well on your way to preserving the perishables as best you can, focus your consumption of food first on anything perishable that hasn't been preserved yet. Space it out so that you consume it all just before it begins to go bad, so as to minimize waste.

Use the stove for as long as the electricity or gas holds out, but start planning on cooking in the fireplace or on the wood stove (or on whatever you build to substitute for one). In summer, the wood stove can be kept cold until you need it, but know that it takes longer for it to reach normal cooking temperatures. In winter temperatures, keep a constant slow fire going except for those times when you're cooking, when you can stoke things up to get the temperatures you need.

Once the perishables run out or start spoiling, begin to shift to the canned and moderate-term stored foods. With these foods, take out those things closest to their expiration dates and only take out or open what you will use that day. Make sure that by the end of that day (or perhaps the end of the very next day), it's all gone. Make sure it's completely empty before opening the next of its kind (vegetable, meat, bread, whatever).

Only start dipping into your long-term food supplies if there is no other canned or fresh good of that kind available. Follow the instructions carefully. Only have one of each type open at a time and be mindful of any spoilage.

In all cases, leave all canned and long-term foods hidden and in storage until you're ready to use it and only then take out just what you need for that day (or at most, the next).

Finally start looking ahead to the next spring. Check your seed supplies and begin planning for what will be planted, where, when and to watch and see if the planting area will be secure enough. Coordinate this part of it with your trusted neighbors.

Keeping An Eye Out

It won't do you much good to simply stay locked-up in your house all day long. You're going to have to go outside sometime (as long as it is safe to do so) and you'll definitely have to start talking to your neighbors (which we're going to cover in just a little bit...) The concept here is to take notice of what's going on around you and in the world, as best you can. Whether or not you have electricity, the radio will be your best bet, especially if you have a shortwave radio that can be hand-cranked. Getting news from outside the country will give you a well-rounded picture of what's going on globally and is a handy way to compare what others are seeing versus what you are seeing.

Forget the television. Even if it is still functioning, it is likely to be packed to the rafters with propaganda and whatever news you do get from it will be the same that you can garner from the radio. Also, televisions eat more power and even a battery-backed radio will be more useful for those times when electrical service is spotty at best.

The Internet might (or might not) remain alive and active, but you may have a very hard time reaching it. If however you have an independent power source (or some powerful batteries) and a means to connect via satellite (yes, it is possible with sufficient skill and the right tools, but we're not necessarily talking about a standard commercial satellite Internet provider service), you have a wider means to get the news, information and to reach out to other people remote from you. On the other hand, this is going to require specialized skills. Most modems and local routers will likely be without power and it is also possible that your local Internet Service Provider may be disconnected. However, if you can rig a satellite modem and a means to connect to an offshore satellite service, you may still have a shot at getting online (albeit the speeds will be extremely slow at best).

Otherwise, you're likely to have the best luck with a hand-cranked radio capable of receiving shortwave frequencies. It's cheaper, easier and will be the most reliable, for as long as there are stations still emitting a signal.

Another means of getting distant news is via ham radio. They're still around, quite active and still plugging along. You don't need a license to get a receiver, but it will take a bit of time and practice to get the hang of using one. You do need a license if you want to use a transmitter, but that will in turn allow you to speak out over the airwaves. Just note that ham radio sets are going to need a power supply and that you will want spare parts to continue using it over time.

For more local news, you could get hold of a CB ("Citizens' Band") base radio, or better still, a police scanner. This will allow you (to some extent) to gather news on a local level, though CB is restricted to a short range (and requires someone on the other end to say something) and police have increasingly begun to encrypt their messages over data channels. All that said, a CB base station and a couple of portable mid-range CB hand radios can be useful in your neighborhood nonetheless, at least until the power goes out for good.

If you are technically inclined and have (or can generate) a reliable source of electricity to multiple homes, you could consider setting up a LAN ("Local Area Network"). This involves a few WiFi routers capable of bridging connections, a spare computer desktop to act as a server of sorts and a bit of coordination. You will pretty much require a server and/or network technician to string it together, but it is an excellent way for the homes in your neighborhood to link together without running cable and provides (if built properly) an encrypted channel over which you can communicate. This will however require electricity to be either made or available to each and every home that uses it. Typical ranges for a WiFi router over open air is about 100 meters (barring any walls, obstructions, etc). That said, this is a bit extreme and should only be attempted if you have both the means to generate your own electricity in multiple places, the components necessary and the skills to set something of this nature up, as well as maintain it.

Other means of getting information involve setting up a network between your neighborhood and others. The trick is to exchange only that information which you intend to pass along, about happenings in your particular area (or from other neighborhoods). Initially, the only information that you want to pass along are things of a regional, state, or national level and nothing about how your neighborhood is doing generally (you best bet there is a non-committal something, like: "we're hurting, but we might just make it if things hold up") and say nothing at all about how your neighborhood is fixed for supplies or defense. However, you will want to learn and pass along any new information about possible ambush traps, areas and neighborhoods that have turned into war zones and to relay any news and discoveries from those neighbors who have left but managed to return.

One final thing to consider: You're going to need a really big 'BS Filter' for most of what you hear, no matter what the source may be. Take it all with a not a grain, but a full-on block of salt. For awhile after things crash, assume the worst. While some of it will be horrendous or miraculous in its own right and even absolutely true, most of it will be heavily exaggerated and more often than not is based on guesses and baseless rumors. You will hear of things that likely do not exist at all or never happened. You will likely not hear of things which did happen. You will likely end up cooking up a few lies yourselves to encourage outsiders to stay away, or to attract something you want or need.

The funny part is, what you hear may not even be a lie at all, but mis-heard information amplified and embellished on... so approach any news with caution.

Whatever you do, do not base any important decisions on what you hear or read from others at all. Packing up and risking life and limb to reach a rumored relief camp or new farming settlement is a near-certain way to end up milling around a remote place waiting to die, along with a whole lot of other desperate people who got suckered into it. Burning off scarce or critical supplies in the belief that a government relief truck will arrive in a few days or weeks with more? That's a surefire way to end up quickly running out of supplies, with no more coming, ever.

Keeping an eye out is more than just getting and disseminating news. It also means keeping a watch and perhaps a few covert day-time patrols (in this case? never at night – you may end up getting shot at), to see how things are going out there beyond your neighborhood. It means watching for signs of conflict and crime while you're out there. It also means being more careful than you've ever been. It definitely means not doing it alone.

We'll cover more in Chapter 3, but this is a good place to get at least some idea overall.

Rounding Up The Neighbors

Remember all those neighbors you've been busy getting friendly with all this time? Well, now you get to find out what they're really like and what shape they're in and to formulate some sort of plan. As soon as you have your home squared away and can safely go outside, as things are still crashing, start walking around the neighborhood (if it is safe to do so, naturally).

Begin with those neighbors you may have been preparing with. Odds are good they may contact you first, but start with them anyway. If there are multiple neighbors that you've been prepping alongside (you lucky dog!), then everyone needs to get together at one of the homes and start talking. Begin by making sure everyone is okay. Then, start going over who in the neighborhood is still around and who left town. Next, figure out the overall security situation and figure out how (and who) in the neighborhood overall is going to (or can) participate in patrolling the area. Also, take a few minutes and sort out the overall situation. Find out who among you should be the spokesperson for the rest of the neighborhood and what roles everyone else will take part in (one would hope that you have already sorted this out, but just in case...). Find out who among you will be staying put in light of events and who is thinking about leaving. If you are thinking about leaving, or if any other party happens to, get it out now and let your trusted circle know about it. This group, if you have one, will most likely be the core leadership for your neighborhood.

If you have neighbors you can trust but that you don't think has been preparing, get together with them next (if you have fellow "preppers" and met with them already, then either all of you should invite the trusted neighbors together, do it in stages, split the load, or whatever seems appropriate). Sort out who has what in the way of supplies, people and skills and don't tell them anything more than vague responses about you that sort of matches theirs. Stress to one and all that any empty homes in the neighborhood are to be left alone for at least two weeks if you can, a week if not. This will give any neighbors who fled a chance to come back and have a place to live, with their stuff undisturbed.

With both fellow "preppers" and trusted neighbors, get together (or while still together), start going over supplies. Figure out where the nearest retail stores and other places are within easy walking distance. Start planning on salvaging these places once it becomes obvious that they are going to remain unstaffed and that there are no curfews, police patrols, or other impediments. Start figuring out who has what skills and begin to sort out which neighbors have the most critical skill sets. Get them together and start sorting out who will lead which duties and when.

The duties you will ideally want are:
- Security (armed if you can do so) – everyone but the doctors are going to take turns helping here.
- Medical
- Water procurement and purification
- Food scavenging (both from hunting and from existing commercial outlets)
- Firewood scavenging
- Specialty supply scavenging (medicines, weapons/ammunition, building materials, seeds, livestock if any, etc.)

Once you have the leadership sorted out, it's time to get all the neighbors together at once. Meet somewhere public (the street works) and make certain that only neighbors are attending (ejecting anyone who doesn't actually live in your neighborhood.)

In this meeting, everyone is informed that we're all pulling together and that the plan you and your trusted neighbors formulated, is how things are to get done. Make it clear up-front that anyone who refuses to participate will be left unarmed and forced to fend for themselves and will not share in any scavenging or other community efforts. Make it clear that anyone who causes any trouble for the rest of the neighborhood will be forced to leave. Try to make it equally clear that the reason for this is so that everyone can survive in the long-term and that all efforts will be made to insure that survival. Together, you have a chance. Stress the point that separately, most of you will die by starvation, exposure, disease, or violence.

Establish some basic rules. The following are recommended suggestions:
- Everyone who is able-bodied will have a duty to perform in keeping the neighborhood safe and supplied and is expected to fulfill that duty to the best of his or her ability.
- All empty homes are to be left untouched for two weeks (one week if that's not possible), to allow for any neighbors who fled to return. After that period of time, all homes that remain empty are to be stripped of all useful goods and supplies to be distributed evenly and one or two will be left furnished for any guests with critical skills that are worth taking in.
- Everyone will have a voice in how the community is run, but emergencies will be handled and decided on by the neighborhood leadership.
- Everyone has a right to speak their mind without being hindered by anyone else.
- Everyone has a right to voice grievances and ideas at any community meeting.
- Everyone has a right to tell their side of the story and defend their granted national rights if accused of any crime.
- Anyone who is detained for any reason, no matter who they are, is entitled to food and water at the same ration levels as the rest of the community.
- Crimes will be dealt with by the community. A fair trial will be had (as fair as possible) and the punishments are either expulsion from the community, or execution of the offending party (get that second portion cleared with the inner circle first – otherwise just stick with expulsion).
- If there are any intruders from outside the neighborhood, a warning system is to be established so that others can come together.
- Expulsion will consist of being forced to leave with only sentimental possessions, a vehicle if the subject has one and 3 days' worth of food and water. Everything else will be redistributed to the community.
- Refugees will be allowed to join us on a probationary basis if the community has available housing, if they have brought their own supplies and/or if they have skills deemed critical to the community. They can be allowed to stay after a majority vote by the full community at any time after two weeks have passed.
- If you take from anyone in the community, you must also give back equally or better.
- All disruptions and disagreements between neighbors will be settled by the community.
- Anyone who refuses now can either voice their concerns, so that we can modify or change whatever is needed by community voice vote, or the refusing party and his/her family have three days to leave with as much as that family can take with them. Once ratified, these rules will not change unless the situation demands it.

As noted, these are just suggestions. Have the community ask questions, voice opinions and together, work out a set of rules that are easy to remember and agreed on by the majority of the neighborhood. Give those who refuse to abide by them 3 days to sort it out and if they refuse, you can ostracize them or expel them as the community decides, but they should certainly be allowed to take as much as they can with them.

Make sure that, once the duties are sorted out and everyone knows what they need to do, that everything gets done. Take this time to put together a small leadership council that consists of 3 people: Two trusted neighbors and the third will be a neighbor whom everyone votes on. Note that leadership does not mean not having to avoid work. All leaders are expected to perform their ordinary duties on top of any leadership work (distributing goods, sorting out problems, running meetings, etc). We just need to get something together to act as a lead role, that everyone can turn to for answers and decisions that affect everyone. These positions will be temporary for now, but they are needed.

Once everyone is on board, have the head of each household sign a statement to that effect, or voice agreement to them. Then, scribble them up and post them somewhere.

From here, you'd think that you could move on to Chapter 3, well, unless you want to know how to get out of town and not get killed doing it. You see, even if you're all cozy and ready to ride the storm out, something may come up anyway, which will require you to pack up and bail. Keeping the car gassed-up and in a safe place with some supplies ready to roll wouldn't be a bad idea. This way, you can at least be somewhat ready if something goes splat.

Therefore, you may want to read on...

Run To The Hills!

Let's face it – even the best-laid plans go sour and you may have to pack up and go. That, or you may be living in one very sub-optimal neighborhood, where living there long-term is a veritable death sentence. You could be living in a neighborhood that has become mostly or completely abandoned. It could also be that you have a sanctuary somewhere out of town and really, really prefer to live there. Either way, you find yourself getting out of Dodge and trying to do so in a hurry. This is where our chapter on bugging-out comes in...

Stop And Think (Yes, Again!)

You're about to take a step from which there is likely no turning back. While I recommended that folks who stay behind leave your property alone for two weeks if they can swing it, not everyone is going to have read this book and odds are perfect that in a collapse situation, anything you leave behind will quickly become the property of someone else. With this in mind, you had better be certain that you really need to leave.

The advantages of staying behind in a coherent community are abundant – communal defense, strong knowledge of the local area and a place to stockpile supplies. If you leave, you're only going to be able to use whatever you take with you and even then you have no guarantees. Unless you know exactly where you're going and how best to get there under the circumstances, you're going to have danger possibly lurking around every corner, every tree and every building. Each mile you travel will become more fraught with danger as the overall chaos of collapse begins to pick up steam in earnest. Unless you have someone or something awaiting you, or you're an ace doctor/engineer/soldier, don't expect to be welcomed anywhere. Leaving your home and the only (however tenuous) grip on property and belonging that you have is a very large step.

All that said, there are times when you really, really do not want to stay put and stick it out. If you live in a large urban or suburban area that has become completely (in the objective sense) overrun with criminals or gangs and your neighbors are gone or won't even help defend the street you live on, it's time to get out. If you have a location (cabin, second home, remote family, whatever) that you would be better off living at, then getting out should be at the top of your list of things to do as soon as things begin to get really bad. If a fire breaks out and your house is right in the path of the blaze, you're going to have to get out. If a civil war breaks out and your neighborhood is in the path (or even close to) the destruction? Leaving immediately is a very good idea.

Either way, unless the situation is bearing down on you *right this minute!*, take a little time and think this through before you go. Do you have a general direction that you're going? Do you know of more than one way to get there? Are you doing this quickly, or do you have time to pack the car/truck/whatever? Is there someone who knows you may be coming and will welcome you? Give it a good, hard thought before you decide.

Packing Up

Unless the situation requires that you grab your bug-out bags and run like hell, you'll want to pack up a vehicle if you can with as much of the vital things you do have. If you have the luxury of a little bit of time (you likely will), try to arrange things so that you leave at night, at least two hours after the sun has set, but preferably any time between midnight at 4am. It sounds strange, but we'll explain why in a bit.

In order of priority, here are the things you want to make room for in that vehicle (your passengers are obviously the top priority…)

- Your bug-out bags – all of them. Keep these where their owners (the passengers) can grab theirs and run - even the driver. For those in the front seat, if your bag is small enough, you can place them under your legs, even while you drive. Backseat passengers can carry them on their laps.
- Water – at least two gallons per passenger.
- Weapons (and spare ammunition to go with them if they use any)
- Food – at least two weeks' worth per passenger, but as much as you can pack in otherwise.
- A coat for each passenger.
- At least four changes of sturdy clothing and make it warm clothing if you can.
- Blankets – at least two large blankets per passenger
- Tools – just a few basic tools here, but include at least a bucket, a large screwdriver, a hammer, a rough saw and a small shovel. A small automotive tool kit would also come in handy here as well.
- Lighters and some basic fire-making tools.
- More food – as much as will fit. We mentioned this before, right? Worth saying again.
- Some spare rope – whatever you use to lash things onto the car roof, plus at least 20' of 1/2" nylon rope.
- Spare gasoline, if it will fit anywhere strapped down to the roof and covered with a spare blanket.
- A small, tough garden cart or even a furniture dolly if you have one. Something with large tires, capable of rolling easily in rough terrain under load and is at least large enough to carry a good share of supplies. Make sure you get a lot of small straps for it.
- Bicycles and a rack? Pack them if you can instead of that wagon.

If you intend on using a boat and it is trailered/ready-to-launch, then you can quietly pack it in the boat, in the garage with all of the above and more of the same if there is sufficient capacity.

A Word Of Warning Before You Go

Two words, actually: Caveat Viator ("*Beware, Traveler*").

Criminals are going to be practically everywhere along the way as things deteriorate and the later you leave, the more of them there will be. Many of them will be disguised as police, National Guard soldiers, you name it. The air of authority is to lure you into stopping so that they can take everything you have. You will find roadblocks of almost every variety being hastily set up – some looking official, some just a a crude pile of cars or a large tree. The idea in these cases is to get you to stop for just long enough to incapacitate you and get at your belongings.

Something else to consider – not all ambushes are static. Sometimes, a small criminal or gang will just try to run you off the road. If they can get you into an accident, especially at moderate speeds, you're going to be injured and unable to fight back while they ransack your vehicle and its contents. Just food for thought.

You've read earlier that you should avoid any military or police checkpoints and for the love of all that is holy, avoid any main roads. It still holds true and deception by criminals is one of many good reasons why.

Tips And Helpful Items Before You Go

If you're missing anything from your supplies (hey, it happens), see if you can find an **abandoned** house nearby that you can sneak into. Take only what you need and any extra gasoline you can find, but no more. Leave everything else alone.

Plan to leave in the early morning (as in some time past midnight while still dark), if at all possible. Leaving during the day will attract attention you do not want (and signal to the neighbors that your house is now open and free for the picking). Leaving during the early and mid-evening will mean that you're leaving when criminal types are most likely on the prowl. Note that you may not have time to wait, but if you do, do it at the best time.

Have your spouse, partner, some other adult relative rest up and take a forced nap before you leave and get them back to napping once you leave town. This way if you get tired, you can swap drivers and keep going.

Spend a moment before you go to breathe, speak a small prayer, or whichever it takes to have one last look around for about 3-5 minutes, if you can spare it. This will get your mind ready, allow you to look around and gives you one last chance to change your mind.

Getting Out Early

Depending on the crisis, you may want to leave early, *especially* if you have somewhere to go (a "bug-out location", family that expects to see you, etc). Unless the masses are already moving, plan to leave that night, sometime between 1 am and 5am. These are usually the times when activity is the lowest and even the criminals start going to bed after 2-3 am or so. This usually leaves the roads at their emptiest and gives you the best chance of getting out via the back roads that you should have scouted out (as we mentioned earlier).

Drive out quietly and at a somewhat slow speed – say, 10 mph below the speed limit. This gives you more reaction time, keeps the engine quieter and makes sure that you have time to do something about any trouble up ahead. No matter the temptation, do not use the high-beams. If there's a full moon out, use only the parking lights. Keep the AM radio quietly going on a news station and keep one ear on it – you will get valuable traffic information along the way. Keep the map handy. Be alert for any snarls in traffic and always have you and the passenger look for avenues and streets that shoot off to the side that you can use for alternate routes if necessary. Keep an eye out for any knots of people, either walking, or especially any which are standing around (if you see the latter – get by them as quickly as possible, or turn around and find a way around them). Keep a good eye out for fellow travelers and try and drive so that you can always turn if possible, or at least jump to another lane. On roads with more than four lanes total (only if you can't help it - you shouldn't really be on one), always keep to either the outermost lane, or the innermost one. Do not drive the center lane of the three, unless you enjoy being trapped in traffic.

If you see any clogged roads with people standing around, immediately turn around if you can, go back a few blocks and start working on one of those alternate routes you were mentally looking for the whole time. Even if that road is clogged with church buses and the people are almost all children, you do not want to be stuck in any one place for any longer than absolutely necessary.

Once you're out in the countryside, keep an eye out for trouble (tough to do at night, but try). The idea is to be at your destination (if possible) by the time the sun begins to come up. Once there is enough daylight building up to see ahead for more than 100 yards, kill the headlights (if you needed them) and start extending your vision outwards. Keep an eye ahead for the same things (and any stray animals, since many of them start moving at this time of the morning), but this time you can go a little faster.

If you're going by boat, then do it at night. Quietly get to the marina, or lower the boat at the ramp after dark. Once you get out to a safe place on the waterway, stop, kill all the lights and let your eyes adjust to the darkness. If the moon is full or there is enough natural light to see ahead, carefully navigate your boat out and away from town with the lights off. Once you are out of town (or if it's too dark), navigate carefully with the running lights on, but do not use any flashlights or searchlights if you can help it.

Getting Out Late

If you leave early enough, or well enough ahead of any mass evacuation, odds are good that you will be able to get there by car as far as it will go (and perhaps even add fuel to it along the way, but only if you have a need to). However, if you leave a bit too late, or leave after things have completely crashed, then things get a little ugly. You will end up driving a bit slower than usual to get around obstacles and road blockage. You will have more obstacles to go around. There will be a far greater danger of criminal activity and lots of ambushes. Things are going to be a whole lot tougher.

The closer you can get to the countryside or open road without having to abandon the car, the better off you will be. Once in the countryside, you stand a better chance of getting farther (or if you have a specific destination, closer to it). One thing though – if you leave late, never drive a main road at all if you can help it, since those will contain nearly all of the ambushes.

However and unfortunately, odds are good that at some point, the car simply isn't going to get any farther, because you most likely left a bit too late. It could be due to traffic snarls from other refugees, or it could be obstacles, or it could be that the road simply becomes impassable. You can do a fair bit of off-road driving in an ordinary street car (assuming it's not 'lowered' or any such nonsense) and with a true off-road capable vehicle you could get even farther. Don't do anything stupid, but see how far you can get that way. A proper jeep or sturdy SUV can knock over small fences and use (reasonably dry) fields to get around obstacles and traffic snarls, but don't go too crazy and keep an eye out for anyone who may resent you knocking down their fencing. Sometimes, even in a normal car, you may be able to get past a gentle ditch or rise from a highway to a frontage road and use that if it's open.

Getting out by boat will be a lot harder to do at this stage, but may be possible, so long as you are able to get to a launch ramp at night, or the boat you have tied up at the marina is still there. If you're launching the boat, keep the bug-out bags in arm's reach of the passengers in the towing vehicle.

As you get towards the launch point, be on the lookout for anyone that may decide to sneak up, stop you and either unhitch your trailer or take an unauthorized ride on it. If you can swing it, have an alert adult riding the boat with a shotgun and spare shells. Advise him or her to use deadly force to keep anyone and everyone away from the vehicle and boat.

As you launch the boat, have someone standing guard from the boat, weapon at the ready and alert. Keep your own weaponry at the ready if you can. Place anyone not doing any work in the boat just before you launch it and make sure the bug-out bags go in there as well. Don't be too fussy about how deep the car goes under or how gentle the launch is, or in some circumstances whether or not you pull the truck and trailer back onto dry land. If you do have the luxury of time, quickly pull the vehicle and trailer off the ramp and off to the side, then strip it of anything useful (especially the battery). Otherwise, just get the boat into the water quickly, get into the boat quickly, shove off quickly, leaving the rest behind.

Once The Car Stops

All said though and especially if you left late, there will likely be a point where the car simply isn't going to get any further. This means that you're going to have to get out and start walking.

If you can take your time about it and you packed a small wagon or dolly, take the time (with someone keeping watch) to load that wagon/dolly with as much food, tools and clothing that it will comfortably hold. Don't overload it, because doing so will make dragging it a miserable experience and will likely cause you to just as easily lose everything. Once loaded, shoulder your bug-out bags, weapons if you have them and head for the nearest cover (trees, a small copse of woods, a small hill to hide behind, etc).

If you have the time but instead have bicycles, again have folks taking turns looking out, unload the bikes and load what you safely can on them. Then, strap on your bug-out bags, the maps and prepare to get moving. Head for the nearest or safest hiding spot before going any further.

If you do have time, either way do one last thing before you go. Take a large screwdriver, a hammer and go around to the back of your car. Use the tool to puncture a hole in the gas tank, letting the gasoline run out. This will prevent the car from becoming useful to anyone else. If you're abandoning the car in a traffic jam and you know you'll never return to it? Just leave the car running... it'll shut off eventually.

If You're Being Run Down

If you don't have enough time (you may well not if there are criminals bearing down on you), things will go fast, but keep talking to everyone as you think of where stop the vehicle. Before doing anything else however, *Do Not Panic!* Part of the trick to successfully running you down is to get you to panic, do something dumb, then get yourself in a wreck. What you need to do is, as calmly and carefully as possible, speed up as fast as you safely can and try to get away. While doing that (or trying to), look for a hiding spot in the distance that you can get lost in quickly, away from the threat. Everyone (except the driver) puts a weapon in one hand and their bug-out bags in the other. Go as fast as you can, but try to be safe about it and only go as fast as you can while remaining in control of your vehicle. Look for side roads and other places where you can quickly turn off if the terrain allows you to do so without being seen. After all, you should have driven this route at least once before, back when everything was peaceful, right? You should know at least some of what to expect, road-wise. This puts you at an advantage over the typical refugee, who won't have a clue and would likely end up in trouble.

While running away from chasers, you do have some advantages. If the chasers are in an SUV or truck, but you're in a car, you can do corners far sharper than they can and probably even drive the straight sections of road at higher speeds. This means that you can out-run them, or turn off into side-roads quickly in places where they will have to slow down or risk rolling over. In such a situation, you also have better acceleration (unless you happen to be in a completely crappy low-end econocar, in which case, do what you can). If they have a higher-clearance vehicle, you can take advantage of any low-hanging obstacles. If they have a car and you have a truck or SUV, you can use your vehicle's mass to ram them, or shove them off the road (just know that they can do the same if the situation is reversed and they will probably try to). Also, if they're in a car and you have 4x4 capability, then by all means use it. If there is a small town or off-ramp to one nearby, duck into that small town, where you can more easily lose someone.

In hilly country and heavily-curved roads, you can use those temporary blind spots to quickly turn off onto a side road, or to duck into a hiding spot, such as turning off a small off-ramp and hiding under a bridge.

Even if the road is straight and there is literally nowhere to hide all the way to the horizon, you may be able to out-distance them, for many reasons: One, most criminals who rely on chasing down their prey will only do it for a few tens of miles at the most, lest they get too far from their own base, or too far away from the main picking grounds of fleeing refugees. Second, they risk running out of gasoline, thus placing themselves in danger of being stranded and then picked off by bigger and stronger criminals. Third, if they cannot catch you within a few dozen miles, then they risk letting easier prey slip by, back where they started. Finally (but only as small comfort to you), there may well be known competition out there and they don't want to risk getting into a turf war with other criminals.

Something to consider: You're spending your gas getting to a destination far away. The chasers are spending theirs trying to get to you, or at least trying to drive you into a waiting trap. You, therefore, have more fuel to spend than they do. This is why you're going to be busy looking for somewhere to turn off as soon as you safely can, so that you don't end up being corralled into the trap and so that you can stand a chance of losing the chasers.

If you can't shake them after 10-20 miles but can swing it (and have or gained enough distance to allow for it), an alternative is to get everyone out and running for a defensible hiding spot while firing at the vehicle (to discourage them being followed), then you speed off for a bit in the car. This will almost always draw away the attackers. Don't worry as much about the car's condition at this point (feel free to red-line the engine if you want), just draw off the attackers until you yourself get to a place where you can either safely hide the car, or ditch it somewhere. Odds are nearly perfect that they don't want you - they only want what you have... feel free to insure that they keep chasing you and the car full of stuff and not the rapidly fleeing passengers. Charge off onto the first dirt trail or gravel road you find – lose them in any way possible, away from your party. If you can lose them, great – hide the car somewhere, grab your bug-out bag and weapon and quietly (using concealment) sneak back out to the road to see where the chasers have gone to. If they're still looking around, then leave the car hidden (perhaps camouflaging it even more or finding a better hiding spot very close by) and quietly walk back to the folks you dropped off. If they have left, be certain they have left, then quietly go back to the car, grab whatever you can from it quickly and hide that somewhere else, then quickly (but quietly) go looking for the folks you dropped off, staying away from the road while you do it. You might be able to re-join them, bring them back to the hidden stuff and car, reload it all, then continue on your way later, preferably after dark.

If you can't shake them, are alone and happen to find a convenient cliff or rock wall, slow (way) down enough to safely jump out with your bug-out bag and weapon and let the car roll off (or into) the landscape, while you quickly look for and run to a hiding spot. With a bit of luck, the car will be wrecked, becoming useless to the attackers. They will of course want to salvage what they can, but at least they won't have an operating vehicle from which to attack others. It also means you have time to escape and quietly get back to your family and passengers hiding a ways back down the road.

Okay, Now You Can Stop The Car, So…

If you have to all get out and run, have the passengers strap on their bags before you even stop the car, shout directions to them as to where to run (left, right, forward, towards something, etc) and then find somewhere to stop the car. Make sure that you only stop when there is enough distance to get away from the vehicle before any attackers arrive (this may mean turning around if you can, backing up rapidly, etc). Once you've stopped the car, grab your bug-out bags, weapons and start firing (or brandishing) in the direction of the threat to provide cover and everyone runs as quickly as possible for the nearest hiding spot that is well away from the car. With a little luck, the attackers will probably be content with taking the contents of the car and leaving you alone.

If you've been running in a boat all of this time? You may come across something you didn't expect, or find that the boat somehow won't get all the way there. You may run out of fuel. You may end up taking on water (from a puncture or collision?) at a rate that eventually you won't be able to keep up with. No matter the case, find somewhere safe to pull off to shore and unload what you can. Odds are slim that you'll come under attack from another boat, but if you do, try and go somewhere the other boat cannot go. If you are powered and the attacker is a sailboat, power upstream and find a place to pull to shore. If you're the one in a sailboat, get out towards the center of the large lake or go out to sea, moving up-current if at all possible to force them to burn more fuel - they'll start having to watch their gas consumption before they catch you. If your vessel is smaller, head for shallower water. If your vessel is capable of handling rougher waters than theirs, then head for the turbulence. If the attack comes from shore, stay towards the other shore (or go out to sea), so that even if you have to ground the vessel (or drift back in later), it will be out of reach from the attackers (just keep a very sharp eye out on the shore you're tending towards to avoid a trap). In all cases, any attack from shore means that you need to push the throttle up and move as far downstream as you can from the attacker.

No matter how you left the vehicle and reached your hiding spot though, take the time to get out the map, figure out where you are, where you're headed and how you're getting there. If the time is well after noon, find a better hiding place nearby, deeper in the woods or hills, to spend the night. This brings us to...

About Traveling In Large Groups...

Whether or not to travel as part of a larger group (especially of refugees) is a sticky question. You have some measure of protection from predators (especially the two-legged variety) in large groups and if you have sufficient leadership skills (and they are equipped well enough), you can even turn them into a whole new community once you all find a place to live. (Example: say there was some ugly global pandemic... the group of you can find an empty small town whose residents have pretty much all died off and simply occupy it). As one macabre benefit, as fellow refugees die off, you can salvage and put their gear to good use.

On the other hand, here is the reason why you shouldn't: You're going to have a hard time eating meals when there are hungry refugees looking at your dinner longingly. Many of them will be wanting to know where you are going and more still will look to you for protection since you're liable to be one of the few with weapons. There is also the risk of criminals (latent or otherwise) in the refugee group who will try to rob you the first moment they think that they can safely do so. There is also the larger risk of you and your crowd being spotted by governmental authorities, rounded up, stripped of your gear and escorted to a governmental camp – whereas in a small group you would have been otherwise left alone as being too small to bother with.

Like I said – a sticky question and highly dependent on the situation you're in. Personally, I'd avoid any moderate-sized groups if it were me, especially if I didn't know them. With a bit of skill, you can keep a sharp eye out for trouble at a distance and have better luck getting to your destination alone, or at most as a small group of like-minded people.

Habits And Headway

As you travel onwards, you will want to get some sort of routine going and an idea of how to approach your traveling habits...

Whether in your car or walking, as the sun begins to fade (or in the desert, as day approaches), start looking for a place to spend the night. An ideal place is somewhere you can hide and at the same time keep a good lookout. If you're in a car, keep going until you're out in the countryside, if you can. If you are in a car, you should have reached your destination by now, but sometimes things happen. Try to avoid any detours, but don't be afraid to go around potential trouble-spots or bottlenecks. Hopefully you have your whole shopping-list filled out and it included the maps. Use only the maps you need in that spot and keep the rest packed away safely in a waterproof plastic bag. At any rate, every 3 hours or so, be sure to stop in a safe place and figure out how far you've gone, where you've gone and how far you have left to go.

Along those routes, try to avoid walking along any public freeways, highways, well-known hiking trails, or other places where folks are bound to congregate in their travels. You may not be able to avoid certain places, such as bridges and mountain roads, but try to. If you're on foot or bicycle, the best routes to take are those forest and country hiking trails that get you to where you need to go without too many other people around. This will also help to insure that you can avoid being preyed upon by criminals and avoid any round-ups or forced relocation actions by governmental entities – as both groups will go to where it is easier: on the roads.

A good idea as you awake each day is to get your personal business done, then figure out where you want to go that day. Plan your route to avoid any sizable roads, population centers and other, similar choke-points. Following waterways is a good idea, but be careful about it. Do however try to plan your daily route to come close to sources of fresh water. This will give you time to filter out drinking water as you go. If you are stuck crossing a bridge, see if there is a railroad bridge nearby and see if you can safely sneak across that instead (that is, make sure that the bridge is wide enough to step aside of any trains). This helps avoid any natural ambush points or traps.

As you stop for the evening, go over the routes you've taken and get a good idea of how far you've gone that day. Get a quick overall view of the route ahead, but try not to spend too much time on it, as morning may bring new developments.

If you come across anything that forces you to change your route (say, a checkpoint appears, or you see something ahead that you definitely do not want to be a part of)? Stop, backtrack until you find a safe place and start planning a route around the trouble-spot. If you cannot change your route or go around it, then keep an eye out until things either calm down, the problem leaves, or you find a detour. Even if you lose an extra day or two of travel in waiting or in going around the problem, it is much safer than trying to run a blockade, or foolishly thinking that you can shoot your way through a situation.

As you approach your destination (assuming you have one), don't push yourself or think that you can jettison your extra gear. Whatever you do, do not start consuming more than you've rationed out so far, thinking that there's more waiting for you when you get there. You may just end up having to travel further than you anticipated.

When you arrive, save the maps and save everything you have. You may need them later on.

Other Routines

As you're traveling, remember that this is not a pleasure trip. You cannot afford to drop your guard, or to keep at least one pair of eyes open. Your best hope is to get to wherever it is you're going.

With enough luck, you'll arrive by vehicle in less than a couple of hours, to a location that is welcoming and secure. For the other 99.98% of us, getting there will mean having a routine that is designed to maximize our chances of living to get to that destination.

The biggest part of it is choosing and keeping to routes that stay away from the masses, from the criminals, from government and military and from any choke-points along the way, as covered earlier. The best way to do that is to plan ahead, look at possible snags and avoid them wherever possible. Look for places along the way that provide water, natural shelter, vantage points and an abundance of wild plants where edibles may be found.

A good routine is another means to help improve your chances. Start with having a pair of eyes always on the lookout, if you can. Sleep in shifts, spreading the shifts among the adults. Keep out of sight as much as you can and away from crowds. Keep a good eye ahead and at least half an eye behind you and to the sides.

Stop at least once every couple of miles and scan front, back and sides, looking for evidence of other people. This also gives the party time to sit and rest for a bit (though do not start dropping packs or anything like that). If you spot any, step off the trail a bit to a place where you can get a better look and study the situation before proceeding. If they're minding their own business, proceed with caution until they can no longer be seen.

As you approach any ridges, approach it carefully, stop just short of seeing over that ridge and find a good hiding place for everyone on your side of it. Then, find a concealed vantage spot from which to *carefully* scan the valley ahead – you're looking specifically for any signs of people and/or trouble. It may take 20-30 minutes to do, but do it. Always keep weapons handy and loaded, with safeties on (and enough practice to know how to clear the safety switch by muscle memory).

Never pass up the chance to stay watered and to replenish your water supplies. Hopefully, you have the water filters in your bug-out bags and are able to use them. The only time you should pass up a chance to fill up on water is if you've just filled up and have plenty left until the next stream or creek. Keep sipping water on occasion as you walk, even if you're not sweating. For bathroom breaks, always walk a few yards off the trail and have someone stand guard close to the person doing their business. Take turns if you have to. Always bury any solid waste and toilet paper, covering the results up so that it appears as if no one was there. Don't litter, or leave anything – not even footprints if you can help it (though usually this is unavoidable). Anything you leave behind also leaves a trail for others to follow. Bury any and all garbage or defecation and cover it with leaves, sand, or whatever makes the spot blend in. Choose hidden places to do this, places not visible from the trail. Also, if you can, walk in a line, so that the footprints don't give away the number of folks in your party.

Keep all noises to a minimum and preferably make none at all. If you must talk, go up to whom you wish to speak to, then whisper and gesture.

When it comes to fires, avoid them if you can, but for cooking and winter heating, keep them as small as possible and put them out as soon as you can. This may be unavoidable in the winter time, but for warmth, huddle together while sleeping, so you can share body heat. Before you leave a fire, always make sure it is out cold and camouflage it completely before leaving.

When it comes to standing guard, do it a good distance away from the party, but close enough to throw something at them (such as a stick) in order to quietly sound the alarm.

Traveling in the winter will be very hard - especially in northern and colder climates. Know up-front that you will be lucky to make half the speed that you would in the summer. Wear clothing in layers, so that in, say, 30-degree weather, you have five layers on: undershirt, t-shirt, thin sweater, insulating jacket, windbreaker. For extra warmth, put on a thick sweater underneath the jacket. Same with the legs: underwear, thermals, pants. For your legs, a thin pair of socks followed by a thick pair, then your boots. If you have rubber over-boots, use those as well. You can strip off what you need as you feel a bit too warm and definitely avoid sweating at all costs. In case of snowstorms or worse, build a shelter and huddle up in it with a fire at the front. Keep your weapons clear of any snow or ice and avoid touching the metallic portions of them with bare skin.

Traveling in summer will be very hard in desert and arid regions. If you can, under those circumstances travel at night and find a place during the day with shade. Some areas (e.g. canyons) won't allow this, but in those cases, avoid traveling from between 11am and 5pm, staying in the shade if possible. Keep yourself hydrated at all costs and drink as much water as you can hold. Every morning, drink enough water to bring about urination before going to sleep. Never stop looking for water sources, but do be careful of alkali water, which many backpacking type filters cannot properly process. In such regions, have a means of either chemically separating the minerals, or of distilling sufficient water each evening.

If you ever have to fire any weapon, for any reason, leave immediately and move along as quickly as you can. If you have the luxury of time, pick up any and all shells and casings before you go, but your first priority is to get as much distance as you can between you and the place where you fired the weapon.

As you walk along, keep an eye out for edible plants that you can positively identify and harvest only the ones you can grab along the way, taking as few detours as possible. Before eating any of them, double-check with a reputable field guide to avoid any unpleasant results.

If you come across any *unarmed* individuals or small groups, or travelers who are sufficiently supplies and that you know for certain do not pose a threat? Well, let's spend some time on that...

Friends And Enemies Along The Way

As you're walking, you will likely come across other folks with the same idea that you have – getting out of town. Before you start traveling in packs, however, take the time to size up these other people and insure that they can be trusted.

First and foremost, if you see anybody moving, try and keep a respectable distance until you can size them up. If they are unarmed, you can edge your way closer, or let them catch up, though you will want to keep an eye open on them nonetheless. If at any time they appear armed, readying an attack, or suspicious, get as much distance as you can, keeping your guard up and weapons trained if necessary.

If however they are still harmless, send one person out to meet them. Have that person talk to the other person or persons and ask who they are, where they're going and how well they're equipped. If they appear at any time to be sizing up either you, your other members and especially your gear, then everyone leaves immediately. If you are traveling in the same direction, then travel loosely together for awhile. Again, if they appear at any time to be sizing up you or your gear, break away quietly and leave them alone. When you all meet, lay out some simple ground rules: "Keep up or fall behind, your choice." "You can follow or go your own way, but we will not change course or slow down." and "You will not be trusted until that trust is earned." "We leave nothing behind, at all and expect you to be just as careful." If they reciprocate in kind, it is a good sign and should be agreed to.

Conversation will happen, but keep it to a whisper and keep a very sharp eye out in all directions. At no time is anyone to say where they are ultimately going, what they have, or what precise routes are being taken. Do not rely on the new people to handle any of the watching or lookout for at least a couple of days – instead do it in pairs during the sleeping periods and together during the walk. Trust at this point is earned, not granted. Over a few days (and more importantly, over a few dozen miles), you can begin to increase your trust, but for quite awhile, quietly verify and insure that nothing untoward is going on. Always keep a lookout for any armed groups following or shadowing your own party. Always look for signs that your new party members are leaving signs for others to follow – stop and confront any attempts immediately, or force that person to leave – at gunpoint if necessary.

Be aware of any and all possible traps. If you stumble upon a pretty young woman who appears injured or hurt, immediately be suspicious. If you come across a woman and child who appear to be injured, then worry about it and keep a very sharp eye out before ever going close to the situation. If you come a cross a "lost child", do the same... stay well back and do not approach the situation unless you are certain and have confirmed that there is no trap waiting.

These are the same traps used today, in peaceful times, by criminals who like to ambush the ignorant and the traveler. The idea is to get a person or two who appears weak and harmless to play the lure. The lure then comes up with some way to either slow you down until the attackers catch up (or so the attackers can position themselves) or lead you to a place where you can be attacked with the least amount of effort. If you come across such a situation, then circle around the person(s) at a good distance, looking for any sign of an ambush. Don't even let them know you're there while you do it, move together and keep your weapons drawn.

If you see any signs of an ambush or anything suspicious, leave the situation if you can, by as wide a berth and with the highest speed you can all quietly muster together. On the other hand, if you are satisfied that there is no immediate danger and wish to be charitable, send one person up to confirm that any injuries or situations are real before proceeding any further. My suggestion is to leave and find another route immediately.

Some of the means by which a lure (or lures) can set you up is to fake an injury or illness, suggest alternate routes, start suggesting that there are better places to go, or even start filling your mind with the possibility of riches, sex, or whatever they think you would want the most. Firmly refuse any and all attempts on their part to change your course or to slow you down. If they try a second time, warn them. If the person becomes too insistent or things become too fishy, remove any supplies from that person, tie their hands together, hobble their legs (which allows them to walk, but not run), bind their mouth to avoid having them shout for assistance from hiding cohorts and simply leave, getting as far away as you can.

In any situation, when you have new fellow travelers, at no time is anyone allowed to wander off alone, period. Not only does this prevent anyone from reporting back to confederates, but it increases security for everyone. As funny as it may sound, this is going to be a lot like meeting a future spouse. You want to be absolutely sure that you can trust that person before you actually trust them. This time around, it's going to be more than just a broken heart that you're flirting with... it could cost you your life.

If you do reach a point of trust and it's been more than just a few days of traveling together, then you can start taking about the future a bit, if the other person is up to it. If you have a specific destination in mind to hole-up, carefully weigh your options and estimated supplies before even thinking of saying anything about it. If you only have a general idea in mind, then say so. In that case, you can start to discuss the options with your new fellow travelers, but any radical departures or suggested changes need to be discussed in detail and should not include going back, under any circumstance.

If you do see anyone who appears hostile and you've been careful to travel as I've mentioned so far? Then you should be able to see them in the distance and either circle around them (we're talking a circle at least a mile wide if you can swing it), hide in a defensible place until they pass, or turn around and go back until you can find an alternate route.

That said, if you find a situation where it looks fishy, or you cannot safely go around at a wide berth, keep all weapons drawn and carefully pass the suspicious person or group with as wide of a berth as you can manage until you are safely past. Keep someone looking back for as long as it takes until you can no longer see them. If they do the same, keep calm, but do not run away or let your guard down until you're safely distant.

In all other cases? Use your head, remain alert and wary and always try to avoid a fight, but never back down from one if you have no other choice, no matter how bad things look. Your big priority is to always keep moving towards your destination, all the way until you get there.

When To Stop Running

If you already have a specific destination in mind, you already know when to stop... when you get there. For those among us who didn't realize that, let's all stop and sound out a lighthearted "d'oh!" However, if you left in a hurry, or were forced to leave, then you may not know offhand where to go, or how far to travel.

Something that may save you some grief in this department is to do your homework in advance, checking out potential destinations to fall back to when things get too ugly in your neighborhood. Some good ideas? Check out islands, small towns near large national forests and parks, small towns centered around mining, small agricultural towns and basically towns that have an abundance of natural resources, or preferably, towns where labor will be needed. We'll cover how that happens in a bit.

Another reason that you would stop running is if you stumble across a place along the way that has potential. These places would have things that include an abundance of natural or agricultural resources, places where the population is still friendly and charitable. If you happen to be among a crowd of refugees, look for places where you can all get together and have a decent chance of either building a town, or in helping to re-populate one that happens to be abandoned or evacuated, but is somewhat isolated and has those natural resources we mentioned a sentence ago. This is going to be as rare as a finely-cooked unicorn steak, but something workable along those lines is still at least somewhat possible.

Sometimes, you have to stop running temporarily. Extremely rotten winter weather, a swollen river you cannot possibly ford, or a distant battlefield that blocks your path are all good reasons to stop wherever you can find enough resources to right things out. If you are fortunate enough to still have a truckload of supplies, you stand a good chance of being able to camp it out. If by some cosmic luck you're in an RV full of supplies and have fellow travelers similarly situated, you should have no problems with this. No matter the case, find a place where you can remain well-hidden and set up for as long as required. Keep a good guard going and keep the usual precautions going as well. Keep a good eye on your supplies, especially if you're carrying them all on your back. Ration as much as you safely can, cutting back the calories a bit as you will be largely sitting around and waiting. Meanwhile, look for as many wild edibles (both plant and animal) as you can while you're waiting. Even if it means eating bugs (most are quite edible and provide excellent protein), small mammals, birds, lizards, whatever... live off the land as much as you can instead of your packs and supplies. Just be on the lookout for any small animals that appear diseased or injured, as they may cause you to in turn get sick. Also, that field guide we mentioned awhile back is still quite required if you want to be safe about it.

No matter how it comes about, the best time to stop running is when you've managed to get out far enough ahead of the fleeing streams of humanity, or at least enough out of their way, to have a good shot at making a life for yourself and your party. If you're a very small group, try to find something close to other, smaller towns, if you cannot join that town completely.

Group Sizing

If you have like-minded neighbors and fellow preppers bugging out with you, then stay in a group of no more than a dozen or so – break the groups up evenly after that. Large groups tend to be logistical nightmares and you can have one group support the other.

If You Get Turned Away

Depending on the scenario, odds are likely rotten in favor of your joining some distant small town. You may have family there, but they may decide to turn against you and demand you not stop there. A town may be fearful (and perhaps justifiably so) that they do not have sufficient resources to keep you fed in addition to the folk already living there. There may be racist or other unsavory reasons – either spoken or unspoken.

In such cases, it makes no sense to try and reason your way in, or to even protest. In a state of general collapse, people aren't going to be inclined towards reason, logic, or even common sense. You could be a top-notch trauma surgeon, but if your family happens to be Black or Latino and the town is full of rednecks, they'll just as likely turn you away as let you in. No matter the case, you do not want to stay there in either case, since doing so will mean living with suspicion and even hate. This is something you do not want to support, foster, or even be any part of.

You may even convert a few, but the majority (or at least those in power) will still make life impossible for you. You may be able to talk your way in, but you'll be the first ones denied in a shortage and the last to be defended in an attack. You'll be the first ones blamed for any calamity or crime. You may even be robbed of everything and driven out of town at first opportunity. It is far better to move on with your things, find a group of people you can get along with and start a new life with them.

If things are dire and you cannot move on, then settle outside the town on any known abandoned or government property (e.g. "BLM" property, national forest, etc) and only go in to barter. You may be able to gain partial protection by proximity to the town and eventually, perhaps folks will change their minds about you.

About Refugee Camps...

These are also commonly called "FEMA camps" (FEMA meaning "Federal Emergency Management Agency"). Avoid them if at all possible. I say this not out of some fevered tinfoil-inspired theory, but because in a state of civilization collapse, the last thing you want is to be stripped of your weapons, excess supplies and then left idle. You do not want to remain unarmed and at the mercy of a dying government, in a camp that can eventually turn deadly to its residents.

While at first it seems like a good deal, it is not. Let's explain why: At first, you'll be fed regularly, given access to fresh water, shelter, perhaps even showers and other amenities. It will be far more comfortable than trekking through the woods looking for the next ambush to avoid. It will be far safer than trying to rough it in a makeshift shelter during a Rocky Mountain snowstorm. It will be healthier than trying to figure out if you correctly back-flushed that water filter. A doctor will at first be a regular visitor and needed medicines will probably be available.

However, that safety and security is going to be short-lived. As civilization grinds to its inevitable end, so will the government benefits. The food and water supplies will start appearing less frequently. Your fellow camp residents will start to get restless and the line for food (when it does arrive) will start to become less orderly. Crime in the camps will begin to rise, eventually skyrocketing. Guards will stop defending you against the outside world and start defending themselves against you, with curfews and ever more regimented rules and regulations. When the food and water stops coming for long stretches of time, the crime will rise to open warfare and the boisterous food grabs will turn into riots. Eventually, the guards will be overpowered (there are going to be far more of you than there are of them and now you have a mob in full-blown riot mode. Anything that can be stolen, will be, as the mob begins to grab anything they can, whether it is theirs or not and run for the hills with whatever they can hijack or carry.

Where does that leave you? Unarmed, without supplies save the clothes on your backs and forced to walk wherever the crowds didn't go. To top that off, you're now weeks (or perhaps months) behind all the other refugees who skipped the camps and have already found a place to live. This puts you and yours in a very, very bad situation.

This is also why you sincerely want to avoid any kind of military of governmental checkpoint. The main goal of the government and military is one of two things: keep order, or put down uprisings. Their idea of keeping order will be to round up as many refugees as possible (this also means you) and keep them in places where they can be, for lack of a better term, controlled, then kept fed and warm until "things get better". The problem is, once civilization begins to collapse, there is likely no stopping the crash. This means that the original good intentions will be left unfulfilled and the people they're trying to serve will be left in a position far worse than if they had just been left to their own devices. You seriously do not want to be in that position.

Best bet? Avoid any sort of government refugee camp, or a refugee camp of any sort. While a sharp and knowledgeable mind will survive nearly every situation, there is no need to needlessly put yourself in a worse situation than you would otherwise be in... had you just avoided it.

Chapter 3

Immediately Afterward

(From Collapse To Six Months After)

Load Up While You Can!

Once you have met with everyone and laid down some ground rules (see the Section "Batten Down The Hatches!" for details), it's time to take a bit of a survey of your new community for stuff. It doesn't matter that you've lived in the same house for 15 years, things will definitely change after civilization drops. Now is a very good time to go around with your leadership and see exactly what you've got and to discuss what your community is made of... but most importantly, to start getting as many supplies as possible before they're all gone – since they will disappear very quickly.

First, figure out who has left and what has become of their property. While it is good manners and common sense to let the abandoned property sit for a week or so (to allow any returning refugees to reclaim their homes), immediately get into these places and take out any and all foods from refrigerators, freezers, pantries and the like. Distribute them among the neighbors who need it the most, or who have the least amount of food (if you're sufficiently prepared, this will likely not be you). Make certain at least three or more people do this at the same time, so that nobody decides to take a bit for themselves. Make sure you get a little so that folks don't think you have a hoard stored away. If the power is still on, the foods can be consumed as is. If the electricity is out, immediately start drying and preserving the meats with salt (no, not your salt) and advise everyone to consume the uncured meats and highly perishable vegetables and fruits first.

Once the food situation is taken care of (or while it is being taken care of) begin to organize a couple of scrounging parties. Make sure they are armed (they will need to be) with whatever weapons thy have or can muster. Each party will get a list (we have them at the back of the book – print out a couple dozen of each) and will each go on their separate missions. This should be done as soon as you can – the very first evening you can possibly muster.

The scrounging parties will be divided into the following and will have the following missions:

- Food: The mission here is to get out to all nearby abandoned stores and retail outlets, break in (quietly!) if necessary and collect as much non-perishable food as possible to bring back. At least a couple of trucks or vans will be necessary, as will a couple of trips.

- Water: This party's mission is to find sustainable and clean sources of water and report the locations of at least two separate sources if possible. While it doesn't have to be completely potable (you'll be boiling and/or filtering it anyway), it does have to be close by and at least somewhat clean. It should be close enough to walk to and back within 5-10 minutes, maximum. This will only be temporary, until a proper well can be sited and dug, if possible.

- Wood: Odds are good that wood will be burned and used for heat and cooking. This party is going to need a truck, trailer, wagon, whatever they can lay hands on. They will be looking for anything that will fit in a fireplace or a wood stove. It doesn't matter if it's too big or too long, just make sure they bring back as much wood as they can.

- Tools and Hardware: This group is out to look for hand tools, hardware and any pre-manufactured materials such as sheet metal, barrels and etc. This group will be scrounging from stores, warehouses, or any other nearby retail or wholesale outlet.

- Clothing and shoes: This group is to go looking for clothing. It won't matter too much what size or style, just that the clothing has to be somewhat sturdy and useful. Same with the shoes. Slinky dresses, high-heel shoes, tuxedos... those things can be ignored in favor of jeans, sturdy socks, work shoes (and tennis shoes otherwise), T-shirts and long-sleeved shirts, coats, work clothing and the like.

- Weapons: This one is going to be the toughest. The idea here is to head for any shops which have outdoors/hunting equipment and anything else which can be useful as weapons. Try to put an emphasis on archery gear over guns, but don't pass up an opportunity to collect firearms and ammunition wherever you can. A word of caution though - these stores will likely have armed owners, so you'll want to be very, very cautious.

- Medical: If you have any EMTs or nurses for neighbors, put them in this group. This group is going to go after local dental offices, clinics, pharmacies and any other close-by abandoned facilities to load up on antibiotics, medicines to treat common ailments, bandages and all applicable medicines to help treat any chronic illnesses or conditions the community may have. Note that pain medication can be taken if it is available and the opportunity is there, but avoid taking too much of it with you in order to avoid becoming a target for criminals.

- Odds and Ends: This group is going after all the sundry bits that can help out: Craft stores (especially sewing and knitting supplies), Outdoors/Sports stores (especially for fishing and camping gear) and any other stores of opportunity to pick up things that are useful to the community.

- Fuels: This group is looking for gasoline, propane, coal, or any other petroleum-based fuel. Give this party as many containers for holding the stuff as possible.

As each party returns, everyone helps them unload quickly so they can get right back out there. Any time the weapons party returns with a load, unload it, get them into a working state, load the weapons and distribute them to the other parties as they return. In between unloading jobs, the materials can be distributed as evenly as possible, but set aside the weapons.

Keep doing this for as long as you can, until all stores within 3-5 miles (and even further in the suburbs and country) are empty. Instruct the parties that if there is any resistance by more powerful groups, they are to leave immediately and move on to the next store a safe distance away. Unarmed crowds and mobs of a small size can be worked alongside, but any moderate or large crowds are to be avoided. We'll go into detail on how to scrounge in earnest later on - right now, time is of the essence.

Once you have managed to bring in everything that you safely can, insure that anything not already distributed is is sorted out as equitably as possible. Sort the clothing and shoes out by sizes. Sort the weapons out and insure that each household gets something useful to defend themselves with. The medicines are to be distributed both to those who need them (e.g. diabetics get all the insulin) and to the medical personnel. If there is anything left over (odd sized clothing or shoes, etc), distribute those evenly (keeping in mind those with kids who grow) since you can save them for barter later on. Some items (barrels, ammunition, etc) will need to be kept as community property, so that they can be used by everyone.

The idea here is to insure that everyone gets an even share of the take and that everyone gains a stake in the community by doing so. Once this is done, set up patrols to guard the neighborhood and get ready to do some serious analysis of your neighbors.

Note that a more detailed analysis and description of what to do can be found in Appendix B, Scrounge Lists.

Getting To Know Your Neighbors, Again

At first this seems redundant. After all, you've already gone out long ago and met everyone, got to know everyone's strengths and weaknesses and you've become a pretty popular and beloved figure, right?

Well, not really. Given that people always change and that you've all just experienced a complete collapse of civilization, you may really want to get a close look at your friends and neighbors again. People change and under this kind of stress, they tend to let the more extreme elements of their personality begin to rise to the surface.

People will be taken up hard by religion, they will become completely nihilistic, or they may fall into complete denial of practically everything. People will cry in public, will laugh at the most gruesome sights and will fly into a blind, frothing rage at even the smallest irritations. On the other hand, such a massive event will cause heroes (both bold and quiet) to rise, people to become thoughtful as they realize what really is important and creativity to spring from a never-ending well in order to forge solutions for problems both trivial and dire.

The trick is to foster the good qualities and discourage the bad. Given enough time and a team effort, the shared experience will bring most people together and will begin to bring out the best in them. However, in spite of best efforts, there are going to be some folks that you do not want anywhere near your community, under any circumstance at all, even if they are family members. So let's start with the unpleasant tasks first and get them out of the way...

Bad Elements

This sounds a little like some 1950's exclusionary routine, but the situation here is the total collapse of civilization and not exactly perfect odds of living long enough to die of old age in one's own bed. You will be spending most of your time trying to secure the basics of life and will have to spend a lot of time trying to hold your community together. This means that you will have to face and remove any potential problems and disruptive you have sitting in your neighborhood.

While this is certainly none of my business at this point and how you run things is your own prerogative, there are some types of folks who will cause internal problems due to the inherent nature of their own personalities, be they natural or chemically-assisted. To this end, you may as well face these folks now, or you will most assuredly have to face them later, under far less than ideal circumstances. You may get lucky and not have to deal with this, but just in case, you may want to do a little investigation and get the community leaders together about it for consultation and decisions.

People to consider for outright refusal and/or ejection:

- Any current drug addict or severe (and active, not reformed) alcoholic. If someone is addicted deeply into a substance and that substance is no longer easily available, they will get very ugly, very quickly. An addict in such a situation will do anything, say anything and betray anything to get the next hit, the next bottle, or the next high. They will take your supplies and happily hand them over to their dealer just to get another fix or another bottle. They will disable your defenses and allow raiders in if there is the promise of the substance they crave coming to them as a result. Any medicines you thought you had will disappear down their throats, whether they need it or not. As such, they represent a very, very serious danger to your community. It cannot be emphasized enough that such folks need to be positively identified and then either treated (if possible) or thrown out. If they have their own home and refuse treatment, let them leave with whatever they can carry, but do not tolerate their presence for any longer than it takes for them to pack up and leave. There are certain exceptions and gray areas (if they are currently going through rehab and making positive changes), but otherwise be prepared to either throw them out, or physically strap them down and keep them hydrated/fed until the withdrawals are over for good.

- Sex offenders. Your community is going to be drawn very close together under the circumstances and having to sort out who should be kept away from the kids/women/whatever is a chore you do not need added to the list. You will have more than enough problems keeping your family safe as it is, without the added burden of trying to live under the stress of a known convicted rapist or molester living down the street. During peaceful times, police and a strict set of laws keep most of them in check and they know full well to behave. This is because in peaceful times, if anything goes wrong concerning a kid or woman, the sex offender's house is the first one visited by the police. Without police, or even civilization to keep things in check? Many of these people will be emboldened and on the prowl and there's no time or even a way to tell who will and who won't. There will be no shortage of potential victims either, due to the ever increasing number of orphans, runaways and other refugees, all of whom will be rather desperate for food and warmth. After the very first meeting, have a couple of neighbors quietly inform the sex offenders among you (if there are any) that they have two days to pack up and get out, else the whole community will be informed and a lynch mob will insure their departure. (before things collapse would be a very good time to get up a list of such folks). There may be exceptions (depending on circumstance and nature of the offense), but they will be very few.

- Hotheads, Incurable Drama Makers and Violent Headcases. Let me be perfectly clear on who fits into this category. Anyone who has a history of outright physical harassment of neighbors and/or the neighborhood, has repeatedly threatened physical violence against innocent parties, or who has gone out of their way to present themselves as an aggressive enemy to the neighborhood. The last thing you need is to try and foster a community where one or more of the residents are busy trying to play out their fantasies of being the penultimate tough guy. Drama Makers will cause discord and trouble for no other reason than to enjoy the suffering of others. A collapse will certainly give them opportunity and motive to amplify that feeling. Some of these may realize that they are no longer able to rely on the civility of their victims to insure against being beaten to a pulp, but these will be very few. In the case of the hotheads and headcases, you will have to remove them at gunpoint and as a group. Allow them to carry whatever they can, but get them out immediately.

- Domestic Abusers. This one is tricky, but if you know of anyone who regularly beats their spouses or kids, they need to go. However, the rest of the family needs to stay. Life is going to be very rough as it is and the added stresses of survival are only going to make things worse for the spouse or kids who end up receiving the brunt of even more abuse with each crisis that arises. Make it very clear to the abuser as a community and at gunpoint, that he or she is leaving immediately. The abuser is to also understand that there is no recourse or chance to return and that the spouse and/or kids are staying behind. Shoot (or attack if there is no firearms) to kill at the first sign of any resistance or any attempt to return.

Removing these people, if no other option is given them, should involve physical restraint (tie their hands and hobble their legs), blindfolding, removal of shoes and transportation by vehicle to a point as remote as you can safely go in an hour's drive (gasoline permitting). Drop that person off there, drop off whatever charity you wish to provide them (food, water, clothing, but no weapons or footwear) and leave them out there. They can untie themselves after you leave and the effort to do so will give you even more time. The lack of shoes will make it nearly impossible for them to simply walk back. Many will probably try to return, which means the patrols will have to be especially sharp. On the other hand, the idea is to make it harder for them to return than is worth the effort.

This may indeed sound harsh and for most it will be a death sentence, but this is the minimum required to prevent their return and subsequent attack on your or your community. As an added insurance, the newly abandoned property and materials left behind by that person should be somewhat distributed, but a large portion of it can also go to the first honest able-bodied or skilled refugee to come along seeking help. The refugee will then take on the role of the evicted member and will have a shared stake in helping to keep things secure. Be sure however to fully inform that refugee what he or she is getting into and make sure they are agreeable to the arrangement.

There are certainly going to be some gray areas here. Some folks are bad enough to be almost worth throwing out, but on the same token, there are likely mitigating circumstances and reasons why you won't want to. Quietly discuss such cases with fellow community leaders. Afterward, but before making any decisions, quietly interview the person being discussed and let them know why you're discussing it – don't bring up even the possibility of ejection, but instead mention the concerns you're worried about and that the subject will never come up again without a good reason.

There are also situations where you may want to keep an eye out, but more towards caring than rejection. However, there may be extreme cases that more may be required.

The folks you may want (but not necessarily need to) discuss and possibly talk with are as follows...

- Ex-convicts. While there is always a danger of an ex-convict returning to a life of crime, especially in a situation where civilization has collapsed, a surprisingly large number have reformed their previous ways and in many instances, the skills they once possessed and put to use may be useful in a world where the rules have changed radically. If there is a family involved and the person has been law-abiding ever since he or she left incarceration, there should be no reason to suspect anything beyond possible temptations and what the stress of survival may do to amplify them. The longer an ex-convict has been free and productive, the less chance of problems.

- Recently (under 18 months) rehabilitated addicts of alcohol, methamphetamines or heroin. While not true in all cases, this is something to pay attention to. Most drug addicts, when rehabilitated, are able to remain clean for the rest of their lives. However, certain drugs have such a hold that even months later, the temptation may return and become quite strong – especially under extreme stress. If the person has been clean for more than 5 years or more with no relapses and hasn't turned to it yet with civilization collapsing, then any worries can be safely dismissed. However, if that clean period is less than a year, then you may want to help that person steer clear of such temptations and make clear to them that a full-blown relapse will be met with ejection.

- Schizophrenics and Formerly Violent Mental Patients. In peaceful times, these folks can control their impulses and actions with the help of psychotropic medicines, counseling and ongoing therapy. However, when civilization collapses, none of that will really be available. As time goes by, withdrawals from the medicines may bring out dangerous tendencies and actions. The idea here is to know if your community has the resources to help these folks (which will require a bit more than ordinary effort), or if you have to remove the unfortunate soul for the sake of safety. If person in question is a minor, then the whole family will have to be counseled on where the community stands and why.

In the case of these folks, take the time and tend towards helping more than shunning. Be more aware of it than do anything about it immediately. In more extreme cases, you may want to consider taking action, but play it by ear.

Overall however, these are going to be somewhat rare cases. We just want you to be aware of it and to be ready to do something about it if you come across it.

Personalities

You've probably heard the cliché that people are like snowflakes – that no two people are alike. This is largely true, but people will fall into general groups under immense stress. These groups are not necessarily good or bad and each has their strengths and weaknesses. Hopefully, you will already have an idea of what the neighbors are like, for good or ill. All that said though, personalities usually break down into the following groups, post-collapse:

- In Denial. These folks are going to swear that things will be "back to normal" in just a few days, then a few weeks, then a few months... as the reality slowly sinks in. They will quite honestly believe that things can only get better and that everything will soon return to what it once was. In most cases, these folks will slowly begin to realize the full horror of what is happening and their reaction may grow to stoic acceptance, hysteric panic, or even both in alternating waves. These are otherwise good people and once they accept the inevitable, most of them will be solid, reliable and kind people. Treat them with patience and do not go out of your way to shock them, or to force them to accept reality. Just be sure they take actions to help them insure their continued survival until they wake up on their own. Some may never do this, but do what you can while being kind.

- Legal To The End. These people fully expect the government, or any recognized authority, as the only means of relief and may reject any efforts you make to help them out. They will be the first to point out that scrounging is illegal (in spite of the supplies about to be owned by gangs or sitting abandoned) and will refuse to take any part in it. Eventually, their empty pantries and freezing house will convince them to consider that maybe the laws have changed drastically (from city to jungle), so don't deny them, especially not out of spite. Whatever their share may be, if they refuse it, hold it for them until they ask for it, which they eventually will, then help them take their full share home.

- Passive In Extremis. This will include a surprising number of people. When society collapses, a lot of people will refuse to do anything to defend their neighborhood, or even help themselves. They will be largely passive and will look upon the idea of violence as repulsive. This is a good trait when held in moderation, but there will be times in this post-collapse world when pretty much everyone will be forced to fight for what is theirs. In extremes, a portion of these people will end up dead – either from starvation after being robbed, or from violence in spite of refusing to participate in it. Until these folks wake up and gain the courage, give them duties that don't involve a weapon, but gently suggest that they keep one handy and know how to use it. Don't push them into it, since that will only cause a defensive reaction that will only harden their resolve to avoid violence at all costs. Natural instinct usually takes over for most of them when it comes to an impending fight, which will wake most of them up. With a bit of luck, these people will realize what's necessary and will begin to participate in defending the neighborhood. Note that some religions require a total avoidance of violence, so do not try to push the folks who adhere to them.

- The Selfish. Some people will, unfortunately, demand that their needs are taken care of before all others. They will feel entitled to anything they find and may even come up with creative ways of demanding more than their fair share. They will sneak around and hoard found supplies and never tell you about it. They may even steal from others when they feel it necessary. They will live by the credo that their needs are above anyone else's and charity is a near-completely foreign concept to them. They may even refuse to help others. Dealing with them is going to be tricky – be polite but firm. Kindly warn the person once and no more. If there are any thefts and that person or his/her family are the culprit(s), bring it out into the open, in front of the whole community and deal with it. If the person refuses to pitch in and help, consider cutting that home and family off from help and supply distributions – especially if they live on the edge of the neighborhood. You treat them as any common raider would be treated if they decide to steal after that. As they realize that there's no way they can survive on their own, give them exactly one chance to join back up (if they ask), but make it clear that they will be watched.

- The Rapturites. You may have a family or two who, upon seeing the entire world go down the crapper, have decided that *The Rapture* is here and they are about to be taken up into Heaven any minute now. They firmly believe that they will be freed from any chore or requirement of having to survive, defend themselves, or to do much of anything except pray during their every waking minute. Many will proceed to abandon their homes and head for church, in the expectation that they will be taken up from there. Others will stay mostly in their houses and await what they believe will be their imminent departure from Earth. There really isn't much you can do with folks like this, but you can gently remind them that until such a time as they depart, the rest of us sinners and slobs can use a little help down here. If they refuse, remind them that a lack of charity means they're not going to Heaven and then leave them be. The eventual realization that they're stuck down here with the rest of you will sink in, but their reaction to it will differ from person to person. Most will realize with disappointment that they will have to survive and will pitch in. Some may decide to get all extreme about it and you will have to keep an eye on them.

- Too Eager By Far. There will most likely be a couple of people who took one look at the world gone mad and decided that they definitely like it. These will usually be people who have previously felt and lived as if they were a cog in some gigantic machine, unable to take charge and forge their own destinies. A large share of the current conspiracy theorists and apocalyptic types fall into this category as well. In extremes, these are folks who will have an AR-15 in one hand, their scrotum in the other and will be frothing at the mouth to push for revolution. The good news is, these type of people will be quite eager to pitch in and help with the scrounging, defense, gardening, or what-have-you. The bad news is, they may start seeing some of your neighbors as problems, or will want to start conquering neighboring areas and expand the community prematurely. They will need cautioning, a large dose of logic and reasoning and some time for the realities to sink in.

- The Princesses. To some extent, a lot of people are going to fall into this category to some degree. After all, humanity in the civilized world has rarely -- if ever -- had to eviscerate a freshly-killed animal, has had to dig an outhouse hole, eat a snake, or has had to fight to the death for any reason. Most people will take one look at such things and step back in revulsion, fear, or both. In extremes, some may even rebel against having to get dirty for any reason. These are people who are quite used to modern conveniences and consider them a right, not a privilege. This may hinder community efforts, but everyone will have to get over it sometime. With most of these folks, once they are hungry enough, or scared and angry enough, they will overcome their built-up revulsion and do the things that need to be done.

Note that everyone will have some of these qualities in them. Most will, by and large, get over them quickly and after an initial period of adjustment, will start thinking and doing the right thing.

Settling Disputes

No group of human beings are going to get along perfectly for long. There will always be disputes, both petty and grave. There will always be people who simply do not like certain other people and if you're part of the community leadership, you're going to have to deal with it.

The best way to prevent these disputes from getting too large is to have a weekly meeting, for no other purpose than to get it all out in the open and to settle any troubles then and there. One adult member of each family is to attend and participate. For emergencies and disputes that involve crime, get everyone together as soon as possible, but for the regular stuff, get it out in the open and do it at least once a week. Have the people in dispute agree that no matter the outcome, the decisions will be binding to both without question. For some, you have the community decide how to settle it. For others, the leadership deals with it. For still others, you should suggest that any parties in contention should find a settlement privately, as it'll likely be more amenable than the one the leadership comes up with. You could even select a jury of sorts from the community on a temporary basis if need be. Each family (or single householder if they live alone) is to get one vote for those disputes which require a vote to decide. Give each side of the argument a chance to air their grievances, uninterrupted, for a period of time – say, anywhere from 5 to 15 minutes. The one accusing goes first and the one defending has an equal amount of uninterrupted time to state his or her case. Each side will not be interrupted for any reason. Once both sides have had their say, everyone either votes, the jury votes, or the leadership votes. Make sure up front that the decision is fair, that it is not too severe to either party and make certain that it doesn't involve putting either party in a position where they become critically short of food, water, or means of warmth.

If any party feels the decision unfair, give them a week to think it over and present it at the next meeting. Any appeals of this nature are to be voted on by the community at large, with each household getting a vote apiece.

Fostering Teamwork

In this humble author's opinion? Teamwork (or blind cosmic luck) is the only way that anyone will survive the coming mess and blind luck is going to be non-existent. A famous phrase from the Revolutionary War comes to mind: *"Gentlemen, if we do not hang together, we will surely hang separately"*. What is meant by that is if you do not stick together and work together even if there is a risk of doing so, you're simply going to die separately, one by one.

To this end, everyone in your community must know and learn that the only help they can count on more than any other comes from themselves and their neighbors, working together. To this end, have most major projects and missions occur in teams and unless it cannot be helped, never have anyone do any major mission or project alone. This not only insures that theft and injury is kept to an absolute minimum, but that everyone pitches in. This especially goes for things such as scrounging and defense. The last thing you want to have happen is for people to think that favorites are being played, or that someone is doing all the work. At no time do you want to confirm any suspicions of individuals thinking that the whole community would fall apart without them.

One very solid way to foster teamwork is to spread the knowledge. If there is only one medical professional, team him or her up with a promising and bright young adult or adolescent. This way, your community 'doctor' isn't doing all of the work, has help and can teach the younger person as much as he or she can. Same with any other vital skill that resides only in the head and hands of one person. Doing this also benefits you by insuring that if that one person is gone, that there are no massive gaps in the skills your community has.

Unless there are obvious physical disabilities or injuries which prevent it, have everyone take their turn doing the grunt work. When everyone has to take a turn at digging or cleaning up, no one gets the idea that they are privileged, or that other individuals may be privileged. This especially goes for you and any community leadership as well, by the way.

From time to time, if it is safe to do so, organize a dance, or a talent show, or some form of get-together. This helps people relax once in awhile and a little morale is a very good thing. Try to celebrate major holidays together and if there are folks who, say, do not celebrate Christmas or Easter, find an event they do celebrate and make it a community affair. Note that this won;t happen for quite awhile, but if the opportunity presents itself, then by all means double-check that things are safe and peaceful, then do it. You'll thank yourself for the time spent relaxing for just a moment.

Looking Out For Each Other

Once you have a post-collapse community going, you will become your brother's keeper. Stress to everyone that they need to take the time and make sure their neighbors are in good shape (or relatively good shape) and that there is no one starving, depressed to the point of suicide, or any other such maladies. If a neighbor has a roof that has sprung a leak, get together and fix it. If someone hasn't quite built a good working system or garden, help them improve it. If you know something helpful that others can use, do it for them if requested or teach them how – even better, do it together while teaching.

This also means looking out for each other on an emotional level. If someone wants to talk and you're not busy, listen – even if you don't know what to do or how to help. Sometimes, an open ear is worth more to a person than coming up with a solution to whatever problems caused the talk. You want to do this because there will be members who are dying and their families will suffer very hard. Having a shoulder to lean on (or cry on) is not a sign of weakness and providing that shoulder is a far greater gift than you may imagine to someone suffering particularly hard.

You can help out in ways both large and small – babysit a neighbor's kids for a day, to let those neighbors have a break to themselves. Give gifts (even if homemade) on holidays, even if you yourself never observe them. Cheer someone up who is feeling down. Speak aloud about future hopes and dreams with each other. Invite a family over for dinner if you trust them sufficiently (just don't give away how much you do have stored away). Even if they do not ask, if you see them doing something that needs assistance and you aren't doing anything at the moment, pitch in and help out, even if the chore is minor. Unlike pre-collapse times, you now have to become more than casual acquaintances and isolated families in some impersonal subdivision. This is going to be more than just comparing cars or seeing who has the prettiest yard or the biggest 'toy'. Looking out for each other is the only way you can hope to survive long-term.

Now mind you, this does not mean snooping into everyone else's business. If they don't want to talk about it, or don't want the help or attention, respect that. However, do let them know that you are there for them and do be there for them as much as you can safely do so.

Establishing Initial Trust

"Doveryai, no proveryai"
("Trust, but verify")

Ronald Reagan is credited with the phrase, but even he presents it as originally a Russian proverb. Trivia aside, it it likely the only way you're going to establish the one basic thing you need to build and keep a community.

If you've been bugging-in all this time and are still in your original neighborhood, then you have a bit of an advantage. If you've been spending all that time before the collapse getting to know your neighbors like you should have been doing, then the job will be much easier.

However, as civilization is beginning to tumble downhill, you know by now that you should be taking that time to organize the community together even further, calling neighbors together and planning together what you should all do. The sooner you can organize folks into a coherent community, the easier life will be once it all does come crashing down.

Without trust, things are never going to get done, people will refuse to work together and you will all individually be picked off, one by one. If the criminals don't get you first, you will be picking each other off as supplies run low. There is often a saying among uninformed people that in such a catastrophe, they would simply use their firearms to procure what they need. Such people are liable to end up among the first to die by their proposed targets, who will violently protest being robbed of their own goods.

You do not want this. When things have finally gotten bad enough to get everyone together and start working on keeping each other alive, those who refuse to join in need to be watched very closely. If they don't come around and pitch in after multiple attempts, you may eventually have to get together and expel such folks, as they will present a danger to you as their own supplies run low (especially if they see all of you gathering things up and distributing them).

With anyone who refuses to join in, don't just drive them out immediately... try and bring them in bit by bit. When your parties come in with goods, make sure they get a share initially, with the caveat that *'we're all in this together and we'd appreciate your help'*. The gesture alone may well be enough to drive the point home and gain you a new community member. Encourage them, but not to the point of treating them any better than anyone else – just let them know that they are appreciated by everyone else and vice-versa. If they refuse to pitch in and continue to refuse to, then start cutting off the benefits. If they complain, gently remind them that if they want to reap the benefits, they're going to have to help out.

Another type of person you'll have to look out for are those who try to maximize the benefits they receive (even to the point of stealing), but minimize the amount of work or effort they put into the enterprise. It is something that most folks usually grow out of once they leave adolescence, but in far too many cases, a person will realize that they can sit back and let the world bear the weight while they hang back and make it appear as if they're doing their share of the work.

You can prevent a lot of most bad types of behavior by always having at least three non-related people working in teams. This way if someone steals, the others will catch it. If someone is slacking off excessively, the other teammates will instantly notice and put a stop to the slacking-off. Teamwork isn't perfect, but it does help reduce a lot of the headaches and problems that tend to arise when you have imperfect humans trying to survive together.

We will be covering a lot of dealing with internal strife later on, but you will have to keep a sharp eye out initially and watch for folks who are out to screw everyone else. Being a society where people are a bit insulated from each other, coupled with a mass drive for *'keeping up with the Joneses'*, it will take a lot of time for many people to learn to work together, instead of competing against each other.

Scavenging And Procurement

In this section, we will discuss how to get what you need, when to get it and how to avoid getting into trouble while you get it.

This collapse thing is going to be both easier and tougher than anything you've done before. It will be easier in the respect that you no longer have to worry about such things as budgeting, finances, interest rates or balanced on credit cards and the like. On the other hand, it will be tougher because you're going to have to get these things without getting shot at (by rival survivors and/or gangs who think they "own" the goods). It will also be tougher because the clock is ticking on the perishable items and the supplies will be forever gone once the stores are empty (though in later times you could barter for some of them).

Is It Scavenging Or Looting?

Many people will look at these pages and recoil in immediate disgust. I want to stop right now and tell you that yes, there is a vast difference between scavenging and looting, but that strict adherence to morals are going to be, shall we say, a bit fluid. In a post-collapse world, stealing is the act of taking from a person or persons who has and needs the item(s) in question to remain alive. Stealing is something that should be punished as severely as you can humanely tolerate. It is and always will be, a crime. On the other hand, items that are owned by corporations that are dying or are no more, items that are sitting around (somctimes literally rotting) that can be used to sustain life and limb and items that are more useful to a community than to some distant and probably dead corporate entity, or some distant board of directors, will do no good to anyone if unused. If not taken by the community, these items will be taken by criminals and may likely be used against you.

I want to make it perfectly clear that scavenging is going to be a necessary part of long-term survival. Unless you have at your house a small hardware store, large grocery store, pharmacy, clothing store and a magic way of keeping it all in one spot without ever worrying about needing more? You will have to go scavenging.

When collapse becomes clear and obvious, what is going to happen is this:

- Corporations will be dead, dying, or at least ineffective at asserting ownership over anything. They will be the last thing you have to worry about.

- Criminal elements (alone or in gangs) will, if even halfway intelligent, begin securing and guarding things like warehouses, stores and pharmacies in order to assert domination and control over the surrounding area. They will of course sell these items to you in return for everything you value: gold, jewelry, ammunition, supplies they do not have, your daughter, allegiance to their gang and cause, whatever. What they will not do is be fair or equitable about it. They will regard these stores as the means to gain power over the population around them, to claim themselves as the kings of the new world that they see. The less that they can control, the less power they have over you and yours. Unless you like the idea of being the peasants in a world in which they see themselves as kings, it is strongly suggested that you deny them the supplies the best way you can – by getting to and removing thee supplies first, or in insuring you have enough so that they cannot assert any control over you due to lack of supplies on your part.

- Abandoned property is simply that – abandoned. If you were unemployed and someone visibly threw down $100,000 in cash on your doorstep then walked away from it, how long would you let it sit there? Would you pick it up and use it, or let it sit around until someone else picked it up and used it instead? Same story here. Why let a refugee family die of hypothermia when there is an abandoned home in your neighborhood that they can use to keep warm in? Why let a sailboat rot in the marina due it its absentee (and probably dead) owner never coming back for it, when you and the community can use it to build a fishing vessel?

- Most non-canned food items are perishable. If there are fresh foods sitting in the newly-abandoned grocery store, why not eat and preserve them instead of letting them sit there and rot, especially if you and your family are hungry and your community short on food?

Long story short, in a world where government is dead or almost dead, corporations are in the same boat and when the community needs it to survive, your best bet is to load up as much of the vital items as you can and put them to use in your community.

A Question Of Timing

The first question and the trickiest, is ...when is the best time to scavenge from a store, hospital, warehouse, clinic, or what-have-you? Too soon and the store employees will obviously raise a fuss and may even be able to call a somewhat functioning police force to their aid. Too late and it'll either be all gone, or the new owners will drive you away with firearms.

The best thing to do is to play it by eye – if crime is at a rate where your neighborhood is having to band together to defend itself and there are buying frenzies going on at the grocery stores and gas stations in your neighborhood and (most importantly!) if police response is non-existent, then this would probably be a very good time to assemble scrounging parties and head for the stores en masse to get what you need as a group, as soon as possible.

Take as much money and credit cards as you can pool together and as many vehicles and gas cans as you can. Leave one armed person out by the vehicles, then get in the stores and load up (use the scrounge lists you built from Appendix B as a guide). If you are challenged on the way out by any employees that have stayed behind, hand over a wad of money (since money is going to be useless by now or very soon anyway) and everyone continues on out to load up the vehicles while the employee is counting it. Don't hand over the whole thing however and leave as soon as everyone else in your party is out the door. Odds are good that you will have given them more than enough and if things are that bad, the employee may just stupidly pocket it anyway. If there is an armed security or police force present (there may be), then go to the checkout counters as a group, split up the money and credit cards and take whatever you can buy.

If the store is deserted, simply go in as a group and load up. Break a window if you have to. Park very close to the store entrance. Keep two armed neighbors (or at least one if you can't spare the manpower) outside the store standing guard. As soon as you get a cart load, roll it to the vehicles and load them, then go back in and fill up again.

After you get home and unload everything for the other neighbors to divide up, see if you can go back for another trip or two and do so as many times as there are supplies to bring out, or until the vehicles are full.

For gas stations, you'll need the credit cards to activate the pumps, assuming the electricity is still on. Fill the vehicle tanks, then fill the gas cans. If the power is out and there is no one there, be sure to bring along any kind of pump you can (it's in the Shopping List, but not otherwise critical). Look for the miniature in-ground filling points – they'll look like tiny manhole covers.

Get those open (avoiding the one labeled "Diesel" unless you use diesel fuel) and use that to fill the tanks and cans. Surprisingly, this can also be used long after civilization collapses, since most people won't know how (or even know) to get gasoline and diesel fuel out of underground tanks.

Note that if there is any kind of military attempts at order (there likely won't be, but...) avoid the place. It is easier to find another store close by than to try and go through the rigmarole. Also, if there is any kind of martial presence, you will likely be severely restricted as to how much and what you can buy.

Commercial Places To Consider

Given that you're wanting to go in there to scrounge and not to steal, it's pretty obvious that many of the other places you would normally think of would be quite empty by the time you can 'legally' scrounge for them.. However, you may still find many of the same supplies if you know how to look. Any store will have some basic hand tools in the back rooms. Neighborhood drug stores often carry a surprising amount of almost every non-drug thing you can think of. Grocery stores often have pharmacy departments. The local hardware store will likely carry a supply of snacks and light foods. Don't get so fixated on any particular type of item that you end up looking over potential goldmines of supplies of other types.

One thing to keep in mind first and foremost, though – if there are any warehouses close-by, go for those first. Take as many trips as you possibly can. Odds are good the employees there are busy ransacking the place too and by then likely won't care too much if you're helping yourself. The only thing to keep in mind is that warehouses are going to be the first place guarded by military patrols and the like, so use your head.

In my opinion, unless you're extremely lucky or have no other viable choice, avoid the big-box stores (e.g. Costco, Sam's Club, etc). Certainly they are big and obvious targets that probably won't have soldiers guarding them, but because they are big and obvious unguarded targets, those are likely the first places that the local gangs will try to take over. There is likely to be quite a bit of gunfire and conflict in those places for awhile... then again, you might get lucky.

Private Places To Consider

Let's get one thing clear immediately: If the owner of a property, store, or large item (boat, truck, whatever) is close by and still alive, don't touch it – it's not yours. However, if the owner has abandoned the property (by becoming a refugee), give it a week or two (barring squatters or the like) and it belongs to the community. If it's been abandoned for a long time, hurry up and make use of it!

Exceptions and Oddities

Note that in the case of stores or other 'mom and pop' outlets, as long as the owner is still around, leave it be, but invite the owner and his/her family into the community in exchange for the contents of that store. Respect the decision no matter what it may be, but try to reason with him or her.

If the stores are in a small rural town, try to properly compensate the owners if they are local residents, but otherwise, the critical contents of that store (food, warmth, etc) should be divided among the community, period. After that initial distribution, anything the store owners collect/make/trade-for/whatever belongs to them, but initially and especially during the worst of things, no one should starve because the store owners want to turn a profit. The store owners will likely understand this and may even participate in distributing the food and blankets, but if not, they should be persuaded into it (gently!) by town authorities.

Is It Really Empty?

Just a small note: Most stores always have a storeroom. If the local grocery store looks ransacked, see if there's anything in the back... you may be pleasantly surprised.

Factors To Consider

- Time: You will have very little of it, so make the best of what you have. Sometimes you may not have a choice as to how long you spend, but when you do, get in, get as much of what you need as possible, then get out. Don't worry about leaving anything behind – just pack as much as you can and get going. The more loads you can get back to the community, the better off you all are.

- Hostilities: Not everyone is going to leave things laying around until you decide to go pick it up. There may be other 'shoppers' there (especially during any panic buying) who will take things from your cart, or from you. If you go as a group and cover each other, there is less chance that individual panic buyers will try anything stupid. There may be a couple of criminals, perhaps armed with small weapons (knives, chains, etc) who think they can hold ownership of the store's contents (your solution depends on how many there are, what weapons can be brought to bear against them, etc). There may even be employees who are either holding on until the bitter end, or are using the store as their new home. Take these into account and act accordingly.

- Selection: Focus on the list(s). do not chase after anything else, unless there is a very good reason for doing so. You won't have a lot of time. If you find yourself in a sporting good store or such, try to focus not on the firearms (they'll likely be all gone anyway), but on archery equipment and any edged weapons (and any ammunition if you can find it).

- Transport: Get together and use the vehicles with the most cargo capacity. You're sticking pretty close to home (or should be!), so gasoline consumption shouldn't be too much of a problem. If you have pickup trucks, make sure that you have a passenger and also two people (if possible) capable riding in the back. All passengers should be armed with weapons drawn to prevent hijackings. They should be prepared to fire at the first person(s) to clearly attempt a hijacking and the passengers should focus on anyone who gets too close. Ideally, with each large vehicle, you should also have a car with teammates in it.

- Rotation: If you can and especially if you don't have specialized teams going at the same time, try to rotate the destinations. Go to one store, load up, go to another if you have space, but otherwise drag the goods home and go hit another store. If you go back to the same store over and over, you may be observed, studied, then hijacked (or followed!)

Always Go In Teams

Never, ever, ever go alone, unless you have no other choice. A team will insure that any minor trouble you come across doesn't turn into a wasted trip or worse. They will also help get the stuff picked out and loaded faster and it is always good to have someone who has your back.

A minimum should be at least four people per team: A driver and three passengers. You can get by with two passengers and in extreme circumstances one, but in those cases you're going to want to approach your activities with more stealth and won't be able to procure as many supplies.

Before anyone goes forth to scavenge, a leader should be appointed quickly and roles assigned. Someone should drive. Someone should stand guard by the vehicle(s) – this person can also be the driver. Everyone else should load and unload. Everyone but the guy standing guard outside needs a copy of the relevant list(s).

What To Leave Alone

- Armories, bases and government installations. Seriously, does this need to be said? Unless and until such facilities are completely unoccupied, they will not be safe places to go anywhere near.
- 'Big Box' stores and "supercenter" stores. These places will be mobbed and by the time they are not, it will either be empty, or guarded by criminals. During Hurricane Katrina in 2005, it only took three days for literally everything (food, furniture, clothing... everything) to be gone.
- Any store that specializes in selling weapons (gun and archery stores). Most of these are privately-owned and the owner is likely sitting on the stockpile, heavily armed. If you think you can persuade the owner to sell or barter some items give it a try, but don't be too hopeful and don't do it with a ton of stuff that he or she can see.
- Hospitals. Odds are good that they're overwhelmed anyway (especially in pandemics!) and they are prime targets for junkies and criminals looking to stock up on drugs.

What To Focus On

In the precious few days (or even hours) you have to scrounge, get as much of what you need as possible and don't think of bothering with anything else. Let the less-informed load their vehicles up with furniture, television sets, video game consoles and cheap jewelry.

Places that are not obvious but are most likely to contain vital bits are as follows:
- warehouses
- industrial complexes (which often contain warehouses)
- distribution centers
- 'luxury' and non-obvious medical clinics – plastic surgery, eye clinics, dentist offices, etc
- veterinarian offices and local pet hospitals (they'll be largely ignored, but many of the medicines are the same for pets or people...)
- restaurants and fast food stops
- auto and truck repair shops, as well as specialty repair shops (e.g. television, vacuum, computer) which can contain small but useful tools.

- movie theaters, skating rinks and small sports and event stadiums (it may be junk-food and popcorn, but it's still food and often stored in moderate quantities)
- strip malls (which may still have been ransacked, but look in the non-obvious stores).
- Parks and recreation department maintenance sheds (tools)
- Farm/Livestock supply houses (for medicines, tools, food, seed...)

If Someone Else Gets There First

Rule #1: Don't spend too much time in any one spot. If it is empty, move on to the next store or place. If it is mobbed with humanity and there's no possible opening to get in, skip it and move on. If it has an armed guard, ascertain how many there are, but only keep it in mind – don't attempt any actual attack, since time is too short right now.

Rule #2: If the crowd size is low enough, you can still get in as a team and load up. Stick with each other in pairs.

A Note About Scrounging From The Dead

In dire situations and the collapse of civilization is certainly one of those situations, do not be squeamish about scrounging needed supplies from the deceased. They obviously will have no use for it and you will. The only thing to keep in mind is to (respectfully!) move the body off to the side and leave its clothing and apparel (shoes, coat, etc) alone. This is not only out of respect, but because anything the corpse may be wearing may be contaminated with bacteria or worse by the time you find it. The only thing useful to search the body for would be vehicle keys and (possibly) glasses. Just note that you'll have to disinfect the glasses first and possibly the keys.

If you are in the house of the deceased, then anything in the house that isn't on the body can be scrounged. While there is a rather large taboo that the dead and his or her possessions should be respected, in a survival situation the dead cannot use those things that you need. Sometimes, using that which belonged to the dead will help you and your loved ones avoid becoming one of them. If it makes you feel better, respectfully lay the body (or bodies) in sheets and tie the sheets. Place them in the basement, or temporarily in a side room with an open window for ventilation. If time permits, return to bury them properly – especially if they are in a home that can be put to use later by you or by refugees.

If you come across young children (that is, orphans) of the deceased on the scene that are still alive, the rule is simple: if you take anything at that point, the child comes with you and you are to care for it to the best of your ability. Use your judgment and consider the situation, but know that it is the only truly moral thing to do.

If there is an adult or near-adult relative/spouse/etc who survived and is on the scene, then the scene is at that point off-limits and the possessions are considered as property of that (or those) person(s). Return anything taken so far if this discovery is made after you and your team has begun.

Scrounging Long-Term

After the initial grab (by pretty much everyone), the act of getting further pre-collapse supplies will be reduced to either fighting for it, getting lucky, or bartering for it. However, as time goes on, you'll be looking for things that are the basis of building other things, as opposed to finding ready-to-consume or ready-to-use items. These will include things like tubing and pipes, sheet metal, lumber, ducting and the like. We'll cover more of how to do this properly in later chapters, but do know that the tactics and methods will be pretty much the same.

Another difference is that long-term, the items you'll be seeking will be less likely sought after and fought over by the criminal elements, mostly because the bits you're seeking will probably not be as obvious, or be considered as valuable. Also, it will be because these items are going to be far more easily obtainable. These items will also be less of a finished product and more of a raw material from which you will be building new things.

Because of this, periodically get a party together and do some scrounging for things that you may want or need. Odds are good that the obviously needed things won't be available, but the less-obvious ones will be. For instance, things like timber, insulation, nails, copper pipes, carpeting, wiring and more... these can all be scavenged from abandoned homes, offices and other buildings. Need tires? Look for the nearest abandoned vehicle with intact tires at the right size you need. Need to build a fireplace? Sheet-metal ducting from an abandoned house will work perfectly.

Finally, as people begin to die off in larger numbers, there will be less competition for the resources that remain. This will bring periodic opportunities to scrounge from the homes of now-deceased people and eventually whole neighborhoods and towns.

Conclusion

You will have a very short window of opportunity and taking advantage of it will have its risks. On the other hand, every little bit will count, so get what you can while it is still there. Never do it alone – do it in teams and involve your neighbors when you do it. Go armed, but do not go looking for a fight. Look for non-obvious places and locations, where most folks won't look (or even think to look).

Securing Home And Community

No matter if you're staying put, or starting a new life in a new home, there are a few things that you must do to keep things at least relatively secure and safe. Doing these will make your life much easier in the long run and will make life much easier for all who live in it.

Visibility

The ability to see all around your home should be your first and most primary consideration. If you cannot see outside in all directions, you're going to have a very hard time seeing if anyone is trying to sneak up on the place. Either have a home with sufficient windows to see outside in all 360 degrees of view, or plan to do watch outside the home. As an alternative and if you have the building materials to do it, you may be able to create a couple of viewing nests or ports in the attic of a home, by carving out portions of the roof, find a way to keep water from entering those new holes and using those as lookout posts. Lay sheets of plywood across the ceiling joists as a floor to stand on and lay them out in a pathway from the attic entrance to all 'nests'. A better arrangement would be a house with a flat roof and an easily accessible means of getting to that roof from the inside, but not the outside. On such a roof, be sure to set up places with sandbags and/or bricks/blocks to hide behind and shoot from, if necessary.

If you cannot see at least 100 yards from your home in all directions, then find out what is blocking the way and see if you can minimize them. This may mean clearing brush, perhaps cutting down nearby trees if needed, or otherwise making clear the space around your home. Doing this will also create a 'defensible space' for fire protection. If you cannot move anything, then find out where the blind spots are and rig them with alarms of some sort.

Given that there is no perfect situation where you can see all around your home in all directions, you will likely need to set up some sort of alarm system. Given that commercially, electricity is likely going to be gone and powering things up will be problematic at best, this means using old-fashioned means of sounding the bell when intruders try to enter. This can be as simple as strings attached to old cans (and hidden well enough to not be seen by intruders), or at least having a dog or two on the premises.

Locking Down The Windows, Doors, And Walls

The weakest point in your home is likely going to be the windows. One way to reduce this is to board up the windows on the first floor and on any ground-accessible second story. This does not mean covering the windows entirely, but it will require screwing or nailing some heavy lumber across the windowsills (using large nails or lag screws sunk into the wall studs), leaving only enough space to see out of easily and to aim a rifle. If you have any sliding-glass doors, try your level best to replace them with something sturdier, or to barricade them as tightly as you can muster. If you can place a wall of mortared cinder block behind the sliding glass door leaving just enough space to walk in and out of, do it.

Leave at least one or two windows so that you can easily and quickly remove the boards if you have to in order to escape in case of fire – use brackets and have the boards act as bars. This is also why you leave the second story windows alone if you can. Also, if you can possibly do it, put up a layer or two of sandbags, bricks, or blocks inside the wall and under the windows, up to the height of the windowsill. Stretch it out a couple of feet to either side. Any sheet steel you can lay between those bricks/bags and the wall would be very useful as well. Done right, doing this will give you somewhat secure firing positions that will stop most small arms fire.

For most other doors, insuring that they can hold up to a decent bashing is your goal. While having the luxury of replacing the doors, deadbolts and frames would be nice, odds are good that you won't be able to. Instead, build a sturdy old-fashioned way to bar the door, by using a couple of strong metal brackets, a length of 4x4 timber to drop between them and the door and some very long lag screws. Since most doors to the outside open inward, this should make things sufficiently sturdy.

After this, your only weak-points will be the walls. If they're made of brick, you're doing well. If they're made of solid stone or logs, you're doing better. If they're simply vinyl siding, insulation and sheet rock, well, you may want to consider bolstering things up a bit. Try to get at least one or two rooms lined with bricks or blocks along the inside of the walls, so that you can have a safe place for sleeping. If you have a basement, that works as well – just try and insure that the exit from it to the main floor is secured.

Any other odd entrances or exits (casement windows, basement/cellar doors, etc) should be secured similarly.

Appearances Are Everything

If you live in a densely-packed subdivision, or a neighborhood that is mostly empty, make your home blend in. If there are multiple houses in the neighborhood that have been ransacked with things strewn about the yards, make sure that, at least at first, you and your neighbors have yards with things strewn about it. If your neighborhood has a bunch of 'quarantine' signs and FEMA spray-paint all over it, be sure to figure out which ones indicate massive contagion and happily decorate your front doors and yards with something that looks just like it.

The idea here is to look like nothing worth robbing, either because it's already been ransacked, or because it looks like a guaranteed death sentence to any criminal thinking of entering. The best way to deter criminals after all is to make it look as if it isn't worth their while to try.

As time goes on, things will begin to get a bit more desperate. If you still have a functioning post-collapse community, you may want to consider even more gruesome and drastic acts to deter the criminal element. This could include lashing dead criminals to a tree with a warning sign hanging from them, or simply gaining a reputation of being one place where any non-organized criminal element will not want to go.

Water In And Out Of The Home

Without a reliable water main coming into the home, you're going to have to secure a means to bring clean water in and let dirty water out. We'll go into more detail on water systems later, but as far as security, this is going to be a big consideration. There are times when stepping outside could mean putting you or your family at risk, which is something you do not want. A means to minimize this worry is to have a largish water supply indoors and buckets set aside for waste water disposal (you could use the sewage system for awhile in towns and cities if you had to and in any home with a septic tank, just continue using that as normal). Fortunately, most homes come with a water heater, which can work very well as an emergency water storage tank. Initially, you will want to use its contents for drinking water anyway, but try to keep it full of water once in awhile, filling it whenever it is safe to procure the water to do so. This will leave you less exposed to trouble and allow you to 'hunker down' for up to weeks at a time if it becomes necessary.

The Garden

For at least the first 12 months or so, you're not likely going to need to worry about it, but eventually you will have to start growing your own food. In approximately a year, the population should have been reduced enough to allow for more room to breathe, so to speak, but you will have to plant your crops with security in mind. To do this, plan your site to not only have the best soil available, but to keep it close enough to home to keep an eye on, or at least close enough to allow for easy patrolling by the community (and yourself).

The best site would be a nearby park, a series of large backyards with the adjoining fences torn down and the yards strung together, a nearby field, or one of numerous other open areas that can be put to use.

For livestock (assuming you manage to procure any), the backyard of an abandoned house would make for a great pig sty (just be sure to put in some sort of shelter for them) and for grazing animals (e.g. small goats or sheep), a block of houses with the inner fences all removed (but the outer ones still intact) makes for a near-perfect pasture, since most of those backyards will probably already have grass in them. Your only other consideration would be in making sure there's enough water available. With you and your neighbors all living in the outer ring of houses and the livestock in the inner ring, you stand a decent chance of securing the livestock from casual theft. The same concept would go for (the more likely to procure) chickens or pigeons - where you would, say, convert a garden shed into a coop and let them live in that. This insures that the food stays inside the center of the protected community. And yes, I said pigeons. They're actually quite tasty when cooked, are considered a delicacy in most nations, breed quickly, eat practically anything, survive under most weather conditions, easy to snatch eggs from and... well, we'll cover all that in a bit. Right now it's just about security.

Security Patrols And Night Duty

Unless you have more than four adults in your household, get used to spending part of the night awake. You're going to have to take turns doing shifts, each about four hours long. If there is only one of you, then you're going to have to rely on dogs and alarms to keep things safe. If there are two of you, you can split the night – one of you going to bed early, the other staying up in late, splitting the hours of 10pm to 6am into two 4-hour shifts that each of you will have to pick from. If there are three of you, you can be a bit more merciful about it and split it three ways. Four of you can split if four ways if you'd like, or rotate out at least two people who will spend the 2am-6am shift on duty.

Just get used to the idea of someone being awake at night and being allowed to go to bed early or sleeping in late. The idea is that you will need 6-8 hours of sleep each night if you can possibly help it. Also try to fix the schedule so that each person can expect to go to bed at the same time and get up at the same time.

If you're in a functioning community, you can guard the whole community in patrols (at least two groups of two people each at all times) and let the folks indoors sleep unless there's trouble. Otherwise (say, on a homestead and such), each person on watch will need to keep him or herself concealed in a place for a given period of time, then walk around, settle in a watching post somewhere else and spend some time there keeping an eye out.

Whatever you do, the one and only thing you should have regularly is the start and stop times of the guard shifts. However, do not simply have one group go in while the other comes out, or simply have everyone stand around and talk. The protection should instead be continuous.

For example, at a shift change in a community patrol, have one person from each new shift's patrol unit go out to each patrol unit in the current shifts *as they pass close by* and speak quietly together while still looking out for trouble. Once the new guy is caught up on events, one guy in each unit from the old shift goes back and tells the others what the latest news and events are. Then everyone else on the new shift slowly goes out and joins their units. Once the new shift is fully in place and on patrol, the folks on the old shift who are still patrolling can quietly break off and head home together, or as the unit passes a safe place for one person to do so. This shouldn't take any more than 15 minutes.

A shift change for individuals is much simpler... the new guy goes out, patrols with the existing guy for awhile, then the existing guy goes home and gets some sleep. The idea in either case is to have continuous coverage. It also allows the new shift to take a walk with the old shift, doubling the protection while being briefed on what to expect.

Another thing to consider is that you should never take the same route, or have the same habits. Breaking the habits is easy: Stop on irregular occasions at a secure spot (behind an abandoned car, a low brick wall, fencing, etc) and have the unit (or individual) look around from there. When you stop, you (or the leader will) pick a random number from 1 to 12 and that number will be how many minutes you (or the unit members) stay put and look around, until you get up and walk towards your next point. Breaking the rut of having the same patrol trails is as easy as walking further out sometimes, staying further in at others. Sometimes going counter-clockwise, sometimes going clockwise. Cut across the secured area quickly and come out from the other end to resume your patrol. There are many ways to throw off any external observers from seeing what you're up to. Once in awhile, turn back and re-trace your steps to the last secure point, stay there for no longer than five minutes or so, then go back to where you were originally headed. Doing things this way may seem a bit complex to some, but it prevents criminals and raiders from figuring out any kind of schedule or path that they can time an attack with.

Something else to consider during your patrols is to never, ever, ever just stand around in a place where you can be easily shot at. Move quietly from one defensible firing position to another. If there are long stretches without these, then build some positions out of rubble or whatever is handy. Make it big and strong enough to solidly get between you and any bullets that may fly your way. When there is a team guarding as a unit, have first one person move with the other(s) covering that movement, then another, then another... until everyone makes it to the next point. This way if someone takes a shot, the others will see where it comes from and can fire back.

One other thing – if you're protecting a community, have at least one person standing at the main entrance (or what passes for one) in case anyone comes up to do business, raise an alarm, or whatever.

What You're Looking For

While it may seem obvious that the point of guard duty is to look for something unusual and keep it from entering, it is in reality a bit more complex than that. The idea is to keep a watch for any movement that is viewable outside of the community or encampment and verify what it is. Odds are good that you may well be watched without you knowing it (which is why we suggested doing the patrols in a semi-random fashion above).

Guarding Without Lights

Most ordinary people can see quite well in near-darkness without the aid of lighting. The one thing you do not want to do is to ruin your night vision with flashlights, fires, improvised spotlights, or the like. Doing this will make your eyes dependent on the lighting, which can in turn reduce or even destroy your ability to see at night.

The way to get used to the idea is to stand in a completely dark room for about five to ten minutes before going outside. This gives your eyes a chance to ramp up their sensitivity and doing so will make walking around outside seem relatively useful with only the ambient light available. If you can, have at least the outer homes and portions of the community in as total darkness as possible, so that there is no indication of lights to show occupancy or movement.

In a post-collapse situation, you will not have what you may be used to when you think of walking around in the dark. There will likely be no "light pollution" from the big city (or even town) to lend illumination. You will likely be able to see every star in the sky, which is good, because even starlight can give off enough to see at least somewhat. In most darkness conditions, you can even use binoculars, which will make things seem just a little brighter, but still useful.

The trick to seeing in the dark is to not look directly at something, but just off to one side of it. The center of your retina (that lining in the back of your eye) is a bit night-blind, so looking directly at something in the dark means you won't see it at all. You can however look just barely off to one side or another of something and be able to see it. In all cases, you should, under most conditions, still be able to see movement, which is what you're looking for first and foremost. When you take your stops, you can then take the time to look around and see more than just movement.

The worst times to see anything at night is on completely overcast nights, or on nights with a New Moon (where there is no moonlight at all). The best times are during periods of full moon, which can allow you to see quite well and for quite a distance. Try and get used to all conditions, though.

If you have night-vision goggles or glasses (note: many video cameras have this ability too – often called "Night Shot" or similar), you can put them to use, but only one person in a unit should do it, since their eyes will become dependent on that light (if using a video camera, use the eyepiece, not the flip-out screen). The thing is, you really should get everyone at least used to the idea of guarding and seeing at night without the aid of light, since eventually batteries die, cannot be recharged and because fire (via torches) is a lousy way to try and see where you're going (it tends to give your position away far too easily).

An advantage of doing this in the dark and being used to is is that you are going to be better able to see what's around you than most attackers will. This is most definitely what you want – that is, to get every advantage you can.

Raising The Alarm

This one is a tough call, because when trouble strikes, you want everybody up and ready for action, but on the other hand you don't want to let the intruders know that you know they're here. Those precious few seconds or minutes will give you a good advantage, by allowing your people to get into position and be ready before the shooting starts.

One thing you do not want to do is to make it appear as if you found out there's an intruder. If you see one and you're in a patrol unit, *quietly* whisper to the others that you see something. Make sure that nobody reacts, however. Stop at the nearest secure point, then look to see what's coming and more importantly, what may be behind what's coming, in case they have some sort of sniper positioned... then start firing as the opportunity presents itself. That will raise the alarm. Also, be sure to teach everyone in the community that those closest to the firearm going off should come out quietly and commence to helping that person out. *However,* those farthest away from it should first look outwards directly away from the center of the community to insure no trouble is coming from there, *then* roughly half of them should go to assist, the rest guarding their position just in case.

If it's just you and your family, screw it – quietly get into a good position a little bit away from the home but within easy reach of it, then start firing as the opportunity presents itself. Everyone inside will wake up and know what to do.

When The Alarm Goes Off

How will everyone know what to do? You teach them in advance.

If it is just one household, you have a plan in place to keep the kids and the sick in a safe place, with at least one able-bodied late-juvenile or adult person inside, armed and ready to defend the home. Nobody gets in without recognizing the voice asking to come in. The rest of the adults get weapons and go outside *opposite* the side of the home where the gunfire occurs, or can go to the windows and look outside to see if they can fight from indoors.

If it is a community, the gunfire will wake everyone up, but here's the twist. Teach everyone to consider for a moment where the gunfire came from. If the gunfire came from close by, on your side of the community, then you quickly but carefully poke your head out the window (*keep the room dark!*), determine what's going on, then get out and help defend the neighborhood, or fire at the intruders from your position and help defend the neighborhood. If the gunfire came from the opposite side of the community and your home is on the edge of that community, keep an armed watch outwards away from the center of the neighborhood. Those in the middle of the neighborhood gets out and joins the firefight in progress, carefully insuring that they move from one place of cover to the next as they get there. If you're not sure where it came from, exit the side of the home closest to the inside of the community and figure it out from there. Once outside and you've determined that the trouble is not opposite your side of the community, go join the fight – otherwise keep a watch out for any trouble coming in from where you are. Meanwhile, in each home, everyone gets up. All kids, sick and elderly are to move to a 'safe room' (more in a moment), with as many able-bodied juveniles and adults armed as possible.

Make sure there are designated people in the community (preferably towards the center) to act as medics. Set up a room or two in a larger home as a place to tend to the wounded. Keep your medical professionals in there during the fighting and if you can spare it, have a designated guard or two standing by with them.

As for the rest, keep fighting until they've been killed, driven off, or they are so numerous and well armed, that you have no choice but to grab your bug-out bags and run.

Post-Attack Security

Once it is over, double the patrols and guards and have parties search for any wounded attackers. Once a wounded attacker is found, kill that person on the spot, preferably without the use of a firearm (e.g. use a machete, knife, club, etc). This avoids raising any false alarms and saves ammunition. This sounds quite harsh, but you won't have the resources to care for them, imprison them, or the like. You also don't want anyone making it back to their encampment knowing what kind of strength you have, how you stack up, or etc. At least, not if you can help it. Again, this sounds very harsh, but this is how warfare has been conducted for millennia and quite frankly, it works very, very well.

While the wounded attackers are dispatched and the situation analyzed, have other (armed!) parties go out to collect all weapons and useful gear that may have been left behind by the attackers. Distribute anything you find, but make sure the community leaders get any maps, radios, or other gear that may be useful to determine what the attackers are up to.

If your community is large enough, you may want to consider assembling a party of those among you who are soldiers/veterans and have them sweep the area for any attacker encampments. If they find any and if the number of attackers is small enough (or the encampment insecure enough) attack them, with the intent of driving away or killing off the invaders. If successfully found and then driven off/killed/etc, have that party hold the ground and send two runners back, then have a second party come out to salvage the gear and supplies found and destroying (or at least rendering useless) anything that cannot be brought back. This means a few more supplies for your community and a few less for the attackers, if any of them managed to run away without injury.

Booby Traps?

Other means of setting up alarms would be to rig booby traps, but be careful in doing so, especially if you have children in the home or living nearby. Preferred styles would involve covered pits and dead-fall traps. Spring-loaded traps tend to require maintenance, which can be tedious and in some cases even dangerous. Try to use fishing line as a tripwire and conceal it as best as possible. On the other hand? The best kind of trap triggers are those that are stepped on, because they are more easily concealed and are harder to see, even during the daytime. A large but concealed spring-loaded plate (the spring holds the plate level with the ground) that moves downward at a footstep, which in turn pulls a partially-buried tripwire, will be far more useful than some string dangling across a pathway.

Try not to be obvious about where you put these traps, either. Look for smaller and game trails, or pathways that aren't obviously trampled down by human footprints. Most of all, though, try to rig them not so much as killing traps, but as (or with) means to raise the alarm should one be sprung.

All that said, do put some bite into your traps. Wooden spikes covered in feces works wonderfully towards injuring someone to the point where they will at least yell out in pain.

No matter what you build or how you rig them, instruct everyone you care about to avoid the areas at all costs. However, do not make any maps, or leave any 'safe' pathways that people can walk through in those areas. They are to be simply left alone until later, when things finally settle down enough to carefully make the traps harmless.

Corrals And Pushways

One way to help improve security is by putting up obstacles, obvious booby-traps and other means of forcing (or at least subtly persuading) intruders to take specific paths and routes to get to your home or community. This can take the form of obstacles (natural and otherwise), semi-obvious booby-traps (hidden just enough to make them appear like they were supposed to be hidden) and even the use of creative demolition (e.g. destroying a small bridge or embankment).

The idea here is to make any paths into your secure area (other than the ones you want) so unappealing or dangerous, that potential raiders will be forced to either take the paths you want them to take (preferably places where they cannot fight from effectively), or to risk life and limb just getting to you.

It doesn't even have to be obvious – aside from the hidden-but-findable booby-traps (to spook any oncoming raiders), placing a new graveyard in a certain spot (and piling animal or raider corpses on the outer edge of it) may well deter intruders from passing through it in order to get to you. Same with any garbage heaps where you and/or your community burns trash – if it's on fire and full of ash and goop (which leaves tracks), a raider will probably think twice before sneaking through it. Other ideas include impromptu man-made ponds, septic pits or dumps and other items which will make sneaking through them successfully to be impossible, dangerous, fear-inducing, or all three.

This of course does not mean you can stop watching those places which you have effectively blocked off – there will always be people desperate or crazy enough to try them anyway and many will see right through any of the more obvious attempts at deception. On the other hand, it does help even the odds that (at least the stupid) attackers will come in where you want them to, instead of from directions where you would have the hardest time defending against them.

Guard Dogs?

If you do go the dog route, try to stick with smaller dogs that are loud. Dachshunds, Collies, Miniature Pinschers, Jack Russell Terriers and other dogs of this type are a bit high-strung and will bark like mad (or be trained to bark like mad) when they see or smell something on their property that they do not know about. Being smaller, they are harder targets to hit, less likely to be seen, can be brought indoors during extreme weather and they eat only a fraction of what a larger dog would.

While large dogs are certainly very useful for things like hauling some supplies, providing a fearless defense and a nice warm spot for the kids to curl up with, large dogs have a couple of problems that tend to make them unsuitable for post-collapse survival. First off, they eat an astonishing amount of food. In a bug-out or traveling situation, the majority of what that dog is going to carry will be its own food. Secondly, they tend to have shorter lives, which means that once the dust settles and you have a somewhat permanent home, your dog will likely have reached old age. Third, large dogs that are not properly and perfectly trained are going to be hard to control if they get a wild hair and go do something dumb, or something that can endanger you or your family. Finally, a smaller dog is easy to pick up and carry across a small river, carry over an obstacle, or most other situations where a larger dog will be nearly impossible to transport.

If you're going to have a good alarm dog though, I would like to recommend the common Dachshund. These little guys were originally bred to hunt badgers, boars and even wolverines. Consequently, most of them have absolutely no fear at all - of other animals or of people. When excited, they are as stubborn as any dog could ever hope to get, even refusing food as a distraction from the object of their attention. This trait is perfect for those times when an intruder thinks he can wave a steak in front of the dog in order to get him to shut up. The long body and deep chest mean far larger lungs for the dog's size, which in turn means more bark per pound – this is why a large (25lb) wiener dog can easily sound just like a 90lb attack dog. Incidentally, this breed is statistically the most aggressive towards strangers (and other dogs) – more so than the Rottweiler and more than any other small dog breed, by huge margins. This means that any intruder on the property will not only get their attention, but you'll in turn hear about it quickly. This is what makes this breed the perfect guard dog. Due to their smaller size, they are often overlooked, or are not seen at all until they start barking and even then maybe. They are very easily trained to hunt small prey such as rabbits and ground squirrels and love nothing more than to dig into a burrow to root one out – a rather useful skill for a pet to have if you're short on food. Finally, this breed can make for an extremely loyal dog, is not afraid of work if you train it and will happily follow you to the ends of the earth and back. With ordinary care, they can live for up to 15-20 years. Since they crave human companionship with their masters (and no one else), coupled with a love of burrowing, they make great four-legged hot water bottles for those times when the weather is too cold and rotten for anyone to survive outside. The only caveat is that you can't let them get fat (especially as they get older), else you put excess strain on their long spine – but in a post-collapse world, a fat dog is going to be about as common as fat people (that is, you won't have to worry about seeing either).

All that said, no matter what breed you choose, always go for something that is somewhat small, fearless, loud and able to withstand most kinds of weather. In this regard, mutts are preferable to purebreds and tend to be far more energetic, especially when excited.

Periodic Security Reviews

You should go over all minor incidents at least the same day. Major incidents and attacks should be analyzed immediately after they occur. Everything else should be gone over once a week. In all cases, go over what happened. Were there any problems with detection and raising the alarm? Did any holes in your security get exposed? If so, how can you fix them so that in the future things are better for you and your community?

Take the time to look over (or have someone look over) the security. Find weaknesses in the layout and strengthen them until they are no longer weaknesses. View your home or community from the eyes of an attacker. Analyze any weak points you find and try to find ways of either removing them, or in making them strong enough to help defend the place better.

Note that in the case of weaknesses being someone's home or similar, you will want to either extend your perimeter outwards a bit, increase coverage of those areas, or reinforce the offending home (try to talk the occupants into it) so that it isn't such an obvious security hazard.

One final bit - don't go overboard to the point where you risk increasing fire hazards and the like. The point is to make it more secure, but not (too much) at the expense of safety.

Looking Ahead

While a bit early considering where we are in the book, you can take comfort in the fact that as time goes by, the problems will become fewer. Initially, you're going to have your hands full. You will likely be attacked repeatedly for quite awhile, especially as easier pickings begin to dry up and as people (especially the criminal types) begin to get desperate. However, the idea I wanted to get across here is that as time goes by, ammunition supplies will begin to dry up, the population will begin to die off in large numbers and those people who aren't living in a community, are either lucky, or are extremely skilled? They will either be too weak to put up much of a fight, or they will be dead.

To this end, always try to spare the ammunition wherever possible. Take whatever weapons and ammunition you can from those invaders who didn't manage to run away fast enough or far enough. Don't use any battery-operated stuff (e.g. night-vision gear) unless you have good reason to.

Health And Healing

Unlike peaceful pre-collapse times, where a clinic is either nearby or relatively so, you're not going to have much in the way of backup if something goes wrong. While you shouldn't suddenly start going paranoid over every scratch and sniffle, you do have to keep an eye out for anything that starts going bad and take a few precautions that you otherwise wouldn't pay attention to.

Keeping Yourself Healthy

The biggest thing to keep watch over is the chance of infection. This can come from even the smallest break in the skin - scratches, bug bites, pimples and minor cuts. While regular washing of the skin can keep the chance of infection low, most situations post-collapse won't present too many opportunities for taking hot showers, or to have antibiotics on hand for cuts and scrapes. Bug-bites will be the most preventable, but the most common potential sources of infection (and worse). In areas with heavy bug populations (especially mosquitoes), you'll want to take special precautions, such as insuring that the windows and door screens are kept in good repair, killing bugs inside the home and attending to any bug bites immediately with alcohol (either rubbing or grain). The idea is to sterilize the area quickly and cover it with a clean bandage if there is any blood drawn. If you let any pets into the home, keep them clean as well. If you don't have flea repellant stocked up, you can use kerosene in a pinch (just apply it like you would any topical flea repellant… run a small line down a dog's back, or on the back of a cat's neck where she can't lick at it).

Wash all minor cuts and scratches as well, with soap and water if nothing else, but definitely with alcohol (either rubbing alcohol or high-proof grain alcohol). For larger cuts, laceration and the like, clean it out as best you can to avoid infection. Since you won't have an infinite supply of antibiotics (either pill or ointment), use them only if a cut or scratch gets swollen, or if the cut is large enough to require a butterfly bandage or stitches. Keep any wounds covered with clean gauze or bandages until it has closed and a scab has covered the entire cut.

Keep yourself, your clothes, your home and your utensils as clean you can. Whatever you can't wash with soap, boil in clean water. Be aggressive in keeping things as clean as possible, whenever possible. It won't always be possible, but try.

One common thing that will crop up in most people post-collapse will be… acne. A high-fat diet all those years has made the skin of most people a veritable playground of acne-generating bacteria - germs that aren't going to go away anytime soon. Pre-collapse, this is mostly kept at bay by daily washing and even regular applications of astringents. Post-collapse, washing daily may not always be possible, your diet will take a bit of a roller-coaster ride, stress will be ever-present for at least a little while and you're going to be a bit more preoccupied with the business of survival than with the state of your skin.

While it sounds silly to discuss acne, the reason why you have to keep it in check is deadly serious. Most boils (that's what a pimple basically is) will eventually burst, each leaving a smear of pus that can form more boils and a small open sore that in turn leads to a potential for infection and disease. Those on the face tend to pose a greater danger, since most folks tend to touch their faces *a lot* during the course of the day… with dirty fingers and hands. You're going to have to pay special care to avoid doing this and if you cannot bathe completely on a daily basis, at least wash your face at least once a day and wash your hands as often as you can (or at least as often as you need to).

Women and adolescent girls should take special care for problems that are particular to them, mostly centering around yeast infections and other gynecological troubles. Many of these problems can arise from lack of regular washing, but also from pH imbalances that can arise from stress, diet and many other factors. There are herbal and natural remedies to keep such things at bay and studying up on these (and more importantly, using them) will be useful. However, daily cleaning and basic hygiene will keep the worst of it out of your life.

Speaking of which, both men and women should always wash up after any sexual activity, period. The last thing you want is discomfort, irritation and worse that arises from lack of cleaning up fluids that deteriorate and harbor bacteria fairly rapidly. It should also go without saying that, well, any of the more offbeat deviations from ordinary sexual activity should be avoided if they involve any bodily fluid or secretion. As an obvious example, this also means avoiding anal or oral sex if you can.

Avoiding disease also means keeping yourself in top shape. Don't intentionally skip meals unless you have no choice due to rationing. Don't run yourself ragged. Get as much sleep as you can to eke out 7-8 hours in a daily cycle. Keep as positive an attitude as you can. Always seek things to look forward to and if time and resources permit, work towards them. Keep the exercise up if you're not doing much physically, but at the same time don't go overboard.

Keeping your home clean is paramount as well - this includes getting rid of as much trash as you generate, while keeping things washed and tidy. Meanwhile, no matter the temptation, don't skip doing the dishes and laundry and don't do either of them half-hearted… use very hot water for rinsing wherever possible (some clothes may not tolerate this, but use your judgement and err on the side of cleanliness).

If someone in your home does come down with a disease, quarantine that person into their own room and nobody is to go near that person except for a parent or spouse (for comfort), someone who is designated as a nurse and/or any medical personnel. Each person who comes into contact with the sick person is to wash up every time they go in *and* out of that room, without exception. Be sure that all disease-contaminated materials are either boiled, buried, or burned. You don't necessarily need filtered water to do the boiling, but it helps. Be sure that for any boiling, you boil the item for at least 30 minutes, to be sure that all bacteria or viruses in it are dead. An alternative for smaller (especially non-porous) items is to soak them in pure grain alcohol or regular rubbing alcohol for at least 2-3 hours before re-use.

For more serious wounds, stitching and cleansing the wound is going to be important. You can also make impromptu butterfly closures from any kind of waterproof tape if you've run out of the commercially-made variety. For deep wounds and worse, seek competent medical help immediately… only if you cannot do so should you try and fix things up yourself (at which point you need to rely on all the studying you've done on the first-aid and medical books in your library. You've been reading those when you get time, right?)

Vitamin supplements from your supplies will help your immune system fight off disease, but don't soak yourself in supplements… take only as much as you require, even cutting them in half if possible (or perhaps taking them once a week instead of daily). The reason why is that your body will simply pass off any excess vitamins and minerals and that expensive and valuable resource will simply wind up in the outhouse and not in your body. If your urine is a bright yellow, you're taking in too much and should cut back. Get some sun on your skin as much as safely possible, as it is a great source of Vitamin D. There are numerous sources of Vitamin C (both in the natural world and in your supplies), so take them frequently, as it will help your immune system. During those times when you're not exercising much (for instance, in the winter), try to institute a daily routine during those times to get some sort of vigorous exercise.

Once in a while, especially if you're religiously inclined, be sure to get some prayer in. Keeping yourself mentally and spiritually healthy is just as important as keeping your body healthy. Depression may well be rampant. It will be even worse for those who suffer clinical depression but are completely out of medications that were prescribed to combat it. To this end, we take a small side-step from self-care, towards caring for your family and neighbors. Take some time to express kindness to them. Give someone a hug once in awhile. Let them know they're not alone. Organize activities to bring folks together. Try to cheer people up when you can. It's hard to keep your own spirits up if everyone else's spirits are dragging you down like anchors.

Keeping Your Community Healthy

This is going to take a lot of coordination, but few rules. Let's start with the rules:

- All trash is to be recycled (after cleaning), buried, or burned, period. Trash left out will only fester and decay, harboring diseases.
- No one is to dump anything into any waterway, lake, ocean, river, pool, or stream. Folks downstream or down-current could contract diseases this way.
- All liquid waste is to be dumped in a hole or in a natural depression, but not near any known well, water source, or the like. Again, the reason is due to limiting the spread of any disease, be it potential or known.
- All human waste (urine or feces, blood, or whatever) is either to be buried or incinerated, or put into a working septic tank (for liquids) if possible.
- All dead people are to be buried or cremated within 3 days of death whenever possible in summer and within 2 weeks during winter (as long as the body is kept cold).
- If possible, everyone is to tend to their personal hygiene needs as frequently as possible.
- Anyone who has come into contact with any contaminant or hazard (of any type) should clean themselves up immediately and should not come into contact with anyone else until they do.
- Anyone who is sick is to be quarantined in their own room and only medical personnel (or one parent in the case of a child) shall come in contact with the sick person. All who do come in contact with the sick person shall take appropriate precautions and will wash up before and after contact.

This alone should take care of the vast majority of your problems when it comes to preventing the worst diseases, since most diseases are spread by way of contact with trash, lack of hygiene and by contact with infected people.

In cases where there are a lot of people living under one roof (or close together), insure that everyone in such a situation is able to wash themselves on a regular basis and that each home has a room set aside which can be used for quarantine if necessary. If anyone in the community can manufacture usable soap, try to have some set aside for those who are without.

It is well worth the cost in materials to prevent the whole community from passing around disease, since post-collapse even minor diseases may well become deadly. After all, diseases (and various forms of malnourishment) are the root causes for a high mortality rate. This was true in the 17th Century, is true in the Third World today and will be just as true in what used to be the Industrialized World (hint: where you live, or end up post-collapse). A lot of this can be kept at bay with knowledge and attention to cleanliness, but it won't be perfect. Many diseases are spread through airborne transmission and by the time someone knows they're sick, the bacteria or virus has already incubated for some time and has spread about for quite a few days.

In those cases where someone is sick with any serious and contagious ailment, the news should be given to the whole community. Inform everyone (gently!) to keep themselves clean, keep their nose and mouth covered in public until the danger has passed and any other sane precautions you think may be necessary during any threat of contagion. Be open and frank about it, since hiding details or denying it will only feed rumors and inflame people into doing something stupid. The idea is to eliminate ignorance as much as possible (to prevent rumors or worse) and to keep everyone informed of how things are going.

One thing you will want to do as a community is to hold classes. Have either a medically-trained person do this, or someone with the right access to the right books and resources on the matter of disease and first-aid. Start by teaching the folks who aren't defending the community basic and then advanced first-aid techniques and tips. Then start including classes on hygiene and on how to identify various disease symptoms, as well as how best to treat them given a lack of modern medicines and facilities. The more equipped they are with prevention, the better equipped they will be to help keep things from getting ugly.

As a community, you should have a group initially (and periodically) go through and bury any dead people or animals that are in homes, nearby fields, etc, as soon as possible. Sometimes a person died alone, without family or loved ones to do the burial for them. Burying these folks is not only the morally right thing to do (if you can spare the time and energy), but also a means to limit any sources of disease. Leaving bodies to simply rot will cause troubles for anyone who stumbles upon them later on. The same goes for anyone you or your community killed off in any attacks… bury or cremate the corpses ASAP and do not let a freshly dead person remain un-buried or un-cremated for more than three days if the air temperature is above freezing.

As a community, you should also take periodic sweeps of the neighborhood and drain or eliminate any stagnant water pools or ponds that you find, be it man-made or natural (e.g. empty swimming pools, ponds, ruts, etc). These places fast become breeding grounds for mosquitoes and molds and because of that will potentially harbor disease. Also look for and clear out any trash piles, or places where community members may have created potential hazards with piles of waste (or worse). Help them clear it out initially and help set them up with proper habits and methods of disposal. Consider setting up a community dump for those things which cannot be burned or buried individually.

To prevent anyone from coming in with diseases, you're going to have to do a bit of checking of each new arrival. Anyone that appears sick (coughing constantly, covered in boils, pale, dark circles under the eyes indicating lack of sleep, listless, etc) should be asked how he or she is doing and potentially quarantined until proven healthy or not. This is going to be more of a judgement call than anything else, but it is something you will definitely want to keep a good eye on.

All that said, even if you cannot screen everyone coming into town (for awhile you can do this easily, since you'll need to restrict who moves in and who doesn't), you will want to screen their stuff. Merchants and traders coming in with goods that have fleas, that are obviously spoiled, or that appear as carriers of disease in some way should be turned away, or at least notified that the offending goods or items are to be left outside of your community before being allowed in.

Finally, make it a point for the community leadership to periodically check in on everyone to catch any potential health issues before they get ugly. Perhaps (if it can be spared), have a couple of volunteers stop by someone's home once a day, to inquire how things are going, perhaps bring by some small but badly-needed things as gifts and to inquire about the health of everyone in the home. The reason you'll want to do this is because most post-collapse survivors will be of the more independent sort and in turn aren't going to be as likely to complain to anyone if they get sick on a minor scale, catch a cold, feel under-the-weather, etc. Only when something is major (or threatens to) will they be likely to seek help (otherwise many will deem it a waste of supplies), so discovering something when it's minor may help prevent something major.

In Case of Outbreak

This one can be tricky. On the one hand, you need your community to be close-knit in order to survive. On the other, you don't want the uninfected to catch whatever the infected have come down with.

If you catch it early enough, you can quarantine the person or family affected. If you can't, you're going to have a rather big job on your hands.

No matter what, at the first sign of a contagious disease (or one that affects multiple people at a given point in time), find out the following things immediately:

- How many people are either infected or has come into close contact with the infected person(s) (especially if the disease transmits via coughing and sneezing)?

- If the disease is fatal, who and how many have died so far? This will give you an idea of how much time you have and how quickly you must act (as in, get the word out and start quarantining the infected folks *right now!*)

- Can it be confirmed? That is, do you know of can you find out what the disease is?

- If it is a water-borne disease, where did the infected drink from and where is their waste going? This tells you what water sources to shut down or put off-limits and lets you inform the community that they should boil all water for all uses (as opposed to only boiling it for drinking/cooking purposes)

- Did it come from any particular type of food (foraged or otherwise)? If so, then make sure everyone else knows to avoid it.

- Did it come from any type of animal (especially pets)?

- Who in your community is pregnant right now? Who has infants or old folk in their homes? Those people get notified first and foremost, since they will be the most vulnerable.

Once you have the answers, it will help you formulate a more useful response to the problem. Whatever you do, do not panic and immediately start locking down the whole community and always take at least five minutes to stop and think things over before doing anything (aside from quick and near-always fatal diseases, such as Hantavirus, Plague(s - there are three kinds), or etc.

Odds are pretty good that if civilization collapsed due to some sort of disease pandemic, people will be very nervous and very easily prone to panic. They will also be very keen on the subject of disease. This is why you should always (even if pandemics never happened) announce the name of the disease first. Then you can tell the community how it is transmitted, what the signs are, how bad it is, what to do to avoid catching it and any treatments they or the medical folks can provide… and do it in that order.

Be certain of the information you give out, as any mistakes can have people screaming "Ebola!" (Or some other unlikely disease) when the reality is a case of mild influenza or food poisoning. Have your community report any and all diseases strong enough to make a person weak or sick to the community leadership, no matter what.

About Drugs (The Legal Kind)

Let's face it - you're only going to have a limited amount of supplies, but a large amount of chances to use them.

Things like antibiotics (if you have any) should only be used when they are the most needed and not for every little cut and scrape. Avoid using them entirely on children over the age of 6 and under 18, unless the wound is really deep, the wound has fecal or other infectious matter in it, or signs of infection develop (this is because you want young immune systems to experience and build defenses against the more common stuff). Instead, insure that the wound is immediately cleaned and sanitized (e.g. with alcohol, tincture of merthiolate, iodine, or similar generic antimicrobial remedies).

For drugs used in the treatment of contagious disease, use them as early and as often as you can. The sooner you treat it, the less chance it has to spread. Meanwhile, identify and remove any sources of the particular disease, so you don't have to deal with it again and again and again…

So what about drugs that are "expired"? They likely won't kill you (with few exceptions - insulin and tetracycline stand out), but do not depend on them with your life. This is because the drugs begin to lose potency after that date, so dosage and potency will start to become big, fat question marks with each month past the expiration date. While most drugs contain up to 50% of their potency even 10 years after the expiration date, some will drop to practically nothing less than a year past the date. Internal testing by Bayer has shown that their aspirins hold up even 4 years past the date.

On the other hand, there was a study done by the US Military that showed 90% of the top 100 drugs in their drug stockpile were safe and effective even 15 years past the expiration date (Laurie P. Cohen, *Wall Street Journal*, March 29, 2000). They didn't say which drugs offhand, just that they were the ones which were most commonly used in combat… so be careful before being too confident on this one.

Find, cultivate and use natural drugs whenever you can. This is not only because you will gain skill in finding/using them, but because they will be a lot easier to get than pills. You should have a couple of good (and reputable) herbal remedy books in your library and you should be identifying and even using them (as needed) by now.

So what about illegal drugs? Now I'm not saying that you should, say, go out and grow marijuana - that is an act which is illegal under federal law. On the other hand, I will say that the plant does have strong medicinal properties and if carefully administered only to those in severe pain, it will do a lot to alleviate that pain. To that end? After collapse, with no DEA or coherent police force to stop you from doing it, I think *in that particular context* you may find it useful as a medicine. On the other hand, doing this carries its own danger as well: some folks are going to want to use (and steal) it for less-than-medical reasons, so you'd better be able to guard it very well and make sure that absolutely no one outside of the doctor and you ever knows of its source.

Withdrawals

This is ugly enough to deserve its own section. A whole lot of people depend on a whole lot of drugs to function in a civilized society. Post-collapse, there aren't going to be many, if at all to go around - and even less as time passes by, until they run out entirely. We covered a lot of this way at the beginning and of the importance of weaning oneself (and one's family) off of as many medications as possible before civilization collapses.

However and unfortunately, this will not happen for most. Millions eat Prozac and other anti-depressants. Far too many millions of children are swallowing drugs to treat ADD and ADHD (all too often as a convenience more than as treatment). Drug addicts everywhere rely on Meth, Crack, Heroin and various other drugs and rely on them at the deepest biochemical levels. For some, the medication simply cannot end for them if they with to remain alive. Diabetics need insulin daily. Paranoid Schizophrenics rely on psychotropic drugs to prevent doing harm to themselves (and sometimes to others). Certain cardiac drugs are needed to prevent further heart attacks.

As you can see, this breaks down to two groups of people: Those who need medicine to stay alive, functional and/or sane and those who don't need the medicine for that, but will experience a very nasty period of withdrawals just the same, sometimes with violent and disruptive results. Sadly, you will probably get to deal with both types in your community.

Helping those folks who need medicines to stay alive or sane is not going to be easy. You will either have to find a viable and renewable replacement for those drugs, or you will have to comfort these folks as they slip into the Great Beyond (that is, either death or insanity... equally unappealing, truth be told). You may get lucky and the patient may be able to wean themselves off a drug (e.g. a Type 2 diabetic being able to slowly wean off of insulin as his or her weight drops). On the other hand, for all too many folks, this is simply not going to happen.

Your best course of action for those who are about to die is to make them as comfortable as possible, to spend time with them and to carefully write down anything they ask you to. Pray with them. Get hold of a preacher or priest if you're so inclined (and if you are able to). Quietly make preparations (out of their sight and earshot) for burial and respect any wishes they may have in that respect. Whatever you do and no matter what the patient says or does, do not consider performing a 'mercy killing'. This will end up mentally scarring you far deeper than anything you could ever hope to think of. Let the person die naturally, but as comfortably as you can make them. Give as many sedatives as you can spare and alleviate as much pain as you can, but do not kill them.

For those who need medicines to remain sane or functioning, this is going to be the toughest in the long-term. You're going to have to make arrangements to restrain anyone who is violent. You will have to arrange things for anyone who may go catatonic (that is, completely unresponsive). You will have to get to know the illness enough to anticipate when episodes may occur and know what actions to take. Then, you will have to act upon those episodes, in both normal times and under stressful ones. Everyone in the patient's family will have to be involved and perhaps even everyone in your community.

For those who are merely dependent on drugs for chemical reasons, things are going to range from uneventful to downright ugly - depending on the drug involved, the person involved and the circumstances involved. For instance, if we're talking about anti-depressants, then for quite a while the person will experience irritability, anxiety, headaches, dizziness, fatigue, insomnia and a possible return of any clinical depression symptoms. Most doctors recommend slowly easing off of antidepressants over time, so you may want to do that in this case.

Once the pills (of any type, for any reason) stop flowing in, count what you have left, then ration them out to slowly decrease the dosage until you're completely out. For instance, if you have only 2 weeks' supply left, then slowly start decreasing the dose - take ½ the labeled dosage for two weeks, then take only ¼ of the labeled dosage for the next two weeks after that and so on, until you're out. This is going to be a lot easier on you than suddenly stopping and hoarding what's left, or taking them as normal until the last one disappears. The idea is to get your brain (and body) slowly used to the absence of the drug. It won't be easy at all and the urge to go back to taking your normal labeled dosage (especially for pain issues) will be intense. However, keep the discipline, else you'll just feel it all the worse once the pills are well and truly gone. If you don't think you can do it, have someone trusted ration them out to you.

For those cases where a natural/herbal remedy may replace the drug's function, perform the same rationing as above, but slowly introduce the natural remedy in the same proportions. It won't be perfect and you may find that you will need more of the natural version to have the same effects, but it is pretty much going to be your only option.

Throughout these writings, I have shown very little sympathy for the person who abuses illegal drugs (or even legal ones), especially those which create a chemical dependency. In the beginning, I have gone out of my way to get you to reform someone of the habit before civilization collapses and have had some rather harsh punishments proscribed for anyone still hooked once things do collapse. The reasons why we have also covered and they still stand in full force, even now… these folks are the most likely to get you killed, pure and simple. However, there are always going to be exceptions to this and you may well find out that you have a junkie in your midst and for some odd reason or other, you may find that instead of expulsion, you're going to have to help that person get on the wagon. This is not going to be easy and I guarantee that you most likely won't have the resources to do it properly.

This is going to require a lot of tough love, long story short. It is possible, but it will require a lot of time, effort, resources, love and attention.

First, you're going to have to physically restrain the individual without hurting them and you need to keep them from hurting themselves. Provide for their every critical need (food, water, warmth), but no matter how loud they scream, threaten, cajole, bribe, whatever… do not give them the drug they're going to crave. It will take days, sometimes weeks before their system is clear of the drug and longer before the cravings die down to the point of automatic self-control. The restraints can come off after a week at the most, but close monitoring will be necessary for up to six months or more.

Note that in most cases, the symptoms will be psychological more than physical. There are however exceptions - alcoholism and dependency on benzodiazepine drugs (e.g. valium and some tranquilizer/sedative types). A sudden stop and withdrawal from alcohol can cause delirium tremens ("the shakes"), which is sometimes fatal. A sudden stop and withdrawal from benzodiazepine type drugs can cause seizures or worse. In these cases, wean the person off as slowly as possible. Most illegal drugs can be weaned "cold turkey", but note that in the case of heroin, methamphetamines and crack cocaine, this will definitely require restraining the person physically. Also note that it will be very hard on the person being weaned off of the stuff.

The problem with ending addiction to abuse (as opposed to prescribed medicines) is that the cravings never entirely go away. This means that you will have to equip your patient with the mental skills necessary to help cope with those. If you know or suspect someone in your family or community of abusing drugs or alcohol, be sure to have (or get hold of) any competent and reputable self-help books on the subject before attempting any sort of detoxification.

Zombies And Refugees

No, we are not talking about the dead rising from the grave. We are not talking about some super-virus that turns people into flesh-eating monsters. However, you will still see plenty of folks that are commonly referred to as "Zombies". We discussed them before, but this time let's dig a bit deeper…

So What Is a Zombie, Really?

A "Zombie" is someone who, woefully unprepared for the worst, is stuck wandering about, looking for whatever food, shelter and water they can find. These are commonly people who are quite used to living in relative comfort and prosperity. In addition (and in contrast to refugees), these are people whose self-esteem is more important to them than any other consideration. These are people who insist that they have a right to be taken care of. Zombies are, for lack of a better description, professional refugees – helpless on their own and unwilling to learn if they can avoid it; a state of being that is gained either through ignorance or attitude. Zombies are people who will most likely end up willingly living in a government camp… and when that camp finally breaks down, they will be on the prowl in various states of demand for food, water, shelter, etc. They will be most likely to travel in large groups.

The difference between a "Zombie" and a raider or other criminal is that a zombie isn't (yet) actively out to cause harm or mayhem. On the other hand, they will steal when they think you're not looking. Their big goal is to stay fed, warm and as comfortable as possible and by any means necessary. The difference between a "Zombie" and you and I is that most zombies will have a too-strong sense of entitlement. They will look to government or military resources, appealing to both (and possibly demanding of them) to stay warm and fed.

In the initial stages of full-on civilizational collapse, these folks will be *very* plentiful. They will travel in very large groups, though if there are government-run refugee camps within reach, zombies will gravitate towards them for as long as those camps contain food and shelter. Failing that, they will gravitate towards any sign of civilization and demand to be let in.

The reason these folks are referred to as "Zombies" is because like the movie version, they are in effect the walking dead – their bodies just don't know it (quite yet).

So What Are Refugees?

Refugees describe anyone who is presently homeless and on the move – be it permanently, or staying in temporary camps. If you are in the process of getting out of dodge, you yourself are a refugee during that time, until you find a permanent home.

What distinguishes refugees as non-zombies is a couple of factors: They are capable of learning and prefer to procure for themselves by way of nature or by scrounging (not stealing, but foraging from abandoned goods and/or supplies as they are found). They are often prepared somewhat and have at least some supplies, tools and the ability to use them. Other distinguishing characteristics include creativity, the ability for rational thought, a strong sense and practice of morality and honor and a willingness to work hard for what they want.

Riding The Waves Of Migration

They will arrive and perhaps pass by in waves. If you live any kind of distance from the main nests of potential refugees (read: big cities), there are a few calculations you need to make here. First, how fast are they going to come? Let's tackle that... the fact is, you're going to see refugees in waves. For the sake of argument and simplicity, let's assume everyone has an urgent desire to just get out of town and moved in all directions equally, shall we?

The first wave will be those who got out quickly or early enough, had a full tank of gas and either have some destination in mind, or simply blew out of town with no regard for packing anything. As a whole, as a rule and for the most part, this group will have little-to-no supplies and maybe they had even left family behind (yes, people are selfish and heartless SOB's at times). A few in this group will be prepared and well-supplied, but these particular folks will most likely have a specific destination in mind and will likely just want to get to it without stopping. The hallmarks of this wave will be easy to spot, since they will be the ones who finally stop about 40-50 miles from town, start waving money/credit-cards/etc around and demanding to buy everything they can lay hands on at any price (that is, assuming they realize that they don't even have basic survival goods, let alone sufficient supplies).

The second wave will be those who also were lucky enough to get out first, but had to stop and pick up gasoline (if they could find it), get the kids from school, pick up Dad from work, or other such small delays. They had some nasty traffic issues along the way, but they somehow made it out. This will be a lot larger than the first wave, but still not the largest, not by far.

The third wave will start with folks who had some serious traffic problems, maybe a flat tire they had to quickly change, had trouble getting gas, or spent some time loading their vehicle up at the last minute with everything they think they needed. However, they somehow managed to still get out. Amongst them will be refugees on motorcycles, small 4x4s and other drivers/riders who managed to get around the worst of the traffic snarls by way of simply going off-road around them, or by taking back roads, etc. In and amongst these folks will also be the first of the criminal elements – those rare few criminals and con artists who were either lucky or clever enough to realize fully what was going on and decided to 'set up shop' outside of town. They may also start picking from the refugees stuck in traffic alongside them. These folks (for good or ill) will likely stop and seek shelter where you are, assuming you're well out of town and in a small community. They may stay temporarily (e.g. for the night) if they have enough fuel and supplies to keep going.

Of course, these three will be nothing compared to the next one. The fourth wave, the largest and longest, will take awhile to arrive and the further you live out from any major population center, the longer it will take both in arrival time and duration of passing. At 50 miles from town, this wave of refugees will take at least two or three days to arrive. At 100 miles out, you may not start seeing them for a week or more. These will be refugees who have decided to abandon their stuck/broken-down/crashed vehicles and continue on foot, or by some miracle or awe-inspiring skill, had managed to drive around the mess, albeit slowly. The earliest elements of this wave will, like the previous, arrive on motorcycles, off-road vehicles, or by dumb luck. They will also start arriving on bicycles. Then, the vast majority will arrive on foot. This wave, more than any other, will very likely stop at your front door and will either desire or demand to have some food or a place to stay. The good news about this particular wave is this: The further out of town you are, the less of them you will see. This is because a large number of them will have camped out closer to town, will have veered off into a nearby government camp, or quite likely will have died along the way of accident, exposure, or malice.

Where Are They All Going?

Another calculation you will want to make is in figuring out where they are all going. Far, far earlier in this book, we had you make up some maps. Now is the time to get those maps out. These will be the routes that most will take – the highways, the freeways, major hiking trails, you-name-it.

What those maps don't answer is... where are all these people actually going? We saved that part of it for now, because until now you had no idea as to what precise calamity would cause people to flee (and at time of writing, well, we still don't). However, if you're reading this in the future, while things are definitely beginning to break down, then you will definitely have a far better idea. If there are any zones of heavy infection or pandemic activity, assume that few-to-none will be fleeing towards those zones. If Yellowstone national park just became a supervolcano, you can safely assume that nobody is going to flee to Wyoming or Montana (or Utah, Colorado, either of the Dakotas...) If there is a civil war or invasion in progress (hey - you never know), then you can pretty much write off any migrations towards battle sites or any regions that would constitute a border between the warring factions. If the calamity is the final breakdown of society without a specific cause, assume that everyone is going to move in all directions - well, unless it is winter, in which case you can assume a bias towards folks moving directly southwards, where the promise of warmer climates can be found.

How Are They Getting There?

Most of this was figured down far, far earlier in the book, when we had you draw up the migration maps. But... there are a few other things you will want to keep in mind.

Odds are perfect that the majority aren't going to know how to get from point A to point B (wherever point B may be) without GPS or maps. Some may have road maps, but won't know how to use them when away from the road. Few if any folks will have (or know how to use) compasses. Some may have hand-held GPS units, but the majority will be quite lost if they leave the roads. This means that most of them are going to be walking/driving/whatever along the roads and any boaters will be sticking to the main navigable waterways. You may have hikers and other outdoor hobbyists who will stick to well-known hiking trails however, so don't think you have the countryside all to yourself.

Expect most of these refugees to travel first as a stream, then further out as groups. They will do this for a couple of reasons – safety in numbers, an instinct to follow someone (or a small sub-group) that has some idea of where they are going and because people are generally hard-wired to travel in groups anyway.

They're Heading For My Town! Now What?

First and foremost, resist the urge to go out loaded with weapons and expect to turn them all away. If they are coming in small, manageable groups, you might be able to do this, but if you live somewhat close to a big city, odds are good that they will be coming as a horde. You simply do not have enough bullets to enforce such a thing, a disturbing number of them will be armed and they will still keep coming.

Figure out how the first waves have been traveling. These first waves consist of folks who will likely blow right through your town and may not even stop. It's that last and absolutely huge wave of people that you have to worry about, since they are the ones who will be arriving on foot, tired, hungry and liable to get very angry if turned away, not fed and sheltered, etc.

If your town can afford the charity (say, you have grain silos nearby), set some food out with the caveat that the refugees are to take what they need and then immediately keep moving. Even better – set that food and tents out a good safe distance away from your town and put up a big sign at the major roads leading to your town that tells all refugees the government set it out for them out there (at the safe distance away). The idea here is to have the main stream of refugees go *around* the town, use whatever food is there, then realize that there is no more, but be in a place where another town is closer, or that it would be more profitable for them to keep walking on. This means choosing the site carefully.

If the cause for collapse involves a pandemic, make (or steal) quarantine signs that will make the crowd think that the town is heavily contaminated and post armed-but-fake military guards wearing fake hazmat gear. It's almost guaranteed that 80% of the refugees will turn away at whatever distance they can read the signs.

If your town is somewhat isolated, or better still, can be made that way, you may have a chance at setting up a solid roadblock, but it will take a community effort. Cram semi-trailers and cars sideways onto a key bridge or two, use rockslides to block mountain passes at key points, set up a floating armed guard at harbor entrances and things like that. Just know that any remaining governmental or military assets out there can and probably will remove most of them if someone feels the need to.

If all else fails? Quickly hide/bury/lock-up as many supplies as you and everyone else can, then simply let the refugee waves through town, using your weapons to guard individual homes from anyone that gets a bit too nosy or close. Meanwhile, sporadic gunfire throughout town will keep the crowd moving onward (this doesn't mean actually shooting anyone per se, but just a few distant-yet-audible gunshots at random intervals to give the hint that staying is a bad idea – a few random screams and loud pleas for mercy from the resident ladies --and even the kids-- will help the effect). Another tip is to make the town look temporarily ransacked by tossing debris and non-critical goods onto yards and parking lots. Setting an old car or two on fire helps complete the effect, as is breaking a few windows along the main drag. Perhaps you can even find empty or closed stores on Main Street, carefully scavenge them of any useful items, then smash them up and make it all look properly looted. The idea behind the charade is to give a strong psychological impression that no refugee really wants to stop here, because not only would be appear dangerous to, but because it looks 'picked clean'. You will have to keep this charade up for a few days and nights, though, at least until the stream of refugees from each wave begins to thin out.

If you do present guards to try and keep the refugees away, make sure the local police are the face of that blockage, backed up by a small group of the strongest, most well-armed men you can find. The police uniforms give an air of authority (which amazingly, many refugees will recognize) and the muscle behind the police will enforce the point for those who don't listen.

When Will The Migrations End?

Hard to say, really. It may be one big clot of humanity fleeing a massive natural disaster or outright warfare. It may be a slow but steady stream as the world slowly strangles on its own economic bile. It could be almost non-existent due to a global pandemic that kills most folks off. The truth is, there is no way to predict this in advance.

However, there is good news. The migrations from any given point can be more accurately predicted. The first three waves you're going to see nonetheless once things finally get to the point where a given metropolis can no longer supply itself. That fourth wave may be quick, or not, but after a month, the numbers will be less than 10% of what it was initially, due to starvation. After two months, it will drop to less than 1%, if that. In desert climes, it may drop even faster due to dehydration.

This drop in numbers is mixed news though. As time goes on, there will be less of them, but the refugees you do see will get harder and harder to intimidate or drive off. The initial waves will be comprised of (mostly) civilized people, most of whom will quietly do as they're told by any perceived authority and will not really question why. They will be 'soft' and any actual resistance they may put up will be haphazard and disorganized at best (though do beware of mobs, which can effectively tear down nearly anything in their path). However, as time goes on, those refugees who are still alive later will be far more savvy and far more able to fight for what they need. They will also be a lot more cunning and a lot tougher overall. There will be very few, if any, of the "zombie" type of person left alive by the third month or so – government refugee camps notwithstanding. The good news there is that by that time, you may well have some room to spare in your community as well, depending on the circumstances and cause of the collapse.

Harvesting Refugees

Wait - it's not what you think! We're not talking cannibalism here. What we are talking about is finding those refugees with critical skills and taking them in if you have the room and resources to. Doctors, nurses and EMTs, engineers, metalsmiths, schoolteachers of hard subjects (think science, math, history... and not things like "feminist studies"), infantry combat veterans etc... whatever you need. You'll be looking for people who can really contribute to the community. If there are any among the refugee waves and if you can do it, quietly take them aside and invite them to stay. Let them go if they refuse, but at least give them the option.

It will be tricky to do in most cases, but if the refugee trickle is slow enough, or if you are able to enforce a blockade, you can interview each family as to their skills and take it from there. One thing to keep in mind: People are going to lie their butts off, especially if they're smart enough to know why you're asking them questions in the first place. If you're not careful, you'll suddenly find that every refugee is a family of doctors with chemist children and have spouses who are professors in engineering. Find someone in your community sharp enough (or knowledgeable enough) to know what to look for in a specific skill (licenses, membership cards, degrees, perhaps administer an impromptu skills test, etc).

Oh and before you even think it: While yes there are likely residents you know who are worthless, or are unable to contribute much, existing residents all get to stay as long as they don't commit crimes, aren't already troublesome, or don't cause extreme trouble as outlined earlier in this book. They have the right to stay by already living there. Do not start throwing folks out just because someone better comes along... doing that will create distrust and is guaranteed to destroy a community faster than whatever calamity touched off the collapse in the first place.

So, What's For Dinner?

Ah, yes, food. Hopefully, you have a ton of it stockpiled away and are quite cozy and ready to ride the storm out for a couple of years before you start planting a soon-to-be-bountiful harvest... right?

Well? Sort of. Before you settle down to the notion of re-hydrating your long-term food stores and having the same things for dinner every night, you might want to stop and think about that for a bit. You see, not all of your neighbors are going to be similarly supplied and the last thing you want to have happen is to have your community leadership confiscate all of your food, shoving it into some community store so that everyone starves equally.

If everyone else in the community is getting lean, but you and yours remain nice and fat? If delicious odors emanate from your home every night while the neighbors are all staring at empty pots and empty bellies? This is a bad thing and will turn the community against you in very short order. What we want to do here is to provide you with a means that you can teach anyone (and everyone) else how to use, so that you and your community can all eat at least something... and this way two good things happen: One, they're not going to look at you as a hoarder (in spite of your hidden hoard) and two, you can save the long-term food storage for those times when there's absolutely no food to be had at all.

Okay - so does this mean starving like everyone else? No. It does however mean using some discretion and concealment. Avoid cooking foods with strong odors. Forage as much as you can, but use your food storage as a supplement, not as the mainstay (unless necessary, of course). Mix in the stored foods with foraged foods when making stews and soups. It also means keeping the meals down to healthy proportions and avoiding excess. This finally means keeping your stored food still well-hidden, locked up and only used when there is no other choice. Meanwhile, you do have choices, both for you and your community.

Aside from the scavenging trip results, collection of government-given relief supplies (if you're lucky), any community-confiscated retail or wholesale foods from stores and perhaps the contents of a few grain silos? You actually have quite a few options outside the home. Let's go through some of the major food groups and see what we can come up with...

Fruits And Vegetables

For awhile, get used to the word "salad". This is because you can scrounge a whole lot of edible greens (during most times of the year) from as close as your front yard and definitely from nearby fields, parks, woods, or what-have-you. This is also why there is at least one good, solid edible plants listing book in the shopping list for your library. Personally, having one local to your area and one general field guide can be very, very useful.

Before you start thinking that you're going to be stuck with eating nothing but weeds, you may be surprised to find that, depending on your region and local climate, you could find and get wild onions, mushrooms (be *extremely* careful and certain in identifying them!), wild strawberries, clover (seriously), blackberries, wild blueberries, wild grapes, mulberries, persimmons, watercress and a whole host of other healthy and tasty plants that can become a part of your veggie intake.

The only real downside to this is that many (if not most) of these edibles are going to have limited seasons. You can however dry most of the fruits and mushrooms, preserving them for later use. You can also take edible wild tubers (roots, y'all) and treat them like you would potatoes, both for the short term and beyond.

Grains and Beans

Grains and starches are going to be the absolute major part of your diet for awhile. Odds are almost perfect that most bulk foods to be found in storage are going to be grains. Again, your handy, tried-and-true field guides to wild edibles can bring you a surprising amount of grains and grain substitutes. To this end, I sincerely hope you have a grinding kit as listed in the Shopping Lists... because most of it will be in a whole grain form and turning it into flour is going to take some doing. As for the wild edibles, you can even get grain or grain-like goodness out of those as well - a good substitute for wheat flour for instance is the lowly cattail. In the bean department, there are actually quite a few plants with edible beans, but always consult the guides before eating any of them.

Meat (or rather, Proteins)

Ahh, meat. The stuff of tasty dreams. That said, beef and pork may be in extremely short supply (unless you happen to have a large ranch or farm nearby with an owner willing to barter). Most of your meats are going to come from either poultry, fish, shellfish, or some rather exotic-but-perfectly-healthy sources.

Before we begin, let's talk food safety. Only eat animals or their meats which you yourself have killed, which do not look injured, isn't discolored or sick-looking and have not been dead for more than four days without some sort of food preservation (salting, smoking, canning, drying, etc).

Now, let's talk about pigeons for a moment. This will be the only time you read the cliché "tastes like chicken", because these little guys pretty much do taste like chicken. They are also fairly easy to catch (you can use a net), their nests are relatively easy to rob for eggs or for breeding stock and you can eat three of them in one sitting without feeling too full. Other, similar birds are also fairly good eating and provide a lot of needed protein and fat, but pigeon is going to be high up on your list, at least until you can get hold of enough chickens to make a good coop-full.

Other good sources of protein and fat would be small ground-burrowing animals (ground squirrels, beaver, moles, rats and mice, prairie dogs, etc.) Don't turn it away if it is healthy, fresh and chock-full of protein and fat. Same with the insects. Ants contain more protein per pound than beef does and about half as much fat.

For you desert-dwellers, we haven't left you off the menu either. Lizards, birds and insects are perfect targets for your palate. They all contain proteins and fats and many species make quite healthy eating. One note though – avoid the Gila Monster (look it up) unless you really know what you're doing. Same with most poisonous insects.

A Word About Rabbits

Yes, this animal is what most folks think of when they think 'survival'. They can be rather tasty when cooked right as well. However, you do have to be careful of two things:

- One, insure that any rabbit or hare you killed does not have white spots on its liver. If you see any, burn or bury the whole carcass and *do not eat it* – you risk dying of Trichnosis if you do.
- Second, if rabbit is your only meat and you don't/can't eat much of anything else, be double-plus certain to eat the animal's brains, any and all fat that you find in it, the bone marrow and some of the entrails (clean brown liver, heart, any abdominal fat, etc). You see, there is this little thing called protein overload, or more commonly, "rabbit starvation". Rabbits have very, very little fat and rabbit muscle will have practically no fat at all. However, your body needs and craves at least some fats, else your body goes into protein overload and eventually it kills you as the excess proteins turn toxic. While avoiding fat at all costs is a nice and trendy way to keep (or reach) an ideal weight in peaceful times, in a survival situation you're going to want and need all the fats that you can get into your mouth – because your body will be burning it up at a powerful rate.

Snakes?

Certainly! Snakes are rather tasty and quite healthy to eat! They are somewhat easy to catch, cook well and the bones are often soft enough to eat right along with the meat (which is a good thing, because the bones are rather plentiful in there). You do have to take a bit of special care in killing a snake, however. Try to catch it with a forked stick, or get a forked stick after you catch one. Hold the thing's head down with the fork of that stick, make sure no one is on the head side of that stick and then cut *behind* the stick, on the other side of the head. This way the head (which is notorious for flopping around trying to bite someone or something) can be flicked away from everyone until it stops moving. This is especially important when it comes to the poisonous snakes (cottonmouths, rattlesnakes, etc). You then skin the critter, gut it and you have a long stretch of edible goodness you can chop into sections for cooking. The best way to eat snake is deep-fried, but you can be forgiven if you don't want to spare the oil. Roasting or slow-cooking in water and vegetables will do the trick without the meat becoming all rubbery (which is what happens when you simply boil snake-meat).

Shellfish?

You do have to be a bit careful here. The old adage about months ending in "R" actually hold true for the most part, but you yourself can keep an eye out over the ocean and look for signs of algae blooms ("red tide" conditions). Crabs and crayfish are safe to eat pretty much any time and are very easy to catch with the right gear. Crabs can be caught with a trap, or caught with a string tied to a large scrap of meat – you slowly pull it it up at the slightest tug and use a net to procure it once it's close to the surface. Crayfish (aka crawdads, mudbugs, etc) can be caught by scooping one (gloved) hand behind it while scaring it backwards with the other hand. Clams and shellfish can be caught (either by rake or by shovel) at low tide on tidal flats, tidal pools, harbor/bay sandbars, etc.

Insects?

Defnitely! You will want to catch a few and at least try it out. Just try and use your handy field guide to discern which ones are poisonous to eat. To this end (if you don't happen to have a guide), avoid caterpillars and certain types of butterfly. For safety and sanitary sake, do not eat flies or mosquitoes (both easily carry human diseases). Otherwise, help yourself to all the ants, grubs, worms, spiders and termites that you can get hold of. Just be sure to cook what you intend to eat.

Cats And Dogs?

Now if things get really, really, really bad, start eyeballing the cats first. Dogs too, but dogs are useful as guard animals and for defense, so keeping them out of the stew-pot is pretty justifiable. Cats in a post-collapse world, on the other hand? Most of them are perfectly usable (albeit furry) bags of protein and fat mostly sitting around and just waiting to be eaten – so unless you have a really big mouse or rat problem, consider the cat as a last-ditch emergency source of protein, right after the insects are all eaten.

In The Winter

Finding food in the winter isn't nearly as easy, but it is still, believe it or not, almost as plentiful, if you know where to look. Most edible perennial roots and tubers are still just fine, locked away in the ground. Many other wild edible perennial plants still have perfectly edible dried seeds and even stalks. Insects are largely gone, but many do hibernate, or carry on just fine underground, just below the frost line. Small animals are still out and about, or are hibernating in small dens underground (or in trees)... you just have to know where they are (speaking of which, do you remember my suggestion of a Dachshund as a near-perfect post-collapse dog? Guess what they're excellent at finding --and getting at-- by instinct? Burrowed protein! Small terriers are good at this too, though not nearly as energetic or passionate about it without sufficient training.)

Anything else is going to be tough going, but with just what was presented, you should be able to scrounge enough food to keep you and your family reasonably well-fed, especially when supplemented by the long-term food stores you have hidden at home.

So Why Can't I Just Shoot Deer?

Well, it isn't going to be that easy for starters. Anyone who has hunted for real, outside of special deer farms (you know, "hunting preserves"), will tell you that living almost exclusively from venison is going to be almost impossible in most places, especially out in the Western half of North America. While you may get lucky, you probably won't. Now, add a couple of million suddenly avid hunters with empty bellies out to feed their families and you can probably see why deer is going to be almost impossible to get on your dinner plate. Besides, shooting at deer for food is going to be a stupendous waste of large-bore ammunition (unless you're just that good at bowhunting). That's ammunition you could instead be using when it really counts - like, say, keeping raiders and criminals away from your family and dinner table.

Same with any other popular game animal, really – Elk, Moose, Duck, Geese, Bear (!?), Wild Pig/Boar, you-name-it. It is a safe wager to bet that most of these animals will be very well-hidden, or almost all dead, within eight weeks of a full-on post-collapse event.

Now if you get lucky, or the opportunity presents itself on a silver platter, then certainly – go for it! Just don't expect to get that kind of luck, is all I'm saying.

So What About The Baby?

No, we're not talking about eating the little guy. We're talking about feeding it. But hey - you should be breast-feeding the little bundle anyway. If for some reason you can't (there are rare but good reasons) and there's no formula to be had? It's going to be kind tight here, but there are some alternatives. Goat's milk can stand in as a ready substitute if you boil it for five minutes first to kill off any germs – same with sheep's milk. Cow's milk is a bit too rough for infants, but in a real pinch, boil it for five minutes before feeding. Another option for those babies who can breast-feed is wet-nursing, where you find a willing mother who is already nursing, or has perhaps lost her infant child in the collapse.

Final Notes

Note that for the most part, this information can be freely shared with your community. Encourage them to go looking and procuring. Just one note though – if presented with a share of the kill, or harvest, kindly take only what you think you need, perhaps a bit less and eat that. Same if you stumble across a bounty, especially wild fruits and seasonal bits... this will cast you in a favorable light among your neighbors and will help foster a sense of community.

Anatomy Of Post-Collapse Criminals

Because we are in an environment where food (among many other goods) will be too scarce to support everyone, you can pretty much count on being attacked at least a few times, no matter how remote or secure you think you are. The question is, what kind of attackers can you expect and how will they try to hit you? More importantly, what will this bit of crisis look like?

Good News And Bad News

We'll start with the bad news first... yes, you will have to fight. You will likely have to even kill in order to defend what's yours. You will be assaulted by young men, old women, children, maybe even clergy and, well, you-name-it. You will have to treat them equally when they attack, or face losing what little you have and starving to death yourself.

But – there is good news. The good news is that the period of time where you will have to expect this will be relatively short. As time goes by and the population normalizes, there will be less and less attackers and attempts. Eventually, barring civil war or some sort of conquest attempt, it will drop to a roughly the same level that you would expect of a normal pre-collapse neighborhood.

Armies Or Large Gangs

(50+ fighting individuals)

If you look at most websites which deal with this topic, you will be regaled with prophesies of how street gangs, biker gangs, or what-have-you will start swooping across the countryside, raping and pillaging as they go, *Mad Max* style – and nobody will be safe! (cue ominous music here).

Well, if you live in an area with a low population density, don't expect large gangs or well-organized armed groups swooping down on your neighborhood at after the fifth year, if at all. In areas with higher population densities, you'll likely have to put up with sporadic incidents for the first three years and in major metropolitan areas, likely for at least five to ten years with decreasing intensity. However, few if any will last beyond that. Unless some army forms for the purposes of conquest, civil war, or similar reasons (and in my opinion even if it does), it will become rare at best. Why? Because math and logic prevent it from happening. Seriously.

You see, a large organized criminal organization will be exceedingly tough to keep together post-collapse and even tougher as time goes on. This is for a few rather logical reasons...

- Without a *sustainable* base of operations, any large group would quickly fall apart due to a lack of supplies. They cannot rely on or expect to keep fed from raids, especially from an ever-decreasing number of targets with ever-decreasing supplies. This is why you might see some sort of large raiding force initially, but that over time, they will dwindle and then break up into smaller groups.

- Any traveling group of raiders will consume fuel, food, potable water, ammunition and medical supplies. They will consume more than usual because of the effort and exertions required to travel. The bigger the group, the more supplies will be consumed and the more quickly they will be consumed. This means that such a group cannot possibly move too far before running completely out of supplies and the countryside will have less and less to offer in the way of readily-consumable materials as time goes on, so it won't offer that much help.

- Any large group of active criminals will have their own internal problems, similar to that of any organized criminal enterprise. The biggest problems will be at the leadership level, where leaders will fear their own subordinates more than any external person or force.

- As each individual raid is conducted, massive amounts of ammunition and other resources are used and the results may not produce enough supplies to replace what was spent. The chances of a raid successfully netting more supplies than was spent (let alone enough to carry a large group onwards) are going to be pretty low. As time passes, the odds become worse.

- Let's say you run a large criminal organization. In order to get a large group of selfish criminals to do what you tell them, you have to keep them supplied with loot, resources, or what-have-you. Fail to deliver, or even fail to do it in sufficient quantities and the whole thing begins to fall apart. If the 'troops' feel they can get a better deal elsewhere, they will leave you in a heartbeat, likely stealing 'your' supplies on their way out.

But wait, you ask – certain mafia organizations have been around for generations, so what about those? Well, organized crime relies on one thing above all others in order to survive: A constant supply of resources and victims. Post-collapse, both will initially be in very large quantities, but will rapidly decline to nearly nothing and in quantities that are not enough to support a large criminal organization.

Okay, but what about folks like Attila The Hun? He managed to rampage through Rome and come away with a ton of stuff and *do it multiple times*. Attila had the exact same problem that most large criminal organizations post-collapse will... you have to keep the troops fed and full of loot, else they will desert you. Attila also had one other advantage: He conducted his raids on a (by then) poorly-defended but still relatively prosperous nation (Rome). This meant an endless supply of goods that renewed itself somewhat rapidly. Post-collapse, no large criminal organization will have that option for at least 50 years or more... likely more.

Now initially, organized and individual criminals alike will have a very easy time of it. There will be literal hordes of potential victims, a whole lot of resources to pick from and the chaos will be such that any organized defense and/or law enforcement will be essentially non-existent. But things will change over time and in favor of the smaller organizations. Let's look at how that happens...

Large criminal organizations will initially thrive quite comfortably alongside the small-time and individual criminals in such environments for awhile and might even grow from smaller groups coming together, so long as there is a strong leadership to form around. However, this is what starts to happen: The larger criminal organizations will likely settle into one or two spots, concentrating their resources into a base of operations and will then try to control or co-opt the smaller groups. The reason why is simple: When you start getting a stockpile of resources, it becomes very tough to move the stuff around and at the same time defend it against other thieves. The larger the organization, the more vital it becomes to keep the defenses up and keep the supplies coming in. This in turn limits mobility, forcing the larger group to use its size in order to impose its will.

However, this doesn't mean that large criminal groups can simply settle in. Over time, everyone will out of necessity start consuming their hoarded resources (be they criminal or not). As these supplies dwindle, more will be needed, so raiding parties will start moving further from their base to collect more goods. At first, they will avoid outright any areas that are defended, because there will be so many non-defended areas around to pick clean. However, once the low-hanging fruit is stripped, the group will have to start doing two things: travel further for the easy pickings, or start fighting to get at the supplies which are better-defended. The problems in doing so? Going further means burning resources for the round trip and/or risking casualties from those raids which involve hitting defended targets.

As time goes by, supplies are still being consumed, by everyone. The rate of consumption might slow down, but can only decline so much for the criminal types before manpower and combat effectiveness is affected. This means the demand for new supplies will never cease and getting more via theft will require more resources spent to get them, in return for ever smaller results (because the victims are also consuming the supplies that are being targeted). This can be alleviated for awhile by co-opting or controlling smaller groups that go out and do the 'foraging', but because criminals will just as readily screw each other over as anyone else, the supplies will begin to get 'cut' (watered down, mixed with inert ingredients, mixed with goods of questionable quality, etc) and the quality will drop right along with the quantity, so this also becomes non-sustainable.

Eventually, the number of targets becomes too low, the quality and quantity of supplies captured will be too low and those supplies which are not stolen will become either too heavily defended or too remote for any worthwhile return. This will cause serious problems and as soon as critical supplies start running out, it will cause the disruption and disintegration of any large criminal organization.

Long story short, expect any large criminal organization to start breaking up after a relatively short period of time. Best bet will be that it fails from internal strife, most likely touched off by lack of sufficient supplies to keep it all held together.

The Medium Criminal Enterprise

(20-50 fighting individuals)

This size will be the largest group that you can expect to see in the distant future. It is still too large to move very far naturally, but it is small enough that it can move if it had to. The ability to field up to 50 raiders at once can be done in an emergency, but won't be too likely, unless the raid is on something deemed worthwhile, such as an armory or warehouse complex. The reason why lies in transportation – unless the raiders intend to remain in the place being looted, it ends up costing more in expended resources than can be gained. The reason for that is logistics: You have to burn a lot of fuel to transport that many 'soldiers' there and back; you have to burn a lot of ammunition to attempt taking something that is probably (and will be over time) heavily defended; you have to burn even more fuel to drag all the loot back to your home base. So, if that many people are coming at your town for your stuff, expect them to stay if they win.

Another reason why you won't expect to see all hands on deck, so to speak, is because someone still has to stay back and guard the stuff they do have. See, a group this size will still have a lot of the same problems of the larger group. One of them is a need to store and guard sufficient supplies to keep from 20 to 50 people fed and warm for any period of time longer than a couple of days. The last thing a budding criminal empire wants is to have all their troops in the field, steal some small amount of supplies, only to come back and find that their own private stash is gone. This means that unless the group is migrating, expect to see only half (at most) that amount of attackers coming after you and your community.

The good news here is, a strong community of, say 75-100 people can field a defense of equal or larger size if the community is fully armed and follows the practice of scavenging from dead and wounded attackers after a fight. A small town of 3,000 people, armed about as much as a typical rural town, can easily field enough defenders to handle even a full complement of 50 attackers (the example being a migrating criminal group looking to live in your town). If you as a community can take out half of an attack force of any size, the other half will likely start thinking about retreat. Because large groups will fall apart quickly due to lack of supplies (especially if traveling), you really shouldn't expect a criminal group larger than perhaps 25 people or so and in rare cases up to 50. This is not set in stone, but should be the maximum in most practical situations.

Because of this, criminal groups of this size and smaller will likely prefer to ambush anyone who wanders out of the community, attack outlying homes in a hit-and-run fashion (staying would only invite reprisal from the closest community) and conduct raids where they sneak in for a smash-and-grab raid, preferring to run instead of fight and only fighting when they have to.

The Small Criminal Enterprise

(5 - 20 fighting individuals)

This will be the most common type of "gang", because once you surpass two dozen fighting individuals, you start collecting hangers-on, 'camp followers', families of these guys and the even without them the logistics begin to get too large to handle in a post-collapse world.

As discussed earlier, larger groups simply aren't able to travel well and staying in one place won't allow a criminal to remain one for very long without dying. On the other hand, a group of between five and twenty fighting individuals can move quite effectively. This is because they would need only a (relative) handful of vehicles to transport their gear and themselves and don't require the supplies that larger groups do, making it somewhat easier to transport all of their supplies without being forced to starve between raids.

The 'best' size would be somewhere around 10-15. This means enough people to conduct an effective raid against single, isolated homes without fear of too many casualties, or a quiet raid sneaking into some small community's food supplies. At the same time, the size is sufficient to be assured of enough loot to keep the gang fed. It also means that once a home is successfully taken over, the gang can all stay in it for a night or two before moving on (assuming the target's home is isolated enough to not attract attention from neighbors). This is part of the reason why, if you want to survive, you need a community and don't try to go it alone... one or two adults trying to hold off a force of 5-10 determined individuals might be possible with a bit of luck, but don't count on it.

A well-armed and well-disciplined community can very easily fend off and even hunt down a group of this size and it would be well worthwhile to hunt them down if it can be done safely. However, beware of being drawn out into any traps or ambush, should you decide to pursue any retreating forces.

This size of group (5-20 individuals) is likely going to be the second most common size you'll have to worry about and certainly the most dangerous.

The Independent Operators

(1 – 5 fighting individuals)

This is going to be the most numerous type of raider group: Enough people to fit into a vehicle, carry out isolated attacks and smash-and-grab raids, but without the need for hardly any supplies over the long haul.

The down side of this size is that there are too few people to really do too much. Five individuals attacking at once may be able to pull off a few smash-and-grab raids if they are sneaky enough, but as the number gets smaller, the chances of success shrink rapidly. A group this small, if combat-trained, can overpower a house however. This means you really shouldn't expect any direct confrontation against a community. These folks will however lie in ambush on the roads, will be the ones trying to sneak up on your place at night and may even try to con their way into your supply stores. The smaller the group, the more cautious they will be.

Most single-family homesteads with at least three armed adults can, with enough warning, actually fend off a group of this size, so long as the attackers aren't combat trained or the like. A group this size will probably run at the first sign of trouble anyway - unless there is something very large at stake, they are that desperate, or if they think they are close enough to their prize to seize it.

Common Types

Now that we've got the basic sizing down, let's look at the types for a moment. You're going to have a few basic ones here...

- Street Punks are going to be a whole lot of hot air, a whole lot of bravado, but very little skill in actual full-blown combat. Sometimes you may be surprised by their creativity, but overall, they have very little skill with slugging it out in full frontal combat. When faced, most will blindly throw lead in your direction while retreating as quickly as possible. They will only press an attack if they think the odds are in their favor. Decent at hand-to-hand combat, but lousy with weapons and tactics. Mostly the smash-and-grab types and are only truly brave when not faced with adversity or a reversal of fortune. Common in larger metropolitan areas and suburbs and rarely seen outside of these areas.

- Bikers differ from Street Punks in the sense that they often do have the skills to back up the talk, though most of these skills are centered in hand-to-hand combat and combat with non-firearm weapons. They also differ in the sense that they will not back down as easily from a fight. These are more common in smaller towns and rural areas and tend to be a lot more mobile than any of the other groups. The downside is that they are more of a frontal-assault type of group, but can and do practice deception well if needed. They will sneak in and out for a smash-and-grab raid when they can, especially if the numbers are not in their favor.

- Desperadoes has nothing left to lose. It may be a group of folks trying to keep their own families fed, or any otherwise ordinary group of people who have come to the conclusion that they have no other means of staying alive. As such, they will slug it out and will only reluctantly leave a fight. They can be reasoned with under some circumstances, but as time goes by, the chances of that happening will drop to nil.

- Ex-Soldiers are going to range from wannabes or has-beens (who were the toughest quartermaster aides to ever go to war) to actual uniformed killing professionals, whose skills are amplified by a factor of 10 due to a thorough knowledge of tactics, combat experience, discipline and training. If you find this kind of criminal, they will be the absolute most dangerous to fight and fend off. However, before you automatically equate everyone in a uniform with a sharp combat killing machine? I have some good news: they will be far fewer than you think, due to something called the "Tooth to Tail" ratio. For example, each actual combat US infantry veteran in Iraq or Afghanistan had anywhere from 10-20 non-combat military personnel to keep them fighting. It breaks down to roughly 1:10 Marines, or 1:20 US Army veterans of Iraq/Afghanistan that had actually went into a combat area, let alone pointed a rifle at another human being and pulled the trigger. That said, the number will shrink further... once you remove those combat veterans who are at home (or at another place they will eventually call home) defending their families and communities, are stationed overseas and unable to return home and the like. Consider further the reduced effectiveness of permanently injured veterans, discharged veterans who have let their physical training lapse to that of any other civilian and the usual attrition due to death, age, sickness and whatever? The actual number of the high-end soldiers you would have to worry about drops to a very small relative number indeed. It gets even smaller when you factor in the higher morals that the typical solider does have.

One exception to this would be a small group of soldiers (most likely national guardsmen) caught in the whole collapse far from home and have banded together to keep themselves alive by any means possible. However, a group of this nature will most likely try to win confidence and use deception to gather or "confiscate" supplies, instead of simply going in and taking it by direct force.

- The Ex-Cops may be the biggest problem, though these will likely not congregate in groups, but may be part of another group. Like most people in any profession, most policemen are going to be with family. The skills to watch out for here is the exercise of authority (but this time used in a bad way), knowledge of how to manipulate groups and hand-to-hand combat skills. They will most likely try to gain supplies by deception rather than force, pretending to be an authority figure of sorts.

- The Con Artist is an individual or small group of people who will appear to be harmless, or even benevolent, but their true nature is known once they get their hands on the supplies and the opportunity to steal them comes along. You will probably not see this type of individual or group in a direct confrontation unless there is no other choice. They rarely congregate in large groups (perhaps 10-15 at absolute most). However, they can likely be found as part of a larger group.

- The Hanger-On is similar to an arctic fox... Oh - arctic foxes hang around Polar Bears (or in our case, stronger criminals) and scavenge whatever the stronger of the two doesn't eat. They're not really a part of the main criminal group, but may participate (or at least be tolerated) to bolster a group's size and apparent strength. You will rarely if ever find a group entirely comprised of this type of criminal.

- The Colonists describes (very loosely) a group that attacks a community *en masse*, with the full intention of not just taking supplies and running off with them, but to completely take over the community as a whole. These will be large groups with the intention of either driving you off or killing you off, so that they can move in and make a life for themselves. Initially, these groups will be very few indeed, since no one is going to have much of anything worth taking over. On the other hand, as time drags on, these groups may become more and more prevalent. They will likely be comprised of either refugees who have banded together, or of another community that was driven off and is now facing failure and starvation.

    This will be the toughest and in some ways the easiest type of attackers to defeat. They will be easy because they have families and non-combatants nearby that will need defending, but will be the hardest due to sheer numbers. If the attacking group is small enough, they may try to first attempt to reason with you, to try and join your group. If you have the spare resources and can use the manpower, there's not a lot of harm in first insuring their intentions are benign, then letting them join. However, verify everything first and take a good hard look around to insure there aren't larger numbers waiting nearby. Otherwise, if you do not have resources or do not trust the intentions at all, you must act quickly. Promptly disarm and drive them out if they object too much to any refusal and perhaps send a large party to attack and drive off any encampments, to give yourself some distance and therefore time. Even if you have no intention to attack, send an armed party to ascertain their size, position and fighting ability. This is not to be merciless or cruel, but to defend yourself as thoroughly as you can.

Colonists can run the whole gamut, but will likely include families – children and elderly, as well as adults.

Professionals Versus Amateurs Versus...

The whole spectrum, no matter what type, will have a range from true professionals at assaults and theft, down to complete amateurs. The amateurs will either fight until they're all dead, or (most likely) will run to the hills at the first sign of losing. The professionals will know enough to avoid any targets that they run a high risk of loss in attacking, but will stick with a fight as long as they are able to see victory. You can likely drive away amateurs with as little as 10-20% of casualties, but the pros will stick around until their casualty rate is well over 60% or more, depending on how desperate they are.

The rank amateurs will be relatively easy to see approaching and you'll likely spot them before they see you. The fully professional you will likely not know about until someone gets attacked. You can lessen your odds of being surprised by training everyone in your community to always keep their weapons handy, know when to raise the alarm and to know almost instinctively that no one should ever investigate anything odd alone.

Amateurs will attack you on your terms. Professionals will try and get you to fight on their terms. To this end, never put yourself in a position where that latter part can happen. For instance, be very wary of traps, ambushes and things that either look too good to be true, or will appear too obvious. For instance, if a group attacks and then runs away quickly, never follow or chase them directly. Always send parties after them in a roundabout way. Always go around the direction that the attackers run off in. This way, if a force attacks you and runs off, going *around* their path of retreat means you're less likely to run into a trap or ambush by a larger force laying in wait. As another example, if you're being attacked at an obvious strong-point, keep your eyes peeled (and defenders posted) in all other directions, lest attackers try and sneak up behind you, or hit you from a side where you're suddenly the weakest. Also, if you find yourself fighting off attackers and they retreat, do not immediately let your guard down and call it a day. Stay on high alert for at least 24 hours afterward and send out small scouting patrols to *quietly* look around for signs of waiting attackers. The last thing you want is to successfully fend off a small attacker force, only to let your guard down just afterward and get hit by a much larger force three hours later.

Hitting Once Versus ' Keep Trying'

Most of the raider types will initially try and hit you once and if they fail, will move on to another target. However, as times become very lean, they will likely come back again and again, testing your defenses almost constantly and trying to get whatever they can.

This can take the form of continuous or at least periodic attacks, or can be a one-time thing. If you consistently defeat them, they may eventually stop trying (unless something changes such as any alliances, an increase in numbers, lack of options on their part, etc). The reason why is that they will run out of resources as the fight goes on and they in turn will be forced to decide if you're worth the effort, or if there are easier pickings nearby to go after. Their solution could be a hybrid one as well - they could break off, go pick off an isolated homestead or two, get more allies, then come back once they gain enough resources to make another attempt.

If you do see constant attacks by the same party, especially the smash-and grab type of attacks? It may pay to have (if you can spare it) a scouting party following that raiding party back to their camp (discreetly of course). If you know where they are camped, you can get an idea as to how to flush them out, burn them out, or in general do unto them before they do unto you (again).

Finally, if you notice that all the attacks are of the smash-and-grab variety, you may want to re-situate your more critical supplies to somewhere a lot more defnsible, if possible.

As Time Goes On...

As Darwin's theory takes its toll on the population at large, criminals will also feel the effects of this, though they will likely feel it far faster, especially after all the easy targets are bled dry. The dumb, the foolhardy and the poseurs will be killed off in ever-increasing numbers - by defenders, the elements, or by simple bad luck. As time drags on, your chances of coming across an attacking party that is easy to drive off will become less and less and those you do find later on will be tougher, smarter and more skilled. A criminal post-collapse will either have to learn fast, be extremely lucky, be extremely skilled, or will be extremely dead.

Fortunately, you and your community will also, with luck and intelligence, become smarter, more skilled and tougher. You won't have much choice in this. With time, you should be able to identify weak spots in your perimeter(s), build up defensive positions (yes, even further than you have done already) and become more skilled in combat.

The criminals will eventually evolve as well. They will have to, since those who live solely by raiding will begin to die off or become desperate but unable to do anything about it (and die anyway). Some groups may either try to integrate into any nearby remnants of society, or they may try to become a community themselves, supplementing their supplies with occasional raids on smaller neighboring communities and on outlying settlements.

In either case, always be wary of any friendly overtures until you are certain that the group will not attack or raid from your supplies. I suggest being on alert for any new settlements nearby for at least a full year, just in case.

When The Government Comes Knocking

This is bound to happen to a lot of communities: You finally get something going, things start looking up, there's a small sliver of hope and then? A division of soldiers from what used to be the government shows up. They promptly arrive with some heavy military equipment, take over and perhaps demand that everyone either cough up their supplies for redistribution, or confiscates them all in the name of 'keeping the peace' (especially when it comes to the weapons). What to do? How to prevent it? How to shove them out if you have to? Should you resist? Well...

Let's Get One Thing Clear

Before we go any further, I want to say that the following paragraphs are only going to be true after civilization begins to collapse. In pre-collapse times, the US military consists of the most stand-up group of folks you will ever have the pleasure and privilege to meet. They will go out of their way to help you out if you need it and are definitely a good group of folks to have as friends and neighbors - I daresay among the best choice if you can swing it. In warfare, these soldiers, sailors and airmen have competent training concerning how to keep the peace in occupied towns and cities and for the most part do their best to leave the more peaceful inhabitants alone. Certainly there are individual incidents caused by idiots, but those get punished in short order and far more severely than you would see a civilian face. By and large, unless there is a civil war in progress, these are going to be the good guys. As a veteran myself, I can, without hesitation, vouch for them as they are today - both as a group and as general individuals and I happily and proudly count myself among their ranks.

That said? In a post-collapse situation, things aren't going to be so clear. When supplies run low (or are practically gone), if there is a civil war on, or if there is no central government or coherent chain of command any longer? Things may well get a little hinkey and perhaps even dangerous. These guys have all taken an oath to defend the US Constitution against all enemies, foreign or domestic, but the problem lies in the fact that this hallowed document may no longer be in force (or adhered to) in a collapsing and post-collapse world. No matter your feelings, this chapter is dedicated towards how one deals with the 'active' military when there is nothing in the way of an effective or functioning government (or perhaps even a chain of command).

Rule #1: Don't Be Stupid

Standing your ground and trying to fend off tanks, artillery, large machine guns and the like? That will only get you and your families turned into a bunch of corpses and/or refugees. Besides, it will be extremely hard to tell what shape the government is truly in if you don't have a look first, they can actually be useful to you and unless you know a bit more, you may be passing up a golden opportunity.

Any commander with at least two working neurons will avoid stirring up the local population. The last thing he needs is to have to burn ammunition and fuel trying to keep himself and his troops alive and ongoing guerilla warfare is a headache that no commander wants or needs. Also, even if the government is well and truly dead, those soldiers are not automatically enemies. They can be, certainly – especially if resources are very thin (or gone) and they're just sitting around consuming what's left. But... they can also integrate with your community as permanent residents... though only make that invitation if you have enough natural resources around to allow it. Either way, fighting them off is a one-way ticket to death and I'm pretty sure that death is the exact opposite of our goal to survive, no?

How To Tell What Shape The Government Is In

Now, I don't mean believing whatever the commander says, at all. Take any announcements with a block of salt, because you'll be fed more hope and sunshine than you can stand to hear. Commanders aren't stupid, after all; they want you to be calm, but most of all, they want you to be both calm and compliant. If they're telling the truth, you'll know quickly. If they sound a bit too optimistic, you will find that out eventually. If they're lying, you'll certainly find that out soon enough as well.

The best way to tell how the government is doing is by taking a look at the troops, their equipment, their supplies (including arrivals of same) and their gear. If the troops and their gear look ragged and worn, odds are good they're not being re-supplied or reinforced, which means there's little-to-no central command or sufficient supplies to do that for them. How much equipment did they bring with them, anyway? If they didn't bring their own non-combat supplies (food, shelter, etc) and there's no regular shipments coming in? Guess where they're going to get them eventually? That's right... from you. It is also a sign that the government is well and truly dead. Even a platoon-level complement of soldiers will have an impressive amount of supplies that they bring along and resupply is almost constant while they're on the move. While you're being nosy, try and figure out (*quietly!*) how much ammunition they keep around per soldier. If the load-out per soldier is less than 40 rounds (two spare magazines), things are okay, but not looking good. If it is less than 20 rounds not already loaded in an M-16 (one spare magazine), things are bad. If all they have is what's already loaded in each of their rifles? Start worrying.

If the equipment (tanks, trucks, whatever) isn't driven all that much, missing parts, are cannibalized and are parked immediately on arrival but then not moved, even when it makes more logical (and tactical) sense to move them? Again, odds are good that supplies are lacking and that resupply will likely be very late, or not come at all.

If resupply doesn't come within roughly 36 hours of their arrival, they're going to start running out of what they do have and soon after they will then start demanding that you cough up whatever you can. These folks should be self-sufficient insofar as supplies, needing only space and water from you. If they want anything more, it is a sign that things are bad.

So Why Not Relax, Just Do As They Say?

The reason why automatic compliance is bad? That's simple: The government may be well and truly dead. If it is indeed dead, then the last thing you want is a bunch of combat-trained folks with M-16s sitting around and facing starvation once their supplies run out. Always, always, always retain control over your own weapons and supplies, even if you have to hide them.

How To Respond To Questions (Individual Level)

Q: *Do you have any guns?*
A: Only my .22 rifle, which I use for hunting rabbits and if needed, for defense. Are you asking me to give up the only means of feeding my family? How will my family eat if you do that? Are you going to feed us? If so, where is the food?

Q: *How much food do you have in your household?*
A: Only a couple weeks' worth of canned food left, but otherwise I gather what I can from the woods, garden, etc to stretch it out. We eat every other day or so, but could use some real food on a more regular basis. Do you have any to spare? If not, when will we see some?

You can see where this is going, right? Hopefully, you've been hiding your supplies and weapons (quickly but quietly bury them if you have to!), the moment you've reliably heard about or seen government troops arriving. Let them see and know nothing more than just a poor family (or if just you, a poor individual) trying to survive a bad situation. You can be perfectly honest about most other things, but just like you shouldn't tell your neighbors how much inventory you have, you really shouldn't tell the soldiers such things either.

Try to ask as many questions as they do, perhaps more. Get as much news as you can out of them, but above all, be perfectly friendly and kind about it. Pretend to submit, but always keep your eyes and ears open.

How To Respond To Questions (As a Community Leader):

Q: *blah blah blah, your duty is to blah blah, so can you do...?*
A: Be friendly and (mostly) honest, but...

Never even hint at how many community stores you have and hide as much as you quietly and safely can. Never reveal any specific plans or routines you have for defending yourselves (a good commander will already know anyway).

Always speak up for the community. Negotiate always on their behalf. Speak out immediately for them. Look out for their interests. If anything is amiss, be sure to relay that info and opinion to the community, in any way that you can safely and quietly do so. Stand up to any threats politely and respectfully, but do not back down in the face of them. Try to turn the conversation to negotiation whenever possible. Never speak to any commanding officers alone – always have at least two witnesses from your community present. Do your level best to demand – and get - private community meetings, with no soldiers or military representatives present. Make them up ad-hoc if you're not allowed to do it openly. Explain that this is to prevent intimidation and as a way to digest community complaints into something that can be presented to the commander(s) You may not get what you ask for, but be damned sure to try.

One other note: If you get any offers from the commander that benefits you personally, but will screw over the community in any way at all? Refuse them, categorically and completely. Without your community, you're as good as dead. If you're stupid enough to do it anyway, then without those troops around keeping order you're as good as dead, when (not "if", when) the community finds out. Also know full well that those troops won't always be around to protect you if you do get stupid - no matter how many assurances you hear otherwise. Long story short – don't be stupid. Always look out for your community, no matter what. They will be here long after the troops are gone or dead and you need them more than they need you.

What To Expect Rules-Wise

Expect that some rules will be imposed. This will likely include things like curfews, the ability/right of civilians to carry weapons in town (unless heading directly out for hunting, etc), the designation of certain areas to be off-limits for any civilians and how any relief supplies that come in are to be distributed (assuming there actually are any). They will most likely spell everything out up-front and make sure everyone has a chance to learn about them.

You can also pretty much expect a census, in which everyone will have to give their name, tax ID number (social security #), age, state of health, number of kids, residence (or original residence), etc. They likely won't ask for too much documentation if you don't have any, but what you do have they'll want to see.

How To Hide Things

There may (after settling-in) be a house-by-house inspection for 'contraband' goods, but this likely won't happen initially. This can give everyone a small window of time to hide things as deeply as possible, only taking out what they need, only when they need it. On the other hand, everyone should hide the 'contraband' (food, weapons, ammunition, etc) as soon as you all can quietly and safely do so, period.

One big caveat, though – if there are troops knocking at your door, don't try to quickly hide things... by then it's going to be too late. Just rely on what's already hidden and don't let on that anything actually is hidden. Be as friendly as you can and happily pretend to trust them while you show them what you have that isn't hidden. The idea here is to not arouse suspicion and to give the impression that you're cooperating fully. Doing so helps them to relax and doesn't make them want to tear your home up looking for stuff, which would quite frankly be a bad thing. It likely wouldn't hurt to have a couple of things shoved into a closet or two, just to throw them off a little.

Otherwise, look for places like crawlspaces, gaps in the inside your interior walls between studs, the attic (spread thinly under the insulation), up in the bed box-spring sets, or even buried. Use your imagination and hopefully you've already hidden the critical bits anyway (like we asked you to do this far, far, far earlier in the book, you know?)

What To Expect Otherwise

Justice matters of a non-petty nature may well be taken over entirely by the troops that move in. They may continue to let your community's civilian leadership handle incidents of a non-violent nature, so if that happens, always handle such issues in as close an accordance to existing (pre-collapse) laws as possible. However, definitely expect any violent acts or thefts that occur to be handled by the troops and not you or your community. Note that the punishments may be softer or harsher than you might expect. Instead of being driven out of the community minus supplies for, say, stealing food? The offender may (possibly) be executed outright, or may simply be ' fined', or perhaps nothing at all.

One thing you really should do is to expect that the troops are almost all going to want to go home as time passes. Reach out to them individually and empathize. Lend a friendly ear. Bring them home-cooked meals from what you manage to scrounge up, or once in a very great while make something tasty from your stores (if they haven't tried to inventory them) and explain that you're grateful they're here, explaining that you used, say, the last can of pie filling you had to make them the item... we'll go into detail as to why in a bit.

Your Defense Is Not Taken Care Of!

While you can be rest assured that these guys will defend the land they're sitting on, don't assume they will defend you. They may be more concerned with their own defense and might not care all that much if you and yours gets raided – especially in outlying areas. Then again, they may care. It all depends on how close to dead the government really is, how much of a conscience the commander has, what his plans are long-term and/or how many resources (read: troops, weapons, supplies) there are to enforce the peace.

Communicating With The Commander

If you're *not* a community leader, don't go up to the commanders and do not complain or suggest anything negative or out-of-the-ordinary to any of the individual troops. Go instead to your community leadership and have any complaints or suggestions passed up that way. You do not want to give the occupying troops any reason or even a hint that you are not a unified community, or to give the commander any excuse to dissolve the civilian leadership and replace it with a military one. Always, always, always stand together. Trust me, you do not want the alternative, because the alternative will eventually starve you. Handle all community affairs with only the community - even if you're angry, you're frustrated, or feeling like you got the short end of the stick.

If you *are* a community leader, always take the time to listen to your community. Insure that all complaints go through you. Solve and fix whatever you can without involving the troops, period. Only if you can't solve it yourself, or if the problem is with the troops, do you take it to the commander. Be as calm and rational about it as possible and do not threaten, cajole, or shout. Demand honest answers and be honest about what you're presenting (so long as it doesn't involve divulging any hidden stores). The last thing you want is for your community members to start griping to the commander directly, because that puts your leadership in peril. The best way to avoid that is to be responsive, give honest answers in both directions (to the commander and to the community) and always be on the straight and level, with no weasel-wording or political double-speak.

One more thing – if you're part of the community leadership, go out of your way to get to know not only the commander, but the second-in-command, the third-in-command and basically all the officers you see and come into contact with. Do this both professionally and personally. Doing so means that if there is any trouble with the commander later on, you may be able to have the underlings do your community a favor and get the commander to see things your way.

In Case Of Civil War...

A rotten situation all around, to be sure. If you do find yourself in the middle of a shooting war and the government (on either side of the fight) comes to town? You'd better be prepared to set aside any ideology or passions and do what you're asked for as long as you're in artillery range. This isn't to say you need to become cowards, but it is to say that you as a simple community do not have enough firepower to hold them off and the last thing you need is for them to plow your whole town under, or turn it into a battlefield. While we're at it, do not even give the slightest hint of resentment. Complaints, certainly... those are expected. But don't go overboard on them and instruct the community to be as gracious as humanly possible to the troops themselves.

A good initial rule to follow: Do. Not. Take. Sides. Your goal is to out-live these guys until their government or movement is completely dead and not to die in a flaming blaze of glory that no living soul outside your area will know or remember. If you're stupid enough to take sides and your side loses, guess what happens to you, your family, your home...? Dead from gunshot, bomb blast, eventual exposure, or starvation. The idea is to be as neutral as you can possibly be, at the very least until you can completely size up who is who, how big it is and what's at stake.

If pressed, proclaim your allegiance to the Constitution of your nation and nothing more. Every soldier you meet has (well, pre-collapse) taken an oath to do the very same thing and this is a no-brainer. If pressed to assist the troops in any way, explain that your community is patently unable to fight and that unless the sides are *very* clear, you need to keep them out of it due to the need to care for the women, children, elderly, etc. At this point, it doesn't hurt to have the men and teenaged boys start looking sickly, or at least make them scarce - otherwise they may find themselves conscripted.

If you find yourself in a live battlefield (or are about to), grab what you can, load what you can and get out of there as a community. You have the option of hiding somewhere out of the way of the oncoming troops until they pass (or die off, etc), or else you need to get together, figure out where to go as a group and take as much as you can (but quickly!) off to that new place.

Informants And Spies

You would be amazed at how quickly and easily someone would sell their family (let alone community) down the creek for a few extra MRE packets. If this happens to you, do yourself a favor and at least try not to kill the snitch outright. Doing so only invites reprisals, which you and your community definitely do not want. Instead, wait a week or two until things calm down. Find out who the informant is for certain, but do not let on to anyone outside the community leadership and most importantly, *let the leadership handle it from there.*

An alternative to killing the informant and hiding the corpse (yes, alternatives do exist) is to quietly feed the informant subtle but false information and remove him from any community meetings. This will destroy any report's reliability to the commander and cause him to discount anything the informant says in the future. It also has the benefit of making the occupying troops do what you want them to do, if the fake information is engineered right. This in turn makes for a very powerful tool in skilled hands.

Another alternative is to publicly out the informant to everyone and their dog, which means he will end up ostracized and therefore useless as an informant. If the informant is a community leader, out the person, remove any privileges and authority from that person and as a community put him to the lowest form of manual community labor you can find.

As Things Break Down

This is why the community needs to be on its best behavior the whole time, in order to avoid becoming targets as things deteriorate. From the day after they arrive, have the community reach out to the troops individually and be as friendly as you possibly can. Empathize with them and sympathize. Offer to host individual troops for dinner if you can do it and to give them an occasional warm bed to sleep in if you can spare it (all without compromising anything hidden, of course). If you can swing it without betraying the hidden supplies, make them home-cooked meals and treats. Talk to them about their families. If any romantic interests bloom between a soldier and one of the ladies in town (trust me – it'll probably happen), don't immediately discourage it. Size up the gent in question and if he's an honest and kind person underneath that uniform, just let it happen. You might need those romantic entanglements later on.

You will find that over time, as collapse progresses, that things will change when it comes to the troops. Some will begin to desert, mostly to try and get to their homes. Any relief supplies that were coming in will stop coming in. Many will try to integrate with your community. Others may form small squads to raid supplies (always be sure the commander is aware of any attempts - if you've hidden your supplies right, the most such troops would get are a few token meals). Communications and news will begin to dry up. The military leadership may start getting more and more nervous. Things will likely break down slowly at first, but then will crash rapidly. There's no real way to tell in words when the inflection point will hit, but once it does, troops will desert in large groups and you may start seeing the larger equipment and weapons either be abandoned, or taken along with the deserting troops.

Do what you can to give the commander the idea that maybe he and his remaining troops should either integrate with the community, or to 'leave the nice people alone', or quietly disband his troops and move on. Go out of your way a bit to help make running the town easier, wherever and whenever you as a community can do so. This means that he won't have to and be more likely to stay out of your business while he spends more time tending towards the disorder in his own ranks.

Note that this isn't always going to happen. A commander facing the stark reality of his government being dead may react in a very bad way... and this possibility is why you've been hiding the critical stuff all this time. You will have to keep your eyes peeled for signs that things are not going to go well. If the guy seems to be considering options which may put your town in a bad way, you don't have many peaceful options, but you do have some. First, try to reason with the guy. If he still has his sanity, he will listen. If he won't listen, or refuses to take any peaceful course of action, try to reason with the second in command and so on (you're on good terms with them, right?) The idea is to find someone sane in that small chain of command that will agree to (and enforce) the most peaceful option you can muster and back that person up with the whole community. If all else fails, have the community start working their connections to the individual troops themselves. You could still have a chance to quickly foment a mutiny and hopefully one strong enough to break the plans of any commander who is out to destroy your community, or at least whose plans will destroy it. The reason why a mutiny is possible is because not all officers are going to crack up – at least not at the same rate and many will likely disagree strongly with any plans that involve pillaging or otherwise damaging your community.

The Overall Idea

Just because troops from a dying government show up, does not mean you have to suddenly go all hostile and guerilla-warfare on them. In fact, they are, at base, another potential resource that you can use until it's gone. For the most part, this new group of folks will be full of people who will at first do their job as best they can, but eventually will just want to go home. Take advantage of this - go out of your way to have the troops see you as ordinary people... and not as targets. You don't want resentment, or any sort of us-versus-them mentality going on, especially when the other party has deadly skills and weapons that you don't.

The final reason to do all of this is because perhaps, just perhaps... the government might possibly recover and civilization along with it. It will likely continue to crash, but just in case, in such an ugly situation it always pays to have friends with firepower.

Civil Duties And Tips

While not meant to be too comprehensive, this chapter is devoted to the basic civil duties and principles that a community needs to perform together and keep in mind, if you are going to survive in a post-collapse world. Many of these are going to be in the planning stage now, since you're going to be busy trying to get the community together into something that functions. However, start thinking ahead and putting folks in mind for these roles, because you will find a use for all of them as time goes by. Note that for now, a lot of this is going to be tentative, because up until the six-month mark or later, temporary is about as good as you can hope for.

So let's break it down a bit...

Defense

Pretty obvious, isn't it? Well, almost... What we're talking about is both defense against external raiders and intruders and offensive attacks against those same folks who may be camped close-by. It also means eventually getting out there and hunting down those who have been happily raiding/ambushing the roads and pathways all this time. The ideal candidates for defenders should be at least 16 years old, armed and capable of fighting when the issue arises. However, everyone who can should pitch in, no matter what other job they have (with few exceptions such as medics). Dedicated soldiers (that is, folks who do only defense as a community role) are going to be few, but do find those in your community who are honest and has experience in combat. If they're competent, have those people look over the defenses and work on improvements to it.

Water Supplies

Okay, so you think that if you dig a well you're all done, right? *Wrong*. You have to insure that the water supplies are and remain, clean. This means a bit of engineering, but also vigilance. Keep your water supply well away from any sources of pollution or contamination. The last thing you want is for someone infected with cholera or typhoid to defecate upstream of where everyone draws their drinking water. It also helps to keep any outhouses or dumps far, far away from any water wells, because the ground may not be capable of filtering everything.

What you will need is for someone to keep an eye out for any contaminants getting too close or upstream of your water supplies. This person will also need to periodically check the quality of these water sources (no need for chemicals, just check clarity and, after boiling, taste).

Sanitation

Simply digging a hole in the backyard will not take care of all the trash, excrement, dead animal parts and other garbage that will certainly pile up over time. You will need someone (or a bunch of someones) to arrange for new places to safely dump community trash, find places for compost piles, recycling (no, this isn't some sort of hippie advice, it's common sense – especially when it comes to metals that you can re-use and large items that you can scavenge later on) and the like. By the way, this also means finding a new, permanent home for human corpses - both of those who die among you and those who got killed trying to attack you. We'll cover that in more detail later on.

Police/Justice System

This is (eventually) going to mean a bit more than simply a few guys with handcuffs who round up and punish criminals. You will need people capable of jury duty if you can spare the time and someone capable of being an impartial (as possible, given the circumstances) judge. Initially, it'll be tough to follow due process even partially, if at all. However, as time goes by, you're going to want to take these things into consideration, so you may as well start thinking ahead for it.

One role you will want these folks to perform is to act as a market judge. If you open a marketplace for barter, it is important to have someone on hand to settle disputes (not to set a price, but to settle the basic disputes that will arise in any post-collapse marketplace).

Engineering/Construction/Repair

Okay, so what does this mean? Well, these folks can reconstruct running water. They can provide electricity to those things which are still otherwise working and are highly useful. They can help provide and improve irrigation to gardens and fields. They can help prevent damage from flooding. They can improve how efficiently a home is heated. They can clean out the storm drains, or reconstruct them. They can, in short, provide creative means to reconstruct a lot of the things that make life at least halfway comfortable and worth living for everyone. The idea here is to identify those folks with the skills to get the job designed and built right. If they have time and energy to start doing some of it, by all means let them at it.

Scrounging (Okay, Procurement)

No matter how self-contained you think you are, eventually you're going to need more stuff from outside your little enclave. Not just the obvious critical items, but raw or partially-finished materials and things you probably cannot comprehend needing in peaceful times.

What you will want here is a combination of two types of people: sharp-eyed and intelligent people who know what to look for and tough fighter types to help defend your scavenging party if it gets attacked.

Medical

This means a whole lot more than just having a doctor (or next best thing) in the community. It means spotting trends in health and nutrition, looking out for contagious diseases (especially since those will be on the rise once civilization crashes) and general public education about hygiene in a post-collapse world.

Communications (Internal)

Someone has to get the news passed around. Have a person (or a couple of people) designated as a central point for information to be passed around. How that information is passed around is up to you: a sheltered communal chalkboard or cork-board, news announcements at community meetings, you name it. All non-critical news and updates should go through this central point. This may be a bit corny, but someone with a bicycle can get word around much faster than non-functioning radios, televisions and etc. ever will.

Communications (External)

If there's a means to do it, it's a good idea to have at least one or two people around who can find ways of communicating with the outside world, or at least be an authoritative source of facts as they arise outside of your community, as opposed to gossip and BS. This could involve Ham Radio operators, folks who know how to use a shortwave radio, people who somehow manage to travel from your community outwards to others (and back)... there really should be a means to coordinate all of this information, then let the internal guys pass the news around.

Education

Just because civilization collapsed is no excuse to close the school. Education gives the kids something to do during the winter and keeping them up to speed on education means you won't have to try and get them up to speed later on. At the very least, find someone capable of teaching children under the age of 11 (or roughly, the 6th grade) and if you can swing it, at least another person capable of teaching up to the high school level.

Over time, you'll want more teachers and a more thorough curriculum, but at the very least? Teaching the skills of reading, writing and basic mathematics (up to and including the pre-algebra level) is paramount. Even this early on in a post-collapse situation, identifying existing and potential teachers (especially those who are *good* at it) is a good thing to do.

Religious/Pastoral

I really don't care how much of an atheist or agnostic you are – there are going to be a lot of people in your community that will want to pray together. The idea is to have someone each representing at least the majority religions in your community - to provide consolation, spiritual guidance, hope and prayer. One thing that is highly recommended – no religious figure is to be in the leadership, period. Keep church and state separate... it's the only certain way to avoid having your community turn into a cult. Don't believe me? Look up Jim Jones, Warren Jeffs, or The Spanish Inquisition, if you want an idea as to what can happen when religious figures run a community. I am a rather devout Christian and I'm telling you that you most emphatically do not want that.

Everybody Else

If there are people with no specific duties or are capable of same, then finding small-but-useful roles and duties will help every adult member of the community feel as if they're part of a larger community. I promise you, this will be far easier to do that you would initially think. Usually, everyone who is able will help with defense, help with any projects that need done and certainly help distribute anything brought in by scrounging parties.

Leadership

This one is last for a reason: Leaders are there to serve, not dominate. If you're one of the leaders, you're given that power precisely because it is assumed by everyone else that you have enough brains to get the jobs done in a way that benefits everyone – not just yourself.

In addition to solving disputes and putting plans into motion, leaders have to start playing such roles as diplomat, troubleshooter, referee, accountant, judge, police chief, promoter/marketer and much, much more. It will be thankless, leaders will be verbally abused quite often, but it is a job that needs done.

Internal Strife And Troubleshooting

No group of human beings are going to work in harmony and anyone who tries to sugarcoat that or say otherwise is setting you up for disappointment. You're going to have disagreements, rivalries and a whole host of friction between individual human beings and families.

Given the stressful situation we'll be living through, the chances of problems occurring are going to be pretty high and incidents may well be rather constant. Many of the stresses and points of dispute will also involve some rather unusual (to civilized folk) situations. Squabbles over food distribution, complaints over being 'ripped off' in a barter, accusations of theft... you name it and you will likely have to deal with it.

If left unchecked, or handled roughly or unfairly, such disputes will fester and escalate, even to the point where your fledgling community may well be torn apart. However, let's split the job up a bit - first we want to deal with how each individual can minimize problems and then we need to set up controls and mechanisms for a community at large to handle problems as they arise.

Starting with the individual level, realize that this is only for disputes. Any incidents involving theft, assault, or worse need to be taken care of immediately, defending oneself as much as necessary (if the situation demands it) and then taking the issue immediately to the community leadership for a more final form of solution.

Common Sources Of Problems

The most common source of problems are going to come from something that can be summed up in one word: Privation. People will be lacking familiar luxuries, privacy, a comfortable level of critical supplies (food, water, etc), familiar habits and even many bases of their identity: home, possessions, family, career, etc.

An important cause of problems will arise from a population who was previously used to living independently, but is now forced by necessity to live and work together. Where people were once previously running their own lives, pretending to be masters of their own individual destinies, they are now forced by circumstance to work together and are uncertain of their futures.

One other big cause of interpersonal friction will be a hyper-sensitized sense of self-entitlement and/or too high a sense of self-esteem. Even in civilized times, some people have egos and self-esteems inflated to the point where even the slightest inconvenience will cause a loud and forceful demand for satisfaction. By way of example, I have a miniature Dachshund. One day he slipped his leash, spotted a neighbor who was walking his larger dog and my little dog proceeded to dance around the guy in circles, barking (not growling, not attacking... just barking, with tail wagging furiously in an effort to play with the stranger and his dog) and occasionally stopping to sniff the much larger dog. The neighbor proceeded to shout a stream of obscenities, claim loudly that my dog was literally trying to kill his and in general raising a large stink. Once I retrieved the dog (slippery little critter is hard to catch), the guy stomped down to the office to lodge a formal complaint (I managed to clear it up by taking my tiny dog down to the manager's office and letting him lick their faces until they giggled and dropped the complaint). Now, imagine this guy in a post-collapse situation, in your community. There will be people like this. They will make you and your leadership's lives miserable at the slightest provocation against their egos. The good news is that the shock and confusion brought on by collapse will force most folks to not automatically assume that their egos can be fed by a sense of petty authority and that maybe they should reach out to their fellow person before shouting at him. The bad news is that not all people will get the hint and think things through before overreacting. After all, a near-solid generation of children to date have been taught that their self-esteem is more important than any other factor. A disturbing number of these children have become adults who believe themselves to be the center of the universe and that everything else exists only to serve them.

Finally, there is one other cause to note: If the dying government was a bit too heavy-handed (okay, outright fascist, communist, whatever) on its way down, there are likely to be folks in your community who were sympathizers, informants, or otherwise were people who gained the vehement and near-total hatred of the rest of the community. Depending on the severity of their pre-collapse activities, you may have to make everyone set aside any hatreds or bad blood and accept the person (or persons) into your community, especially if they have a critical skill that no one else has. The reason why is that you're going to have enough to do just staying alive and defending what's yours and you're going to need all the help you can get. Your best bet is to find out just how involved the person was, why they were involved (who knows? There may have been a credible threat or a family member held hostage) and what kind of person got left behind by the now-dead government. It may turn out that you wind up with a vigorous ally. I give as example a gent by the name of Werhner Von Braun. He was a rocket scientist in WWII… for the Germans. A Nazi scientist who used slave labor to build weapons. His V-1 and V-2 rockets were horrific in their reach and strength and his work helped kill thousands of civilians in Allied territory. After the war ended, the US Army helped smuggle him into the US, where he immediately did the one thing that he really wanted to do all of his life - make manned space travel possible. Long story short, his work was directly responsible for putting men on the Moon in July of 1969 - 24 years after WWII ended.

In conclusion, all of these notions will have to end and for most, it will end quickly post-collapse, for better or worse. However, for those offenders who have managed to survive to this point and continue to make life tough for everyone else? It is going to take a whole community or group of families to come together, sit such an undesirable person down and inform him in no uncertain terms that either he gains a bit more tolerance and/or kindness for his fellow citizens, or he ends up on his own. Individually, each person is going to have to realize that their only chance at staying alive lies in how well they get along with their neighbors. Anyone who fails to act on this will learn the hard way. This does mean you too, by the way… we're all going to have to change ourselves a little. I'm not saying that you need to create some sort of commune or commune, but you will have to learn how to get along. The results will be way less than perfect, but do what you can.

Tackling Individual Issues

Each individual and family can avoid problems by just showing a bit of tolerance for each other. The only real way to help things along is to try to understand the other person's point of view, to stop *and think* before speaking (or shouting, or worse) and to go out of your way to understand the other person's point of view. If you've been reading the book all this time, a lot of this will look familiar: you should have been exercising your mind all this time to do just this.

If you see a problem arising, settle it as quickly as you can, before it gets too big or too angry to handle. Seek out the other party and discuss it with them. Ask for their point of view first and quietly listen (bite your tongue if you have to) until they are finished speaking. At that point, calmly point out what you see, ending your turn to speak with a question as to what to do to reach a solution amicably. If you are in the wrong (or a family member is), say so up-front and offer a means to settle the matter. If the other person is in the wrong, don't say it outright, but say it in a way that helps them realize it, then suggest a solution that will settle the matter.

If the solution involves a bit of labor on the other party's part, offer to pitch in and help get it done.

This will require both parties to actually do this, which is obviously not always going to happen. In such cases, set up a later time with the other party to discuss the situation and explain why. If that isn't going to happen, take it up with the community leadership, even if you're the one in the wrong. Doing this will keep individual issues from harming the community.

The best way to avoid individual issues is to prevent them. Teach your family the situation. Be kind to a fault. Let the petty stuff pass. Just like a parent raising young (and teenaged) children, you have to pick your battles. If someone does something dumb (but minor) once, simply let it pass without mention. If they do it repeatedly, point it out. If they still do it, it's a problem that needs to be solved and you need to take the person aside and fix it. If it still goes on, take it up with the community leadership if it's big enough. If it's not a big issue, take steps to minimize its impact.

Tackling Community Issues

As a community, the leadership is going to have to tackle things that are going to seem petty and beneath notice. However, tackling small problems before they get large is something you're going to want to do. The less internal strife you have, the better (and more easily) everyone is able to work together.

First off, you should set aside one day a week (and in larger communities, two) for internal disputes. Set this time aside to allow people to come forward with complaints against each other, against the community, or whatever comes up. Set aside at least three leaders to address the issues and if any complainant (or defendant) is related by blood or spousal relationship to any of these leaders, the leader in question will step aside and another will take his or her place for that complaint. Each side will have up to 30 minutes to explain their side of the issue. Any physical evidence should be submitted at this time. Any witnesses that are presented will be presented by each side during their side of the argument, but each witness should be questioned by the opposite side in a party. After the sides and witnesses are presented, the three leaders withdraw to a private area and discuss three things: who is right/wrong (if applicable), what to do about it and how the solution is to be presented or carried out. If a vote is needed, vote on it and the majority vote wins. If more information is needed, everyone goes back out together and gets that information before making a decision. Once everything is decided on, all three leaders go back out and *publicly* announce their decision and the course of action. If possible, it is also to be written down into records for later reference.

Both parties are to agree with the decision, no matter how unfair it may seem. If information comes up later that changes the decision, then the leaders should reconvene and publicly announce these changes, why the changes were made and remedies made to right any wrongs.

Another thing to have going as a community is to appoint a market judge. Choose someone trustworthy, or choose multiple individuals who rotate out on alternate days. This person (or people) should be fairly sharp, even-handed and knowledgeable about what items are worth. This person is to preside over any disputes during bartering and heads up any enforcement of rules in the market. If the community is large enough, you will also want a couple of armed individuals who will help enforce any of the more angry problems that arise. The market judge should be positioned prominently in the market and easily accessible. To make things easier, market rules should be posted prominently and legibly.

Trials

For non-petty items, such as theft and assaults, the accused is to be restrained and confined as soon as the accusation is reported. A trial will be set up as soon as possible and if you can swing it, a jury of at least 7 people is drawn from the community. The jury members cannot be related to or otherwise blatantly prejudiced, either for or against the accused or any victims. The accused is to be sufficiently fed and basic needs taken care of (food, water, warmth) and should be allowed (with escort) to gather any evidence or solicit witnesses to assist him or her. The judge should be one of the community leaders (again, with no ties to the defendant or any victims.)

The trial is to be public, period. Basic rights are to be observed: The accused is to be considered innocent until proven otherwise. If the accused is unable to speak for him or herself competently, or is a child, then someone should be appointed to defend that person.

The accusers go first. Any witnesses presented will each give their testimony, then be cross-examined by the defendant or his or her representative. Afterwards, the defendant will give his or her testimony. Again, any witnesses are to give testimony, then be cross-examined. After both sides have presented their stories, each side will be given a chance to deliver a summary of their case to the jury. At that point, the jury will go somewhere private and decide the issue. When they do, they come back out and deliver the verdict. It has to be unanimous. If they cannot agree, then a mistrial is declared and another jury selected to run through the whole thing again.

If the defendant is declared innocent, he or she is to go free immediately.

If the defendant is declared guilty, then a punishment is to be decided on and delivered by the judges. The punishment is to be delivered immediately, with one exception: If the punishment involves execution, then the convicted should be given a chance to appeal and tell his/her side of story one last time in public and any execution should be delayed for 7-14 days after sentencing.

You should have or scrounge up relevant guidelines (or books) on how to properly conduct a trial, but what we're presenting here is a rough and ready guide, modified so that it can be put to use quickly, with a limited amount of people.

If the troubles involve a community leader, then that leader is not immune to any of the rules and should be treated exactly the same. Someone to temporarily replace the person in his or her post should be appointed, but that replacement will take no part in any trial or decisions involving the accused leader.

Preventing Strife

If you as a community can possibly swing it, someone should be appointed to go around the community as an ombudsman of sorts - someone who checks in periodically to make sure everyone is okay and to solve any minor grievances that he or she may come across. This should be someone generally liked by the community and should have some diplomatic and managerial skills.

This person is given a bit more leeway than anyone in the community as far as making petty and minor decisions. This person will also act as mediator to those issues which are big enough to cause friction, but not big enough to involve the entire community. This person is also the one folks should go to over any issues of a delicate nature and thus cannot be compelled to testify on those matters unless the community member confiding in that person allows it.

One other thing to note: You can use a local church leader or pastor for this role, but I recommend that you don't - not everyone worships the same way and the idea is to have a separation between spiritual and secular matters. Keep this position secular, so no one feels that the religion gets in the way (or that some residents think they have some sort of advantage because of a similar religion).

The Role of Religion In Your Community

While it is dangerous to place any church official, pastor, deacon, or preacher/priest into a position of secular power, there are vital roles that these folks play in the community and roles that are helpful towards preventing troubles.

In our current peaceful times, there is a whole industry centered around therapists, psychologists and folks who generally get paid to help people feel better about themselves, as well as work out any troubles in their patients' lives. Post-collapse, these folks will likely be too busy trying to feed themselves and I sincerely doubt that too many other people are going to barter scarce or hard-won goods to talk about their self-esteem. On the other hand, the local preachers, priests, pastors, rabbis, whatever-you-call-them have had years of experience in doing pretty much the same thing, but with a religious bent. For example, many local leaders of long-established (e.g. Jewish, Catholic, Lutheran, Episcopal) religions have had formal training in psychology and therapy during their preparations for ordination and have been putting it into practice the whole time since. While praying is a big part of what church entails, there's also the duty of insuring the flock are happy and healthy and of trying to help them become that way if they are not. Note that this also extends to non Judeo-Christian religions to an extent, though the training may be a bit more informal (e.g. Islam, Hindu, Buddhism, etc).

Consolation, advice, a friendly ear, a friendly word, knowledge of who in their flock may have something another needs (and vice-versa) and a keen sense of human insight are all within the scope of a competent religious leader. While obviously this is something that an Atheist or Wiccan is not going to readily accept, it is well worth the effort to get a local preacher working to help the community in ways which the secular realm is not going to be equipped for. As an example, a Catholic man usually considers the rite of confession to be an excellent place to release inner tensions and anxieties (and sins), in a setting that has the utmost confidentiality. On the other hand, a non-Catholic (or even non-Christian) man may also find similar relief by simply talking to the same priest in a quiet but non-religious setting. Similarly, a Christian man opening his heart to a local Jewish rabbi may find an incredible amount of sympathy in return, which works just as well

Any generally respected religious figure with a good reputation can also act as mediators in a dispute, especially between any two parties of the same religion. As long as the clergy in question isn't extremist, or out to stir up any kind of negative sentiments, the results are usually better overall. In fact, you would do very well to inform any and all religious figures in your community that mutual respect will be shown at all times and that nobody is to preach against any of the other members of the community, period. While this is a bit close to regulating religion, calmly explain that riling people up against each other is harmful to the community as a whole and that it should be avoided at all costs. Odds are perfect they will understand this and if not, the reminder can serve as a warning.

Folks in your community should be encouraged (but not required!) to attend church whenever services are offered. You may well be surprised at the number of folks who actually do attend these various services. For those who do not, perhaps encouraging an activity that is secular but open to all, say once a week or so, would do well to bring these folks together too.

Obtaining And Filtering News

At the one time when solid information is most desperately needed, information will probably be impossible to come by. The television (if it's still going) will become full of propaganda and misinformation, all of it designed to either keep the populace calm, or to get everyone to react and behave in certain ways (and not always to your benefit). Same story with the local radio stations, the newspapers and the like. The Internet (which is likely the last thing to remain operable) will be a maelstrom of bad information as well - from government propaganda on the mainline news sites, to misinformation spewed by people on all sides of every issue (pretty much like it all-too-often is in pre-collapse times).

As things get progressively worse, so will the flow of useless garbage from most common news outlets, especially as reporters are increasingly intimidated or bought (and not necessarily by the government… corporations and other entities have their agendas too, you know). Eventually, electricity will start becoming an iffy thing and most sources of news, compromised and useless anyway, will cease to be. This is not to say that everything is to be distrusted. Local independent radio stations and papers (and even some independent local TV stations) may shift allegiances and/or stick to actual fact and news, especially as central governments and other entities begin losing the ability to enforce their collective will everywhere at all times.

So how does one get and distribute news under such conditions? This is actually easier (at least on the local level) than you think:

- Bribe The Travelers: if you can spare the food, host travelers and (individual/small groups of) refugees passing through for a meal and pump 'em for information.

- Shortwave Radio: While most of it will be in foreign languages and likely full of propaganda as well, countries and places that are still operable are capable of broadcasting news and information on an international scale that you may be able to put to use.

- Passing Word Around: Keeping in contact with other nearby communities is going to be nearly impossible at first, but as things get easier to do in that regard, passing news from one community to the next may become a rather efficient means of finding out what's going on regionally. As news arrives from one direction, you send it off with the next person or group traveling in the other direction.

- Ham Radio Network: This may be a bit spotty with the decline of ham radio operators these days, but if you have someone who can work a radio and the equipment is handy, try to get some electricity going for it.

- Recycle Them Thar Wires: If you have intact copper wiring between two communities (and can keep them isolated from any other wires), you may be able to rig up a rudimentary telegraphic or even telephonic system. This will however require a bit of work on both ends, some working parts, some electricity on at least one side and a little maintenance.

These are only a couple of ideas to get things started. Over time, you can get creative and find a way to pass word around on a regular basis. However, when you start seeing any kind of news and updates flowing, you will have to look at the information with a critical eye.

Here are a few things to adhere to and keep in mind as you receive and spread information:

- Any news you get from just one eyewitness needs to be reported with that person's name. Anything witnessed by two or more unrelated/independent people can be reported as tentative fact, but with the caveat of "unconfirmed" unless all witnesses (or a majority) saw it first-hand and are community members. Any controversial or contradictory facts should be reported with the names of all witnesses. The only exception is that which involves information which may put someone in trouble. In that case, verify what you can before spreading the news and use some discretion when needed.

- Always ask for names/witnesses when receiving news. Also ask for dates, times and as many other specifics as you can.

- If there is any portion that you're missing or misheard, or are not sure of, ask to have it repeated. Be sure of what you're hearing.

- Always try to get more than one version of the same story.

- If the story involves movement or migration, try to get some sense of what speed the subject of the story is moving.
- Write it down as accurately as you can and only stick to the facts. Do not add, subtract, or embellish anything. Do not try to fill in any missing information. This is to avoid a game of 'Chinese Telephone', in which each person who passes along a story subtly changes it. Embellishment has a habit of changing the story to the point where it means something radically different from what was originally being reported.
- Avoid editorializing at all costs. The news should be as neutral and politics-free as possible. While obviously no one in a fledgling post-collapse community is looking to run for President of the World? You do have to keep in mind that people, even under completely rotten circumstances, will be looking for any advantage they can find, even to the point of manipulation. If the news involves something technical or includes information you simply do not understand or comprehend, find someone who can clarify it.
- It's been said before in so many ways, but verify what you can. Discard anything that looks or sounds fishy, crazy, or too far out of order. Note that some of the things you see or hear may be designed to get you (or your community) to do something that may not be in your community's best interests. Anything that seems too far out of line should be reported quietly (and immediately) to the community leadership.

Whoever handles the news in and around the community, the news folk(s) should be independent of the community leadership, period. The leadership is not to modify, suppress, or change the news in any way, shape, or form. It is what it is and any attempt to change or censor it will create distrust and possible unrest in any community. If something is fishy, say so. You (the news guy) and the community leadership need a relationship that involves enough trust to eventually know when something is not right and when something may make the leadership look bad, but still must be spread. Any news which happens in your own community should be (if possible) disseminated honestly, openly and without any sugar-coating or spin.

Finally, one thing to always keep in mind: Avoid "FUD". The acronym stands for "Fear, Uncertainty and Doubt". Usually it's engineered (as we mentioned, as an attempt to get you and your community to react in certain ways). However, sometimes it just sort of grows on its own, like a yeast or mold. Your best bet in either case is that anything else should be met with a demand for proof - not hearsay, not guesses and certainly not vague omens, or (laugh, but…) dreams. Stick to the facts when demands and decisions are called for and you should do fine.

Going It Alone

Is This Even Possible?

This is going to be more than a little bit tough. This is because without a community of at least 10-12 able-bodied adults, *at an absolute minimum*, you're going to have a very tough time surviving; I daresay it will be almost impossible. Certainly, in the earliest stages you may be able to get by with hiding and basically staying out of the way, but that's only because there will be so many rich pickings around that involve less effort and time on the criminals' part.

However, there are instances where you may be able to hold out in smaller groups. If you are isolated enough, you may be able to hold out for quite awhile, using distance and isolation to provide the majority of your defense. This works because any criminal groups who travel have some logistics problems - namely, knowing that they may not have enough resources to make it worth the resources spent getting to you and back, plus the risks associated with combat and 'conquest'. This is also assuming that such criminals would even know where you are, let alone how much you have. Because of this, you do have a shot at surviving for quite awhile before anyone even knows your small band exists.

Unless you or your small band is isolated enough, tough enough (hint: you're probably not), or lucky enough, it will be nearly be impossible to survive otherwise.

Okay, So How?

If you cannot find a community to join, or one you'd want to join and you're able to have planned enough out in advance, you and your small band (or family, or just you) can pull it off, under the following conditions:

- *By Sea*: If you stock up enough supplies (including sufficient fuel and fresh water!) and you can equip a large enough boat, you can actually pull off staying offshore for up to six months or more. This will require having a crew that knows what they're doing and is able to weather living in cramped quarters together without going crazy or blowing up at each other. The vessel will also have to be able to endure the extreme weather conditions found on the open water. A sailboat is preferable to a diesel one (but a hybrid would be best), but only because it means less fuel that would be required. This will also mean getting far enough from shore to avoid any criminals with boats. You may have to deal with any possible piracy, but if you're in the middle of, say, the Pacific Ocean and emit no electronic signature (e.g. keep the radar off and don't use the two-way radios), your odds of successfully hiding from the rest of the world are actually pretty good. While you're at sea, you can look for places where you may be able to re-supply your water (good luck with anything else). The advantage is that in most oceans, you can find a few places that are isolated enough to pull it off. Note that his is also useful to a lesser extent in extremely large bodies of water, such as the North American Great Lakes (and to an extremely limited extent on The Great Salt Lake, though in that case, well, you won't be able to hide very well and fresh water will be almost impossible outside of Willard Bay).

- *On An Island*: Forget such shows as *Survivor* or *Gilligan's Island*. Think more along the lines of *Lord Of The Flies* and try to avoid that. You will have to choose an island with its own permanent fresh water supply, enough of a perimeter of water to provide advance warning and defense and enough natural resources handy to provide what you didn't bring along. You will have to keep a tight and watchful eye on all available resources and upon arrival immediately begin to scope out the long-term possibilities. Another consideration is to design and plan for defenses against threats that may arrive from the water. Since that island isn't going to move and you can't swim for hundreds of miles? You cannot simply walk or swim away when things go wrong. You will need not only the boat you arrived in, but start looking for or constructing a second boat just in case. You will also need to find a way to camouflage both of them. You will have to also have a habit of reconnaissance that keeps watch at all angles outward from the island.

- *Out In The Wilderness*: Instead of going the nautical route, you could get you and enough of your supplies out to the back-country - somewhere that is isolated and far enough from any form of human settlement to acquire some form of safety. This is not only a factor of distance, but could be a factor of terrain. In the depths of a desert or mountain range that is nearly impossible to reach by car, you could find yourself a cozy spot that few if any humans could reach without a ton of supplies. The problems with this approach are few, but very important. You'll need a way to eventually (well, someday) get back to whatever replaces civilization. You will need to bring everything with you, or have it all waiting there. You will need to be able to endure the extremes of weather and climate that is common to wherever you end up hiding out. You will also need to secure basic resources on a renewable basis. For instance, if your hiding spot is in the desert, you will need a permanent source of fresh water. If your hiding spot is high in the mountains, a permanent supply of wood and protein will be needed, as well as some place to grow vegetables in the short season that such an area would provide.

- *Smack In Plain Sight*: This one is going to be extremely tough to pull off, but is possible. Hiding in plain sight means finding a place that is concealed or fortified enough to rest safely, in a place where you would least expect it, such as a major urban center or a suburban area. The biggest rule you will have to follow is to learn how to stay hidden at all times. You will be staying indoors nearly all the time, rarely going out. You will need a constant and trustworthy source of fresh water, enough supplies to remain hidden for up to two years or more (before you'd have to leave anyway) and a reliable-but concealed means of disposing of any wastes you generate. You're likely to not get much in the way of local news and you will have to adjust your routines to maximize concealment. This will mean only going outdoors at night when the coast is clear and only for short periods at most. If anyone finds out about your presence (and more importantly, the presence of your supplies) and you are unable to keep them out, the game will be up. The closed-in nature of this option means that *everyone* staying there will have to know how to live together in fairly close quarters without getting on each others' nerves and without losing sanity.

In all, living alone or in very small groups will be extremely tough to do. You will have an impossible time gathering any news and your world-view will shrink to the confines of wherever you live happens to be (which also happens in communities, but the larger pool of people at leaves gives you a larger range of diversion and a healthier diversity of opinions).

Bunker = Mausoleum

There are those who think that riding the storm out in a bunker is the best way to go and are wondering why I didn't mention such a thing directly. There are those who have spent literally millions of dollars to insure their continued survival in literal underground vaults. Some of them look (and probably are) extremely cozy, cheerful and, well, safe from the outside world.

However, the problems with living in a bunker are incredibly numerous. You have an isolated group of people with little stimulation, long stretches of boredom and latent personalities that will eventually explode due to the lack of diversion and external stimulation. You will have bad behaviors and actions arise that would otherwise be dampened or diverted by hard work or new experiences and by the common need for everyone to rely on each other against external forces. People sitting safe in a hardened bunker have none of these things. People in an active community will get at least a trickle of news and even gossip that will keep minds interested and working, whereas a bunker is mostly isolated and most likely unable to receive news.

As time crawls by in a bunker, personalities will begin to warp. Small irritations will become large explosions of argument. Minor disagreements will have the easy potential to become open warfare. Leaders in a bunker will make Homeowner Associations look like college beer bashes by comparison, as Napoleon complexes and a lust for power begin to bare their teeth. Left unchecked, eventually it becomes more dangerous to live inside a bunker than by trying one's chances outside. However, with a common fear of the unknown beyond the bunker walls, most of the residents will remain fearfully inside, while suffering increased chances of claustrophobia, fatigue, disruption of circadian (sleep) rhythms and worse. This and more is why you can safely consider a bunker as an expensive-but-comfortable mausoleum. Very few disasters or causes of collapse will require living in one, so going through the expense and trouble of doing so is more than a little futile, truth be told. The money can be better spent on preparations that help insure that you and your community survives aboveground.

Conclusion

You may have noticed that a lot of (well, nearly all of) the content in this particular book involves community. This is because without one, survival is going to require a massive pile of skill and luck. To this end, your primary goal is going to be gathering people together into a working community and put together the basics of what could eventually become a functioning society. Everything in here is designed to lay down the foundations of what can eventually become a working, self-sustaining community and possibly even more as time rolls on.

It'll be tough to do and you'll have a very hard time doing it. Odds are decidedly against you unless you have enough preparation and skill. At the same time, you're going to have to keep your own individual/family preparations a secret, even if some in the community come up short. In such cases, use charity at your discretion.

This will be easiest to do when you already know most of your neighbors, have a working friendship (or stay on good terms) with as many of them as possible and live in a place where there are enough local resources to forge a new life. Seek people and skills that will help and discourage (or even reject) those who would tear down. Stress tolerance and respect for each other, while at the same time banding together against those who try to only take from your fledgling community.

As long as everyone (mostly) works together, if you are capable of defending yourselves together and you actually get along together? You stand a pretty good chance of making it.

Chapter 4

In The Short Term

(6 months to 2 years)

So You've Made It This Far...

So far, so good? I don't mean reading through this tome - let's assume that you're reading this sentence six months or more after civilization has collapsed. You've actually managed to start a community by and large and there's actually a small hope of making it long-term. But before we start drawing up plans for the new coffeehouse/pleasuretorium in your new city, we're going to have to get some things put together and going first.

By this time, some areas may have settled down a bit. Civil wars may (or may not?) be raging in areas and if your region is one of them, it is hoped that you're at least out of the way of the fighting. It is further hoped that you at least have some kind of a working community and that no one is in any immediate danger of starvation (though this last hope is not a perfect given).

You will likely have seen more than a few bumps and grinds up to this point. You may have even had to fight to within an inch of your very lives, yet somehow survived. You will likely have seen hard times and will likely know of neighbors and friends (or even family members) who have died along the way. You will have learned by now that life has indeed become like the cliché - hard, brutish and short. People will have died from things once considered trivial. You will have become at least somewhat used to being short of something, of rationing and even used to doing without. Your previous routines and habits from the pre-collapse world will have seemed like a distant memory and simple luxuries like television and the Internet will have seemed a lifetime ago. You'll probably dream about them once in awhile.

I won't lie about this - it will dig deep into your very core. You will have re-examined your relationships with even the closest people in your life. You will have questioned the existence of everything you thought and held dear - even questioning your very faith in the Divine. You will have seen and done things that civilized human beings will consider as barbaric. There will be few moral boundaries left that you haven't crossed. You will be at once a hero and a coward. As a small consolation, you will now know exactly what it is that you are made of.

The good news is, you will have found out who your real friends are and will have forged unbreakable relationships with them. You will know for certain if your spouse loves and trusts you …or not. You will know your physical and social boundaries and will be smart enough to stay within them unless circumstances dictate otherwise. You will have discovered just how creative that you can get and realize that you're a lot tougher than you ever thought.

At this point in time, death is going to be a constant, as pre-collapse supplies are exhausted and the bulk of the human population begins dying off in unimaginable numbers. The criminals who are still alive will become far smarter, far more devious and far more violent in their attempts to get what they want. By now, all of the easiest (unarmed) pickings will be gone, dead, exhausted, or too isolated to get at. Simply scaring raiders off won't quite cut it any longer. There are few to no other options now, so the raiders will not leave unless they're either dead, or until they get some of what they came for. As an unfortunate side-effect, no lone traveler will be safe by this point.

As a community, you will have to be more vigilant and more willing to fight than ever. You will have to have a very tight patrol and a very strong response to any attackers. The only hope you have of any raiding party leaving you alone, is that you may see some peace until they're done picking off the smaller camps and homesteads.

This is going to be one very ugly time. Even many prepper types will start running out of stuff at this point. You too will notice that you can no longer rely exclusively on stored goods to do everything for you (nor would you want to). Bellies everywhere will start getting a bit thin and some people (even in your community) will start doing stupid things just to keep fed. This is why you spent a bit of time in the last book teaching folks how to forage, scavenge and otherwise get hold of foods that don't come in a box, bag, or can (and we will be covering even more on this..)

As we've discussed far, far at the beginning, the US has roughly 310 million souls, on land that can only support a small fraction of them. This roughs out to at least 6 of 10 people dying off from starvation… these are not good numbers at all. There might be some semblance of an economy and technology left by this point, but these benefits will at best be restricted severely to heavily-guarded supply lines and selected well-defended cities. The odds of you living in one of those remaining strongholds is going to be very slender, but living there likely won't be a paradise either.

This is the time and you should spare it, to start planning ahead. If it is winter, then planning for the next spring's plantings and eventual harvests will bring hope and vitality. If things are still too troubled, then planning and finding ways to get food and try getting enough to everyone will definitely be useful. Your community should be doing regular foraging and scavenging trips - in armed groups, but still out there and bringing back useful supplies. You should also be looking into building things up a bit further - both personally and with the community.

Around The House

Up to this point, we've been concerned with how to secure the place, hide stuff in it and how to use it all to stay alive. Odds are pretty good that things have been a bit uncomfortable in there. After all, without an operable sewage system, running water, electricity and/or natural gas? Things are probably a bit strange now, to say the least.

However, once you have enough spare time to actually think about things, but before the mountain of work involved with growing your own food, you can actually get a few things done around the house to make life easier on you.

Water

One of the first things to look at is your water situation. If you're still dragging water in by the bucket-load, this is obviously going to get old. You can make things somewhat easier with a bit of carpentry skills and a big enough bucket. Here's how…

Find a clean 55-gallon drum or other large portable tank of sorts that's capable of carrying clean water (hint: you have at least one in your shopping list - did you buy it?) Build a structure strong enough to elevate it so that the top of the bucket is close to or nearly level with (at least) the gutter along your roof (or if that's too tall an order, at least raise it so that the bottom is at least 6' high). Attach your rain gutter to it (and run a hose from any other gutters to it as well, if possible). You will also want to do two other things: first, attach a tap to the bottom of the drum (about 1" up from the bottom) to let water out; second, set up a small but sturdy ladder or ramp to allow you to get up to the top of the drum and dump clean water into it. From here you have some options: You can either take water directly from the tap at the bottom, or you can run a clean hose from that tap over to your kitchen sink, where you can attach it to the faucet directly (after disconnecting the house pipes, that is). You now have running water into your home. Note that it isn't much and that you will still have to purify it (or at least boil it) before drinking any. On the other hand, now you only have to get water once in awhile, until the drum is full. The reason you hooked up the rain gutters is to help things out a bit, especially in regions (and time of year) where rain is more than plentiful.

Now the winter will present some obstacles in colder regions, but this can be slowed down a bit by placing the tank next to a chimney or flue, where the slightly higher temperature will help keep ice at bay. Other useful means to keep the water from freezing up is to pack grass or straw around the drum, or to occasionally put a sizeable hot rock into the drum. The heat will keep the water from becoming ice for at least a little while. In very cold regions, you can likely drain the drum in autumn and eave it empty, instead melting snow and/or ice from clean areas when you need it (just be certain to boil the ice.) As an alternate? Instead of a drum, you can use a large hot water tank (which is already insulated - bonus!)

Another option is for those who have a well handy: Simply run the pipe from the well to your kitchen faucet and replace the faucet with a hand pump. As an alternate to that, run the pipe from the well-head to a hand-pump mounted on one side of the sink (or to the drum if you have one set up).

An option for you folks in a two-story house: Disconnect a large empty hot-water tank, seal the pipes leading to and from it and drag the tank upstairs. Put a funnel and screen to the infeed point on top of the tank and connect the bottom tap to the house piping. Fill the hot water tank with clean water (a lot of it will drain into the house piping) and you're set. Just remember to insure your pipes are sound, the main water feed is shut off, boil/purify the water before pouring it into the system and keep a regular schedule of keeping it full. Like the outside drum, you can run rain gutters into it as well, just have something built to clean the inbound water and to handle overflow so your floors don't get soaked. A large pan or tarp underneath is also a very good idea. On the plus side, keeping it from icing-up is no longer a problem (one, because it's indoors and two, because hot water heaters are by their nature already insulated).

We're not out of options yet, though… if there is a permanent year-round stream somewhere uphill of your home, a bit of piping (okay, a *lot* of pipe) can bring a small trickle down into an elevated settling tank (because you're dealing with sediment here) and from there into your homemade filter, then finally into your home pipes. In this particular case, you'll want two taps coming out of the tank - one at least three inches above the bottom for water going into your home and a second tap on the very bottom to occasionally flush out sediment.

Just because there wasn't a well around before things collapsed doesn't mean you have to go without one now… a few shovels, a few friends, a few sturdy boards, a lot of rope, some stones and a good idea of how deep you have to dig before you hit the water table? In most cases that alone should be sufficient to help you decide whether or not a well is possible. The good news is, it is usually possible in most areas (but extreme mountain areas, deserts and anywhere near an industrial area are obviously excepted). There are plenty of books describing how to do it (and at least one of them should be in your library) and you can certainly share a well with neighbors.

The only things you want to keep in mind with any external water source is security (to prevent anyone destroying or poisoning it), cleanliness (no dead animals, poop, or trash in the well, please), maintenance (to keep it operable and to clean out the sediment) and preventing any icing over (which can damage pipes and worse).

While we're on the subject of water, did you know you can take a car radiator (or better yet, a large air conditioner or refrigerator condenser coil), some metal tubing and your fireplace (or wood stove) to make on-demand hot water? You'll have to find an all-metal radiator/coil and flush it out very, very good (hint: a newer one is easier to flush out than a rusty old one, but avoid any that have plastic components). The radiator either goes in the fireplace, or strapped very closely in contact with the back of a wood stove. The tubing goes from your water source to the radiator and from the radiator to your bathtub, or a tap where you can take buckets of hot water to where you need it (like, you know, the laundry tub?) This saves you from having to boil water one bucket at a time for things like bathing or laundry.

Now if you live in an arid/desert region? This little subject is going to get very critical and very fast. It is hoped (and strongly) that you took the blunt hints way earlier in this tome and located a place with year-round water and that you have a means to secure it, yes? Your particular considerations will be preventing evaporation of what you do have and of storing as much as you humanly can in a (relatively) cool, dark place. On the plus side, you do have a useful source of energy for distillation of iffy water into clean water - though you will have to have a lot of them going on a near-continuous basis.

Heating And Cooling

A lot of this is going to depend on where you live. A home in Florida or New Mexico doesn't require a whole lot of heating - just a small stove or fireplace for those occasional cold snaps. Conversely, a home in Alaska or Montana isn't going to require a whole lot of cooling, if at all. However, you will want to cover the bases, just because well, you never know. Hopefully you've taken the time before the collapse to get a good idea of what temperatures to expect throughout the year and at their worst (for both hot and cold).

If snow and ice are a near-constant (or even expected) throughout the winter, then you're going to want to put a lot of effort into keeping your home warm. If you haven't done this already, choose at least one large room that can be kept heated no matter what. This is usually going to be a room with a fireplace or wood stove in it, or where you will place one. If this is not in a bedroom (odds are it won't be), then be sure to keep extra blankets and comfortable furniture. This will come in handy during the really nasty cold snaps, with everyone huddled around the fire. Be sure to keep at least two to three days' worth of fuel (for most folks, that means wood) handy so that you don't have to go outside just to get wood when things are too cold or too dangerous to go outside.

Now that you've had a chance to break the home in for awhile (either for the first time as a newly-arrived refugee, or in your own home after a good, hard post-collapse live-in), you can now figure out which rooms are vital and which ones aren't so vital to keep warm (or cool). You can take a good look around (and perhaps by now had enough experience) to see how well your fireplace or stove keeps things warm and to improve things as needed. This can be as simple as re-arranging furniture to allow maximum heat radiation, to as complex as building and adding additional wood stoves throughout the house (just a note: be *very* attentive to safety if you do that). Before you get grand visions of having a super-efficient wood or coal stove in every room your home has, keep in mind that someone is going to have to go around to keep them fueled and clean out the ashes on occasion. Ideally, the most you'll have to do is to perhaps build something to bolster the heat in a fireplace that you may discover as a bit undersized or inefficient.

Your heating sources are likely to be your cooking spots as well, which means that if you're going to put in and build a stove to cook on, placing it in the kitchen (or very near it) is probably a very good idea. At this point, you'd think that it would save you a lot of effort to go raid an abandoned wood stove store (such stores usually sell hot tubs as well - go figure), but then again, those things are *heavy*, which is going to make getting one home a trial (and a colossal investment in gasoline or diesel), so keep that in mind if your thoughts drift in that direction.

Finding and building one's own heating source is (relatively) easy to do, but you have to keep a very close eye on two things: fire and carbon monoxide. Even the slightest leak in the exhaust will give you troubles ranging from strong headaches to outright death. If you don't pay careful attention towards isolating the hot bits (and sparks and embers) from the room's structure, you'll end up with your home burning down. If the materials you use aren't study enough to resist the intense heat of the fire inside it, the whole shebang will melt, making a mess at best and burning your home down at worst. In your library list you will find a book or two that will show you how to build something workable and usable… I'd suggest looking into it before making any plans.

Other means of keeping warm in a home without a working gas or electric furnace? Keep extra blankets on the beds and sleep two (or more) to a single bed. The combination of extra body heat and extra insulation will make less of a need for a constantly fast, hot fire. Under such conditions, be sure to keep clean clothes at the foot of the bed, under the covers - your body heat will make them warmer when it comes time to put them on. Stack large stones along the sides of the fireplace or wood stove and along any walls that are very close by - they will slowly absorb heat and then slowly radiate that heat back out into the room as the fire dies down. Hang heavy blankets (or very heavy curtains) over the windows to keep heat indoors (especially if you have single-paned windows or metal window frames). Scavenge extra insulation from any abandoned homes, or from any building damaged beyond safe habitation - then take that extra insulation and pack the crawlspace, attic and any dead-air space around your home.

Cooling one's home is for most regions is less of a critical item and more of an option. In spite of living in an air-conditioned society and becoming quite used to finding escapes from the more brutal heat waves, most people can actually tough it out without undue stress.

Initially however, there will be quite a few number of fatalities in your area come the first serious heat wave. The very old, the very young and the very sick will all be potentially at risk of dying due to excessive heat. There are however things you can do to help reduce the risk to anyone in your family.

Water helps out a lot in this case… a damp towel or rag on the neck, or wetting the hair will do wonders in keeping someone cooler for up to an hour or more. Wetting a t-shirt while it's being worn as you walk about or do chores will also go a very long way towards keeping you cool. If someone is seriously overheating and a basement is handy, send the person down there for awhile and have them lay on the floor. Keeping curtains or blinds closed helps keep the sun out, which in turn helps things out greatly. Cover anything fuzzy (sofas, chairs, etc) with a smooth white fabric (sheets, for instance) before sitting or laying on it. Take a walk outside on occasion, if you can… odds are good that things are cooler outside under shade than indoors where the air is stagnant. Restrict your more strenuous activities for the early morning or late evening hours (or even at night), when temperatures are naturally cooler. Finally, keep hydrated (especially if you've been sweating) and avoid alcohol if at all possible (hint: it's perfectly possible), since alcohol tends to dehydrate you even faster. Overall, unless you live in an area that is well-known for temperatures well above 100 degrees Fahrenheit for weeks on end, beating the heat will be a bit uncomfortable for awhile, but certainly survivable for the vast majority of humanity.

Laundry And Cleaning

Let's face it - post collapse, keeping clean isn't going to be as easy as firing up the washer and dryer, or whipping out chemicals that were once easy-replaceable. This is going to take some creativity and a lot of work.

Let's tackle the laundry first, since it'll likely be a rather grueling process. However, it really doesn't have to be. Nowadays, most folks wash each of their clothes after wearing them only once. Unless you happen to fall into a mud-hole or otherwise get extremely filthy, you can easily wear your non-underwear clothing at least two or three times before things get to the point where you have to wash them up. Things like coats and jackets can go a year or more without the need for laundering if they don't get dirty (and you don't wear them next to your skin). Blankets? If you use sheets, the blankets can go for six months to a year between washes, easily. By the way, a friendly bit of advice: Sleeping in just shorts (or even naked if you're safe enough) will reduce the need for washing pajamas, but then again, it'll increase the number of times you have to wash your sheets, so don't get too creative. Another friendly bit of advice? Aprons and/or smocks will keep your clothing cleaner and are easier to launder.

Either way, no matter how far you can safely hold off the chore, it's going to have to be done eventually. Here's how to minimize it, yet still keep things safe. First off, odds are good that you have a stockpile of laundry soap, right? This is good, but read the directions carefully. Take whatever their recommended usage is and cut it down to 50-75% of that. You may not have a choice if you have hard (heavily mineralized) water, but otherwise you can safely get by on using less, especially when you consider that most people overuse the stuff anyway. Eventually, you'll be learning to make your own soap, but we'll go with what you have for now. Make sure you have two very large tubs that can take a bit of hot water, a stick, a washboard (improvised or otherwise) and clothesline all strung up somewhere outside (in the summer) or indoors (in the winter).

Unless the clothing definitely requires cold water (to avoid shrinkage), you'll want water as hot as you can stand it without scalding. Add soap to one tub and use a large clean stick and a bit of clothing (e.g. a shirt) to churn things up and get the soap all dissolved. Once the soap is all dissolved, let's start with the whites. Toss in all your whites, churn things up for a bit with your hands (or the clean stick if needed), gently kneading the cloth like you were kneading bread… call it 5-10 minutes of doing that. Scrub out any heavy stains against the washboard (or a clean rock) as needed. Take out the clothing, wringing out each item tightly before throwing it into the other tub of hot water for rinsing. Once the first tub is empty, rinse each bit of clothing in the second tub, pull it out, wring it out tightly, then set it aside in a basket. Once done with that load, change out the rinse-water in the second tub with clean hot water and inspect the first tub. If the first tub doesn't look too dirty, you can top it off with a couple buckets of fresh hot water and perhaps a little more soap. This lets you stretch out the soap a bit and lessen the time you're spending by heating up all that water.

Once everything is run through (or perhaps have a second family member do this), hang the clothing up to dry. Give it a few hours (more if it's winter-time and indoors), then put away the clothes. It sounds easy, but trust me - it'll eat the good part of a day running through the laundry for a family of four.

Cleaning the rest of the house is going to take some ingenuity, but you can lessen the frequency of having to do it with a few simple habits. First off, make everyone remove their shoes at the door and walk around indoors barefoot or in sock-clad feet. Perhaps having a pair of slippers for each family member is an alternative that works - especially in colder climates. This keeps the outside dirt mostly confined to the area around the door. Wearing outside clothes to bed may allow you to jump out of bed and into combat, but it makes the sheets and blankets dirty in a hurry.

Other good habits to keep things clean indoors? Keeping a sheet or some sort of cover over furniture that folks sit on keeps that furniture cleaner. Kids are required to stay out of their bedrooms unless they're sleeping, cleaning, or changing clothing. Consider getting rid of that wall-to-wall carpeting, instead using rugs (perhaps fashioned from the now-unused carpeting) which can then be taken outside and cleaned once in awhile (otherwise, you're going to have a very rough time vacuuming carpets without electricity). Get used to using rags instead of paper towels. All eating should happen only in the kitchen or dining room, to minimize having crumbs and food bits otherwise spread all over the house, where mold and mice (or worse, bugs) would be attracted to them. It's a lot easier to keep one room clean of food crumbs, than in keeping every room in the home clear of them.

Disinfecting in a post-collapse era is going to rely less on aerosols or chemicals and more on hot water, soaps, vinegar, herbal antiseptics and even alcohol - whatever you can make or get hold of. This will require a bit of a change in how frequently you clean, how much time you'll spend cleaning (bad news: more) and how effective that cleaning will be (good news: enough to stay healthy if you're fairly vigilant about it).

Dusting and ancillary cleaning should also continue as usual, as time permits - less dust means less allergens and a lower incidence of respiratory ailments.

Clothing

Whatever clothing you have (or can scrounge) is pretty much what you're going to be wearing for quite a few years, so take good care of it. Eventually textiles will return - it'll have to, certainly (even if you make the stuff), but make the best of what you have. Have clothing that you wear when doing hard or dirty work, as opposed to fashionable stuff or the ones you have for just around the house. They may seem like anachronisms now, but things like aprons and smocks are going to come in rather handy towards saving your clothing from stains, dirt and other things that will wear your clothes out prematurely. This means getting hold of (or making) these items and keeping a few around for all of your chores and work.

Kids are going to present a particular challenge when it comes to clothes… not because they get dirtier (they do) but because they will rapidly grow out of them. One big benefit of having a community is that you might have multiple families with multiple kids of different ages, so that clothes that one kid grows out of will be just the right size for someone else's kid growing into them. As any mother will tell you when shopping, when you're scrounging for clothing, always get clothes a couple of sizes larger than what a kid wears now, so that they can, well, grow into them.

When it comes to scrounging or bartering for clothing, always choose something that is tough, able to be laundered without any special work and are easy to maintain. Choose clothes with simple, sturdy fasteners and/or zippers. Pick clothes that are somewhat thick and a little loose-fitting (but not too much so). Forget about fashion, by the way; now you get to shop for warmth and durability.

Until folks get the hang of making their own textiles again and do it without the wearer looking like an extra from some b-grade biblical movie? Clothing is going to be one of those things that will be initially ignored during civilizational collapse, but then get very valuable as time goes on.

Meanwhile, even if you have all the clothes you could ever need, the idea is to collect and care for even more clothing (and something even more valuable - cloth). Scrounging abandoned sewing shops and stores which once sold bulk bolts of cloth is going to be a very good idea. The cloth can always be put to use in making parts of water filters, linens, curtains and, oh yes, *more clothing*. Even if you cannot find ready-made bolts of cloth, taking old large clothes apart and making new smaller clothes is not a half-bad idea. You can even recycle worn but clean cloth into things like water filter parts or other filtering media. If you or members of your family really know how to sew (and make patterns), you could easily build up a ready-made goldmine that you can barter for later on as things stabilize. At this point in time post-collapse, if you haven't already, you will definitely need to start looking ahead. Stockpiling and building inventory that you can sell later on is a definite part of this process.

Shoes

Most folks really don't think about shoes too much. Those that do (ladies, I'm looking at you) usually don't think about them much beyond styling. Well, given that a new pair of shoes will be far, far off in the future post-collapse, you're going to want to take very good care of the ones you do have. You can stockpile materials to repair shoes, but the chemicals aren't going to last for too long; this is why your shopping list should contain a lot of extra pairs of sturdy shoes. A solid all-around pair of shoes will last for around two to three years under hard use, but less than one year if constantly hiked in. This means you'll want to stockpile them up, just like the clothing. For adults, this will be fairly easy to do, since shoe sizes don't change all that much once you reach the age of 20 or so. For kids, you're going to have some work to do, since kids will grow through shoes faster than they'll grow out of their clothing. You can use their old shoes for barter and since they will be grown through quickly, they should last through about 2 to 3 sets of kids' feet very easily.

Needless to say, in your scrounging forays, shoes are going to be pretty big on your list of things to get a pile of. In a pinch, you can make a pair of shoes out of wood (for the soles), leather, hide, or cloth. Eventually, you will have to do this, but hopefully, you'll be more easily able to find (or barter for) new shoes. Incidentally, good shoes are something that you should scavenge from any killed attackers or even victims from within your own community. Yes, it is macabre and yes, it will definitely seem gross, but there's no sense in letting a highly valuable resource go to waste and you can always wash them first.

Cooking

That food isn't going to get hot and tasty all by itself. Also, now that you have some sort of routine going and are likely (or at least most of the time) not having to eat with one hand while holding a rifle with the other, you can start paying attention to things like taste, nutrition and your overall dietary health.

Before we dive into the grist of it all, let's take a look at how you're cooking that food. First, how exactly are you cooking it, anyway? Without electricity or natural gas, odds are good that the shiny new induction stove in that ultra-modern kitchen hasn't been doing anything useful for awhile now. Same story with the microwave, I suspect. So we've been reduced to what, eating out of cans and boiling water for the freeze-dried stuff? We can do better than that…

Hopefully, you've paid attention to your shopping list and have managed to scare up the cookery recommended in it. The reason why is pretty simple: non-stick cookware will be positively wrecked after constant use over an open fire and cooking without a stove tends to bang up the pots and pans a lot harder than most were designed to handle. Now if you have a wood stove with a decent set of cooking surfaces built onto them, you'll do okay, but I'd still keep a sharp eye out for sturdier pots and pans. While we're on the subject, the same goes for all those plastic spatulas and spoons. They can be useful for awhile, but the higher temperatures and less powerful cleaning products (which translates into harder scrubbing) will make plastic utensils pretty useless after awhile and turn many of them into miniature germ labs (same with most plastic bowls and dishes, by the way). Obviously, this is something you will want to fix.

So let's take a peek at what you're using for a stove these days. Assuming that most folks don't have a (by now considered fancy) wood stove to work with, we'll concentrate on the fireplace. Sure, you could use the charcoal grill out back, but the last thing you want is tasty cooking odors wafting all over the neighborhood with an easy path to your backyard (especially if anyone in the neighborhood is short of food, you know? Just creates all kinds of awkward situations that way). A fireplace with a good chimney will carry those odors a bit higher up, helping them to disperse a bit faster and making it harder to pinpoint from afar.

About that fireplace - you've likely rigged up some sort of grill-like surface over it, but you can get away with using dutch oven pots, a tripod and hook arrangement to hang pots with wire-type handles, etc. While we're on the subject, did you know that you can safely make your fireplace bigger (provided you have the materials to do it)? It will require a bit of work, but it certainly can be done. You'll have to remove any carpeting around the new hearth size, lay some bricks down two layers deep for your new hearth (and pack all crevices *solid* with sand or mortar) and then get up a big ol' hood of sorts to funnel all that smoke up into the existing flue (note that this will probably mean carving a big hole in your mantle, so I hope you're not too attached to it). A course of bricks sticking up around the perimeter of your new hearth will provide a handy means of catching ashes and keeping them off the floor. Utilizing existing sheet metal ductwork from the home (or that by now unused range hood) should provide plenty of material for the hood to go over your bigger fireplace, but you'll still have to put in the work. Just be very certain that you insulate or isolate any building material in the wall from the soon-to-be-hot fireplace hood, mm'kay? We don't need you burning the house down just to keep warm.

Doing this will give you something you can actually use - somewhere to cook food. Most fireplaces built today are built to provide ambience and maybe (if the home is rural) heat. A large number of fireplaces rely on natural gas or auger-driven wood pellets to provide the flames. While that's quite convenient when you have electricity, wood pellets and natural gas aplenty, it's a bit useless when you have none of that. The idea of stretching out the floor a bit on that fireplace is to provide an area where you can have a fire (or simply rake large coals) underneath a grill-like surface, where you can in turn put your pots and pans to cook food. If you have any spare bricks and/or blocks after your hearth building, you can use them to hold up the grill.

Now that we have some sort of decent cooking surface to use, let's turn our attention to arranging the kitchen. Since in most cases your fireplace is in the living room and all your food preparation stuff is in the kitchen, something probably needs to change here. Moving the fireplace to the kitchen is going to be a bit crazy, so the only option you really have in such a case is to set aside a bit more room in the living room to use as a food preparation area. This is actually not as bad as it sounds… it can be as simple as putting the coffee table to a slightly different use, or as complex as having cabinetry and built-in furniture that can go from peacetime wet-bar to post-collapse kitchen counter. However, since the hearth is a bit low to the floor, you may be better off using the coffee table (assuming it's made of a material that is easily cleaned and can take the abuse of knives and normal cooking activity). The good news is, this frees up a chunk of the kitchen that you can use to store things, convert it to a workshop, make it an impromptu medical/surgical area, or a whole host of other post-collapse uses. Either way, it'll save a lot of time and effort in the long run. Of course, a small wood stove in the kitchen that one can cook a full meal on would be a whole lot less of a hassle, no?

Now we get to the food itself. Hopefully, you've stored as many spices and high-longevity condiments as you could possibly lay your mitts on before the collapse. If you did, awesome! They say that hunger makes the best sauce, but honestly, a little chili sauce now and then is a wonderful thing. Most long-term storable foods are going to either taste bland, or will be salty beyond belief. What you need is to find a happy medium, to avoid having a stratospheric blood pressure reading, or to avoid being bored to tears by the simple act of eating. Fortunately, in addition to whatever you stored, you can likely gather and grow additional herbs and spices for use around the home. Chili peppers, rosemary, mint, thyme, sage, lavender (yes you read that correctly)… a huge variety of herbs can be kept growing around the house, either inside or outside. Many (especially wild onion) can be found growing wild in most parts of North America. These can be not only used, but excess herbs can be dried and used to barter with, especially with folks who don't have any of these things growing around the place (or know how to properly identify them).

Nutrition is going to become something you will want to pay very close attention to, especially long-term. Now if you did the research and have stored foods that are well-balanced, no problem for now. But as time goes by and especially as you begin to rely less on your stores and more on foods grown locally, you're going to want to plan and eat a balanced diet whenever you can. This isn't going to be as easy as it seems - fruits and vegetables largely come into season only during certain times of year, which means you'll seriously want to store enough of them to carry your through until next harvest. Certainly you can preserve them, but most preservation techniques are going to remove a lot of nutrients from them (which means you'll likely eat more than you may be used to, but don't worry - you'll be too busy to sit around the couch and get fat). A lot of the produce is going to be high in starches and carbohydrates (again, forget the whole dieting thing… you'll be eating smaller portions and getting enough exercise to make up for it). Expect more beans or lentils and more grains in your diet as well.

Meat is going to be less common than you were used to. Before collapse, eating a thick steak 3 times a week could be easily done by any middle-class family, but afterwards, things are going to change… a lot. Fish, seafood and birds will be your most likely sources of fresh meat, while beef and pork will be rarer and usually preserved (which usually means salted). The amount of meat in your diet will drop a bit as well, with perhaps ¼ of an expected typical meat serving on average at a meal, or maybe eating some form of meat only once or twice a week. While we're on the subject of meat, pay special attention and care towards eating the organs and bone marrow of the animals you procure for food. Normally, folks think of meat and only think of the muscles (steak, roast, etc), but eating many of the organs (heart, liver, kidneys, brains, testicles, etc) and bone marrow are going to be highly important to you. This is because the organ meat and marrow will contain a lot of highly essential fat-soluble vitamins and minerals, things that muscle meat simply will not have. For instance, bone marrow provides many nutrients which you would normally get from dairy products and then some. Note that even with fish, you can still gain a lot of nutrition from fats and certain organs (e.g. the liver, brains and undeveloped eggs) and get enough to provide for most of your needs.

Milk and dairy? Not unless you're a real skillful trader, or you happen to have a healthy goat or cow stashed somewhere in your possession. If you do stumble across a source of dairy, it will likely be cheese or butter, since raw milk spoils pretty quickly.

Purely processed foods and sugars will likely still be around for a bit, but don't count on it too much. You will likely be healthier without it in the long run. Your sources of sugar post-collapse will come less from cane or corn and more from sugar beets, tree sap, honey, certain fruits and even certain home-processed flowers (e.g. clover).

This isn't all bad, though, since all of these changes will actually be healthier for you. Your food will be less processed and over time won't contain hormones, dyes, chemical preservatives, or artificial flavors. You'll also be eating less of it for awhile, which will trim excess weight off the body and force your body to more efficiently use the food it does get. For the type-2 diabetics out there, this means hope of being rid of the disease …though hopefully your pancreas kicks back to life before the insulin and/or Metformin runs out.

I'm assuming that hopefully you have a book or two on cooking the old-fashioned way stashed in your library. You will likely find these foods a bit more bland, but over time your taste buds will actually adjust to this (this is especially true for all you smokers out there who have been forced to quit by the lack of tobacco - your tongue will wake up again after about 3 months of going without the ciggies).

Personal Sanitation

Simple rule of digestion: what goes in one end, must eventually come out the other. The question is, where do you intend to let it all out and how inconvenient will that be for both your family and your neighbors? Better yet, how on Earth are you supposed to brush your teeth, or, you know… shave?

Let's get the icky bits taken care of first. Solutions for this most personal of activities will range from digging a trench somewhere nearby, to having some highly complex composting toilet that will likely require maintenance and more than just a little upkeep. Your best all-around solution, if you can swing it, is to find a permanent source of running water and a working septic tank, keep the toilet tank filled with water and carry on as usual. Lacking that (which most of us certainly will), you're going to have to build an outhouse and party like it's 1899.

Now, about that outhouse. Unless you really enjoy drifting off to sleep accompanied by the smell of fermenting human waste on a hot summer's night? You will most definitely want to give some consideration towards where you're going to put this thing, what you'll want to build around or over it and how you intend to maintain it. Let's start with siting it first. Before you pick up a shovel, make sure your proposed outhouse site is at least 50 feet from any and all wells, nowhere near any open sources of water (like streams and creeks, for instance) and somewhere that is downwind of your home for most of the year. Picking a spot that is somewhat naturally private on its own is a very good idea as well. Keeping it at least somewhat close to home (within, say, 100 feet) will make things more convenient, as well as safer (you really don't want to be under attack while sitting on the crapper, then need to make a world record 100-yard dash to get back to safety). Once you have a spot all picked out, find another just like it, but somewhere else… that second spot will be next year's hole. Get that first hole dug nice and deep, at around 4-5 feet deep and small enough in diameter to fit under the bench or outhouse you put over it. You may want to put a small traditional building over it, which keeps the toilet paper and your butt dry when it rains, but just remember to put that structure on sled runners so that you can drag it elsewhere when needed. Also remember to keep the thing well-ventilated, for very obvious reasons.

If you don't want to do the whole outhouse route, you can improvise a bit with tarps, or with whatever is handy to provide some basic privacy and shelter from the rain/snow/whatever. No matter what you do, there's going to be some maintenance involved. Once every couple of uses, throw a handful of quicklime (kiln-fired limestone powder) down the hole to keep the odors down. Lacking that, once a week or so throw a shovel-full or two of dirt down there to put a small layer over the yuck at the bottom, which will also help keep odors down to something tolerable. Once it gets shallow, say, about a foot of hole depth left, cover the entire hole and drag your outhouse (or whatever) over to a fresh hole at the second site you picked out. You can likely return to the first site in about two-three years, or at least after 12 months if you dig a fresh new hole at the first site (preferably at least 5-10 feet away from where the original hole was…)

To help keep the whole neighborhood from smelling like a septic tank, you'll probably want to share an outhouse with your next door neighbors, alternating backyards when you change sites. You could even conceivably have one shared between four homes in higher-density areas… but going beyond that will probably make it impractical (and busy. Busy outhouses are bad things).

Once you've got that whole bladder/bowel thing squared away, you can then focus on the other waste issues, like your normal garbage. Some items you will definitely not want to throw away where others can see them… used cans of long-term foodstuffs rank pretty high up there, as do food boxes, or any other items considered as luxuries. To be honest, you should be recycling a lot of those items anyway. In fact, recycle as much as you can. Whatever food doesn't get eaten for dinner should be eaten the next breakfast. Empty cans can be washed, saved, then put to use for something else later - or even bartered if you have way too many already. Same with any empty box that is intact and study enough to hold something. Old cloth (even socks) can be put to use as filters, or re-purposed as rags. Broken items that are combustible should be burned if they cannot be recycled and those few items which are not recyclable, re-usable, or otherwise of any use can be either burned or buried. The combustible stuff can either be burned in a wood stove or fireplace (paper and wood products), or outdoors. Anything made of plastic that you cannot use, you may as well bury. Burning most plastics will give off a toxic odor and it leaves residue in chimneys and flue pipes. If you're especially creative and know a bit about how to do it, many plastics can be melted and then made into other more useful items. Given the circumstances and a bit of creativity, you'll likely be amazed at how much stuff you end up not throwing away.

Before we end this section, there's one other bit that we will have to cover, because unfortunately you will see a lot of it. When someone dies, that someone's corpse will quickly become a huge repository of disease and odors that you do not want anywhere around the home. Bury or cremate any bodies in or near your home within 2-3 days at absolute most. Any funeral or other rites are up to you, but do not delay in disposal unless otherwise necessary. Putting an existing community cemetery to use is your best bet (most have quite a few open areas for expansion), but otherwise set aside a field that you're not going to use for farming or other activities if a cemetery is nowhere close or handy. It is suggested that as long as someone has been dead for no more than a day and didn't die from disease, you can use pretty much all of the intact clothing and items you find on the corpse. What and how much is up to you (err, leave the underwear behind - you won't want it).

The same story goes for any animals that die or were killed - though in this case, if the carcass is freshly killed and didn't die from disease, you can likely butcher it, eat it and/or cure the meats. Anything that's been dead for more than two days in hot (80 degrees+) summer temps - more than two weeks or more if the temps have been below freezing the whole time) shouldn't be eaten and should be buried or burned immediately (though you could likely use the hide up until the point where the carcass begins to stink).

About the Kids…

If you have children in the house, a couple of things are definitely going to change. Some of these changes will be of necessity due to circumstance and some of them because the overall environment has changed.

Let's start with child labor. In peaceful times, kids aren't expected to get a job or do any work (save for chores around the home) until the age of 16 and usually they keep the fruits of their labors. However, things are a little different now. Instead of doing a few chores to keep the house clean or the neighbor's kids babysat, now those chores are going to directly help keep the family alive. Boys over the age of 14-15 may easily be pressed into service to help defend the home and community, may be out hunting and will likely be participating in repairs and modifications. Most kids old enough to do useful labor (call it around 6 or so) will be doing actual labor - planting seeds, tending gardens, helping with the cooking, foraging for food, digging holes and ditches, you name it.

Education is going to be a bit different as well. If your community has anyone who teaches, there will be something still for the kids to learn from. However, unless you luck out and have teachers handy to cover all subjects and all ages, you'll likely have basic education at best for the younger children. This means your pre-teen and adolescent kids will need to pick up studies on their own, with as much help as you can spare, or tutoring from older kids and elderly folk who can spare the time in exchange for bartered goods. Your library should have quite a bit of books handy to help things along, but the local/nearest schools will also have these and a lot of additional supplies.

One of the greatest gifts you can give to your children, whether a school exists or not, is to teach them to learn on their own. Teach them how to read and write. Teach them logic and critical thinking. Teach them teamwork and a good, strong moral set. Most important of all, teach them to do research and to question everything they don't already know perfectly well.

Children post-collapse will see and experience a lot of events that are going to shock most adults. In such events, try and keep the youngest kids out and away from it, but don't go out of your way to shelter them by too much. They are going to be part of a rather unfortunate generation - one which will have to grow up fast and learn some rather harsh lessons that haven't been taught for hundreds of years, if not millennia.

One thing to consider is that these kids are going to be just like a lot of adults - confused and scared. The only difference is that the kids won't be able to hide the emotions as easily, or understand any of it. You're going to have to sit them down, then carefully but honestly explain to them everything you know about the world around you (with only few exceptions, to be explained later in this section). Don't try to be brutal about it, but at the same time don't sugar-coat the facts. Explain things honestly, but gently. Be certain to express hope for the future, even if you yourself don't have any hope. Listen to them, especially with the older kids. Be certain to hear any ideas they may have and answer as many questions as they may have. Don't try and fill in any details that you do not know and do not try and make anything up. If you're guessing at something, tell them that.

The rules you may have once had about how far away your kids could wander from home, or who they hung out with? All out the window for awhile. Kids that stray into the woods, or in some cases even too far away from the home, may end up kidnapped for reasons that I'd rather not relay in any detail. Suffice it to say that these motives can range from ransom, to sexual assault, or even to cannibalism. I trust that this will be enough of a reason to modify the rules under which your kids operate.

These rules are actually pretty simple and very much along the side of common sense at this stage of things. The younger kids are to never leave the front of back yards and are to always remain within eyesight of a parent or guardian. Older kids are to never go anywhere alone and should always check in with a parent or trusted (by the parents) adult periodically. No minor is to leave the home at night without some sort of armed escort.

Kids tend to have one fault that adults usually do not - they cannot keep their mouths shut when they really need to. If people in the neighborhood is rationing food and your kid is bragging to his little friends about how much he ate (or even what he ate) the night before, you're going to be in for a world of hurt. To his end, if you have kids, you're going to have to perform a bit of operational security even in your own household. This means having rooms where your kids are not allowed to enter under any circumstance at all (these will be where your stores are) and you will have to impress on your kids to not even speak of anything concerning what supplies you have or how many of them you have. You may even have to lie to them, or to disguise the food with foraged materials to camouflage the fact that you are using stored foods.

Finally, one thing to do as often as you can is to take a few minutes of the day out and spend time with the kids. Tell them you love them. Show them that you love them. If you can possibly (and safely!) spare the time, spend a few minutes playing with them, to let them know that even without civilization that they can still be kids once in awhile. It'll be a good thing for them and a good thing for you.

When Night Falls

Without electricity, night time will be dark indeed. Your only sources of light after the sun sets will be as follows: moonlight, firelight, candle light, torchlight and whatever still-operating lanterns and flashlights still exist. It won't matter how many batteries you have, or how big the solar collector is - you may as well start getting used to the idea of seeing with whatever you have on hand. Of course, if you have the supplies, by all means use them – discreetly. After all, you don't want neighbors in the dark seeing artificial light emanating from your windows.

Your eyes will actually become used to the dark over time and while the human eye is fairly night-blind when compared to, say, a cat's eye, you will be amazed at how much you actually can see without a ton of artificial light around. Earlier, we've covered how most folks can see fairly well in full moonlight. Not enough to really read by, but certainly enough to see where you're going and to see what others are doing out to around ¼ mile or so…once you get used to it.

You can make your own candles, but until you get the hang of that, you can also improvise light from other means:

Oil Lamps: For this, you will need an open, hollow, but shallow container, such as a shell, small bowl, or a small depression (or dent) in something fireproof, such as stone, some small pottery, or similar materials (note: don't use metal unless you like burned fingers). Procure some fat, paraffin, wax, or oil (corn oil, olive oil, canola oil…) - anything that burns slowly. You can even use pine resin (that is, the stuff that oozes from cuts and breaks on pine trees). Make a wick out of scrap cloth, yarn, thick string, tree bark fibers, or similar materials. Put the fuel into the 'lamp' and lay the wick into it, with a little bit of the wick sticking out over the side. If the 'fuel' is solid, melt it down a bit with the wick in place and make sure the wick soaks a bit of it in. Let the wick soak a bit otherwise before lighting and you should have enough light to see by.

Torches: While obviously unusable indoors, torches are useful outdoors in the open. Most commonly, you can use a stick, cloth and some sort of petroleum fuel or other oil to make one. Bind the cloth to the stick tightly with baling wire or similar fire-proof binding material. For fuel, you can scrounge pretty much any type of flammable liquid. For example, motor oil salvaged from disabled or destroyed cars work just fine for this purpose.

It was mentioned once, but deserves to be mentioned again - only use these things outdoors. Always keep the torch tip higher than the stick to avoid catching the stick on fire. Have a bucket of sand or water handy by the home to put out the torch and try to avoid getting the torch fire too close to overhead things that are flammable, like low tree branches and such. The best way to use a torch is to have one person hold it, while others use the light to see by, or in a group have one torch for every 2-3 people in the party.

Candles: you can improvise candles from used candles, bee hive combs, paraffin, rendered animal fats, or any other solid flammable substance with a low melting point. To recycle a candle, melt some wax or paraffin and make a wick out of string. Pour the wax into a small can (or paper cup, or something similar) with the wick suspended through the center… just like the school craft projects you may have done.

Lanterns: Just like making an oil lamp, the same principle is at work, except instead of using an open shallow bowl, you'll use a metal can with a large hole cut into one side (high enough to put a supply of oil on the bottom) and a short wooden stick. You carve the hole in the can (about the size of a golf ball), get a wick and nail the can to one end of the stick, aligning the stick and can so that you can hold the stick parallel to the ground while it is lit (attaching it is simple: drive a medium-sized nail through the can wall opposite the hole, into the butt-end of the stick.)

Note that these are going to be mostly short-term solutions, though a good oil lamp can last for a couple hours before refilling. Given the new lifestyle, staying up late is going to be hard to do unless you're on guard duty/patrol, so you'll likely be looking to turn in shortly after the sun sets during most times of the year. An interesting fact to bring up: In Medieval times, most people had a 'first sleep' and a 'second sleep', where someone would fall asleep shortly after sunset, wake up at midnight or so (to take care of a couple of personal chores, take a walk, have a snack, hit the bathroom, get romantic with the spouse, etc), then go back to sleep until just before sunrise. I suspect that folks may by and large go back to such a routine after a decade or so of not having any real reason to stay up late on a nightly basis and in some ways, it actually seems to make sense.

Food, Food, And More Food

By this stage of the game, the food situation is going to get very dire for a lot of people. As mentioned at the beginning of this book, a whole lot of people are going to be a whole lot of hungry. Most people who give any thought at all towards preparing a stock of food will have a month or two stocked up at most. Many folks who have managed to grab food supplies in the panicked shopping rushes may have anywhere from a few weeks to 3 months' worth of food stored away. Folks who are out away from the cities may have up to a year's worth of food packed away. Some of the more faithful members of the LDS (Mormon) church will have followed their doctrines and have up to two years' worth of food stored away (which is what I hope you have managed to do, no matter your religion or lack thereof.) Some relief may have arrived from governmental sources, the more aggressive and intelligent criminals will still have some stocks left and there are likely some lucky individuals out there who have managed to fall into something that will keep them fed for decades on end.

Now here's what's going to happen to the rest of humanity: Six months in, the vast majority of people still alive will either have nothing, or will be running out of food rapidly. This is going to start presenting some problems as people get hungry. You may also want to look into not only helping your neighbors find food (so they don't start trying to get at yours), but to help stretch out your own food supplies, which is always a good thing.

Foraging

Yes, we've been talking about this since a hundred pages back or so, but it's time to bring it up again with more detail, especially this far in post-collapse. Foraging for plants and insects is a good way to start, since a surprising amount of things are edible on this planet. From worms to steak, protein can be had quite easily.

The best places to forage will be wherever you find open fields, copses of trees, woods, coastline, riverbanks, streams and many other places. Basically, you want to go where the concrete and steel isn't. Even in the deepest urban areas, you can find edible weeds, insects, worms and grubs and a whole host of things that turn out to be quite edible. Those in the country will have the easiest time of it, but even those in the city can forage in places like public parks, rooftops (especially of old buildings), industrial areas, parks, docks, any underground area and other places where there usually wasn't any vigorous landscaping going on during peaceful times.

Insure the plants you find are edible (check with your education and field guides) and nearly any insect you find in North America is edible - except for flies, mosquitoes and certain spiders. You may have to remove the stingers from wasps and bees (and if you find any bees, try to follow them back to their hives first, where you may be able to snag some honey). The idea is to gather enough not to necessarily fill your bellies, but to provide enough protein and vitamins.

The best times to forage for plants and insects will be during the day from around mid-morning onwards, when there is the most activity (insects are cold-blooded - they need warmth from sunlight to get moving). Also check during the twilight or nighttime, as some insects mate or migrate during nighttime hours. The best time to forage for any underground insects will be right after any hard or prolonged rains, when insects such as earthworms and snails surface to avoid drowning and over-saturation.

Hunting

If anyone reading this has visions of bagging deer left and right, or thinks that they will have a constant stream of big game or waterfowl to keep a steady supply of meat? Well, sit down, because I have some bad news for you: This simply is not going to happen, period. During the initial weeks and months of civilization's collapse, most of the easy-to-catch game species will have been eaten and whatever is left will have long since become masters at hiding, or will have long ago left for places that you won't be able to get to. Let's put it this way: If you're a deer and every two-legged creature with a rifle, club, or knife is banging around in the woods trying to kill and eat you, would you hang around? Of course not - you'd either be long gone, or you'd already be dead. Trust me - everyone with a rifle, shotgun, pellet gun, club, or slingshot is out there trying to bag a deer (or a cow, horse, sheep, bear, rabbit., whatever.) Note that there are no longer any game wardens, any seasons, any bag limits, any minimum sizes or ages or prey, or basically any rules at all. If it moves and has meat on it, it will likely be on the business end of someone's gun, bow, spear, stick, rock, or bare hands.

You do have one initial advantage however, if you're willing to consider it. Most people will be chasing traditional animals: deer, rabbits, elk, moose, wild pigs, ducks, geese, doves… what they're not aiming at (yet) is raccoon, squirrel, seagulls, turtles and basically everything else. In some parts of the South and Appalachia, folks do seek out these non-ordinary prey and eat them, but everywhere else? You'll likely have them all to yourself for quite awhile (that is, until starvation begins to set in and folks get really desperate).

So, when we talk about hunting here, we're not talking about your ordinary big game, the type of animals that you typically think about when you think hunting. No, we're talking about animals that either can't easily run off, or are plentiful enough that you won't have to worry so much about extinction after everyone tries to kill them off. We're talking about pigeons, rats and mice, common birds of all description, snakes, frogs, groundhogs, opossum, wild pigs, frogs, lizards and a whole host of other small animals, all depending on time of year and region.

Also note that hunting won't necessarily require unlimbering that giant big-game rifle with the massive 2000-yard scope, either. You can do most of this type of hunting with a BB gun , a small air rifle, a net, or even just a big stick. What makes these smaller weapons even better is that they are inexpensive, require little to no maintenance, the ammunition (if any) is dirt-cheap and they can be stored up pre-collapse in huge stockpiles - sufficient to last a lifetime if necessary.

You can also hunt with tools as primitive as a small bow and arrow, a net, a blow-dart gun, a slingshot, or even just a stick with a sharp point on it. You can use snares and dead-fall traps as well, or even a humble classic mousetrap baited with peanut butter (just keep a watch over the traps and snares periodically - as in, on a daily or even twice-daily basis). You can capture pigeons quite easily with stale bread crumbs and a net. Catching mice and other rodents is as easy as burying a large open (and smooth-walled) plastic container into the ground, buried just up to the lid and putting some sort of bait in the bottom of it. The idea is that rodents jump in to get the bait, but the smooth walls of the plastic container won't let them get back out again. Those large storage tubs they sell for storing Christmas ornaments are perfect for this.

You will have to do a lot of work for not much return, but sometimes you get lucky. If you come across any extra meat, always try to share the extra with neighbors - you'll never know when you may come up short but your neighbor came into something good. Fostering a sense of sharing now will also help you out in the long run. If there are neighbors that don't share in return, focus instead on those that do. If you've got your preparations, you don't want to necessarily waste your salt and preserving supplies on things like mice and rabbits in either case, as they will be relatively plentiful. Save the salt and spices for those rare bigger kills, such as venison or other animals which are pig-sized or larger. If you can't give it away, eat it as quickly as you can and anything you know that you won't be able to eat within three days, you dry into jerky or perhaps salt-cure if you have enough salt to spare.

Almost anything is edible if it walks, flies, or crawls, so don't pass it up so easily. Do however avoid hunting down cats and dogs if you can help it, since these animals may still be owned by someone and you really don't want to make the owner(s) mad by barbecuing Spot or or making stew out of Fluffy, you know?

One last thing: Be stealthy about cleaning and bringing home any of the kill. Don't laze around it or celebrate. Once you've killed the animal, quickly drag it somewhere hidden to clean it. Quietly gut it, bag the organs you want to keep, quarter the body if needed and then get the hell out of there with your kill (trying to avoid leaving any kind of trail when you do). Bury or otherwise hide (as best you can) anything left behind from the kill - guts, heads, feathers, whatever. (Honestly, though? Take the head back home with you too - there's lots of good meat in there and the brains provide a wonderful source of fat.)

If you use a firearm to hunt with, that shot may drive off other wildlife, but it will very likely attract other hunters - especially those people more interested in taking the fruits of your labor from you than in getting their own. Even if they don't attack you directly, they may come closer for a chance to catch and take the wounded animal before you can track and find it. This is another reason why you really shouldn't bother with large game unless you are certain you're alone (preferably alone with a partner or two), or with a large enough party to help enforce possession over what you kill. Honestly, though? Learning to hunt with bow and arrow or a small air rifle will make things much less hazardous for you in the long run and save ammunition for those times when you really need it.

Fishing For Real

Unlike more peaceful times, you won't need a license and won't have to obey any size or catch limits, or any such thing. You can use any kind of bait, trot-line, net, seine, or device of any kind you want to catch fish. You don't need to throw the little ones back.

Any kind of tackle will work, as long as you have live bait or a good net. You can even use a shirt and some pantyhose if you had to. The subject of fishing of all types - sport or survival - fills way too many books, so let's concentrate on how this will happen post-collapse.

Unless you live along the coast, in a swamp, or along a very large river or lake, fishing spots are going to be somewhat limited and everyone is likely going to try their hand at it, especially in urban areas. There will likely even be gangs and groups who will try to control the banks in more densely populated areas. They may charge 'fees' to fish there (either in goods or in a percentage of the catch), or will routinely attack anyone unlucky enough to be out there fishing alone, unarmed and/or unaware. In the country or along a coastline, you have a few more choices and areas safer to go fishing, but keep in mind that the closer to a populated center you are, the less choices and safety you'll have.

Don't get me wrong - for the most part, folks will likely leave each other alone, especially if everyone is catching something. On the other hand, as food gets scarcer and as things get tighter? You may find people asking, then begging, then stealing and then assaulting you for caught fish (and other foraged foods) by force. To this end, you'll want to do most of your fishing at night and as mentioned before, keep it quiet and stealthy.

Fishing off of or from a boat is best done quietly as well. On inland rivers and lakes, avoid using any motors so that you can not only save gas, but you can also avoid attracting attention from the noise a boat motor makes.

For larger-scale fishing, *if you can spare the fuel* (or have sail, or have the stamina to row, etc)? You'll likely want to make it a group, family, or community affair. It'll be a great way to bring people together and larger catches will mean a better outcome for the community at large. Families who succeed can barter those parts of the catch that they themselves don't use. In this particular case, it helps to know how to use nets and have information on their construction, use, etc. This is especially useful in coastal areas (for most coasts - some coasts will require the vessel to go very far out, perhaps beyond the horizon). Other coastal benefits come from shellfish - crabs, lobsters, oysters, clams, etc. Either personal or 'commercial', these will bring more than a few benefits, but it helps to have the gear. On a personal level, crab traps and small rakes should definitely be in your supplies. If you live on a coast and are near large, known fishery areas, this will all help you out greatly, so you will want to make it a priority to get up to speed as quickly as possible and start bringing in fish. The sooner you can do this, the less chance your coastal community will have to face starvation.

At this stage of the game, around six months post-collapse or so? This is a good time to start thinking ahead community-wise. To pool a lot of the less-critical-but-still useful resources. To teach each other skills. This doesn't mean automatically sharing everything, but it does mean sharing whatever you safely can. The skills you teach each other means less chance of a neighbor without those skills coming to you for food (either by asking or by gunfire).

Livestock And Crops

For those of you lucky enough to have livestock, you have your work really cut out for you. You have to keep those critters alive in spite of a sudden lack of vaccinations, high-quality feed supplements and all the other amenities that you were once used to buying. For those of you who have crops, you have to maintain, harvest and re-plant a whole lot of seeds. It also means that you won't be as easily able to manage the sheer numbers you may have.

This leaves you with a couple of options:

- 'Hiring' some help. This means taking on neighbors and refugees as workers to help you keep and protect the herd/flock/crop/whatever, in exchange for giving them a share of whatever you cull or harvest from it. You'll have to help set them up with housing if you're remote (unless you really want everyone living in your house) and it will mean working together as a team.

- Giving Away The Excess. For those animals you have way too many of minus petroleum and electricity, you may want to literally give away the extra to neighbors and townspeople. Doing this keeps the extra animals from going to waste and is actually in some cases a lot more humane. For instance, if you own a dairy farm, there's no way you'll be able to milk 400 head of dairy cows on a daily (or even weekly) basis with no electricity… failing to milk a cow practically engineered to give mass quantities of milk? As a dairy farmer, you already know the results: that cow will suffer quite heavily. For crops, it really goes without saying that there's no way you'll be able to harvest that large of a pile without machinery… and there's no way to store it all for any appreciable length of time (depending on crop) without it spoiling.

Now certainly you can collect, cull and barter off a lot of your stock or crops as well and set yourself up nicely. Just be prepared to lose a lot of it to spoilage unless you share it.

Something to consider: don't get too high and mighty about having livestock or a large stockpile of crops. The nearby townfolk have something you don't - protection, manpower, firepower and communications. If you get stingy and un-neighborly with your livestock, the townspeople aren't going to be in much of a hurry to defend you against raiders, or to come to your aid when you need it… they'll save their energy and resources helping out the neighbor who is helping them out in return. If you're particularly nasty about it, they'll simply wait until you're dead (it won't take long) and then come scavenge what the raiders didn't run off with. Just something you may want to keep in mind.

The best way to get the best of all worlds is that first option - have folks come out as hired hands. You won't be paying them in money, but instead you'll be paying them in a percentage of the crop, livestock/meat, or even a percentage of the land that they can call their own after a few years of labor (just avoid the 'sharecropper' solution - that is, don't 'rent' land… it leads to a lot of wasted effort and nobody wants to end up a peasant. Even in these times, you'll find people moving away the very moment that better opportunities arise). If you live close enough to a town, then most of your hired help can come directly from that town. Offering a share of the crop or herd in exchange for work is plenty of incentive to get them to come work for you.

Be certain that whatever solution you choose, you do your share of the work. If you appear to be slacking off and not working as hard as those you've hired on, some folks may get bad ideas. This in turn leads to bad actions. Always hold up your end of the bargain and share any bonuses or windfalls, because appearing to be too miserly is a sure road to disaster. Be sure everyone knows their roles and knows how to perform them. Some of those you 'hire' should be assigned to security patrol, or at least make that a part of everyone's duties (including your own). Herds that scatter, or crops that are out of visual sight can quickly become easy targets for raiders, desperate neighbors, or pretty much anybody.

You're likely going to have to perform a bit of education for those you do take on and you yourself will likely be learning right alongside them, considering that you'll be revisiting agricultural techniques last used by your great-great grandfather. The good news is you will have advantages that folks in the 19th century didn't: Knowledge.

By the way, with the lack of petroleum (and thus, well, horsepower of the mechanical kind)? If you have actual horses (that aren't geldings), you stand to come out way ahead in any future community if you start breeding them. It will take a year or two before you have enough new horses to start breaking and bartering off, but the sooner the better, As things began crashing down initially, you could have been sharp enough to start that breeding program earlier, insuring that some mares have produced foals by now, getting a good jump on providing a massive source of post-collapse income for yourself and your family. Similarly, those with donkeys can do similarly well and you can even train juvenile cattle as draft animals to be bartered off (or train a few for your own use as plow and wagon pullers). Overall, if you have livestock, you may want to get a bit creative with what you do have and more importantly, how you use it.

Food Preparation And Storage

Okay, let's say you get lucky and bring in a regular supply of food to feed yourself and your family. There's no sense in going through all that effort and work, only to end up curled on the floor due to food poisoning, or watching in horror as your hard-earned produce and meats turn into science experiments. There is also that one little bit of psychological hurdle you'll want to get over as well, especially when your meat intake consists of animals you may not be used to eating. What I mean is, it'll be easier to get your kids to eat meats in hamburger form than in a form that is instantly recognizable as having once been a squirrel, rabbit, mouse, pigeon, etc.

Let's start with the safety bits and a few do's and don'ts.

Dos:
- Always cook all meats and insects to at least 180 degrees (F) internal temperature before eating. Take the temperature at the bone with a meat thermometer, or at the center of mass for the meat.
- Use as much of the animal as you safely can, including muscle organs (heart, tongue, gizzards, etc), livers, brains, eyes, kidneys and other parts you wouldn't ordinarily think of.
- Always wash plant materials (leaves, berries, whatever) to remove dirt, contaminants and worse.
- Always check field guides and positively identify all vegetation, insects and even certain animals (frogs, lizards, snakes) as safe for eating before you chow down on them.
- While you're at it, if there are any special preparation methods to safely eating something, follow those instructions to the letter. If you cannot do that, then whatever you do - don't eat that item.
- Just like in peaceful times (but even more so now), always keep raw foods separate from cooked and do not use the same tools or surfaces for both raw meats and vegetation.
- Eggs (of any type) can be kept for up to a week or two if kept cool, but realize that unless you got that egg from a coop, the egg is likely to be fertilized, so expect it to be a bit crunchy once in awhile (but you can eat it anyway).
- Preserve whatever you don't plan to eat in the extreme short-term. For meats, this means 2-3 days, for softer fruits this means a week or so (depending on the fruit) and for vegetables it will depend on the vegetable in question.
- Most metal-canned foods that are not dented, bulging, or rusted can be eaten well past the printed expiration date. Don't pass up a good meal due to the date. Home-canned foods can generally be eaten if the lids are not bulged or rusty and if the contents are not severely discolored or moldy.
- If you're not certain about a food and you're really, really, really hungry (as in, facing danger of actual starvation), have one person in your party nibble a small bite of it, then wait for two hours. If no ill effects of a tell-tale taste of spoilage, then eat a small (as in, less than 4 oz.) meal of it, then wait for 12 hours. If no one has gotten sick after that, then it is generally safe to eat. Use this test as well for anyone who think that he or she may have an allergy to a given food.
- Keep your un-preserved foods (especially meats) as cool as you can.

Don'ts:
- Never let meat go un-eaten or un-preserved past the three day mark.
- Never eat any vegetable or grain product that appears wilted or moldy.
- Avoid the urge to 'cut around' the 'bad parts' of meat products if you can possibly help it. Molds (and spores) and bacteria may extend well beyond the visible portions. You can however get away with it in certain vegetables and fruits.

- If the animal looks sick or infected, do not kill and/or eat it. Avoid it entirely.
- Don't barter for or use pre-processed dairy products that are well past their expiration date, even if the package is perfectly intact.
- Don't accept home-brewed alcohol for consumption unless you can perfectly trust that it was made safely, or if at least two other trusted people have consumed the same batch without ill effect. This is because initially, most folks trying to brew their own booze will most likely screw it up, or make wood alcohol and pass that off as safe.
- Don't even consider eating roadkill animals unless yours was the vehicle that killed it, or the carcass is fresh.
- Do not use the same utensils or dishes for raw foods that you use for cooked ones. Do not use the same utensils or dishes for raw meats that you use for raw plant matter or fruits.
- In the first year, do not eat anything found growing along a once-busy roadside. The sheer amount of county-applied herbicides and pesticides over the years, combined with a steady diet of auto exhaust and worse, will make such plants a potentially hazardous thing to eat long-term. If you're truly hungry though, eat themin small amounts.

When it comes to cleaning freshly-killed animals, here's the rules… first off, gut the animal quickly and get out of there with it (it cools the carcass faster and the sooner you get out of there the less likely you are to lose your kill to predators, be they two-legged or four). While gutting, take enough time to inspect the animal for signs of disease… if you find any, discard the animal. Keep the hide on it until you get home (to keep the meat cleaner and to provide a useful hide from larger animals). Grab those eviscerated organs that you'll want to keep (generally, the heart, liver, kidneys and perhaps the lungs and small intestines) and put them in a bag. Anything you leave behind should be buried, or at least shoved into brush and covered with leaves. Quickly leave the area with your kill (for larger animals, you may have to rig a drag-pole or a carry-pole for it, quarter the animal, have your partner help out, or whatever you can… just get out of there ASAP).

Once you get home, then you can take the time to do the butchering work. Hang the animal for a day or two if the weather is cool, to let the blood drain. Carefully skin the animal and set the hide aside for later preservation. Try to keep the cuts as large as possible, since you want to expose as little surface area as possible to the air. Set aside what you can eat within a day or two and set aside whatever you intend to give away to neighbors the next day, or for barter within the next two days. Start working to preserve everything else. Be sure to prepare the organ meats for consumption within 24 hours. Meanwhile, package the bartering cuts and wrap them, storing them in a cool place until you are ready to go to market with them. If you won't be trading these for at least 2 days, work on preserving these as well, or at least brining them in a cool place. Keep in mind that you can get a better price for preserved meat in either case. For those cuts you're giving away, proceed to give them away as soon as possible.

You'll notice that I constantly mention charity. To best be a part of your community, your priorities for procured meats is for yourself (and your family), then for charity, then for barter… in that order. Like I've written before (a lot), you never know when you'll come up short and if you're known for being charitable whenever you get plenty, then others will be more likely to give you charity when you have none. This action and reaction will help bring you and your neighbors together into a tighter community, which is vital if you all want to survive long-term. For most Judeo-Christian religious folk, 10% to charitable purposes (or a 'tithe') is sufficient, but you'll probably want to give away more if you can spare it. Seek out those who are most in need when giving things away.

A final note for preparing meats… if the animal is small, or recognizable as something that will freak out your kids, spouse, whatever? Make hamburger or cut it into fajita-sized strips, to make it unrecognizable. This way it will be more palatable to everyone once it's cooked.

Preparing vegetation and fruits is easier… far easier. If it is on your property, harvest only what you need that day unless it's in danger of spoilage (then you harvest like mad and preserve it all if possible). If it is on someone else's property, leave it alone. If it is from the wilderness or from unclaimed abandoned property, harvest as much as you need. For those plants which you intend to eat that day, pick what you need and leave the rest be. For those which you intend to use long-term (say, it's autumn and you want a supply for winter), take the whole plant (leaving any mature seeds behind if you don't need them) and use it. Do be sure to leave some plants behind to finish seeding the ground for next year. This leads to something else… try to avoid stripping every trace of edibles if you can avoid it. You want more for next year in case something goes wrong with the crops.

For any food, keep them all whole for as long as possible before preparation, bartering, or preservation. For shellfish and seafood, keep them in wet (and alive!) for as long as you can before doing the same. Insects? Same deal (but using a cage this time instead of water, eh?)

Try as much as possible to get a balance of foods into your gullet. You'll want proteins, fats, leafy green vegetation, starches and as much variety as you can get.

Beyond getting fed, there are a lot of creative things you can do with food to make life easier for yourself and for your community. For instance, you can find/harvest/grow local spices, or make oils from certain seeds (pine nuts, walnuts, peanuts, soybeans, sunflowers, un-hulled rice, hemp, etc) - vegetable oils are going to be hard to come by anyway, so making a usable/edible source of that will be rather valuable.

Planning Gardens And Farms

This would be a good time to look around and see where you want to grow food long-term. I would strongly caution against growing anything just yet - you will want to give it at least a year post-collapse before trying to do that. The reason why is simple: nothing draws a hungry crowd faster than a garden with ripe vegetables. In spite of the dangers of actually doing it, you should still plan ahead here.

While at the beginning we figured an average of 14 acres needed to support each person, we did so assuming that many of these crops would spoil before being transported, that livestock would die off during transport and we were counting a statistical average overall at worst-case. The good news is that at the extreme local/family level, you can feed two adults and two kids at a subsistence level with as little as 3-5 acres, depending on the ages of the kids and depending on sustenance from foraged or kept livestock elsewhere. Any claims of doing so on less land will often require a huge stockpile of fertilizer, equipment, or structures that will be fragile and/or irreplaceable (something you seriously shouldn't trust your life to), or require exotic or special-strain plants which in turn will be hard to come by or potentially unusable at some future point.

However, in order to pull this off, you need to find the right spot. Let's get going with that bit first, shall we?

First we need to know what kind of land we want to use:

- Ideally, this land should have a clear and full south-facing view of the sun. This becomes especially important the further north you are from the equator. What I mean by south-facing is the southern slopes of hills (or mountains), southern sides of buildings and homes and similar. This also means being clear of obstacles on the southern side of the plot (trees, buildings, other homes, etc) - basically so that you can get at least 6 hours of full sun each day from early spring to late autumn. This also means that the plot is among the first to thaw and the last to freeze.

- The plot should have soil that is well-balanced if possible. It should be a good mixture of sand, clay and humus (humus is all those organic bits that make soil brown - basically rotted organic material.) Speaking of soil, ideally, the best land should be easy to dig into and should be easy to 'fluff' up, working air into the soil until it is soft and easy to dig with your bare hands.

- The plot should also be well-drained and clear of known seasonal flood plains or other water-hazard areas.

- The plot should be clear of any overly-large rocks and/or boulders - the smaller stuff you can remove, but the bigger stuff will be impossible without machinery.

- It needs to be close enough to easily defend and to keep a very close eye on from your home. If all else fails, you should be able to walk no more than five minutes to it.

In order to find the right place, we also have to know where you are, so…

- In the countryside and wilderness, odds are good that such perfect spots are already covered with existing crop- or pastureland. It can be as simple as asking the nearest farmer to use a couple of acres for barter (or in exchange for helping defend his home), taking over an abandoned farm, claiming a plot of unclaimed/un-claimable land as your own and clearing it as farmland, or similar activities. Not that you won't have too many (if any) options in remote desert/arid or mountainous areas, but with a bit of creativity, it is still possible.

- In small towns, you can utilize existing garden plots, extremely close-by (as in observable from home) farmland, any open parks, empty lots and nearby empty fields.

- In suburbia, you actually have a lot more options than you realize. Besides parks and sports fields, most suburbs in non-mountain/desert regions were once farmland. This means basically any patch of land that you can make clear will do just fine with a bit of work. This can even include something unconventional, such as a concrete parking lot cleared with a sledgehammer. Just note that when you initially till suburban dirt, it will be full of all kinds of garbage - gravel, building material cast-offs, pipes, bricks, electrical conduit and even old trash.

- In urban or high-density areas, you have a ton of work to do. You will likely have to clear a patch of land not with an axe or shovel, but with a sledgehammer (hint: start with old parking lots where the concrete and asphalt will be thin and easy to break up). Vacant lots can work, but many will have the remains of foundations, will be full of buried trash and will likely need to be enriched heavily with compost before it can be really productive. This is why cities are going to be very hard to farm on.

Once you find the right spot, you will have to be ready to do a few things:

- Insure that the land to be planted is clear of large rock, stumps and other obstacles - both above the soil and under it to a depth of at least 8-12".

- Work compost into the land as soon as you can pick it out, claim it and clear it. You can also bury the inedible animal guts and animal (not human!) droppings into the same land. (Avoid human manure at all costs - you stand a perfect chance of spreading E.Coli, Salmonella and worse through the food that way.)

- Be ready to become hyper-vigilant over the plot after you do plant crops there. This is for three reasons: One, you want to insure that varmints and pests don't eat your food (and we mean two legged, four-legged and multi-legged varmints). As a bonus, any meaty varmints that show up can be harvested via pellet gun (or net, or spear, or whatever) to add some meat to the table. Two, you're going to have to weed the thing with near-paranoia. Third, you will want to pick the goods as soon as they're suitably ripe and begin processing them (canning, dehydrating, smoking/salting, brining, etc).

- In drier areas, you'll want to have an easy means of irrigating the crops if you have the water handy to do so.

- You will want to insure that your crops aren't easily viewable from a distance. In most areas this is no big deal as terrain, buildings or trees can conceal it, but in open and/or hilly country? Some of the best plots of land will be visible for miles away.

Once you have the spot picked out, one option is to go in with multiple neighbors and work adjacent plots. This means everyone can more easily work together and there's (perversely) less competition for prime plots, as everyone thinks they have a piece of the best farming bits. On the downside, it also means that unscrupulous neighbors might sneak over to your plot and snitch your new produce when no one is looking (then again, that's a possibility anyway, so…)

There are some areas in which farming will be a nearly impossible thing to pull off. Three immediately come to mind: coastlines (sand, salt), mountainous regions (rocks-a-plenty), swamps and large wetlands (seriously - do I have to explain why?) and arid/desert regions (lack of water and potentially poisonous mineral deposits). In three of these cases, this is offset by harvesting seafood (coastal), or by a far larger variety of foraging and even fishing due to the sparsely-developed regions you're in (mountains, swampland). In arid and desert regions, you're going to have the hardest go of it.

Before you despair, all is not lost here. Coastal areas become quite arable less than 1/8th of a mile inland and even as close as 50-100 yards upland from the sand dunes. Mountain regions often have lakes, meadows and geologically older lakes that have level ground around them - these can, with enough work, be placed into usable production (just note that you will likely need more land to do the job than you would in the warmer lowlands). Swamps are going to be tough - mostly because dry land will be used for housing and everything else, in addition to farming. It is however still possible to eke out enough land to grow food on. Deserts and arid regions can be put into use (depending on rainfall and water sources) by irrigation, or by judicious planting along dry riverbeds if there is enough moisture and time in between flood stages to pull it off.

As mentioned before, don't go planting things just yet - wait until things have calmed down a little - and until the majority of humanity in your area has died off. It seems very harsh, but as stated repeatedly, the reason is simple: you'll want to be able to grow enough to feed yourself and your family and not watch your own future means of staying alive disappear into a horde of desperate refugees' mouths.

Before you get too defensive about the needy, I want to introduce a thought that you yourself should bear when doing this planning. Always set aside a bit of land and/or produce for charity and for seeds. In times of plenty, it can happily help those in need. In lean times, it might just offer enough of a safety margin to keep you and your family alive (without having to devour seeds to do it).

Safety And Fire Concerns

Post-collapse and at this point (about six months in) the fire department will likely be, shall we say, permanently unavailable. If a fire breaks out, you're pretty much on your own. It'll be up to you, as an individual and community, to put it out as quickly as possible. We'll start by discussing fire because, well, it's going to be your biggest hazard. Let's get some indisputable bits out there right now…

- Once a fire starts outdoors, it is extremely hard to contain. Even in civilized times, wildlife fire control personnel are trained to save natural resources, not individual homes or properties. Why? Because it's often the only option they have, even with massive resources.
- Lack of a water pumping system (or even sufficient water) will cause even more troubles when putting out fires.
- A fire can consume a home in minutes. Un-fought, it can consume a whole community in hours, often sooner.
- Raiding parties can use fire as a weapon (even if they don't intend to burn down the whole neighborhood).
- Because folks post-collapse will be improvising their cooking, heating and lighting with fire, by people with little skill in using fire for these things, the risks are orders of magnitude higher than you think.
- Homes are, in most suburban and urban settings, perilously close to each other, making the spread of fire just that much easier to accomplish.

Let's face it… the situation is just *begging* for someone to accidentally burn down the whole community. It is not enough that you keep from burning your own house down, but you're going to have to keep your neighbors from burning it down too.

Keeping Your Home Safe

The first and most important building block to avoiding fire and other hazards, is to get your own affairs (and home) in order. Our post-collapse home is going to present a rather new and interesting list of hazards and I assume you don't want anyone getting hurt.

Before we go into anything else, you're going to have to be careful around fire, period. Your family is going to have to be careful around fire, period. You're going to have to re-arrange your home a bit to accommodate this whole new dynamic of using candles, lanterns, fireplaces (on a constant basis instead of occasionally) and wood stoves. For instance? Awhile back, I mentioned how to enlarge a typical fireplace to make it usable… and you should really over-engineer the materials and distances used to keep flammable materials clear of it. Never underestimate the heat or temperatures involved, or over-estimate the construction of anything that sits next to a fire… always include generous safety margins.

Here's what you need to do beyond that…

- Keep a 5-gallon bucket or two of clean, dry sand next to all fireplaces and/or wood stoves. This is useful for putting out nearly any type of common household fire. Keep a 5-gallon bucket full of water next to that.
- Always use pots and pans that have close-fitting lids and keep the lids handy when cooking, so that any cooking or grease fires can be quickly smothered by putting a lid on the offending pot or pan, then removing the pot or pan from the heat.
- Always have an adult start, handle, or maintain fires - including candles or lanterns. Adolescents and children should only be near them with adult supervision.
- Have a plan in place for evacuation in case of fire. Each person should have their individual bug-out bags either in a common place to grab them quickly, or sitting next to their beds. Practice this plan once a month if you can.
- For each adult person keep a shovel, rake and buckets handy by the front or back door and insure that everyone in the home knows about it and why it's there (to quickly fight any small, localized outdoor fire). Keep at least one ladder handy as well.
- If possible, clear out all small trees and brush within 30 feet of your home.
- Keep your firewood at least 30 feet away from your home as well, save for a small stack next to or just inside the home.
- If you can help it, do not burn trash in an open pile. Use a steel drum (with holes punched around the bottom to allow air in), an outdoor brick barbecue, a pit, or some similar semi-enclosed device. Keep everything clear around it (including the grass) for at least 10 feet in all directions. Keep a shovel and a large pile of sand nearby when burning, in case something gets out of hand. Always have someone tending a burning trash barrel or pit until it is out cold.
- Do not burn anything outdoors in high winds or in high summer unless absolutely necessary.
- Near any fireplace or wood stove, always have fireplace tools handy and only use those tools to do anything with the fire. Do not improvise any kind of fireplace tool unless you have no other choice. Always be sure of a clear vent or chimney and be sure to clean out that vent, flue, or chimney at least once a year (in pre-collapse times, you could get away with once every few years, but post-collapse, odds are perfect that you'll be using the thing daily, so an annual cleaning is a must).
- When you put out a fire, always make sure that it is not only out, but that it is out cold. As in, cold enough to touch directly with your bare hands.

Aside from fire, you're going to have a lot of hazards lying around the home and if you have children, you're going to find out rather quickly that you either keep the place safe for them, or you end up with tragic consequences. Let's look around our post-collapse home, shall we? There are firearms within easy reach, knives and sharp objects nearly everywhere, various other weapons, tools and implements, fires and candles burning and a whole host of troublesome things that present a hazard. Mind you, this pile of hazards isn't just for small children… adults can slip or trip and get hurt as well. So, in addition to all the firearm safety, you and your family are going to have a bit of other habits to acquire…

<u>In General</u>:
- Keep your home clean as possible.
- The same goes for your property outside the home. Don't leave things lying around that can present a hazard.
- There should be a place for everything and everything should be in its place when it is not being used.
- The more organized and tidier a home is, the less chance of hazards or accidents from things lying around.

Firearms/Weapons:
- No one is to touch a firearm at all unless he or she has been taught to properly use it
- In the home, all guns are considered to be loaded unless otherwise personally checked by the one holding it and re-checked every time by anyone picking it up.
- All weapons will be within easy reach of their owners at all times.
- No one is to touch a weapon that isn't theirs without explicit permission, unless circumstances obviously dictate otherwise.
- Never point a weapon at anyone unless you intend to immediately kill that person with it.
- Weapons are to be kept as clean (and if applicable, as sharp) as possible.
- Store all ammunition and firearms away from open flame of any kind.
- Guests should always leave their firearms on a table or designated area by the door they entered. This prevents arguments from turning into deadly fights and it allows the homeowner/residents to control their own home.

Tools/Utensils:
- Treat all tools and utensils with care and respect.
- When you are done using it, clean it and then put it back where it belongs.
- Do not use a tool to do something that it wasn't designed or built for, unless life-or-death circumstances dictate otherwise. If the tool breaks, it could injure you and you won't easily (if ever) find a replacement for it.
- Treat all metal surfaces with a light coat of oil on occasion (or wipe them with an oily rag). This not only prevents corrosion, but if you get injured by it, you won't have a ton of bacteria-harboring rust to deal with. For eating utensils and knives, a good washing and dry storage is all that is necessary.
- Keep all sharp tools sharpened. A dull tool is more dangerous to the user and to bystanders than a razor-sharp one.

Candles, Torches, Lamps and Lanterns:
- Always set a candle or lantern down on a fire-proof surface, or in a fire-proof container.
- If you have or can make a candle-holder, then get it and use it.
- Always keep an eye on what is above the item. Remember that flames travel upwards and anything flammable above the flame's tip may catch fire, even if the flame isn't touching it.
- If you're not using it, make sure that it is out cold.
- Make sure that whatever you used to light the item (match, kindling stick, whatever) is out cold after you've lit it.
- Be especially careful when using any of these in a barn, outdoors during dry weather, or anywhere there is a lot of flammable material nearby.
- If you cannot find a fireproof surface to set it on, either make one (e.g. find a ceramic plate or bowl and use that), or don't put it down.
- Always keep torches generally upright at a moderate angle away from you and off to one side, away from the body. This will prevent cinders and burning material from falling on your head, body, arms, or hands. Keeping it angled off to one side prevents the flames from coming back to your face if you start running, if you trip, or if the wind suddenly shifts.
- Be especially careful when walking through archways or low-hanging branches with a torch. Try to keep as much space as possible between the tip of the torch flame and whatever is overhead.
- Be aware that parts of metal lanterns can remain hot for quite some time after they've been put out. Let them cool completely to the touch before putting them away, or before setting them onto any surface that is flammable.

Common sense tends to take over at this point, but always remember (like it'll be hard to) that a post-collapse home is going to be a bit more hazardous to everyone in it than the pre-collapse modern home with all of its amenities.

Keeping Everyone Else's Home Safe

As said earlier, it is not enough to prevent your own home from burning down, or your neighbor's messes from harming your kids or someone else's. To that end, it is up to the community to insure that the following things happen:

- Always make sure there are always clear paths for security patrols to walk, especially around the perimeters and insure that there are clear paths through every yard.
- The streets should be kept (and made) as clear as possible, with only defensive barriers and checkpoints in place (usually around the perimeter). Be sure that there is enough room for small vehicles to pass.
- Keep an eye on everyone's yards, to make sure that debris or other things don't create piles that can cause a hazard. This is not only to keep kids safe, but someone walking in the dark will appreciate it as well.
- Take all of the aforementioned home safety tips that apply to the outdoors and apply them to everyone (within reason). When you do, explain carefully why they're being enforced - don't just push dictates from on high. Make sure that there aren't folks who are causing situations that can be dangerous to the community after gaining agreement from everyone. For those who initially refuse or are unable to do so, inform, educate (if needed) and help implement these measures for them if you can.
- Organize and train a fire brigade. Make sure every able-bodied person over the age of 12 knows what to do and how to do it. Unless you have possession of a fire truck and sufficient fuel to run it, plus a solid water supply? You're likely going to have to organize bucket brigades. To this end, collect and scavenge as many water buckets as you can and keep them in stacks at strategic locations.
- Locate all moderate and large sources of water that can be used to help put out fires. Swimming pools, water towers, rain barrels, ponds, streams, rivers, whatever… and organize bucket brigades from those points nearest any potential fire. Do note that stagnant pools will present a health hazard anywhere that mosquitoes are present though, so arrange to occasionally dump bleach into these places as needed, or drain them.

One last bit - yes you have to be scrupulous about safety, but no you do not have to be an overbearing pain in everyone's butt about it. Use your head, some common sense, pick your battles and don't come off as a local petty authority. Calmly explain the most important bits, explain why something should be the way it needs to and nearly everyone will get the idea and will, more or less, make the place safer overall.

Scavenging For More

It doesn't matter how much stuff you have. It won't matter how much stuff you can get hold of in the early chaotic days of civilization's collapse. It certainly won't matter how much you and your community can keep hold of in spite of attacks from desperate people. You're guaranteed to be out of something that you really need, or someone (or group) in your post-collapse community will need some vital part or resource in order to complete an important project and no one will have it.

However, that's going to be a bit of a problem, especially this far in. Well, I should instead say that there is good news and bad news. The good news is, there will be a lot less people around each day to contend with when you go scavenging for whatever items you're after. The bad news is going to be a bit more detailed…

- If the item is useful enough all by itself, odds are excellent that someone has it and they may not be amenable to bartering for it, or will demand more than you're willing to pay.
- If the item is fragile, perishable, a gas, or contains any liquid, it will likely be broken, spoiled, leaked away, or spilled out.
- If you live out in the country and the item is not a commonly manufactured thing, you're going to be traveling long distances to find it - if any still exist.
- What items of the type that you do find likely won't fit, will need (but not have) yet another item to make a work, require a tool or implement that you don't have, or will be non-functional itself. The odds are one hundred times worse for any electronic devices, components, or parts.
- If it's coffee, ammunition, manufactured medicines, or chocolate? Well, get it out of your system now, because quite frankly, you're probably screwed in that regard.

It's better that you know this going into the idea, than to get your hopes up and finding it out the hard way. You may get lucky and stumble across the needed item(s) in an abandoned building, vehicle, or perhaps on a corpse, but unless we're talking about the collapse event being a global pandemic that wipes out 90% of humanity? Well, just don't get your hopes up. Far better to see about coming up with a solution that involves only those things you already have on hand, or which you can build yourself from readily available or common materials.

Before even thinking of doing this, you really need to sit down and think this out. If the items you need are critical, it won't take much thought. But, if it isn't (and merely important, or if it is crucial to future plans and actions), then you'll want to think this through and be certain that it is what you must do. At this stage of post-collapse, there are going to be a lot of people still desperate, still alive and still more than eager to take whatever you have from you. They might also follow you home and discover your community where they never knew if its existence before. Therefore, consider all of this before coming to any decisions.

So let's say that you want to give it a go anyway. Maybe you've heard that such-and-such a place has what you or your community will need. Maybe a nearby community died off from the plague or something, leaving all that stuff there for you to get hold of. Maybe you heard through the grapevine (trust me, there will be one) that someone at a distant location has what you need, but demands that the buyer (that would be you) come get it. Or, maybe it has been awhile since anyone has heard of anything out there and it might be worth sending a scouting party out to take a look. Either way, for some crazy reason you want to go have a look anyway. Well, let's get started then.

Planning And Preparation

First and foremost, the last thing you want to do is to just trot out there, even if you're just wanting to take a peek. Always, always, always plan ahead, because this is going to be a new and far more dangerous world than you'll be accustomed to, in spite of everything you've experienced to date.

The rules are going to be real simple. Before you plan anything…

- Never go alone. Always go with a team, preferably comprised of at least three well-armed and able men.
- Have enough fuel to get there, back and perhaps take a little more just in case. Only in sheer desperation should you ever consider going with 'just enough' of anything. Same with ammunition and weaponry.
- Use a vehicle with enough cargo capacity to carry whatever it is you're looking to get, plus enough for any opportunities you find along the way.
- Speaking of vehicles, take at least two if you can. One vehicle can break down, lose a tire, etc. Two vehicles can provide cover for each other and an alternate means of evacuation.
- Modify those vehicles if needed so that armed guards can ride in them and be able to fire freely while moving (this is more for providing covering fire than anything else…) You may also want to consider reinforcing them, but note that this will cut into fuel and speed.
- Make sure those vehicles have spare tires, spare water (radiator) and a few quarts of oil handy between them. Also insure they have hoses capable of siphoning and some gas cans.
- Have a good, solid map of the proposed route, destination and (if applicable) the return route.
- Have more than one route there and back.
- Set up communications protocols and signals with all team members - to use while there and while traveling.
- Make sure you have hand tools. Two shovels, a couple of good crowbars, bolt cutters (at least two), two small mechanics' toolkits (containing wrenches and socket sets), some good 20' lengths of rope and any specialized tools that may be needed to extract the item. Also include two tow straps for the vehicles.
- If you can find working flashlights and spare batteries, include them - one for each person. Better yet would be night-vision gear, but these would be optional.
- Have each person wear clothing that would make them blend in with the typical survivor at the destination site. This is no time (and there is definitely no need) to get stupid and go all camo and face-paint when no one else there is… doing that will only make you a target.
- Have enough first-aid gear on board to handle a combat injury or two.

Once you have those basic supplies and personnel, *then* you can start planning. Get everyone together who is going. Make sure the leaders of the party are known and that everyone knows to obey orders from them. Explain carefully what it is you need to get, what it looks like and how to identify it. If it is a mechanical part, then everyone should learn exactly what the item looks like in working condition, what size it is and what to look for in it.

Map your routes. First, find anyone who knows the destination well enough to give you a good initial idea of what the place is like, where everything is and a good idea of what to expect from the terrain there. Take the time to get as familiar with the place (and any maps) as you can. Have maps of the destination only drawn up and make sure everyone has one. Do not mark any obvious rally points on them, just have the map drawn up and have everyone memorize it. A good idea is to use symbols for each step - an innocuous symbol (circle, triangle, whatever) marking the destination, a different one marking places to sniff around if you have time and a different one entirely to mark any rally points. Draw this detailed map and put it on a table and walk everyone through it, step by step - repeatedly if necessary. This map stays behind, by the way. When it comes to getting to and from the destination, those are separate maps and they are not to be marked at all, save for the place(s) where you intend to scrounge. You will have to go over them with everyone, laying out the main and alternate routes. These maps will go with the drivers of each vehicle, but that's about it. If the maps are lost or captured, then it is hoped that you had everyone memorize the route.

The next part of your plan is going to involve timing: How long it takes to get there, how long you have to spend while you are there and how long it will take to get back. Go over each of those carefully and make sure everyone knows by heart what the deadlines are. Be sure to give yourself enough time to account for possible delays, but not too much time. You do not want to remain in one place for too long. You may be able to be a bit flexible about this, but not by too much. The idea is to get there, get in, get what you need, maybe pick up some opportunistic items (only if things are calm enough), get out and get back. At this stage of the game, strict timing will do two things: First, it will minimize the amount of time your team spends exposed to the dangers outside of the protected areas of your community. Second, it will give you goals and a purpose, which in turn affects your attitude while you're in that remote destination. That sense of purpose will make you walk with a more confident demeanor and will discourage the more petty criminals from bothering with you. Something to keep in mind when discussing timing is to go over contingency plans if someone is late or delayed. Don't get too complicated about it, but set up a second rally point elsewhere - somewhere relatively safe, that everyone can go to if a pair or trio is late meeting up at the first one. If everyone is going in together and no one is splitting up, then this won't be as necessary.

Communication protocols should be next, even if you're not splitting up the team. You should have separate visible and audible signals that determine when there is trouble, when to abandon the mission and just return, when to come together to the leader, when each goal is complete and it's time to move to the next one and when to call the whole thing done. If you have walkie-talkies that work or similar gear, this is a good thing, but make sure to agree on a specific channel. If your destination means going to a market and interacting with people, have four code words: one for all-clear (that is, one that means "everything's good!"), one for caution ("keep a watch out - potential trouble"), one for immediate danger ("weapons out - prepare to attack or defend") and one for, well, *'run like hell!'*. Make sure everyone agrees to them and knows them well enough to use the correct ones for the correct situation.

Once you've sorted all of this out, then figure out what other items you want to search for while you're out there. Be sure that first, you have some chance of finding them and second, that doing so isn't going to take too much time. Then, figure out if there's anything special about loading them or bringing them back if you do find any and if anyone might oppose you doing so.

Finally, figure out when you want to leave. If you're going to a market to sell or barter, then this is going to be relatively easy. If you're going strictly to scavenge, then you want to time it so that you arrive at your destination sometime past midnight and leaving the place before daybreak.

The only real difference between going scrounging and going to market is your behavior once you're at the destination. If you're scrounging, you want to be quick, efficient and most of all, quiet. You'll be weapons-out the whole time and your main goal is speed and stealth. If you're going to market instead, then you will want to be calm, cool and collected - you'll be supporting each other and you're going to look ordinary and impoverished.

Now that you have a plan and a time set to implement it…

Getting There

Once you leave the relative safety of your community, it is time to get into combat formation.

For one vehicle, every pair of eyes should be peeled outward looking for trouble, weapons at the ready. The driver keeps his eyes on the road, occasionally checking around. The passenger looks along the sides of the road ahead for trouble and navigates if needed. Anyone in the back keeps their eyes peeled out to the sides and rear, looking for signs of trouble or attack.

For two vehicles or more, it works like this: The lead vehicle keeps watch forward and to the sides, with someone keeping an eye out for the vehicle behind him. The rear vehicle keeps close watch behind and to the sides. Any additional vehicles keep watch to the sides and keeps an eye out for the vehicle directly in front and directly behind. If anyone spots trouble, a warning is given to all other vehicles discreetly. If you have walkie-talkies or CB radios, this can be done verbally, but otherwise you can have the passenger wave a piece of paper in the windshield or back window, flash your headlights or running lights briefly, or simply keep a close eye out.

If the trouble is mechanical or otherwise non-threatening, everyone quickly stops as soon as a safe place to do so is found (if possible) and everyone gets out into a defensive position, aiming their weapons in all practical directions. A couple of people in the troubled vehicle gets out and proceeds to fix whatever is wrong, while the leader(s) assess what to do from there.

If the trouble is an obstacle (fallen tree, wreckage, whatever), everyone immediately stops upon seeing it, backs up, then turns around and heads for the alternate route ASAP. Stop only long enough to turn everyone around and that's it - if possible, do all of this without stopping at all. Here's how you do that: If the road is wide enough, simply turn around in a chain and head for the alternate route. If the road isn't wide enough, then the lead vehicle immediately empties of all but the driver and everyone who gets out should assume defensive positions with weapons out and ready to fire. Any middle vehicles should have weapons out and ready to fire, but remain inside the vehicles they're in. The rear vehicle turns first and quickly - then the other vehicle(s) turn from there, one at a time, from back to front. The rear vehicle should drive a few yards back down the road (towards the alternate route), stop quickly and everyone in it gets out with weapons ready to fire in anticipation of an ambush and to cover the other vehicles. Once all vehicles are turned around and past the stopped (rear) vehicle, the lead vehicle loads up quickly and leaves.

Get a safe distance down the road (and be sure you're not being tailed) before even thinking of stopping to discuss what to do next. I know it sounds rather paranoid and militaristic at this point, but honestly, any obstacle on the road, especially those on narrow roads with few turnouts or means of bypassing, will most often spell ambush and you really don't want to be in one.

Another and more effective means of avoiding ambushes is to stop occasionally at high-points in the road where you can see ahead and actually have a scout or two get out and carefully scan the road ahead. You stop just before the very summit of the rise and have scouts get out and quietly (and with concealment) go to the rise, look over it and do their scanning from there. Another still is to only choose routes that don't have narrow roads that are easily blocked-in, such as tunnels, mountain passes, bridges and the like. If you can avoid these spots, you'd be better off, but you cannot always do this. If you have walkie-talkies, you do have the option of having a small and fast scouting car go out head of the group, but be sure to build some sort of light protection into it and be sure that there are at least two armed men inside the car (also, be sure the driver has at least a sidearm in case things get too bad).

With a large enough and well-armed group, traveling with some common sense to it, you will likely scare off the smaller 3-5 man ambush groups, since they won't have the firepower to deal with something bigger than they are. On the other hand, they do have the advantage of concealment, fortification (to an extent) and surprise, so don't be too sure of yourself.

If you do get caught in an ambush, your only immediate goal is to fight while retreating. It makes no sense to try and stand there taking out the group. If you know your party outnumbers (and out-guns) the ambush group, then pull back anyway, so that if nothing else, you can launch your counter-attacks from better tactical and strategic points. Your best bet though is to think twice before attacking and if you have an alternate route, use that instead. You can send expeditionary forces out later to clean things up if the ambush point is too close to your community. By then they may well have moved on anyway.

Opportunities may abound on your way there (abandoned vehicles, wrecks, etc), but keep one thing in mind - many of them may well be bait for an ambush. Approach them carefully only after thoroughly searching the immediate area for ambush, inspect the area and item of interest carefully for any possible traps, strip the thing quickly and then just as quickly get moving on your way.

Finally, do not stop for anyone, period. It doesn't matter how pretty, how desperate, or how useful someone along the road may appear… do not stop. Such people are often set out as (willing or unwilling) bait for an ambush.

While On Site

So, you've arrived at your destination safe and sound. The ride was probably rather nerve-wracking, but you've only just now gotten to the most tense part of it all. Now you actually have to put your plan into action.

If you're going to market, you always leave someone (and in any group of five or larger, at least two or more men) behind with the vehicles to guard them, period. The last thing you need is to have a successful mission, only to come back and find your vehicle(s) stolen. Always seek to park them in a somewhat camouflaged place that can be defended if need be. If they have working locks, by all means lock them up. Get your smaller barter goods out to take with you as long as they fit in your pockets. Leave the bigger stuff with the guys guarding your vehicles. Then, go shopping. While you're there, get as much news as you can, having someone writing it down in a small notepad if you can do so. Bring your bartered goods back to the vehicles immediately as you get them, then go back into the market. If you break up your shoppers into smaller groups, then have everyone check in at the vehicles or at a common spot in the marketplace periodically, at least once every 30 minutes to an hour. Odds are good that any decent-sized market will require you to be unarmed. Respect the rule, but never let the folks guarding the vehicles go unarmed, even if you have to park just outside of whatever zone is set up that prohibits weapons. Do not stay for too long. Have a set deadline and everyone meets up at 1 hour before the deadline (no excuses!) to get ready to go home. If anyone is late, you have an hour to search and/or to find out why, but when the deadline arrives? Your party leaves, period. Anyone left behind can wait until you return, or can get home on their own. This sounds incredibly harsh, yes… and you can be flexible if you like by a few minutes here or there, but delays in getting home will throw off the timeline and in some circumstances may put you in greater danger.

If your goal is to scavenge for materials, then things get more interesting. Kill the headlights before you get close to the rally point - needless to say, you'll be doing this one at night, or early dawn at the absolute latest if you're only going to spend an hour or so at the place). You will have to park the vehicles as quietly as you can, in a place where you can most easily defend them, but can get them out fast if you have to. Half of your force is going to be defending them (unless it's a 3-man team, in which case one person will be defending it, so find a very good spot for him). The scavenging team then gets to work, going to the place where the materials are. While they are getting there, all eyes are out looking for signs of trouble. You get in, search only as long as you have to, get what you need, then get back to the vehicle, dropping the goods off there. If you have time after that and there is no sign of trouble at all, then your scavenging team can go back out and do a little sniffing around, picking up whatever useful items they can. Just be sure that there are at least two people (one on a two- or three-man team) with weapons out covering for those who are doing the actual scrounging (that is, those procuring and carrying out the materials). Make multiple trips to and from the vehicle if you have to, but never walk the same route to or from the scavenging site twice in a row (or twice at all if you can help it).

If the vehicles are full, the material you seek is gone, you got all that you needed, or the allotted time is up, then you're done. Everyone gets in the vehicles and leaves once the parties have all returned. If men in the scavenging team do run into trouble, they are to drop what they have (if it's bad) and all fight together while retreating to the rally point. If the team at the rally point hear trouble and the group is large enough, there is the option of sending men out to help rescue the scavenging team in trouble as well. You have the same option if someone (or a team) is more than late getting back. Everyone else at the rally point gets in the vehicles at that point, or stands close to them with weapons out.

Once everyone and your scavenged stuff is loaded up, get out of there as quickly as you can if there's trouble, or as quietly as you can if there's not.

Getting Home

The same rules apply when getting home as they do when getting there. The only real difference is that if the mission was long, or successful, then the men may be prone to relaxing their guard a little. This cannot happen and must not happen. You must remain as vigilant and on alert as you have been the whole time. Another thing to consider is that you may be tailed… that is, someone may be following you home. If anyone credibly thinks this is the case, then relay the information to the leaders and quietly look for ways to shake the tail, so to speak. This may involve literally shooting at the person tailing you, or quietly leaving behind a trap of either men or an obstacle to block the ones following you. If you leave men behind, they are to disable the vehicle(s) and get out of there ASAP. They are then to run down the road to where you have your vehicles hidden and waiting. Your best bet though is to quickly drop a tree or boulders (or some sort of obstacle) into the road behind your last vehicle (if someone truly is following you, they're going to hang back a good distance away). This will delay the followers long enough to prevent them from catching up and especially from knowing where you're ultimately going. You may want to do this anyway along your route home, just in case. It always pays to be a little more paranoid on the subject. Just make a note of where you dropped the obstacles.

Once You're Home

Take stock of everything. Split the materials up evenly, or distribute them as agreed upon before you left. If individuals in your team did their own bartering without compromising your team's security, or they picked up something useful along the way, then try not to begrudge them for doing it - they risked their necks out there, after all.

Take the time to go over the mission. Figure out what worked and what didn't. Try and find ways of fixing what didn't work, so that there is less risk next time. Solicit and listen to suggestions and comments from everyone who went out with you. They may have ideas that you can use next time.

Electricity

Ahh, electricity. A wonderful and vital component to modern living. With it, so many things are possible… light, heat, entertainment, communication, computation, transportation… nearly all of you reading this book has never known a life without it. No, I don't mean camping trips and the like - those are temporary things and odds are good you had at least a flashlight anyway, as well as a vehicle to get you there and back.

Post-collapse, you're likely going to have to find ways to get along without it. But, this doesn't necessarily have to be a permanent thing. Certainly, you see and hear a lot of talk about generators, but honestly? Generators require a constant supply of gasoline or diesel - something you're not going to have a whole lot of to start with and will have even less of as time goes by. This doesn't mean giving up on having one, as you can always use the parts in it later… just don't rely on it to provide you with power from here to eternity. In fact, odds are good that you'll likely have run out of fuel by this stage in the game.

You can generate your own electricity without gasoline or diesel, if you have the right materials and know-how to put it all together. This is going to require a bit of planning and forethought, but if you do it right, you just might make out very well. Without diving too deeply, let's start with the basics of what you'll need.

Making your own juice is going to require five things:
- a reliable and constant source of energy (heat, light, or kinetic)
- a means to convert that energy into electron flow by using it to distort a magnetic field
- a means to transmit the resulting electron flow to the devices you want powered and back (that is, to create a complete circuit)
- a means to control that electron flow within the circuit
- a safe means of shutting things off if they get too hot, or too much out of control

Procuring all of these will be easier than you think, except for the first one, which determines how you will be taking care of all the others. You actually do have options, though - read on in Chapter 5 (the section called "Engineering") and you might find which ones are right for you. Before we do, note that you will want to procure and study each type more deeply as you edge towards any decision you may make.

Note that we dive much, much deeper into electrical power generation in the Engineering chapter of Chapter 5, but we should go over all the other parts here.

Using Power Wisely

This topic is actually two-fold: One, using power to only run things that you really need, occasionally using it as a means to boost morale. Two, not advertising it like a beacon to those who would get envious, or to outsiders who would immediately make it a priority to take what you have, since it would stand as something that best resembles what used to be, before civilization collapsed.

Let's tackle the first bit that we mentioned. Just because you may have a full solar and hydro setup that can provide a massive amount of kilowatts to your home, does not mean you should simply call it good and live like you did pre-collapse. First of all, having the entire home lit and using all the bulbs will mean that you won't have spares when they all begin to burn out one by one. It is far better for you to restrict lighting to one bulb in the living room and maybe one in the kitchen at most, removing the others and saving them as replacements for the one or two you do use. If you have nothing but CFL (Compact Fluorescent Light) or LED (Light-Emitting Diode) bulbs throughout the home? Their long lifespans mean that you can potentially have enough spare bulbs in a typical home to keep things lit for as long as you can supply electricity to the place. Even with old-fashioned incandescent bulbs, you can start scavenging around for enough bulbs to last for perhaps a decade or three.

The same story goes for appliances and gadgets. While things like refrigerators can last for decades on end, most modern electronics will definitely die off after awhile - for example, LED, Plasma and LCD televisions are rated to last for only 5-10 years at the most. By contrast, the old "tube" CRT televisions can last for up to 30 years or more. Computer desktops, if they're of sufficient quality and kept with proper care, can last for upwards of 10 to 15 years, though the operating system on it (especially Windows) may corrupt into unusability long before the hardware wears out (to that end, it pays to have at least one computer in the house that runs a UNIX or UNIX-like operating system, such as Linux or Apple OSX). However, never expect a laptop or any kind of cheaper computer to last for more than four years, unless you have plenty of fully-compatible spare parts on hand and are skilled at replacing and repairing them.

Looking at the more vital bits of electronic gear (radios, walkie-talkies, night-vision goggles if you have them and the like), there is going to be one constant: No portable battery-operated device will last longer than 7 years, unless there is a way to run a power adapter into it and power the device off of that. This is because the batteries themselves will wear out. Most brand-new and unused batteries (of any kind - disposable or not) can be stored for up to four years before their charge begins to wear down. Rechargeable batteries begin to lose their ability to hold a charge after about a year or two of constant use, unless you are extremely careful about how it is charged and discharged. Specialized batteries such as those found in a laptop (Lithium-Ion type batteries) are even more finicky and may be rendered useless in as little as a year if handled roughly, but can last for up to 5 years - though note that the amount of charge it can hold by then may be cut down to as little as 30% of its original mAh (milliamp-hour) rating.

At the larger end of the battery scale, you get a little more time. Top-end automotive and marine batteries can, with care, last for up to 7-10 years before they are completely unable to hold a charge, but can possibly be 'refreshed' if you have the sulfuric acid and know how to do it. Taken in a combination of top-end car battery plus a couple of refreshes, you might get a bit over 15 years of dependable use out of it. There is also good news on the hybrid car front as well. The latest batteries to be manufactured for this purpose have been able to hold a charge for up to 10 years, though these batteries are going to be very large and very bulky. The good news is that you can possibly find one car-load of these, scavenge it (and its charging circuitry) and use it as a home battery backup source for a wind or solar electrical plant… just keep in mind that you will still have only a relatively limited amount of time to use it, long-term.

The second bit about using electricity wisely has nothing to do with making things last and everything to do with security. If your community is dark at night and your house is lit up like a giant Christmas display, then a whole lot of neighbors are going to be more than just a little envious of you. The community leadership may well become envious of it too and may start thinking of confiscating it for community uses if they don't have their own source going. There are two ways to prevent this - one, by hiding it (good luck with that - it'll be almost impossible), or two, offering to either build something similar for critical community use (for example, keeping walkie-talkies charged, using it to light areas where raiders may sneak in, or for the local hospital or clinic), as well as offering to build it (for barter) to the community at large; this allows you to use it as a valuable post-collapse skill that keeps you and your family better supplied and better looked after. As a step further, you may also want to have some sort of goodwill gestures going on with it. For example, you can build and run a setup where you play movies on a big-screen TV in a communal meeting-place on occasion, or to store any remaining (or even homemade) insulin and refrigerated medications for the diabetics among your community.

While at first a lot of this is going to sound crazy and naive, think of it this way: If you generate goodwill among your neighbors and are known as a person that can bring electricity to those who are in the dark? Those neighbors are going to be a little more eager to defend you and yours against raiders, since any losses that you get hit with will translate into losses which they will surely feel. The alternative (being overly stingy and secretive about it) will only generate hostility and make neighbors less willing to defend you and yours. Make sense now? I figured it might.

Barter And Markets

Up until now, you've only heard and read bits and pieces of the idea of bartering. You may even know what it is at a basic level - the direct trading of goods and services, without a medium of exchange (you know, money). Well, the definition is correct, but it is a whole lot more complex than that. Before I go into the gritty details on bartering, let's explain why it is that a post-collapse society will rely on it almost exclusively for awhile.

Money is still going to be around, right? Well, yes it is. Money has been used as the best (to date) method of exchanging goods since the dawn of civilization 8,000 some-odd years ago. It has a set (albeit arbitrary) per-unit value that everyone can agree on when setting prices. It is extremely portable. It can be somewhat divided without rendering its intrinsic value worthless (unlike, say, dividing a live egg-laying chicken). It is relatively easy to store and it tends to not rot after awhile. Well, here's the problem: Paper money is only as valuable as the government that backs it. Take a look at the US dollar bill, of any denomination. The only things that gives it worth are these:

- the phrase "This note is legal tender for all debts, public and private"
- it bears the official US Treasury seal
- it holds the signatures of two people: The Secretary of the US Treasury and the Treasurer of the United States.

That's it. Those are the only things that give it any value at all. Without those parts, it's just a piece of printed linen. The only thing enforcing them is the government behind it and without that, the whole thing is pretty much worthless.

So why not shift to something with intrinsic value, like gold, or silver? Well, eventually that's a good idea - a very good idea. On the other hand, at this stage of the game post-collapse, people can't eat gold. They can't use silver to keep warm at night. No one can wear either of the two on their feet. There's really not a whole lot of gold or silver floating around in a standardized unit weight and shape. If the disaster that caused the collapse is anything radioactive in nature (let's say nuclear warfare), then there will be (unjustified, but present) fears of radiation contaminants in the metals. There is always the suspicion of forgery by the use of dilution with base metals (iron, aluminum, copper, tin, etc) or by shaving down the weight ever-so-slightly. Basically, we're still at a stage where people are only going to be interested staying alive and they aren't going to trust a single soul.

To break it down one stage further; it's six months post-collapse. If you showed up at my door to trade gold for something you think I have, I would turn you away. I need my stores to keep me and my family alive, even if I have extra, because I don't know if circumstances dictate whether or not I'll need the extras anyway. Anything I have to spare is only going to be directly traded for something I can use. Now if you showed up at my door with a live young chicken or two, or perhaps some ammunition that fits my existing firearms, or a 55-gallon drum of gasoline? Then we'll talk. And this is where I need to fill you in on what barter actually is…

The Mechanics of Barter

Barter requires a few things to actually happen before it is successful. In order for you to sell me something, all of the following items have to be true:
- You have to have something (or a skill) that I want or need.
- I have to have something (or a skill) that you want or need.
- You have to be willing to part with the item or perform the service that I want or need.
- I have to be willing to part with the item or perform the service which you want or need.
- You have to be willing to pay my asking price.

If any of those factors cannot be satisfied, then the deal is off. It really is that simple.

Finally, there are factors that affect the barter as well. For example, how many of something do I have laying around to give? If I only have two shovels to my name, then the asking price for one of them is going to be a whole lot higher than if I had 30 shovels laying around the house. What is my mood towards you as a person? If you were a lousy neighbor who always treated me and my family like crap before the collapse, then my asking price is going to be a whole hell of a lot higher than it would be if you were a close and trusted friend. How desperate are you for the item? If I were a less-than-perfect gentleman, I'd perceive that desperation and try to get as much out of you as I could, especially if you were in desperate need of something that I could spare. If I perceived that you didn't really need it but I really want or need something you have, then the price would likely be a lot closer to what would be considered a fair market value.

As you can see, there's a lot more to it than just asking a price and paying the total.

Obstacles

There are also going to be two other things that gets in the way of any successful bartering for certain items. First is the secrecy. If I have a lot of spare food, enough to actually barter with, I sincerely doubt that I'm going to let you or anyone else know about it. The reason why has everything to do with my own security. If you found out that I'm bartering off extra food, then word is going to get out and I'll end up with a very long line of beggars at my door - and then have to fend off every desperate and hungry human being who will want to take what they cannot trade for. To top that off, if there is any government left and they found out, they may be busting down my front door demanding to know where I keep my 'illegal hoard' of food, medical supplies, etc.

I (like most same people) would rather not have to deal with that - it's more of a headache than its worth. The second reason that barter for critical items will initially be slim and none has to do with the scarcity itself. We're still at a stage where vast numbers of people are dying of starvation. On a purely objective and logical (but certainly not moral) scale, why trade you for that item when I, by merely surviving longer than you, can just scavenge it from your home/corpse/vehicle later on? Yes, that is very harsh. Yes, it is very much against what a community would stand for. But, it is human nature and in a situation where most of the population is going to be dead of violence, disease, or starvation? It is unfortunately going to be the norm unless you're a member of a working community.

A final obstacle to barter involves the fact that many of those who are prepared sufficiently are not going to be in much mood to interact with those who are not prepared at all. They will be more interested in 'holing up' and waiting for the population to die down (literally), or will have to be critically short of some unforeseen item before they will bother to barter. These are also going to be people who will isolate themselves from the rest of the world as much as possible. The good news is, as said multiple times before and beyond, without community, small isolated groups or individuals are liable to end up being out of goods sooner than they realized, because you can't be perfect and they will be the ones picked off first by raiding parties due to isolation. Very few folks will escape it, so on the plus side you can expect at least some barter to happen.

Goods That Are Always In Fashion

All of that said, this does not mean that barter is going to be dead. Some things will sell no matter how bad or ugly things get. Some of these things will – for good or evil – include:

- Sex. What, you thought the world's oldest profession would disappear? Prostitution has survived for far longer than history itself has. The very first human settlements had prostitutes and I can confidently predict that the very last ones will have them too. It will be even more prolific in a post-collapse world. The reason why is that any passably pretty woman knows full well that there is always one method she can employ to procure food and critical supplies, even when logic says she otherwise shouldn't be able to do so. Let's face it - men think with their penises and a little thing like the collapse of civilization certainly isn't going to stop a guy from wanting to use it once in awhile.
- Alcohol. Almost as old as human history itself, alcohol, be it homemade or from remaining pre-collapse stocks, will always be popular enough to get attention from even the most hardened survivalist. Alcohol after all has three uses - two of which are legitimate: First, it is a very good means of preservation. You can keep calories in liquid form and with few exceptions, it actually tastes better with age. Second, it is a good anti-bacterial solution if the concentration is high enough. Pure moonshine can kill most

germs without killing or harming you. It is why even the strictest Mormon or Muslim teetotaler needs to have it in quantity. The third reason is completely immoral, but completely understandable: The rather overwhelming need to escape reality for awhile. People are simply going to want to get drunk and forget about the miserable state they're in. It also helps such folks to forget that they're hungry, that they're cold and will definitely help them forget that they're scared. Because of all this, alcohol is going to get made and/or traded no matter how rotten things are.

- Drugs. Yes, you knew this was going to be in here. Like alcohol, people will buy and consume it primarily to avoid reality for awhile and to make themselves feel better for a spell. Addicts aren't going to suddenly stop craving the stuff just because they're hungry, cold and tired. Because of all this, there will be people who will trade for it and there will be people who supply it, for as long as the stuff is available. It is also why you want to identify and eject anyone in the community who is addicted to the stuff - they will do anything to get more, even if it puts you, your family and your community in danger. However, before you write off all drug dealers as targets to be shot on sight, consider that a limited number of drugs will actually be beneficial to the community. Marijuana is non-addictive and makes a perfectly all-natural pain killer. Prescription drugs (usually painkillers and antibiotics) are also useful for proper medical uses when correctly administered by a medical professional (or as close to one as you can get).
- Counterfeit Goods. Food that is expired or mixed with inert ingredients (e.g. flour or powdered milk mixed with talc or gypsum) will likely sell like crazy, even if the buyer knows it's been 'cut' or is expired. Hungry people do desperate things, after all and there are immoral and greedy people who will happily sell it to them.

In addition to the obvious, there are goods that make perfect sense to barter with, won't give away knowledge of your critical supplies and are yet quite useful to the buyer. In fact, you would be amazed at how much you can put up for barter to get the things you need:

- Clothing. This one seems not very obvious at first, especially in the summer. However, children are going to grow out of their existing clothes, most clothes will be overlooked in the initial mad rush to get supplies and if you had a family member or spouse pass away, you may well be sitting on a goldmine or items that someone else really needs, but you likely do not (unless you're really into wearing dresses, I suppose…) A smart prepper who bought up a huge pile of well-built clothing at the local thrift stores before collapse, paying pennies on the dollar, can easily find him or herself opening a veritable store once civilization collapses entirely. If you have kids, the clothes they out-grow that you can no longer use can be put to use by someone else's kids.
- Shoes. Same story as the clothing. If you have extra pairs of sturdy shoes that you cannot fit into, you have a loss in your family and extra shoes left behind, or you had stocked up from the local thrift stores before collapse? You have a potential gold mine post-collapse.
- Diapers. No, not the disposable kind - they take up far too much room and aside from a few you'd keep for massive injuries, are pretty much worthless. I'm talking instead about cloth diapers, either home-made or pre-made. A stockpile of these are useful for babies, they're useful as rags and they're quite useful for a lot of other things besides.
- Tools. Be careful with this one and only do this if you have a lot of extras. Tools are going to be very hot items long-term, because most folks will need them but not have them. Certain tools will have the highest desirability: axes/mauls, shovels, hammers, saws, sheet-metal tools, manual woodworking tools and the like.
- Recovered Nails. If you are good at scavenging nails (or have sufficient blacksmithing skills to make them), you stand to make out very well, because it won't matter how much wood someone can get, nailing it all together is going to require, well, nails. A supply of recovered or 'virgin' (never used) screws will carry even greater value.

- Matches and cigarette lighters. It seems silly at first, but someone who bought up massive quantities of cheap matchbooks or lighters pre-collapse may well find him or herself walking in tall cotton in the post-collapse market. This is because unless you smoke, or had tended a wood stove or fireplace frequently for heat pre-collapse, odds are good that you're not going to have matches or lighters handy. For a society suddenly thrust into making heat from burning wood or coal, matches are going to be extremely (excuse the pun-) hot items.
- Batteries. Obviously you won't want to leave yourself short, but if you have (or have procured) a massive stockpile of excess batteries, you may as well trade them off before they expire. At this stage of post-collapse society, batteries are definitely going to come in handy.
- Coffee. If you drink or serve coffee on a regular basis, forget this paragraph - you'll need all you can get. On the other hand, if you don't drink coffee and have a ton of it laying around, then by all means put it to work for you! Just keep in mind that you shouldn't under-price it… the stuff only comes from equatorial regions and post-collapse they likely won't be making too much of it, if any.
- Pornography. Go ahead, laugh. Get it out of your system. All done now? Good. While highly immoral, there will certainly be a market for the occasional dirty magazine and if you have any that you're willing to part with, you may as well get something useful in exchange for it.
- Condoms. Again, silly, right? Well, not silly at all. Unless you intend to use them yourself, you may as well get something useful in exchange for them.

Note that this is just a very small sample of what you can sell at barter if you need to. At this stage of a post-collapse society, I would recommend that you go easy on doing so and only barter when you absolutely have to. Most of the manufactured goods will only go up in value as time goes on. Just keep in mind those things which will expire or wear down after awhile (e.g. batteries) and try to prefer selling those first.

Stuff To Avoid Selling or Taking

Unless you're absolutely desperate, there are a few items that you should avoid selling or taking in trade at all costs and it is hoped that you only stockpiled these things for your own use:

- Weapons and/or Ammunition. Selling someone the means to come to your home and take the rest of what you have is dumb, dumb, dumb. Give weapons or ammunition only to trusted neighbors if need be, but never sell weapons or ammunition, especially not this early in the game. There are many in the prepping community who advocate stocking up on .22 caliber ammunition for bartering purposes or even as a proposed future currency, but honestly, avoid even the thought of doing that. Now if someone wants to give you ammunition in trade for something you're selling, then by all means soak it up - especially if you don't have a more pressing need for something else.
- Jewelry and Gems. Seriously? Save that stuff if you have it. With the advent of laboratory-grown diamonds and fake gems good enough to fool the unprofessional eye, you will find few (smart) takers. While you're at it, avoid taking that stuff in trade at all costs, because I doubt that you'll ever be able to get rid of it.
- Common Electronics. It should be a no-brainer to avoid taking, say, a big-screen TV, a blender, or a gaming console in trade for something. If civilization has indeed collapsed, you'll find very little use for the things. On the other hand, common

household items that contain motors and gears can be useful if you're the hardcore tinkering type who can generate his own electricity.
- Sex. Seriously? Think about this one for a moment, guys. You're giving away something valuable in exchange for what… a few seconds of hormonal pleasure that's over all too soon? If you need some 'strange' that badly, then you should have stocked up on some porn magazines before it all went down. Avoiding prostitution also means you can avoid getting venereal diseases (or worse), which in the post-collapse era is often going to be painful and fatal… and explaining *that* to your wife and kids as you lay dying is going to be no fun at all.

Stuff To Not Trust Immediately

Because we live in a world filled with human beings, a lot of them are going to want to rip you off. This means that you're going to need to do some testing and verification

- Gasoline and Petroleum (without testing it first). Unless you are able to test it (to insure it isn't watered down, rotten, etc) and more importantly, know what to look for, you may be in for a sucker's bet. Believe it or not, this is somewhat common even in pre-collapse times among unscrupulous dealers and gas stations, since a water dilution of up to 10% will likely go unnoticed by most folks and the results are more easily ascribed to a mechanical failure. If you need gasoline or diesel that badly, siphon it from abandoned vehicles at this stage… for awhile, it will be the purest source you'll be able to get your hands on. However, if you really need it, be sure to quickly scoop up some liquid from the bottom of the container (scraping the bottom), dump it into a clear glass jar, then let it sit for a few minutes. If you see layers form, or if it looks cloudy, then you've got water in the fuel (think about how Italian salad dressing works… same principle here. Pure gas or diesel isn't going to separate out into layers of water and oil like that).
- Medication from Untrusted Sources. Just like the illicit drugs, the prescription drugs from sources you do not know and trust may be placebos, may be 'cut' (mixed with inert ingredients) or may be the wrong stuff (e.g. blood pressure medicine put into a bottle with a prescription label that reads "Oxycodone", or insulin injected into an empty vial labeled as Morphine). Unless you have a pill identification guide handy (most pills are marked in unique ways to identify what they are), or you're a pharmacist (or happen to know one who can check), then you may be in for a crap-shoot. This stuff is going into your body, or the body of a loved one and the last thing you want is for nothing to happen (or worse, an entirely bad result due to drug allergies, poisoning, etc). You should be safe if the vials or bottles still have their metal or plastic tamper seals on them and you happen to know how to identify these.
- Ammunition. Unless the seller allows you to fire a random round or two, or unless you trust the seller, don't trust the product immediately. It is stupefying how easy to hand-load a bunch of ammunition without gunpowder in them, or by putting dirt instead of powder in the cartridges (so that you hear it 'shake properly'). A fake round of ammunition will clog a rifle barrel quickly (the primer will push the bullet into the barrel, but that's often about as far as it gets), leaving you in a very bad position - probably at a time when you really need the thing to work.
- Gold, Silver, Precious Metals. If you're looking at pre-1962 quarters and dimes, you have 97% silver. If you're looking at bullion bars or any gold coin, you won't have any idea what you're looking at. There are ways to test precious metals (acids, scratching the surface hard to check for gold plating, using a magnet to pick up iron/steel-based fakes, etc) and you will have to know these methods before trusting

any source of precious metal. It is suggested that you gain this skill, or have a skilled neutral party in the community who can test this for you and 'certify' it.
- Food. As mentioned before, food is too easy to alter and 'cut' with inert-but-worthless ingredients. It could be spoiled. It could be re-packaged from cans which were swollen (botulism), or worse. Long story short? Unless you're running out of it fast and have little choice, don't barter for it unless you trust the source. The seller should be perfectly willing to eat a few bites of it himself, right then and there and without hesitation. You will also want to keep an eye out for any 'venison' that may have instead come from dogs, cats, or even people. While such meat sources do contain calories and certainly won't kill you, human 'meat' can easily carry human diseases and you always want to know what it is you're about to eat. Insist that the person selling it always leave the hide or part of the carcass attached, including the skull and do not buy any meat that you cannot trust or identify.

Running a Market

As a community, the best way to get some sort of trade going is to provide a safe and secure place to perform trades, for both your own community residents and for any traders that come by. At this stage post-collapse you will want to thoroughly inspect and insure that any outsiders are legitimately trading and that any disputes or instances of cheating/fraud are handled on the spot.

The best way to start is to pick a good site. Choose a place just outside of your community, so that outsiders can buy and sell without allowing them to wander around inside your community scouting for weak points and the like. Insure the site is reasonably level with good drainage, has some means of setting up shelters or tents and can be defended against in case of raiders. A grassy or level paved area free from mud (even when it rains) is a good place to start. A store parking lot would be perfect. Your site should be close enough to town so that residents can flee there quickly and a checkpoint should be established to only allow entry and exit for the residents during business hours. Finally, you should put up a set of cork boards and chalkboards, sheltered from the rain, so that messages and notices can be posted.

The site should be equipped with bathroom areas that are reasonably private (e.g. latrine ditches cordoned off by sheets or tarps) and should be maintained as much as reasonably possible. The site should also have a covered area for the market judge(s) to sit and should include a table or two for their use. If you have enough spare tables, you can 'rent' them out to sellers for a nominal fee (payable in working ammunition, fuel, weapons, or other items that can be put to communal use). A place at the outside entrance for checking in all weapons should be next. This should be a fairly sturdy shed or small building and should include a table, some sort of chit or tag system (basically tags set in pairs with numbers, like a coat check stand) some large numbered bins for holding weapons and a place for someone to work the stall as well as two armed guards. The guards at this spot will always have their weapons out.

Manpower is next. Be sure to have enough 'muscle' (armed guards) around to enforce the rules, defend the place if necessary and insure the safety of everyone involved. At a minimum, you should have an armed guard for every 30 people, plus two armed guards at the weapons check-in spot, plus at least one armed guard at the market judge's stall. The Market Judge runs the market, makes decisions on the spot and his or her rulings are final - this person should obviously be trusted by the whole community (or at least the vast majority of it) and be reasonably intelligent and wise. It would also help if you supply this person with things like an accurate set of scales, measuring cups and calibrated weights. Next, you will need someone to check in all the weapons as patrons and sellers enter the market and check them out as they leave. Finally, you might want to have someone monitor the cork/chalk boards, to keep them reasonably neat, tidy and to pick up on anything suspicious or newsworthy.

Once you have all of that set up, it's time to get up a set of rules. These should be simple, easy to follow, iron-clad and posted at all entrances, the market judge's stall and at weapons check-in. For example, a good example set of rules would be as follows:

- <u>Business Hours</u>: Monday-Friday, 10am to 6pm. Saturday, Sunrise to Sunset. Closed Sundays. *(Note: you can set your own hours - just keep it during daylight).*

- Absolutely No Loaded Weapons Allowed In The Market, For Any Reason. All Weapons Trades Will Occur Outside The Market. No Exceptions. All Weapons Will Be Checked In At The Entrance.

- No Gambling Or Games Of Chance Allowed *(Note: this rule is optional, but in my humble opinion a damned good idea).*

- No Alcohol or Illicit Drugs To Be Consumed In The Market *(see also the gambling provision - optional, but you'll want to enforce it, trust me.)*

- The Market Judge Is Located At *{give location- "Center Of Market", "By the Corkboards, etc}*. Take All Disputes There. All Decisions And Opinions Are Final.

- All Sales And Trades Are Final.

- You Buy And Sell At Your Own Risk.

- We Are Not Responsible For Any Payments Or For Items Left After Business Hours.

- Frauds And Thieves Are Not Tolerated. Repeat Offenders Will Be *{list punishment - "Detained", "Executed", etc.}*

You are more than welcome to add other details and rules as you see fit, but this should be enough to get you started. Be sure to make the signs large and legible, so that there is no confusion.

Finally, you should also have signs labeling the Market Judge, the Weapons Check-In point and the 'Restrooms' (latrines). A good option is to also post (in large markets) smaller signs in prominent places, with arrows pointing the way to these places.

When you open and close the market each day, you should make it a habit to have some unique sound, or make some visual cue that the whole community can see. Examples would be ringing a bell, raising a flag, or something similar that everyone for a decent distance around can see or hear. Just make sure it's not to be confused with an alarm. Other preparations can include seeding the cork/chalkboards with news, notices and things of that nature.

Playing The Bouncer

While most trades in the market will go smoothly, it takes work to insure it. This includes the following roles and the way they should be run…

It all centers around The Market Judge. The Market Judge runs the show; he will insure that any scales on the premises are accurate, that there is no fraud going on and is able to marshal any of the armed guards to wherever they may be needed the most. The Market Judge is also the one to settle disputes. If anyone has a question on the authenticity of an item, or thinks that they have been ripped off, they can take it to the judge, assuming the seller hasn't already left the premises. The judge is not to make any trades at all for him or herself, period. If the judge does need to make a trade, then someone else is appointed judge for that entire day.

The guards will be all about the place. There will be fixed guards at the weapons check-in station and at the market judge's booth. The rest need to be in and about the crowd in pairs, keeping an eye on things and making sure everything is normal. They will also be responsible for ejecting any small troublemakers, or in detaining any serious troublemakers.

The weapons check-in station is important. Each person entering the market will leave his or her weapons at the check-in station. In return, they will get a tag with a number on it matching the bin number where their weapons go. Each customer gets his or her own bin - no exceptions (this prevents any mix-ups). When they leave, they turn in the tag in exchange for their weapons. The reason this is important is so that heated arguments do not turn into fatal ones and it will cut down any armed robbery to practically nil. Anyone caught in the market with weapons will be detained and all weapons confiscated without return. You might choose to be friendly about it and escort the weapons-holder to the check-in station, or you can choose not to and forcibly remove their weapons at gunpoint - this depends on the situation, naturally. Now if anyone is trading weapons in the market (or wants to), they still check in their weapons at the station and any interested customers can inspect the weapons there. Everyone going in or out who is not a guard has two choices: Check in the weapons, or do not enter. For everyone's safety, you must hold to this rule.

If there is any serious trouble in the market, at least, say, 2-3 pairs of roaming guards should go to the scene, with the rest keeping a closer watch on things elsewhere (unless you only have three pair, in which case leave at least one pair keeping an eye out in case the trouble is a diversion). If the trouble is minor, then all but one pair of guards should quickly go back to watching the rest of the market. If the trouble is serious enough, the market stops for at least an hour (so that an investigation can occur) and no one leaves or enters except the community leadership.

In case of arguments that get loud, a pair of guards should wander close by, but not too close - call it 10-20 feet at the most. If the arguments get too loud, then the guards should go together to the participants and suggest that they either cool things down, cancel/reverse the trade, or take it to the judge. The guards don't leave until a settlement of some sort is reached, or the guards can escort the participants to the judge's stall for settlement there.

Notes On Dealing With Money

The smartest bet insofar as running a market at this point in time is to avoid getting entangled with setting any standards as to what is or is not money and definitely avoid setting values on any currency that still may be in use (or offered). Leave that to the buyers and sellers, period.

However, in reality this isn't going to happen. The people will demand that you get involved with this and you're going to be stuck with setting some sort of standard. Odds are good that various remnant governments (state governments, city governments, whatever) may start minting their own money, or assigning different values to whatever national money is still floating around in circulation. There may be national government still in existence, even at this stage and foreign money may still have some sort of value attached to it as well.

If you're dumb enough to agree towards setting any standard against any existing printed money, then here is how you'll want to do it. First, set aside a bit of the central cork board area and a part of the market judge's stall to post the latest 'standards'. Be sure to have someone posted at the cork boards to insure that only the market judge changes any figures. Then, get together and base the standards on the value of a given currency against one gram of gold (and you'd better have an accurate scale or two handy to measure such things). For instance, the 'currency board' might read like this:

Currency	Value (per gram of pure gold)
US Dollar	0.5
Canadian Dollar	4.5
Euro	2.5
Boston Dollar	5

These are only examples - they only illustrate and reflect a hypothetical situation. For example, from this chart we can read that Canada's society and government are still somewhat of a going enterprise but may or may not be fading fast, that the European Union is still around but struggling hard and that the city of Boston is nearby and is strongly operating as an independent entity. Meanwhile, the US dollar's shown value indicates that the US Federal Government is still in existence and will be for a little while still, but is likely so weak that it can no longer fend for itself or exert much if any influence out where you are.

The factors in play are the strength and health of the issuing entity in question, the amount of trade that occurs with (and from) its territories, the distance from you that the issuing government's influence may be and a slight prediction on short-term future trends. These are what you will have to carefully take into consideration as well. We will dive a bit more deeply into the subject of money in future chapters, but suffice it to say that this is about as far as you'd really want to go and even then only if you cannot otherwise avoid it. It is best to start things out as just even straight barters, where the buyer and seller determine value, not you.

The Psychology Of Post-Collapse Living

Let's face it: Where once you could live happily and isolated in suburbia and for up to 16 hours a day not have to interact with anyone but maybe your family? Well, those days are way over now. You literally depend on your neighbors to stay alive and if you're not communicating with them, you're fast going to be forced to. Where once no one particularly cared what you did or said (and you couldn't care less about others) while outdoors, you now have to pay very close attention to nearly everything and everybody, if you expect to remain alive for very long.

Your living conditions have changed radically too. Where once you may have lived where everyone had their own bedroom (except for the one you shared with your spouse), now everyone is likely sleeping in the same room, usually huddled around a fireplace or wood stove for warmth. Privacy is a rare thing. Everyone is stressed. Your biggest worry went from office politics, to making sure you can feed your family for the next month or two. Everyone is scared, though many won't admit it. Nobody will have the same daily routine that they once had. You're carrying a weapon almost constantly from the moment you step outside the door. A criminal world that once didn't give a damn if you even existed, now wants what you have and will do anything just short of dying to get it. Keeping up with the Joneses now involves making sure that the Joneses' house doesn't burn down and take yours with it. Where once you had never seen a dead or seriously injured person except on TV newscasts, by now you likely have had to bury a few bodies and not all of them will have been intact. Where before the collapse you likely hadn't been in so much as an argument with a stranger since childhood, by this point you have probably had to fight off an intruder or two and have likely killed a human being or two by now… and for some odd reason you may not seem all that affected by it now.

All of this and much more are going to directly assault your sanity, your personality and your disposition. It will shake your faith in God, humanity and anything else you hold dear. Your biggest goal will be to keep your wits about you, retain your faith and do your best to make the world a better place in spite of itself.

Keeping Your Head

First and foremost, before you can influence anyone else, you'll have to insure that your own head is screwed on straight and make sure it stays that way. It is all too easy to give in to those fears and emotions that will make you angry, paranoid, scared, or suicidal.

The last thing you want to do now is to despair. Yes, everything you know and love has been changed - radically. Yes, you may have even lost a spouse, a child, or a parent - or, well, all of them. You may have no idea how loved ones living far away are doing, or even if they are alive. In spite of all that however, you mustn't give up. Take the time to read The Bible, or whatever books your faith (or lack thereof) centers around. Take the time to imagine better times coming - force yourself to construct scenarios to that effect if you have to. Always try to look ahead, plan ahead, think ahead.

If you've lost a spouse or companion and you have the supplies to support one, then seek another that you can trust, if only to comfort and protect each other. Odds are perfect that person is hurting too and it can't do any harm to help build each other back up again. If you've lost a child and you have the supplies to support one, then find an orphan and adopt the child as your own. The love that child will build for you will in turn help return you to being a normal person again. You will be amazed at what loving and caring for someone else will do for your own outlook. Just one caveat, though - these people are not replacements for what you have lost - always remember that. Instead, enjoy the discovery of this new human being. Unless you intentionally abandoned or killed your spouse or kids when you had a choice to not do so? Don't feel guilty about inviting this new person in - if they reciprocate, then this is a *good* thing that you're doing - for both you and this other person. Always respect what you have lost, but hunger for what you can gain in love, affection and togetherness. Don't expect any of it to happen overnight or immediately (after all, they're likely grieving too), but with mutual understanding, tolerance and kindness towards each other, the odds are definitely in your favor long-term.

Take the time to pray if you have the faith. Find the time to grieve. If there is a preacher or priest handy, take the time to talk to them quietly, even if you're an atheist. Find someone you know and trust (or at least trust) and talk together, being certain to listen as much as you speak.

Yell and curse at the heavens if it makes you feel better. Guys, seriously, cry when it hurts - no one will think of you as less of a man for doing so, especially considering all the crap that everyone has gone through up to this point.

Whatever you do, do not bottle up your emotions any longer than absolutely necessary to save life and limb. As soon as you have a quiet moment, take your emotions out and examine them. Let them out. Look at them. Work through them at these times and do not avoid them. On the other hand, do not let them dominate you. When it is time to put them away, you put them away and wait until the next quiet moment to take them out again. When they have faded, then you can put them away for good. Save them for old age or the Great Beyond.

Don't worry if you don't succeed at first… as long as you try and keep at it every day. Soon, you may be blessed with brief moments of happiness. Treasure them. As time goes on, you will find that these moments happen more and more. Seek them out as often as you safely can.

Helping Others Keep Their Heads

If you are able to keep your head about you (or even if you're having trouble doing it), you need to help the others in your home keep their spirits up. It is imperative that you encourage each other, help each other and most importantly, be tolerant of each other as much as you can.

When there are moments of peace, take the time to sit together, make up silly songs and play games. Take the time to listen to the kids and to answer as many of their questions as you can, being as honest about it as you can (just keep in mind the child's age and maturity level) and remember to be optimistic. Try and carve out some time to be alone with your spouse and you shouldn't expect him or her to be up for sex, since that's not the point of the exercise. The point is to be intimate and close. To spend time with you, your spouse and nothing else. To whisper. To talk. To share a joke and laugh together. If you're interrupted, pick up where you left off the very next moment you're not. Shoot for having a couple of 'special' hours together at least once a week if you can. If you make it a regular thing, it gives the both of you something to look forward to throughout the week. And hey, if you get a little nookie, bonus!

Helping The Community Keep Its Head

The community residents are all going to be scared, confused and on edge. A few of them will likely lose it, with disastrous results. A big part of being a community is going to involve supporting each other. To get together and help each other. To help your neighbor patch a hole in his roof, or help each other clear out debris, or to console someone who has lost a spouse, child, or friend. To help a neighbor dig the grave he may be stuck with digging. To help build an outhouse to share between homes.

Getting together for more than defense and scavenging is a must. Religious meetings and services, non-religious social events, workshops (especially for needed and critical post-collapse skills) and other events are vital towards bring people together. Be certain to organize some events for purely social reasons, so that people can forget about the world for a little while. Organize old-fashioned "socials". These are basically dances and games where young single kids and adults of the opposite sex can get together and meet up in a setting that no one can complain about. School (if you can set one up) helps kids learn and at the same time provides the benefit of escaping from the daily grind (and their fears) for awhile. Appointing someone (or a few someones) to talk to without judgement and in confidence is a very good idea.

As mentioned numerous times before, going around and checking in with everyone, to see if they need anything and to provide them with what you can (as community leaders) is an excellent way to help bring people together.

Close Quarters

Getting back to the home, let's take a bit of time talking about strangers living together. While you really want to avoid it if possible, it will probably happen, so dealing with it is going to be a priority for the folks living in that home. Living together in close quarters, especially if the people living together are not bound by blood or marriage, is going to take a lot of getting used to.

One of the reasons (top reason, actually) that holing up in a bunker with strangers (or even casual acquaintances) is a bad idea? It has to do with personalities under stress being confined together. A whole lot of people have done a whole lot of studies on this phenomenon, from NASA to the US Navy (for submarines), to numerous university psychologists, sociologists and psychiatrists. Long story short - if you can otherwise avoid it? Never, ever hole up in a bunker if you can at all help it.

To a lesser extent and even without a bunker, many of the same things happen to a group of people thrown together in a semi-confined area under hard stress. A common thread among them has to do with the following problems:

- Fatigue. It's hard to sleep when there are a bunch of strangers in the same house or room. Certainly it can be done for short periods of time and over time this problem does diminish as long as each person gets to know each other and is allowed their privacy. On the other hand, under periods of prolonged stress (the end of civilization counts as a cause of that), people are going to want and need more time to delve into themselves without having a bunch of other people around. If you are used to sleeping in a dark and quiet room your whole life, then the noise and movement of strangers will naturally wake you up. Conversely, if you are used to some sort of noise and light when you sleep (example: sleeping with the TV on), then the darker surroundings and lack of the comfortable and usual background noises will also keep you awake. Until you get used to the other folks enough to be comfortable with them, you're going to have a very hard time falling asleep unless it's from exhaustion.
- Pecking Order. People who are naturally used to being in unquestioned charge of their homes are going to have a very hard time compromising and accommodating others in their new home. It's doubly bad when you get a bunch of strangers or not-too-closely known people and have them live together and triply bad if there are any "A" type (aggressive) personalities in the group. A pecking order will try to get established whether you want one or not. Whoever is 'in charge' will try to assert his or her dominance to the point where it starts to defy common sense and it will certainly grate on everyone else.
- Paranoia. If you're all alone in a home full of strangers and you don't go out of your way to get to know everyone else, there is a chance that you're going to start to think up all kinds of scenarios that will automatically put you on the defensive, even when there is no call for it. Taken to extremes, you will even begin to see things when they really aren't there. If any cliques develop among the fellow dwellers and you're not in those cliques, then things will start to get real bad. Even worse, if there is any question at all about supplies and their availability, or if anyone in the home feels deprived of goods that others have (or have received), then all sorts of bad feelings will develop in a hurry.

- Increased Intolerance And Irritability. Initially, everyone is going to be relieved for awhile… after all, you're still alive and managed to escape the hordes. During this brief period, everyone is going to be a bit more accommodating towards their fellow man and especially towards your fellow roommates. However, once the initial wave of relief and contact has passed, things will begin to grate on you - especially others' habits. If someone picks their nose, makes a repetitive noise (clicking nails on a hard surface, say), or does something that you yourself find only slightly irritating, then that irritation will be amplified as time goes on. Things that once escaped your notice, or was easily tolerated, will begin to really grind on your emotions. The longer any confinement occurs, the worse it will get. Things that irritated you but could be endured for a short time (example: a crying baby)? Over long periods of confinement or close-quarters living, these things will quickly become unbearable, to the point of quickly escalating to a heated argument every time it happens. The 'offender' in turn will feel far more defensive about it as well, leading to a bad situation all around.

If you find yourself living cheek-to-jowl with folks who are complete strangers, or who are mere acquaintances, you're going to want everyone to get together and lay some ground rules down - quickly. First and foremost, if it is at all possible, set aside an area where folks can have private time, alone and undisturbed (except for emergencies). Make sure everyone has a time all their own reserved to be left alone, even if they don't want it - using that time is optional, but having that time set aside is not. Second, try and agree on a basic set of house rules and behaviors. Each person should also have a spot that is his or her own private territory, to be left undisturbed except for emergencies (this is usually the bed, bunk, or corner where the person sleeps).

Third, make sure there are at least five or seven people and always an odd number, in a home where strangers are living. It sounds weird, but it works. You see, if there are only three people, two of them will tend to be closer, leaving the third person out. With four or six occupants, it becomes two versus two (or three versus one), or it becomes a milder variant with three versus three, or so on. This isn't any sort of hard and fast rule, but it should bear consideration in any situation where you have more than a single person, couple, or family living under the same roof.

Something else to consider: If at all possible, either make the home unisex, or insure that there are more females than males. Having a home with less single females than single males is a recipe for a whole lot of fighting, especially when everyone starts getting feelings for each other. Certainly, the ladies may start fighting over the pool of men in the house, but it doesn't boil over as easily into rage and physical violence. As a bonus, the presence of women in the home will subtly force the guys to behave in a more civilized manner.

Never, ever let an argument fester. Always try to respect the other person's point of view and always talk it out with the whole group if you can. The more of it that is out in the open, the less of it will sit and fester until it explodes later. If you feel that something is too much to bear, try to get away from the situation for awhile if you can. For those in bunkers, this is impossible, which, again, is one of many reasons why living in a bunker is generally a bad idea.

If someone is suspected of stealing, or of behavior which is clearly detrimental to the whole group (e.g. sleeping while on patrol or guard duty), confront that person openly and always do that confrontation as a group. When doing so, never go after the person, but go after the incident(s). Focus only on correcting the behavior, especially if it is the first time or incident. If the person is a child and the parent is available, then the parent is to take care of the behavior after being confronted (again, by the whole group) and is to right any wrongs committed by the child.

Finally, if something does get too bad, then try and see about swapping roommates with another home. The leaving party takes only what they own in the way of what they brought (and can provably own) and nothing else. If the problem is someone (or multiple someones) exhibiting bad behavior and refuses to end it or amend it, then eject that person (or people). If the person being ejected is a child, then the whole family goes with him or her, depending on age.

Overall, in such a situation, try your utmost to be kind, patient and understanding. Everyone there has lost something, or someone. You're not the only one who could use a little forgiveness for something minor. Help each other out. It should never be 'every man for himself' in the home, but instead you should work as a team and with one goal: survival.

Gender, Romance, And Babies

Initially, this is going to be the absolute last thing on anyone's mind. Low calorie intakes mean decreased sex drive and libido… the body is smart that way and prefers to conserve energy for the more important function of staying alive. If you're just above the level of actual starvation, odds are perfect that you're not going to be in the mood to do the horizontal fandango. If you're just barely hanging on, the last thing you want is another mouth to feed. If your brain is wracked with trying to find ways of eating every day and fending off predators, you're not going to have much time to think about how soft and curvy the ladies in your survival community are.

However, if you manage to survive this far in and have managed to eke out an existence that leaves you with a little bit of free time and an occasional full belly, then odds are good that your thoughts are going to drift to other things…

…and that is where this chapter comes in.

Fantasies And Fallacies

So, let's get this out of the way now… Oh, yes. I already know. If you've read this far along, I can tell you right now what you may have had dancing in your head with regard to this subject.

If you're a guy, I definitely know what you've conjured up. Let me embarrass you a little and spill it now: You've managed to survive this far. You've managed to store up and gather sufficient supplies for yourself and for one other person if you stretch it a bit. You're out one day and you stumble across a refugee out all alone. She's young, beautiful and she's oh-so-scared and lonely! You take her in. Out of sheer gratitude and for some odd reason that defies millions of years of evolution and female personality types, she eagerly agrees to every weird and crazy sexual craving that you have ever had. She becomes a willing servant to your every whim and desire and you live happily ever after.

If you a woman? Oh, this is also way too easy: You're in a similar situation and have managed to get by with enough supplies. You find yourself in danger by raiders. Out of nowhere this dude with rippling muscles and perfect abs appears from the trees and opens up a veritable drum of whoopass on them, driving them away for good. Out of gratitude, you allow him to stay with you for awhile. He promptly begins procuring all these new supplies for you (including chocolate!), helps you clean the house (and in fact insists on doing so) and always leaves the outhouse seat lid down. He's the perfect listener and one night he reveals wounded feelings deep inside, caused from the collapse. You gently take him by the hand in sympathy and your eyes lock for just a second - but that second is more than enough. You feel a shiver run through your body in recognition of that spark of passion. You lean in to kiss him… you experience things you've never felt before with any other man (especially that lousy ex of yours). You lock against each other in hot, torrid passion, then cuddle and whisper and giggle for the rest of the night. The next day, as he's cooking breakfast for you, he announces that he wishes to be your perfect partner as you build a new world together. You spend the rest of your lives together, raising wonderful, polite and strong children… well, when he's not out working the fields as the sun shines on his golden, glistening abs.

Yeah, right. Let me pop those bubbles right there and just sum it up this way: It's all crap and you all know it.

First off, if you see a young and pretty damsel in the middle of nowhere? It's an act designed to lure you into an ambush, a distraction while you're getting robbed, or she's using her looks to use you for your shelter, food, supplies and whatever - at least until a better (or 'wealthier') guy catches her eye. Simple and sweet? The most dangerous damned thing in a post-collapse world will be a pretty young woman.

Let me lay an example on you: You see a pretty young refugee come to your door. You agree to take her in and she gives you a little pickle-tickle (*we'll just go with that term*). You fall asleep and she whacks you on the head with a stick of firewood, then slips to the door to quietly let her boyfriend and his buddies in. Then they proceed to kill you and take all of your stuff (if of course you're not already dead).

Needless to say, when it comes to any pretty girl post-collapse, your life demands that you approach with extreme caution - that is, if you're stupid enough to approach at all. Real women post-collapse won't be wearing makeup, will be too damned hungry and busy to think about romance, will likely have a kid or two in tow and she will definitely not be interested in becoming your sex toy unless you're either astronomically lucky (hint: you're not), or if she has no other choice in the matter.

Ladies? Seeing a good-looking stud-like stranger appear out in the middle of nowhere? Well, ladies - odds are good that he's either going to rob you if he's hungry, rape you if he's horny, kill you if you look like his ex wife, or, well, all three. No need to elaborate further, I suspect.

These scenarios bring up a very important point about the post-collapse world: sexual roles will change in a very big way. If you want a good idea of just how much it could change, go back to the time when Romans walked the Earth. Women were basically property back then and men took (and did) whatever they wanted to before her dad got involved.

Now in a post-collapse society, many men and women will retain the societal mores and gender behavioral roles that we've all grown up with. That is, most of us still-civilized guys will still remember to treat women as we did before everything went splat - with kindness, deference and respect. However, in a world where rules rarely if ever exist and outside of the few communities and towns that are left and intact? Four words: Law of the Jungle.

What may come about is the few folks left that are civilized (your community hopefully being one of them) may have to enforce a few rules and moral codes that haven't seen daylight since the horse and buggy were the most common form of transportation. For instance? Well, let's walk through a few and the reasons why.

Women, Post-Collapse

While proper treatment of this subject would require an entire book in and of itself, we're only going to touch on those things which you really need to know.

Women within your community will have to be watched over and protected - and not because they're frail or mentally weaker. You do it because there are going to be way too many men out there who will rape them at will and take them as actual slaves. Slavery, you say? Yes, slavery. Sexual slavery exists even the pre-collapse modern world and will be a very profitable means of income in the post-collapse world. After all, there will still be a lot of men out there who have a lot of supplies, some sort of lordship over their new territories (either through fear or intimidation) and will willingly part with valuable goods in exchange for a girl that can cook, clean and open her legs once in awhile - willingly or not. If she's young and pretty, she's a target for people who would willfully engage in slavery. This sexual aspect is what makes women a lot more vulnerable post-collapse than men. To that end, men, if they want their community to remain alive and prosperous, will have to protect the ladies among them.

Women themselves will have to be careful as well. Wandering too far off alone (or even as a pair), wandering around unarmed at night, dressing immodestly, or doing things that would be considered risky in a dark inner-city alley? Unless she's post-menopausal or pre-pubescent (and no guarantees on the latter), such behaviors are going to attract danger, period. Ladies, I hate to say this, but you're going to want to either have or find a guy, or find a group of strong and well-armed fellow women who can help you protect yourself.

I know, I know: *"Oh yeah!? Well I'm a crack shot, I have a black belt, I bench-press 350 pounds and I have lots of guns you chauvinist pig ...you Jerk!"* Well, I respectfully beg to differ with you. The strongest woman in the world is physically weaker than a mere above-average male, your ammunition won't hold out forever, ambushes will always leave you at a disadvantage, kidnappings and home invasions won't give you much time or space to unlimber your weapons (or muscles) and there are way more of them than there are of you. So unless you really are a black-belt that can bench-press 350 lbs *and* capable enough to shoot the eyelashes off of a mouse at 2,000 yards (hint: you're not), then you're seriously going to need to drop the tough-chick act and seek out some help. For the 0.001% of the ladies out there who truly are *'Xena, Warrior Princess'* -level material (hint: if you think you are, well, you're not), then you might be good to go. Long story short, you follow the rule that everyone has to follow, no matter what gender: Never over-estimate your abilities, or the situation you're in… and it is only fact with which I say that most ladies among us are not going to have the same level of strength and combat ability. Check any of that militant feminism at the door, because the post-collapse world is far less tolerant or accommodating than you think. It's better to settle with this thought now, than to find out the hard way later.

Women can make it with a group of fellow women (as mentioned before), but everyone is going to have to be well-armed, well-trained, well-provisioned and very alert. Basically, the same rules that you would follow if stuck in the bad part of town at night, should be the same rules you follow *everywhere* post-collapse - even in your own home.

Now - enough about the weaknesses. Let's bring out and let shine those things that make women far more powerful than men in a post-collapse world. Unlike men, women are able to endure phenomenal amounts of pain and still keep functioning, both physically and mentally (see also: childbirth), which helps give them greater stamina than men. Women are better balanced, physically (men are generally top-heavy) and tend to be more agile. Women are, in general, more able to add wisdom to intelligence when it comes to problem-solving, as men are more geared towards logic - this also makes women slightly more creative.

Unlike men, women by and large tend to focus on the emotional and relational aspects of the human condition, which in turn will bear the seeds for rebuilding civilization. They also tend to 'civilize' men in a way that an all-male group would not bother with if left to their own devices.

Something to consider: Women actually have a few more options than you think, post-collapse. Women can (and may well) be more than just homemakers, babysitters and teachers. They can easily be engineers, doctors, leaders and nearly any other skilled job you can think of - including soldiers (keep that last bit in mind, especially if you're short on 'manpower' to help defend the community).

Women also have one means of earning an income post-collapse that men simply do not: prostitution. We've touched on the subject lightly, but only to discourage the practice. It will break up married couples and get people into all kinds of general trouble. It is especially dangerous for the woman - physically, mentally and as a potential for serious medical hazard. Unlike pre-collapse times, a sexually-transmitted disease can't simply be cured with a few injections of penicillin, because there simply won't be any penicillin to go around. Women in larger communities or what used to be urban areas will find themselves quickly dominated and controlled by the first sufficiently armed jerk to get the idea that he should be a pimp. Self-esteem and self-respect will plummet to practically nothing. Every germ and virus that the 'johns' carry will now belong to her. Every 'customer' brings with him the potential of robbery, rape, violence and worse.

Condoms and birth control will rapidly run out or spoil, which means not only is there a risk of STDs, but also the risk of pregnancy. Now under normal circumstances pregnancy is certainly not a bad thing and even post-collapse a pregnant woman stands a fairly good chance of delivering a healthy baby without getting killed in the process. On the other hand, supplies are going to be extremely tight for quite awhile post-collapse and having one more mouth to feed isn't going to help things any. Overall, it's a very bad proposition, especially for a woman who is selling her body for food. In spite of this, well, it's the world's oldest profession and if anything, will likely increase dramatically post-collapse. Let's face it - women are just as desperate to eat as men, are just as willing to do whatever they can to insure that they remain fed on a regular basis - twice as willing if they have kids to feed, since they will need to eat as well.

All that said, if you're a woman trying to do your best post-collapse and especially if you have kids, it is hoped that you have gained critical skills otherwise and have mastered them enough to not have to resort to renting out your body in exchange for food, clothing, or any other basic necessity of life. To this end, the community needs to insure that the ladies have those skills - and teach her if she does not. Gents, the last thing you ever want is to get killed and have your wife resort to renting out her body to other men just so that your kids get fed.

How To Treat A Lady, Post-Collapse

Overall, you menfolk out there still need to treat women as a true gentleman would, period. Just because there are no longer laws against rape and domestic abuse, doesn't mean that you can suddenly start performing those acts. Why? Because quite frankly, women have a new, vast and far more clear-cut set of choices now. If you beat your spouse, the other men in the community will no longer have any restrictions against beating you to a pulp in reciprocation. If you belittle, abuse and insult your girlfriend of wife constantly, she can very easily find and move in with another guy, take most of your supplies with her and that guy is likely going to be far better-armed than you. If he in turn gets stupid, he'll likely find himself in the same sinking boat. Meanwhile, there's the subject of rape. If you're stupid enough to rape a woman or girl, then you will very likely find yourself being executed without sympathy, mercy, or delay.

So how exactly does one become a gentleman (in case you haven't already figured that part out by now)? This part is simple - be kind, be understanding and most of all, defer to them in all social matters. There are instances where you will have to cast that aside and take charge (during attacks, dangerous situations and the like), but any time things are peaceful, you treat a lady like a lady. The only times you will ever have any justification in physically attacking, striking, or otherwise harming a lady is in defense of life and limb, or in the defense of another's.

Notice how I said "lady". If the woman or girl in question is an obvious decoy for an ambush, or is actively part of a raiding party, do not hesitate to use all necessary force - up to and including deadly force. Also, do not fall for the damsel act. That is, the act of playing a helpless and frail girl in the hopes of manipulating you.

Another thing to keep in mind: Never, ever, ever (*ever!*) exchange goods (or worse, room and board) for sex, or withhold promised goods in exchange for sex. Yes, it's been said multiple times before, but it bears saying again. She will resent it, you won't enjoy it and she'll be gone the first moment something better (for her, not you) comes along. Well, that is, if she doesn't kill you in your sleep first and claim all of your stuff as her own. You see, being held hostage for sex tends to kindle hatred and that hatred will only burn hotter as time goes on. If you're doubly stupid and to do it with a spouse in the home (thinking that you're some sort of demigod now that you've got control of all the household supplies), the spouse may well kill you first, or you may one day wake up to find yourself naked and hog-tied on the road somewhere, spouse or not. Either way - don't do it.

When it comes to talking to your lady, what holds pre-collapse will hold post-collapse. Explain things in terms of emotion - ease up on the logic and reason. If something she does makes you angry, point that out (gently!) and explain why you're angry. If something she does makes you sad, same-same. Same story if she does something to make you happy… and in that case, don't just say it, show it (within the situation and propriety, of course.) Take the time to tell her whatever hopes and dreams you have while you're alone. Explain things to her in terms of emotion before laying out logic.

Finally, guys - if the lady in question is living under your roof and you're taking care of each other in some manner, take time out often to praise her for the things she does. Oh and you'd better be sincere about it as well. Choose things that upon thought, you really do appreciate about her and point it out to her. Don't wait until she's sulking, angry, or sad… just do it whenever you can.

Men, Post-Collapse

Guys? Well, guys are going to be a bit different once the civilization hits the fan and splatters into an unrecognizable mess. Without the strictures of law and given a pre-collapse society soaked with moral relativism, a disturbing percentage of guys are likely to decide to make up their own moral code as they go along, if they even bother with that. The relative few men and boys who have been raised with a strong sense of character and a solid moral code on the other hand, will be able to control (and not rationalize) their more primitive desires and urges. Note that this lack of civilized behavior doesn't mean just sexual selfishness, but also the urges to steal, kill, or ignore their fellow man and in general doing whatever it takes to gain personal comfort - that is, becoming an amoral self-centered animal.

You see, for many men, things like empathy, kindness and charity are mostly taught behaviors and are not always innate. We guys are far closer emotionally to the primitive pre-historic man than women are and so we have to learn this civilized stuff from our parents to a very substantial extent. Depending on how we were individually taught, the lessons may or may not stick around - especially when the things that keep us in check (law, society, religion, etc) are no longer around as boundaries.

However, do not despair! There are still many men out there who have not only learned these lessons well, but have made them an integral part of their personalities and are both willing and eager to teach them to their own sons. These are the guys you want helping to form a new civilization out of the ashes of the old. These are the guys you want to help out the most when they're down, because they will be the most likely to return the favor when you're in a bad spot.

Individually men will, more than women, be at once the most devastated and the most empowered at the end of civilization.

Empowered? Well, in a perverse way, yes. Existing power structures will be harshly altered, or they will evaporate completely. Most of the things a man is forced to spend attention on, such as bills, bureaucracies and social conventions, will be burned away or eliminated. Men can, post-collapse, take a far more direct role in forging their lives and the lives of their families. For many men, this can be (perils notwithstanding) a very liberating thing. A man who has spent his whole life playing office politics and worrying about his credit history now only needs to worry about planning ahead to feed his family, keep them warm and to fend off thieves and marauders. While the risks are obviously far greater, the goals and means to attain them are well within the reach of any reasonably intelligent and creative man's mind. Before moving out of this thought, I want to stress that this will certainly not make it easier, nor will it guarantee the outcome. This has nothing to do at all with nostalgia, or some lame hearkening to days long gone. This is going to be damned tough (and hazardous!) work. However, it is a lot more direct and thus easier for a typical man to wrap his head around, especially as most of the trappings and restrictions of the civilized world fall away. In the end, many will find it the most satisfying way of life as well.

In that same perverse way, many men, especially those who feel a loss of masculinity in the pre-collapse world, will quickly rediscover their testosterone. They will begin to feel as if they have regained their 'rightful' place in the order of things: provider, protector and even pioneer. Again, this is not a judgement as to whether or not this is right or wrong, but instead a statement on how a successful man post-collapse will likely feel - with success being defined as staying alive and reliably providing for his spouse and children.

On the other side of the coin, men will simultaneously feel the sting of failure and loss. A lifetime of building his reputation, his standing in society, of achievements and respect gained among peers in his little slice of the civilized world? All of it will be gone down the crapper. That retirement he's planned and saved for over years, if not decades… gone. The house he's been faithfully paying into got left behind and was burning to the ground when he last saw it… just so that his family can stay alive. The comfortable lifestyle of watching sports and indulging small hobbies - dead. And, most disconcerting, the feeling of being somewhat certain of how his life would play out has vanished into the ether and long-range plans become almost impossible.

There is finally the nagging feeling of uselessness that won't go away until the guy in question manages to be reasonably certain of keeping his family alive long-term.

Meanwhile, good luck getting most men to admit to being afraid. At most, they may admit it briefly to their spouses, in a whisper under the sheets… and then immediately follow it up with some sort of qualifier. We're simply not allowed to admit any emotion that can be perceived as a weakness and this will be twice as true in the post-collapse world. Seriously, we're just not wired that way and even in the pre-collapse world, we have some pretty vicious monikers for any man who does admit to these feelings openly (clue: the insults are usually designed to question his sexual orientation and/or masculinity).

To a man, experiencing and/or showing weakness is a very, very, (*very!*) bad thing. It's even worse than failure. We won't admit to it, we won't confess it under torture and we certainly won't entertain these feelings at all if any stranger is present - or even to ourselves. Pointing out failure to a man is the quickest way to get him angry, defensive and in a very bad mood. He already knows he failed in some way, even if he won't admit it. Pointing it out won't solve anything and only serves to stoke anger, shame and a whole lot of other hurt feelings.

In harsh post-collapse conditions, you will also notice that a lot of men have a bad habit of dispensing with what we consider to be "minor" stuff: washing our hands, keeping our language socially acceptable, keeping the toilet seat down, using a napkin… Under stressful situations, a guy tends to hyper-focus on those things which he considers important: obtaining and consuming food, fending off attackers when necessary and if his calorie intake and energy levels are high enough, getting laid. Anything else is considered an inconvenience that gets between him and his goal. Mind you, not all men do this. However, most guys will certainly (and in rare cases gladly) revert to such a lifestyle. Usually a feminine presence will keep these things in check and lend perspective, but not always.

Treating a Gentleman, Post-Collapse

While you wouldn't think this would be necessary, it actually is in a way. Ladies, this does not mean suddenly becoming subservient or suddenly becoming a doormat. However, it will take a bit of understanding on your part to best help the guy get himself back into a mode where he can best do his new job - protecting and providing.

First and foremost, forget appealing to emotions if you want him to do something. He'll have those bottled up tight and won't let them out until he's ready. Be patient though, because that likely won't be for quite a while. Men, especially in a stressful environment, run on logic and reason (well, most men do) - especially logic. If you need him to do something, give him a clear logical argument as to why it needs to be done and done within the time frame you want it done. Otherwise, there is a high chance that he'll either dismiss it, or put it at a very low priority in his mental list of things to do.

A good example would be, say, to do his half of the laundry when it's winter and not much is going on. Present the logical argument (*"you're not busy and this will get done faster if we both do it together"*) and you'll get better results than an appeal to emotion (*"This isn't fair!"*). While this is true no matter what the environment, it is especially true when things are ugly all around.

While no one is saying to flatter or feed a guy's ego, definitely give him a bit of a boost now and then. If something makes you happy, tell him. If something he did or said really helped things out in a way, be sure to let him know about it. Tell him your hopes and dreams, but do it cheerfully and in a way that presents a logical problem for him to solve… giving a guy something to mentally work on *when he's not otherwise occupied mentally* is (quite perversely, I'll admit) a good thing to do. Deep down in our primitive minds, most of us guys love solving puzzles and riddles and even more, we love presenting the solutions to our lady (and kids) in a way that makes her (and them) smile. We're weird like that.

Romance And Relationships, Post-Collapse

Now that we have some idea of what each gender is like (well, roughly), let's see about getting them together, shall we?

A post-collapse world is going to do a whole lot of shaking-up when it comes to romance. Spouses and lovers that survive together will either bond even more tightly than ever, or their relationship will disintegrate quickly. Shortages (and rationing) will likely lead to fights as stress levels rise to the breaking point. Making it past a rough patch in good shape (attack, shortage, whatever) in one piece and healthy will convince couples that they made the right decision in choosing their mate.

The collapse of civilization is, obviously, going to put every love-oriented relationship in existence to the test and do it without mercy, respite, or warning. There will be times when there is only food for one that night - who eats it? There will be times when a spouse may stumble or be seriously injured, but getting away from something quickly means life or death - do you stay behind and help while facing certain death yourself, or do you run on, leaving that person behind to die? Or does the injured partner demand you leave him or her behind? Assuming both parties miraculously survive, what happens if one turned out to willingly abandon the other? What happens if some third party takes shelter with you two and then proceeds to seduce one of you? What happens when you are asked to risk your family by taking in a dangerous relative (say, a drug-addicted in-law)? What happens when you have to turn away an innocent stranger at the door because you barely have enough to keep yourself alive and the partner objects to you turning the stranger away? What if your spouse or partner's response to the collapse is to simply go to church and wait to die (or for "the Rapture") and demands that you come along to do the same? These and many more tests await you when it comes to your relationship in a collapsing civilization.

If you can survive these tests and you and your spouse or partner both make it, then you'll end up doing two things: You will have learned a lot more about your loved one than you ever have before and you will have come to an unshakable decision as to whether or not he or she is worth staying with (in your opinion). This much I can guarantee you and with complete certainty.

For you single folk out there, this whole collapse thing is definitely going to change your criteria in seeking a mate. Once (okay, "if") you manage to get situated to a point where you can keep yourself fed and sheltered on a regular basis and are then capable of doing it for two, your thoughts will probably turn towards finding someone to share it with. After all, as human beings, being alone is not a normal condition and we all have the drive for intimacy and reproduction.

So… what changes, you ask? Well, let's start with glaring generalities, shall we? First and foremost, women are no longer going to seek men who drive a nice car, has a "nice butt", a sensitive side, a beach-house, or a high-paying career. Nope. Women are going to look for things on a more basic and easily-discerned level. The basic criteria becomes this: A strong, confident man who can provide, protect, supply the goods (*yes, I already said provide, but it's that important*), is hard-working, smart, listens well and won't abuse her... anything else is just icing on the cake at this point. On the flip side, men won't be chasing the prototypical girlish 'hottie' any more either. Odds are good that men will be seeking women who are strong, healthy, hard-working, mentally capable (both in stability and intelligence), young to an extent (insofar as reproduction) and even a little submissive. Well, it's much the same things men look for now, but like the ladies, they'll be seeking them on a more basic level - with the superfluous stuff like pouty lips, hourglass waist and balloonish boobs going right to the bottom of the priority list. Both genders will, as top priority, be looking for someone to best help them survive long-term... if they're smart.

Of course, there will be the stupid and the shallow, but this time around, choosing the wrong person could very well kill you, which means that the stupid and the shallow will quickly find themselves alone, dead, or both. An incompetent and wasteful mate isn't just the type who will jack up a bill (power, phone, credit card), but now it means wasting things like food, heat/fuel, or things that will hurt or even kill if you end up being stuck without those basic needs. A mate who is adulterous isn't just going to cause headaches and heartaches - now all of that fooling around can compromise your home's security, leaving you without a mate *and* without food, supplies, or weapons (or all of the above), as the 'other guy' or 'mistress' ends up moving in and you end up being forced out. A mate who is abusive pre-collapse could be left behind or sent to jail when there were cops around (though honestly such a thing is tough to have happen even in pre-collapse times) - and either way the abuser has at least *some* sense of holding back for fear of getting locked-up. Post-collapse, an abusive mate can torture and kill you outright and likely not even fear retribution for doing so (especially if he's quiet about it), because neighbors may be too busy keeping their own affairs in order to inquire too deeply. If you live isolated and any of this happens, you're basically screwed.

To this end, you're going to seriously want to choose very, very carefully before allowing any potential mate into your home, let alone your bed.

You notice that I've made no mention so far of casual dating, casual sex, or the like? There's a good reason for that. In case you haven't noticed, as we've progressed along, you'll find that very little in a post-collapse society is casual or wasteful and romantic entanglements are no exception. Certainly, prostitution will still be around and perhaps even flourish... but only a flaming idiot would let a hooker into his (or her?) home to do the act. Security will demand otherwise, since stolen items will tend to be missed a whole lot more than they would pre-collapse. Casual dating will take on less of the sex-play and more of a serious search for someone to share one's life with. This pairing-up may happen rather quick at first (in desperation), but as time progresses, the single among the survivors will start becoming a bit more critical towards who they want sharing their beds and they're going to want that sharing to happen long-term when they do find the right someone to share it all with.

A lot of the reasons why this will happen are actually simple and logical and for the same reasons that many of our great-grandparents had the same level of criteria and pickiness:

- The phrase "plenty of fish in the sea" won't be true anymore (and will get less true as the population continues to drop and as travel remains impossible), so going into a relationship with the idea that you can always get someone else later if things don't work out? That's going to be a really stupid idea.
- Women will regain the power they once had long ago, as they become the supreme arbiters of who they will willingly have sex with. Why? Because birth control, STD cures and abortion will no longer exist in any practical (or safe!) form. Simply put, there is no "Plan B" anymore if something sexual turns into something unintended.

- Resources will be limited, with no certainty of more to replace them once gone, so both men and women are going to be a lot more careful about who they share those resources with.
- The communities that survive are, out of necessity, going to be very tightly-knit for quite awhile and community approval or disapproval will become a very large factor in who you choose to mate with. If they say "no", your continued survival as a member of the community might depend on your agreement with their decision.
- Where pre-collapse many younger people did what they pleased without a lot of family interference, family will figure very large in the decision-making process now. Why? Because families will need every able pair of hands that they can keep fed, clothed and housed. This means, at least for teenaged people, it's going to be way different when it comes to dating. Where before your parents might not care too much about who you date, now they're going to be analyzing and looking into everything about that young man or young lady you're becoming attracted to. Oh and their opinions will carry a whole lot more weight - your decisions will now affect them too, after all… especially when they get old and unable to care for themselves and you have to care for them instead.

As you can see, a radically different environment means that your romantic and sexual activities in it will be radically different as well.

Instead of a quick hook-up, courtship may well make a comeback. This isn't simply because that's how it was in pre-industrial times. Instead it is because the greater influence of parents and community will demand that you not only get her approval, but that of her family and community as well. Unless you're inherently worthy (by way of resource wealth, bravery, or skill), they will be damned picky. It is worth noting that this will mostly apply to younger girls and women more than the older ones. Older women will still likely have their own say in who they mate with and when. However, smart women will know full well what the risks are and will be more likely to start being rather picky themselves.

You notice how we've focused on the female gender? After all, you might think that guys have equal footing here, right? Well, think again. Guys are most likely not going to be as protected and coddled in a post-collapse society. After all, guys are going to return to the pre-industrial role as the soldiers, the laborers and the enforcers. In other words, the males will return to more direct, aggressive/assertive roles. This means that guys will be the ones doing the chasing due to mere temperament brought on by the jobs they will be performing.

This is not to say that women and girls will shrink into a more passive/coddled role. In fact, women and girls, while of necessity taking on the more domestic role, will begin to wield a lot more power than they currently have. You can easily see women making decisions at the community level just as easily as men, especially in affairs outside of combat or defense (and perhaps even within it). As they regain control over the sexual dynamic, they will realize that they are no longer toys or objects to be used by men and that they have a lot more control. They will realize that they no longer need to attract a guy by appearing (or being) 'easy' and that they can be very choosy in who they share a home with.

Finally, know that none of this is set in stone. However, while there will always be exceptions? By and large, what you've just read is a good prediction of many of the roles in which men and women play. It will look a lot like pre-industrial times, but with a few (yet important) modifications. There will be a lot less ignorance (at least initially) of the capabilities and skills that women and men each possess. There will be a seismic shift in priorities, for both genders. And, when a couple decides to actually get naked together, odds are good that (especially with the lack of birth control) another rather important subject will come up, namely…

Babies, Post-Collapse

This isn't going to be your everyday chat with Doctor Spock, or a study in how to raise a smart kid. In this world, Junior will have to sharpen up fairly quickly, or else he stands a solid chance of ending up a casualty. Your job is to help him get there.

Before we say anything, there's a disturbing but inescapable fact that we're all going to have to get you used to. Post-collapse, infant mortality rates will obviously skyrocket and not just due to starvation. Even as the population balances out, a lot of death will come from things that you thought had gone extinct, or had been cured a long time ago. This goes especially so for babies and children. In the pre-industrial world, it wasn't uncommon for only 1 in 5 children born to see the age of 5. Most died early due to (often preventable/curable) diseases, malnutrition, accident, birth defects (usually brought on by malnutrition, abuse, or injury of the mother) and a host of other misfortunes that an adult could stand up against, but a helpless child cannot. This is assuming the child is even born at all, since miscarriages were rather frequent before modern times. Long story short, death is going to be very common for anyone born post-collapse and especially so for babies.

All that said, many of the pre-industrial deaths were brought on by ignorance, lack of the right resources, lack of hygiene and lack of proper care, all of which we can fortunately do something about. It won't be perfect and you won't be able to save them all, but you can do better than most. Of highest importance is the pre-natal care of the mother (nutrition, workloads, exposure, etc) and the critical first two years of childbirth (sanitation, hygiene, monitoring, prevention, etc).

We'll start at the beginning. Ladies, the moment you realize that you might be pregnant, you've got a lot of choices to make and things to do.

Speaking of "choice"? We'll get one thing out of the way once and for all: Abortion is off the table from now on. This is not due to any moral or ideological reason, but for a lot of practical ones. For the militant ideologues among you, you'd best get it through your heads right now. Here are the reasons why: Unless the pregnancy will kill the mother for certain (e.g. an ectopic pregnancy), an abortion is the worst and most wasteful decision you can make. An abortion procedure will use up much of the few precious medical resources you have left. In a world where skilled surgeons are going to be vanishingly rare, the procedure will introduce a lot of serious risks to the mother that will lead to death - either by infection or blood loss. If you're lucky, you'll only have to put up with injury, infertility and chronic pain. In short, forget the whole pro-choice/pro-life arguments… abortion is strictly a pre-collapse-only option that you simply will not have, period.

So, now you're pregnant and for the next nine months you're going to know all about it. For the last three months, practically everyone else will know it too. Given this, the first thing you need to do is to be a bit more aware of your nutritional needs and workload until you are certain that you are pregnant. Take things a bit easy if you can from the moment you think you know until your second or third missed period, which is around the time when your physiological changes start becoming noticeable to a trained medical person without the aid of chemical tests (by month six, the 'baby bump' will be noticeable to anyone with no medical knowledge at all).

This would be a very good time to spend spare moments reading up on what to expect and how best to prepare for the new arrival. If you're the mother, hit the relevant books as hard as you can and begin to put them into practice. If you're the father, take the time to study a bit of it yourself and support her as much as you humanly can.

Note that especially in the early stages of post-collapse, miscarriages will be all too common. The mother likely won't even notice most of them until after they have occurred, since they'll usually happen well before the 10th week of pregnancy. This happens largely because in extreme stress or during lean times (when few people are taking in enough calories and/or a balanced supply of vitamins and minerals), the body will, on its own, declare the growing child unfit for birth and self-abort, in order to keep the mother alive by saving her nutrients for survival.

Other causes of miscarriage will be things that in a pre-collapse world would be preventable. Examples are pre-eclampsia (abnormally high blood pressure caused by pregnancy), Rh factor incompatibility between mother and child (Mom has A/B/O-neg. blood type, baby had A/B/O-pos. blood type), various minor birth defects, cervix troubles, breech births, etc.

Assuming that things aren't too bad and that the mother-to-be gets to eat enough each day, you'll want to get everything together for childbirth and have it ready by month six. Also stake out a room in which the child will be born. Sex during pregnancy is ordinarily safe in pre-collapse times all the way up until the last few weeks. In post-collapse times, you'll likely want to stop doing that once the kid starts kicking, just to be safe.

Childbirth itself is going to be one of the leading causes of death for most women (usually though hemorrhage), so it will be the most critical time for both mother and child. If there is a nurse or midwife in your community (or within reach), you do whatever you can to secure his/her services - this is no time to be cheap on the bartering goods. While most women will be able to survive and thrive past childbirth (humans have done so for millions of years), a disturbing number will not - at least not without help. If it comes to the point where normal childbirth isn't going to be safe or possible, know that caesarian section births are actually quite doable and safe, even in post-collapse times. Just keep the area and the tools as sterile as possible and both mother and baby will most likely do okay.

In pre-collapse times, birth and childhood were purely events where the parents had a little trepidation, but for the most part it is a joyous occasion. In post-collapse times, there is going to be a lot of worry and a lot of anxiety. After all, shopping sprees for clothes and nursery gear will instead become worries about having another mouth to feed and another butt to keep covered in cloth.

The good news is that babies are pretty much easy to feed for the first 6-8 months, thanks to breast-feeding. As long as mom gets enough to eat, baby will get enough to eat. The rest is a gradual introduction of finely-ground bland, then normal foods. Even more good news is that with enough cloth diapers in your stores (or impromptu diapers made from clean cloth scraps), the whole diaper thing isn't that hard to take care of either. Clothing and blankets? Not really much of a problem either and can be improvised with relative ease. Crib? No pressing need, but one can be built or improvised.

You will have to make more radical living arrangements than you may be used to, especially in the early months. Baby isn't going to have his or her own room, nor would you want to waste the space. You may have to keep the baby nestled between you when you sleep just to insure there is enough warmth for the child (which in turn means you will need to sleep as carefully as you can), or at least have a small crib nearby. The mother will likely needs to be the one doing the primary caregiving and tending - especially if the father is continually growing and procuring food, working on patrol and etc. One potential bonus - if you have grandparents or elderly folk living in the home, they can jump right in and do a lot of the caring for the child when the mother is busy. Another option includes older children (depending on age, maturity, the availability of occasional supervision, etc) picking up the slack and helping to care for the child.

I know what you're thinking - especially the ladies reading this. However, let me put something to rest: This isn't sexism, but instead it's a practical and necessary division of labor. Yes, the mother will work just as hard in and about the home, but tending the home isn't going to be nearly as dangerous or demanding of attention, strength, or stamina. On the other hand, the guys have no excuse to slack off at home when things are quiet, especially in winter. Raising a child is hard work, as many parents already know.

Gents, if you're not doing anything and you have spare energy and time (if you're doing things right, you probably will), then get your ass in there and help out, as much as you can. Be involved. One important thing to note, especially in stressful times: No matter what, do not lose your temper at either baby or mother. It will be far too easy to do that, especially when baby is in the stage where he cries non-stop at all hours for weeks on end.

While we're talking about division of labor, please don't get dumb or crazy. Don't take the little guy out on guard duty with you, or carry it along in a sling while you're working. Yes, you've likely seen pictures of women doing just that in third-world countries, but note that you probably don't want or need to follow the same example, unless circumstances dictate otherwise.

Something to consider when it comes to a crying baby, but only because it comes up in discussion so often when it comes to babies in a crisis situation… there may possibly be times when that crying baby may put you and your family in mortal danger, especially if you're hiding from something or someone. I'm certainly not going to tell you to smother a baby, but you will have to keep in mind that a baby cries whenever it wants to and doesn't particularly care when, where, or why. A quick-thinking mother can breast-feed it, quickly take off its diaper, or whatever is necessary to shut the baby up. On the other hand, consider that such a situation will be pretty much non-existent, so don't waste too much time trying to work that one out.

Unlike most pre-collapse parenting techniques, yours is going to be fairly simple overall. Teach the kid to get up and walk as early as it safely can. Encourage the child from infancy to learn, to look and to listen. As soon as the child is strong and smart enough to do something, give him or her simple chores to do while at the same time teaching the child on other subjects. As they become smarter and stronger, give them more responsibilities as quickly as you can (safely!) do so. Unfortunately, childhood will be short and you won't have much time. You'll also need all the help you can get.

As the baby becomes a child, both mother and father need to take the time with their children (new or teenaged and at all points in between) to teach, to love and to occasionally play together - whenever and wherever possible.

Conclusion

A lot of the same challenges, desires, problems and solutions are not going to go away just because civilization does. Men and women will still fall in love. Babies will still be born. Everyone is still going to have to fight off the same diseases, dangers and despairs. Orgasms will still be sought and achieved. Couples will want to grow old together. Your best bet is to keep the good and work past the bad.

Civil War

We, The People…

…are jackass-stubborn, intensely greedy, insanely competitive, astoundingly ignorant and we definitely don't play well with others. As long as everyone has enough of everything they need, this is usually not a problem. Generally, in peaceful times we keep it all tamped down and folks are (usually) polite to each other. In pre-collapse times, there's a government structure and police force to make sure that people remain largely peaceable towards each other. There's even diversions to help burn off excess energy, such as sports and other competitive events.

In post-collapse times? Without a working government or police and without enough necessary resources to go around (food, clean water, heat, etc), things are going to get very ugly, very fast. Every idiot with a gun and some friends will suddenly get delusions of grandeur and think himself to be the new king of the post-apocalyptic frontier. A vacuum of power will always get filled after all, as long as there are humans around to try and fill it.

We need look no further than history for proof. When the Roman Empire began to come apart, multiple people rose up out of the woodwork, each proclaiming himself the new Emperor. Once the empire was good and (mostly) dead, local warlords would form local armies and proceed to attack their neighbors, simply in a quest to increase their powers. Europe finally stopped doing that (more or less) in mid-1945, roughly *one thousand five hundred years* after Rome came apart and some portions of the former empire are still struggling for conquest, even today.

This age will be no different. As it is, you will have had to organize your local community and therefore create a power structure - just to keep each other alive and relatively free from criminals and raiding parties. You may have even had to reach out to other similarly-organized communities to procure things they had which you do not and vice-versa. Therefore, you will have already seen and participated in organized and independent power bases.

Why can't we all just get along?

Well, let's break it down, from global to local, just so you can see why people get all crazy for conquest. However, just to make it interesting (and more factual), let's remove megalomania from the table and stick to why *groups* will want to come dominate your little corner of the world…

<u>On the global front</u>: As national governments begin to collapse, neighboring governments may begin to take advantage of the collapse and invade them for numerous reasons, including resources, ideology, religion, or raw demand for power. Note that those four reasons are in order of importance and priority. At this level it's simply war and while a bit out of the confines for our chapter, it is something to be watched out for (especially if you live in Europe, Asia, or Africa). The odds of an invasion for the United States, Canada, or Mexico (the continental focus of this book) may be somewhat low, given the widespread freedom to keep and bear arms for private ownership and use, coupled with the sheer mass of land and terrain. But – yes, there's a "but" – as low as those odds are, never write off what any country with a starving population is willing to try. In our (North American) case, it's just something to keep in the backs of our collective mind. For instance? Maybe if you live on a coast, you may want to keep an occasional eye out for military ships bearing down on you. The good news is, as time passes, the likelihood

approaches zero and we'll call it 24-36 months post-collapse (global-scale) when we can safely discount such things happening.

<u>On the national front</u>: As the federal government begins to lose control and grip on power, factions within it may well rise up, each claiming to have the only sole constitutional mandate and solution to "bring back" the "golden age". However, odds are perfect that nearly all faction leaders are lying through their teeth, so rallying to any of their flags will likely be suicidal for your community in the long-run (especially once your favored faction loses).

<u>On the regional front</u>: As the central federal government begins to die, state/provincial governments may begin to fill the role once held by that federal government and begin to assert itself over neighboring states/provinces, doing so by force if necessary. This is especially true of states or provinces which have wealth, large populations and a lot of military hardware on hand. This will be especially true if the neighboring areas have needed resources. Even within the same state/province, the state or provincial government will likely assert itself as the only power worth obeying by its population. This will likely mean a restriction of travel to or from the area controlled, as well as very tight restrictions on intra-state (or intra-provincial) commerce. If money is still worth anything and being used, all taxation and levies (including federal ones) will likely be redirected towards the state (or provincial) capital - at least until a new form of money is minted or printed.

<u>On the local front</u>: As any community gets over the shock and (if) it stabilizes, it will begin to come up against resource restraints. Stocks run low, stuff runs dry, things begin to spoil, the population over-uses it, whatever… long story short, you're going to need some things to keep your community going and you won't have it, but the neighboring towns might. The affected community (yes, yours) will be forced to seek out and control sources of any additional nearby needed resourcc for its own use. If that source is controlled by someone else, the community will have four choices: trade, find an alternative, do without, or take it by conquest. If the resource is especially rare, or the price too high, you can forget trade. Finding alternates may not always be feasible. Doing without may not always be feasible.

For example, if the resource is water in an arid region and the folks living at a large spring are well-armed but barely have enough water for themselves, you will be stuck with either fighting for it, finding/securing an alternate source, or just dying of thirst. Yes, there is someone out there thinking: "…you could also try to join with them!" Not happening. They obviously won't let that happen if they barely have enough for themselves. Then again, your first reaction will naturally be "…they're lying! They have plenty for everyone!" They may be, they may not be. Fact is, you're still stuck with having to deal with it and there's no way around that. Note that this can and will go both ways, as other communities eye what you have with envy and suspicion.

To quickly summarize, people are simply not going to get along, even if we got along at all other social, ideological and religious levels. Speaking of which…

Hatreds And Hollerings

When your belly is full, it is easy to tolerate and welcome folks who differ from you in skin color, religion, ideology, sexual orientation, moral code, or even things like sports team loyalties. In lean times, it's much harder to do. In desperate times post-collapse, it will take the willpower of a bona fide saint.

Unfortunately, modern pre-collapse media has gone out of its way to highlight and aggravate such differences in order to attract eyeballs and therefore ratings (which in turn mean advertiser dollars). All too many people are more than eager to get sucked into these contests, willfully aligning themselves in these rather stupid games. You may know these games by their 'teams', such as "liberal vs. conservative", "black culture vs. white culture", "gays vs. straights", "pro-choice vs. pro-life", "unions vs. entrepreneurs", or the most recent one (at time of writing): "Occupy (insert town name here) vs. The One-Percenters" (think of it as a warmed-over 'poor vs. rich' ploy). As times get tougher, the media cranks up these faux contests, because they need to attract more viewers/readers/listeners, so they can in turn attract more money. Once collapse hits, the media as you knew it will be dead, or distorted into comical propaganda just before it dies... but all of the passions they stirred up will still be quite alive and kicking. With everyone facing the very real possibility of starvation, those passions will be used on all scales, both micro and macro, as excuses to exclude or deprive some lesser-numbered group or other of their possessions and lives.

Mind you, this isn't just a WASP (White Anglo-Saxon Protestant) thing, either. Examples from all around? No problem: A predominately black community will begrudge non-black people living amongst them at best - as long as the minority residents behave themselves, otherwise it's expulsion or execution. The existence of radical groups such as *La Raza* will pressure a Latino community to care for it's own and excluding anyone else (unless you happen to be Catholic, but even that is no guarantee). A small-but predominately Mormon ("Latter Day Saints") community in Utah will likely ostracize or expel the "gentiles" from among their number (this occurs on a social level even today within even the larger Utah cities). A small Southern town will actively drive out any known gays, "liberals" and atheists from among their number. Same with a small community in New England, but instead these folks will be driving out the "rednecks" and those folks whose religion or moral code does not sufficiently conform to the "progressive" ideal.

If you're black or Jewish in Arkansas, you might indeed find some trouble up in the hills. If you're white and find yourself in a post-collapse St. Louis, Detroit, or parts of Los Angeles County, you are quite certainly in trouble. A Catholic or Mormon family in a Southern Baptist county may find themselves unable to get help from their neighbors. An atheist living in Utah may find that no one even acknowledges his existence.

Basically and without any particular prompting, people will slowly but definitely segregate themselves into racial, ideological, or religious groups. The communities most likely to survive will be the ones who actively avoid falling into that particular trap.

All that said? This is only a general assumption based on history and is not going to be true everywhere. However, this is a normal, natural thing for people to do in a crisis situation, which also means that what holds true for smaller (town-sized) groups of people will also hold true for larger (state/provincial-sized) groups. Take for example the whole (and in my opinion, stupid) Red-State/Blue-State thing. The United States has gone to war with itself before - there's no reason it won't again, especially if the federal government disappears or is rendered useless.

Without wasting too much ink, let's just say that given the evidence presented so far, there are a lot of reasons why any given group of people will attack and conquer another group of people. The big question is, how can you avoid falling into such a trap yourself? A second, nearly equal question is, how do you prevent, fend off, or at least survive some other group of people attacking you for those reasons?

Avoiding These Traps Yourself

Tolerance and kindness, like charity, begins at home. Now before anything else, I am most certainly NOT advocating that you embrace ideologies, religions, or sexual proclivities as if you approve of them. What I am telling you is, don't let it matter to you. Every major religion has the same general guideline when it comes to these things and that is to diminish the acts while not diminishing the person. Or, in Christianity: "Detest the sin, but love the sinner" (St. Augustine wrote that), or if you're looking for bible verses, look to the story of Jesus and the Adulteress, where the sin is condemned, but not the sinner.

As for any racial issues? It is hoped that you have not only grown up enough to dismiss them, but that you would actively teach your children that skin color means nothing in regards to the ability, intelligence, or morality of any given human being.

Be aggressive in how you deal with these issues and feelings. You don't have to approve of the way a person thinks, but (in my opinion) God gave us all the freedom to think and opine as we wish (and atheists of all folks should cherish and value freedom of opinion) and thus the most you should do on such matters is to respectfully disagree and do so without lessening your kindness towards the person. You do not have to approve of another's actions, so long as those actions do not diminish you personally, or put you into any kind of **credible** danger (note the emphasis). For example, if your community includes a gay couple, leave them alone when it comes to activities which occur in their bedroom, as it does not affect you in yours. If they preach about how their lifestyle is better and you do not agree, then you can respectfully disagree, state your opposing position as kindly as you can muster without verbal attacks and then (most importantly) move on to a different topic of conversation that is agreeable to you both.

Overall, for those folks who differ from you in some way (be it physical or ideological), always treat each and every one of your friends and neighbors as human (and humanely) as you would treat yourself.

Avoiding These Traps As A Community

Usually, people can control or handle differences when they are alone or in their own family. It is when people get into groups that things start to slide all sideways. Mob rule is a powerful force and like any other endeavor, community leaders must avoid even the slightest opportunity for mob rule to raise its ugly head.

A typical scenario could very easily begin like this:

You've gotten your post-apocalyptic community together. Everyone in it has (more or less) enough to eat and is capable of defending each other against outside influences. But then, some minor crisis arises. It could be a small band of refugees that has arrived and is begging to stay, it could be a minor food shortage, it could be a temporarily bad water supply… anything that raises a threat to your community, where suddenly there's a problem. A local charismatic rabble-rouser starts going on and on about how it's the fault of some minority in town and if only we could be rid of that minority, things would be all better or safe again. Some of the crowd starts nodding their heads while others shake theirs. The guy keeps going on and on and on… about an hour later, most of the crowd is very attentive and some of them begin shouting. Some of the crowd leaves nervously, with a few rushing off to warn the folks being spoken against. Less than 20 minutes after that, the crowd disperses to their homes to retrieve weapons, meets together and begins marching on the people whom the speaker has been railing against, bent on expelling those folks from the community (and keeping the supplies for themselves, naturally). Another crowd has rallied around the besieged scapegoats and are also armed, looking to stop the madness from completing itself. Shots are fired. People begin to die.

Congratulations… you just lost your community and with it, your (and their) only real hope of long-term survival.

The solution is very simple: You stop that garbage hard and you do it back when it's just some lone person (or very small group) calling for action against any innocent group of people. You immediately get the community leadership together and confront the speaker/leader and you confront him or her right there in public. The entire leadership, no matter what their personal opinions may be? They must, as one, stand together and put a solid stop to it while it is still small and easier to control. You do not have to enter a debate about it, you simply state loudly but clearly that everyone in the community is working together and any attempts to divide the community will result in the expulsion of whomever is working to divide them.

You give them exactly one chance to speak a public apology and to never try a stunt like that again. If the rabble-rouser refuses to stop, you expel him or her right there, period… even if it means calling in your troops to expel the person (and potentially the person's family) at gunpoint. Note that it doesn't matter who the rabble-rouser is, or what their standing may be. You simply do it. Even if it's the town preacher who has been a fixture in the town for 30 years, you simply do it. Even if it's the only doctor in your community, you simply do it. No exceptions. The absolute hardest thing you would ever have to face is a mob and you do not want one.

You can also go a very long way towards preventing such a potential situation by integrating everyone as much as possible. Do not allow separate enclaves of any kind to form, no matter how innocent and any group that is to form in your community can do so with one caveat: all groups must accept any community member as a member of the social group being formed. Sometimes you can't help but allow separate groups to meet, such as church services and the like, but always have someone listening in if possible and all gatherings (with very few but clear and obvious exceptions) are to be open to the public. This means most of your own community leadership meetings as well, by the way.

Dangers From Without (the community)

When we say "Civil War", we don't just mean within your own community. We also mean dangers from other areas and other regions, in what used to be your country (or even your state). As noted at the beginning of this chapter, when the federal/national government falls apart, it will be left up to the states and/or provinces to fill the vacuum, if they can. While not all will be able to, many states will certainly try.

Some states will have a strong enough identity to hold together for quite awhile after the collapse of federal government (Texas and Vermont stand out as examples, as would the province of Quebec in Canada). Others (most notably California) will come apart very quickly, due to having various factions within it, or from simply being too big to even try. For example, California could very easily split up into a northern and southern component, or split along racial/ethnic lines. Even states such as Arkansas have two radically differing ideologies and cultures within their borers (in Arkansas' case, we're talking the mountainous and ideologically conservative Ozark region versus the flat and more liberal southern/river region.)

The divides which arise can be ideological, racial, ethnic, religious, or even geographic. A surprising source of division may well be wealth, but more accurately it will be power. What I mean is, the much more powerful cities may begin demanding more resources from the rural or less-powerful areas that are either unwilling or unable to provide them. For all we know, it could be a mixture of many factors, as was the case with the US Civil War of 1860-1865.

Oh, you thought that Blue versus Gray thing was all about slavery? Well, time for a small education, which will in turn help us understand what can come about in our own post-collapse world. The two sides had radically different cultures, different religious outlooks –The South was mostly Baptist vs. the North's wide variety of religious leanings – and the two sides had differing outlooks on where one's loyalties truly should lie (with the state versus with the federal government). They even had two radically different economic modes (The South's massive agricultural base versus The North's massive industrial base.) Given all of this, the issue of slavery merely brought all of that together in a way that covered all the bases (truth be told, the black man got equally abused and looked down upon by soldiers and citizens of both sides alike). This is not to belittle slavery, but to force you, the reader, to look beyond the simplistic sound bites and headlines and to realize that there are many factors that play together when it comes to figuring out what the hell happened. Knowing this leads to…

Everybody Wants To Rule The World…

In either case, none of that is going to help you when war begins brewing in your neck of the post-collapse woods… or is it? Let's face it: Every former governor, general, secretary-of-whatever, mayor, admiral, congress-critter and even gang leaders across town will be doing their level best to seize as much political and influential power as they can humanly grab and keep hold of. Those who are used to power pre-collapse and have the access to weaponry (theirs or others') will be twice as motivated to exert it over as many people (and as much land) as they can possibly reach - including you and yours. This will obviously lead to conflict - both with communities that don't want to be subject to anyone, as well as with other little warlords (or wannabe warlords) who are trying to do the exact same thing. Like it or lump it, this stands a good chance of putting your little community in a lot of danger.

As you've probably figured out by now, this is going to be a problem. Your options are going to be few, but they are worth exploring…

As we've discussed in earlier chapters, simply being neutral will work to an extent, but here things are different. How? Well, so long as the army showing up at your doorstep is controlled by a functioning federal (or state!) government and so long as they're not simply coming around to loot you to the bone, then playing a neutral party is your best option, due to all the other options being very ugly ones that will only get you killed.

Problem is, at this stage post-collapse, these troops aren't going to have much left in the way of a home to go to, no central controlling authority to speak of and what few resources they do have are quite effective at turning homes (and occupants) into smoldering craters. I'll save you the speech and just say that this far in post-collapse, when people are still dying by the millions each week? Unless you are more powerful than the force arriving at your community, playing friendly and neutral to any approaching force is an invitation to get pillaged, raped and stripped of everything you have. Even if you are more powerful, you may not want to risk being nice, unless there is a specific and tangible advantage to doing so and I can tell you now that there aren't any.

So what other options do you have? Well, exactly three: Fight, Flight, or Deception. The first two, once committed to, will be for keeps. Fighting means that even if you lose (and well, die), the invaders will have had to earn the privilege and there won't be much left for them to consider useful. Running away means that even if you could eventually return, there would be nothing left to return to (because you either burned it, or they consumed it all, or they will occupy it). If you commit to either option, you're all-in, period. The third option requires a lot of work, a lot of skill and a willingness to resort to one of the first two options should this third option not work.

But, let's step back for a second and we'll revisit all this in a bit, if it comes to that. After all, you may be able to avert or avoid it all. No, seriously, it is possible, if you do it right. However, you have to know what is actually coming first. This means long-distance communications and intelligence, which in turn means wading through, well…

Take A Ganda At That Propaganda

Even post-collapse, with the whole world going 'splat', communications of a sort will still be there. Problem is, what you get will be all-too incomplete, chock-full of rumor, unsubstantiated at best, definitely slanted to serve someone's purpose and will need to be run through the old BS filter multiple times before you get any useful information out of it. This is doubly true when wars and rumors of war start floating around.

Your best routes of communication will be via traders and any traveling specialists (e.g. doctors) that are brave enough to journey between communities, by whatever electronic means still exist (Ham radio, perhaps a small segment of Internet, etc) and by whatever information your own scouts bring back. Suck in as much information as you can possibly get hold of, but believe very little of it, because a lot of it will fall under one of two categories: engineered news and exaggeration.

Exaggeration is as old as mankind - just ask any fisherman, hunter, or car salesman. In most cases, if something sounds a bit too far off from reality (it'll still be easy to spot, in spite of the whole insanity that accompanies the collapse of civilization), odds are good that the truth is about four to ten notches further down.

Engineered news? Oh, that's going to be tough to spot. It can be as obvious as some pronouncement from a near-dead potentate (governor, president, whoever) proclaiming that golden days are just ahead if everyone were to just {insert some privation here}. It can be as obscure as rumors of a disease outbreak, of troop movements and shortages of some resource or another. It can also be as subtle as rumors of raiders literally raping and pillaging some area, with a far-too-clear description of who they are and where they can be found.

When in doubt, try to get the story from two independent sources if at all possible. Send scouts out to quietly check into something you think is of interest and is local enough to verify first-hand. Never accept anything at face-value from someone you don't perfectly trust and never be pressured to act unless it is something obvious (e.g. someone tells you that your house is on fire and you can see it burning). Take anything involving optimism with a pinch of salt… make that a block of salt.

Priorities And Sorting Out Information

Your job in all of this is to look out for your community and sort out what you can use from the fluff you can ignore. It doesn't matter if you're running the community, or if you're the lowliest member of it.

When civilization collapses, the dynamics of priority tends to change dramatically. Where before a letter from the IRS or from a bill collector may require your immediate attention, you can now (post-collapse, that is) safely discard any loyalties or obligations to any entity which does not have direct and immediate control over you or your community. The local band of raiders at the other end of your valley? That's something you have to pay attention to. Some guy halfway across the continent claiming to be the "new president" of a dead government? Not so much.

If it affects your community, you verify it quickly, then you act on it quickly to the best of your ability and skill. If it affects your immediate neighbors, you may want to act on it, if acting on it is safe enough to do and doing so doesn't put your own community in danger. The farther away the situation is, the less you can worry about it, but always keep an eye out for anything that gets closer. If it's more than 50 miles away and isn't heading towards you, you can safely ignore it. If it's local trouble but heading away from you, keep an eye on it anyway until it is well and truly gone.

It may help to keep a chalkboard or paper handy to keep track of things. Divide it out by distance and severity and update it frequently as new information arrives. Make sure the community leadership at least talks about it on a frequent basis. Keep everyone in the community apprised of anything that floats to the top of the which affects them directly.

What They Want

(...and why you don't want to give it to them unless you've no other choice).

Okay, so let's say you've got trouble coming. You've been keeping up the usual patrols against raiders, right? You've not wasted too much ammunition? Good. But now you have something a bit bigger and a whole lot uglier coming your way and it isn't going to get scared off by a few wounded or dead on their end. This incursion could range from some local 'warlord' gone fanatic and trying to increase his territory, up to what remains of a national army sweeping across the countryside under a leader determined to re-establish the now-dead government. In all cases, the intruders will want the same things out of you and your fledgling community:

- Resources (as much as you can give them, unless its weaponry or fuel, in which case they'll want and take it all.)
- Loyalty (…or else.)
- Soldiers and other fighting manpower.
- Possibly labor, women, shelter (for their soldiers), etc.

The reason they want your resources is to feed their army and enrich their own home community and (if they're smart) they also do it to leave you both defenseless and dependent on them for your continued existence. After all, in a post-collapse world he with the most resources gets to run the show.

The demand for loyalty is obvious. They obviously don't want your community calling the shots, or questioning why they demand half the men in your community take up arms and go fight some far-off battle. Loyalty also makes it easier to get you to send off the work of your labors for no logical reason, though they will usually tell you that in return they will protect you (from what? Well, them, but they'll say it's raiders or the like).

They'll always need troops to help them conquer other communities and areas and taking manpower from you to do it has the added benefit of leaving you, again, weak and dependent on them.

Labor? Of course! They can force you to rebuild roads, build shelter and other projects for them and basically enrich their little infant empire at your expense. After all, the Egyptian pyramids didn't build themselves, now did they?

As for the rest? Well, pretty girls are always in demand and their soldiers will want to stay somewhere warm and dry (and your house happens to be both, so...)

As you can see, the benefits almost always flow towards the new little dictator and away from you - just like peasants giving to the local feudal lord. The bad news is, this will likely not only impoverish and deprive you, but put you in danger of starvation or worse. You really don't want that, so I'm hoping that you've made the decision to either drive them away, or try to fight it out, or at the very least make them pay in blood for the opportunity to dominate you. You see, if you give in, you lose more than just the full use of resources (which is bad enough), you also lose the ability to determine your own fate and to forge your own destiny. In short, you become a slave. Now I realize that the whole name of the game is survival, but do you seriously want to do it as someone else's property?

What Do They Have?

Needless to say, you'd better find this out very quickly. Send out scouts as soon as you can and have them send someone back ASAP with updates. The size and composition of the forces will determine what you do from there.

The best time to know what you're up against is just after some force has taken over a neighboring community and hopefully one far enough away to give you time to do something about it. It will take them time to consume the new conquest and will teach you a lot about what will happen to your community should they conquer it. You can use that time to sabotage the supply vehicles*, gather intelligence and spread subtle rumors designed to draw the invaders away to directions other than where you live. What you do during this time best determines what such invaders will or will not do later on, so make it count.

* Wait - just the supply vehicles? Yes! Without supplies and the means to transport them, everything else grinds to a halt. Even the strongest armies need logistics to move forward. Concentrating on them will deny invaders the means to move, it slows them down and if done right, can even defeat them entirely. If it's flammable, you burn it. If it's food, you poison it. If it's mechanical or electrical, you break it, short it, or jam it. If it's liquid, you spill it all over the place. Whatever you do though, don't steal it... doing so weighs you down and by simply having it, gives them a reason to chase after you in the hopes of recovering the stuff. It's better to show them that they'll lose it permanently, with no hope of getting more.

Keeping Them Away

Now let's be realistic here - you most certainly do not want to be left defenseless and dependent on some far-off self-styled 'warlord' to remain alive and odds are perfect that they'll always take a lot more than you can afford to give. Long story short, you simply cannot afford to give them anything.

If you're unfortunate enough to be square in the way of the oncoming rush, you have two choices:
1) Stand your ground and fight to the death, or
2) Gather as much useful stuff as you can, drag the important stuff off to the woods and bury it, hide everything else you can and burn whatever won't move.

Well, wait: there are other options you can use before then: deceive, delay and divert and distract. After all, it is far better to try and head it off and save all the desperate stuff for when you have no choices, no time and no hope of fending it off.

Let's start with deception. Sure, every tin-pot warlord worth his salt will have a map and certainly they're going to have some idea as to your numbers and strength. After all, unless you're completely isolated from humanity and kill anyone who tried to leave, word will get out as to where you are and what you may have. However, you can still deceive the smaller forces and anyone thinking of attacking. Fake weapons, fake soldiers, a fake disease outbreak (complete with fake funerals and fake graves), fake shortages, or numerous other lies can be told and kept secret enough to fool a potential invader into looking for juicier (okay - "safer", or "worthwhile") targets. The idea is to know far enough in advance to prepare and display the fakery and to make it convincing enough to the right people so that word carries back to the right ears. You may want to get some ideas offhand and play off of relevant recent events when the time comes. Just keep the fakery subtle, so that it doesn't come off as if you're faking anything and don't stop until the threat has passed sufficiently.

Delay and divert them? By this I mean you literally throw up enough obstacles in their way to make getting to you in force so hard to do, that you are no longer worth the effort. Some communities are naturally isolated enough that cutting off any armed incursions is almost trivial. For example, a mountain community can be cut off to the point of impossibility by filling the roads with landslides, which in turn makes vehicular traffic impossible (by car, horse-drawn wagon, or basically anything that doesn't fly). An island-based community is already cut off to all but properly-equipped boats and/or ships. However, you likely don't have any of that, so you have to use what you do have. If you do have terrain that helps you out (wetland, hills/rough terrain, river(s), lakes, known danger areas, or the like), you use them to your advantage.

Now before we go any further, you're going to have to decide: fight, or play nice? I ask this now because the next steps are going to make big messes that won't clean up very easily. If you decide to play nice, the next steps won't really be things you'll want to try.

…but, if you decide to fight? Okay then. Once you've figured out where they don't want to go (all those big and ugly obstacles we listed up there), you'll have to block off the paths where they want to go. Even something small, like diverting a stream to wash out a road or two can do a lot to present an obstacle. What you cannot do in size, make up for in quantity. If an approaching force has to clear out fallen trees every 50 feet along a road, they're likely going to start looking for somewhere else to travel. Bridges? Blow them if you can, take them apart or block them solid if you can't… even a little wrecked bridge over a drainage culvert can become a big obstacle to vehicular traffic, if you're creative about how you take it apart.

In a pinch (and with the wind in your favor), you might be able to launch your own disasters on an approaching force and at least whittle their numbers down enough to handle them. In these cases, be damned careful or damned desperate, because these types of obstacles are going to pretty much act on its own after they get going and it may well turn back on you. Touch off a forest or brush fire, which will definitely drive them away, around, or back . Sabotaging a smaller dam to flood an approaching force will definitely divert them (and if timed right, will certainly drown a lot of them). In hilly/mountainous terrain, landslides are your friend. Is it wintertime? Nothing like hard-packed 3-4' tall humps of snow interspersed with roads slicked with ice to keep an invader on his toes, no? Speaking of snow? In avalanche country you've got a friend in literal mountains full of sliding snow.

While we're on the subject of holding off invaders, I'm going to explore what many others would advise: Set up ambushes along the way and kill 'em off a little at a time, but don't make it look as if you're the ones doing it. The game plan is to have small teams, each with a three-part job: stop, snipe and scoot. Get a couple of good shots in, then get out of there. Hit them from behind, to make it come from a direction opposite from where you are (and to hit the softer but more vital targets like supply vehicles). Have two or three teams a half-mile or so apart (but not at regular intervals - make it almost random), with each team leap-frogging the others as the column gets closer. Each team also keeps an eye out for scouts or patrols, picking them off first if possible. With a bit of luck, you'll draw off soldiers, or they'll start chasing shadows. If you're really lucky, they'll start thinking of easier targets (don't count on it though, at all).

Now this approach is good and bad: Good, because it takes out a few people and vehicles beforehand and spooks them pretty hard. Bad, because if you do manage to make a dent and the force is large enough, they will likely take it as a challenge and the effect could be quite the opposite, in that it will motivate them to take you down. Given this, I say go for it on one condition: if you know full well that you're going to fight to the death, they're getting closer and that there's no way to make them leave.

You do have one really big ally in this tactic if you're willing to use it, no matter what time of year or resources you have on hand. That ally is distance. The further they have to travel (without resupply from communities along the way), the less likely they will be successful. As the miles pass, more irreplaceable fuel is being burned, more food is consumed and more effort is being exerted. Use this to your advantage. If they have/use vehicles? You get your little ambush teams in to sabotage vehicles, which does a LOT towards crippling the advancing force. In this case, you don't snipe at the people; you snipe at the vehicle engines, the tires and especially the gas tanks, spilling or burning as much of their fuel as you can. You shoot the horses and pack animals. You take out headlights to make driving at night impossible. Concentrate on disabling or destroying the supply vehicles, to the exclusion of everything else. Without them, everyone has to walk, which stretches out the amount of time you have. If they have no vehicles and are dragging carts by hand, you shoot the ones dragging the carts, or you shoot to take out the cart wheels and axles. If it's down to just people, then you start taking out anyone who looks like a leader.

Now mind you, sometimes you may come across a force with some serious military gear: tanks, armored vehicles, etc. However, those can be stopped as well by large ditches, landslides, large trees, concertina/barbed wire (which tangles in tank tracks nicely) and all kinds of obstacles which will force them to burn off fuel and slow them down. You still concentrate on the supply vehicles as your top priority. Without fuel and supplies, even a fully-armed tank or military armored vehicle will quickly become useless, so *always concentrate on their supply vehicles!*

Combined, this comes together to distract and grind down almost any opposing force, be it amateur or professional.

One Small Exception...

If you're up against a sufficiently large, complete and actual military force (especially one with any kind of airborne assets such as helicopters), then we're talking about something entirely different here and shooting at them or their stuff would probably be a very, very bad idea. In such a case, please refer to the chapter in Chapter 3 called *If The Government Comes Knocking*, modifying the advice a bit to take into account whoever is running that force.

However, this far post-collapse? It will be rare indeed to find someone or some group that is still capable of marshalling enough resources to keep such top-end hardware and weaponry going on a continuing and effective basis. As time passes, the odds will get slimmer and slimmer that you would meet such a force.

Summary

Civil Wars will be an unfortunate thing for quite awhile post-collapse, even to the point where it may be an almost constant thing. If your community is somewhat isolated, or you are able to ally yourself with other like-minded communities (that's coming up later on) and you are able to see it coming from a far enough distance, you stand a good (but not perfect) chance of fending off the worst of it and enduring that which you cannot fend off.

This also means you have to do your best to prevent internal strife within your own community as well. If you can do this and be consistent about it, you then stand a chance of enduring what may come.

Making It a Permanent Thing

Let's say you've managed to keep a more or less constant supply of food going for everyone in your community. You've managed to defend yourselves well enough. There's some semblance of commerce going. There's no immediate danger of shortage (water, food, etc). You have your community members looking out for each other and you've all even managed to form some sort of identity and loyalty. You may even find your population increasing as survivors and refugees from outlying areas begin to settle in and begin the barest start towards making a life for themselves. It's been at least a year to 18 months since civilization collapsed and the majority of humanity in the region has likely died off, leaving the local ecology at least somewhat sufficient to provide for those who have survived thus far. So... how do we go about making this enterprise a permanent, thriving and going concern?

Well, this is going to require a couple of things to happen. You obviously cannot slack off on the basics. You have to insure that a continuous supply of clean water is there. You need to insure that food is available both now and in the future. You seriously need to insure that your militia is sufficiently trained and equipped and operating in good order. However, there are a few other things you will need to start doing. First, you have to insure that your government is permanent, fair and as child-proof as possible. To that end...

Governance

Up until now, you've gotten by with a council of sorts. You have a leader, someone who runs the militia, others to keep a collective eye out for various health and safety issues and folks to keep the trade and commerce a fair and safe experience.

So far, that's a pretty good thing. However, there is one problem with just assigning someone to do a job that grants power - over time, that person gets a bit too comfortable with the job and starts thinking about making things even more comfortable for him or herself. There is also the fact that you've been doing all of this so far by the seat of your pants. Well, we need to fix that. We need a local government.

First, you're going to need a charter. A charter is a constitution of sorts, where universal rights are written down and the structure and duties of your new government are outlined and written down. You don't need to go crazy with flowery language, but a basic community (let's say "Town") Charter has a basic outline that goes a bit like this…

The Name of Your New Community
   * It would help to describe a bit of why you named the town what you did.

Preamble
   * (a mission statement - what is it that you as a community really want to accomplish? It sounds silly, but I promise that it isn't. Think hard on this bit.)

Rights and Responsibilities
   * A list of basic rights that everyone enjoys, no matter what
   * A list of what is required to be a member of the community

Offices and Terms
   * A list of each office (Mayor, Council members, etc), what each office is responsible for, and how long each term is. Also note if the officers are limited to only serving so many terms, etc. Finally, make sure that when it comes to critical offices (mayor, militia leader, etc), a line of succession is clearly outlined.

Voting and Appointments
   * In here, outline how voting will occur, when it occurs, what happens if there are conflicts or ties, how to appoint someone to fill an emergency vacancy, etc.

Business and Procedure
   * Outline when everyone meets during ordinary (and relatively peaceful) times, how business is conducted, how citizens can raise issues and suggest changes, etc.

Amendment and Modification
   * Describe how this charter can be amended, changed, or modified.

This should be enough to get things started. Remember to keep it simple and useful, because you can always add to it later. The idea is to get something simple and solid started.

When you put this together, make sure as many community residents are involved as possible. Do it in shifts and over multiple meetings if you have to, so that everyone gets a chance to see it and approve or suggest changes to it. Once everyone has had a chance to look and work it over, it goes up for a vote. Simple majority vote by all residents wins it. If it fails, get everyone together and work it over some more. Once it passes, it becomes your guide to how your new town is run.

As soon as you have a good, working charter that the majority agrees on, it's time to hold elections. Anyone who hasn't committed a crime in or against the community is eligible and should last for at least two weeks before votes are held. The candidates with the most votes in each office wins the office and become the new Mayor, Market Judge, Council Member, etc. From then on, power transfers to the new officers and any existing community leaders who failed to make the vote will have to step aside and go find something productive to do for a living.

Setting Up The Town

Once you have a functioning government, now it's time to figure out just what you have. These new professional town officers will need to be paid. Initially, this will be in bartered goods: Food, shelter, etc. It's not important (yet!) To pay enough to cover everything an officer may need, save for the Mayor or any other officer who demonstrably works for at least 40-60 hours a week. How much of what items should be in payment need to be agreed on by the whole town initially, with no power to increase those amounts without the whole town voting on it. The idea is that you should provide them with enough to compensate them for their time spent in running the town. This is an initial form of taxation, so it should be as light as you can make it on the town's citizens.

Next up is to figure out the town limits and where on the map your government's responsibility ends. Anyone living in the limits is a citizen of the town and given all rights and privileges (and is taxed). Anyone living outside those limits is not a citizen - how they are treated while they happen to be in town limits is up to the citizens.

Finally, a few basic laws are in order. These are rules that everyone will live under, *including you*, so don't go too crazy. The idea is to incorporate the basic rules you've had before and to codify them in writing. Once you're done writing up these laws, post them for everyone in town to see, allow for a period of time where any citizen can object to a law, suggest changes, or make comments. Write down everything that's said. Any laws that are objected to are voted on - any which are not rejected are put into effect. The rest go into effect at the end of the comment period.

Once you have the basics of law, it's time to get some order. Now is the time to set up a couple of people whose only job is to enforce the law, direct firefighting, run the militia and other vital functions. It's not (yet) necessary to have professional positions for every conceivable facet of town government, but at least try to have at least one cop (who can deputize as needed), one firefighter (who can direct volunteers) and one militia leader (who in turn commands volunteers).

Reaching Out

Once you have all the basics of a town going, it's time to start doing the things that a post-collapse chamber of commerce would do. If your citizens have goods they want to sell but the local market isn't diverse or lucrative enough, get together an armed patrol and set up caravans to help them take goods into other communities to sell. Put it on a regular schedule, but be sure your caravans are heavily armed, including scouts to look ahead for potential ambushes or other trouble. Why do I keep harping on armed scouts and soldiers? Because nothing attracts raiders and other troublemakers like a regular schedule of useful goods going down a given route. In fact, while we're thinking about this, perhaps it's a good time to map out multiple routes and alternate (or better - randomly select one) each time a caravan goes out to market in another town. Initially, this will be a dangerous undertaking and will require a lot of vigilance and a sharp group of soldiers. However, as time passes and the criminals are either killed or starved, things will (well, more or less) go smoother.

If you can manage to get some good trade going with neighboring communities, you should then try and take some time to have one or two of your most trusted leaders go out to meet their leaders. Initially, it's an opportunity to talk a bit, to compare news and rumors and to maybe bring a few gifts. Over time, if they're doing okay and are like-minded, you can forge agreements, treaties and other means to increase your mutual security and prosperity (perhaps starting with having mutual patrols to hunt down and exterminate any raiders on the roads linking the two communities? I suspect this would be a good start).

But then, we'll be covering these particular topics in far more detail before you're all done, I promise. We just want you to get to thinking about it and maybe start some preparations.

Thinking Ahead

If you're still breathing and you're actually doing far better than expected, then let's get you to looking ahead at your wee home and community. After all, you're not going to live like a desperate slob your entire life, are you?

About The House

Recently, we've been focusing on community and interpersonal survival. However, maybe it's time to come back home for a little while.

If you have the time, set a bit of it aside and take a good, long look around the home. First off, odds are perfect that unless you're Amish, your home was built and geared for a ton of modern conveniences (flush toilets, running water, electricity, cable/satellite, Internet, central heating, etc…) Needless to say, you're likely using the place a whole lot differently by this stage in the game. Now I'm not telling you that you should rip all of that out, because doing so leaves holes and some of it can be salvaged or put back to its original use later on (or even sooner).

What I'm getting at is to look at how you use the place now and to look into a few changes that will make it a bit more comfortable. Not only can you re-engineer your post-collapse home to be cozy, but you can take a good look at how everyone moves about it. You can then start looking at how to better re-arrange things not only for utility, but for defense as well. Let me give you some examples:

- Running Water? If you have a water cistern (that big water storage tank out back that we were talking about a whole lot of pages ago) and it's elevated? you can try to find a way to pipe it into your existing plumbing and shutting off the water feeds to all except the faucets in any part of the house that is lower than the bottom of the water storage tank. If you have a lot of water, enough water pressure and a septic tank? Maybe you can add the toilets as well, which knocks out one really huge headache for everyone in the home.

    If you lack the water pressure or sufficient water flow, you can still get by if you have (or can build) a few manual pumps and know how to sufficiently replace the water faucet in a common area with it (and running the plumbing to your water storage or well, etc).

- Electricity will take a bit of skill, but with a bit of forethought, you can do pretty well. The idea is to disconnect individual circuits from the main breaker that are not going to be in use and then disconnect the main electrical meter leading into your place. You can then connect your AC (not DC!) leads from your power source into the main breaker and use the existing wiring as a means to conveniently distribute power throughout the home.

- Let's face it - the modern kitchen is not built for 18th century cooking. Your refrigerator is going to be almost useless, your stove isn't going to run and the dishwasher? Well, the only working dishwashers will have two legs now. Maybe it's time to take those little beasties out and perhaps store them somewhere just in case a day comes where they'll have a use again (and if it doesn't, you'll have a ready pile of usable parts for other things). Then you can replace the stove with a home-built wood-fired one. The hole where the dishwasher once sat can be fitted with shelves for extra storage. The giant hole where the refrigerator once sat? Tons of storage space can be built into the spot. Some folks may debate a bit on the refrigerator and some may even have a dire need to continue keeping one running at all costs (Type I diabetics for instance, who have a need to keep homemade insulin cool). I contend that you would be better served by building a root cellar, or if absolutely necessary, then at least replace that monster sized appliance with a tiny "dorm" sized one.

- Heating, especially in colder climates, is going to be a very huge topic of discussion. If you're like most survivors, you have at least one wood-burning apparatus and it's likely in the living room, where the most people can huddle/sleep/etc around it. Maybe it's time to start thinking about expanding that a little bit. An obvious second place would most likely be the kitchen, where it can do double-duty as a means to cook food. Another place you'll want to put one is in a master bathroom, for those really intense cold snaps where everyone can bunk in a smaller room that doesn't require as much fuel to keep warm.

- Speaking of heat, the end of the world doesn't mean you get to skip out on making the place more energy-efficient. Find/scrounge heavy curtains to help keep heat from escaping through the windows. Scavenge fiberglass insulation from abandoned/destroyed homes and stuff it into every void, crack and crevice that you can find in yours.

- If you haven't done it by now, go through the home, top to bottom and figure out what is really there. Take the things you don't need (the gaming console, the electric treadmill, toaster oven, etc) and store them somewhere else, outside the home. This gives you more room for the things you do need and it also gives you an idea of the things you can scavenge for usable bits later on. While you're at it, get the family together and re-organize all your stuff. The vital stuff (weapons, cooking/eating utensils, everyday tools, etc) can be put in places where they are readily available at a moment's notice. The not-as-vital stuff (leisure books, etc) can be arranged where they can be reached in fairly short order and the not-vital-at-all-but-nice-to-keep stuff (Christmas Tree and ornaments?) Can be gotten to eventually. I think you get the idea from there…

- Let's take a look at defense. Can you quickly move from one room to another? If not, what is in your way? If you have the tools, perhaps carving a new doorway between rooms will help things (but do try not to just start carving holes unless folks agree that it's critical to do so). If there are things like furniture in the way, move them. Check the windows - are they sufficiently reinforced on the ground level? Do you have enough material to cover and/or barricade them? Take a good hard look at all of the doors. How can they be reinforced to keep bad things out? Before going crazy on bolting up the doors, give a thought to speed - can you get in and out of it quickly if you had to? If you have a door that is rarely used and is not really necessary (the one leading in or out of the garage, for instance), you may want to give a thought towards reinforcing it a bit stronger than the others, such as putting in an internal bar or two across it to keep it closed. Sliding glass doors should be heavily reinforced, even to the point of having half of it bricked-up. Half of any double-door should be considered for nailing completely shut. Give a good look at your walls, too. You can have the strongest doors and windows in the world, but if the house is built from 2x4 boards and drywall, you may want to consider putting up a layer of brick or stone along the outside of it - for at least the first 5-6 feet of height.. Even logs will work, as long as they're securely fastened.

- Take a good look at privacy. As long as security/defense isn't compromised, how can you help your family and fellow occupants gain a bit more privacy? This is especially something to consider if you have more than one couple living under the same roof, since they're definitely going to want some 'alone time' without worrying about you or someone else walking in on them. Even kids will want some place they can think of as a secure and safe area that they can call their own - especially as they reach their teenaged years.

- Don't forget to organize the outside areas as well, but with a twist: keep it defendable and fire-aware. Don't clutter the yard and move aside/store anything that does clutter things up. Clean out debris and crap that may have accumulated. Set aside and set up definite work areas. While you're at it, work on proper drainage if it rains with any frequency, which keeps things less muddy.

- While you're outside, lay out any potential gardens and areas where you can do some real agriculture, if you can. The areas should be fertile, fairly easy to defend (even if not at close range) and relatively free of rocks, trees and other things that would get in the way of good growth. If you have a choice, go for the south-facing and gently-sloping patch of land, with no shade. Now reality will dictate that you likely don't have a nearby field that you can just start plowing, so modify something to get as close to ideal as you can.

About Town

Once you get your home in order, get together with the town/community leadership and take a look around town. A lot of the same considerations for defense are definitely going to be at play here, but now you have time to really do something about it. Some things to look at? Well…

- When looking over the defenses, see what the land offers you. Some approaches to your community may be very tough to pull off for a raider, so not much work needs to be done in order to reinforce that. Make the easy areas harder to attack from, forcing the raiders to come at you from well-defined places that put you at the advantage. Forcing raiders to come in to your community from open roads with little cover, or up narrow roads that have high hills facing down towards it? Far preferable than having them come at you from places that offer a lot of cover to them as they approach.

- Once that's done, you should take the time to clear out defensible zones everywhere that you can, especially in areas that are avenues of approach. A defensible zone is an area surrounding the community that is open and at least either level or downward-sloping as you walk away from the community. This open area makes it easier to pick off raiders and provides a good psychological barrier as well. Don't try to get cute and put in traps or ditches, as they will inadvertently provide cover to anyone trying to attack. This zone also acts as a barrier against forest or brush fires, provided that you keep at least a 20' wide road or other continuous barren strip of land on the outer (and if you can, inner) edge of it. Note that you don't have to leave the whole thing barren - you can use a good portion of it to grow crops if need be (just remember to keep the road or strip on the *outside* of the zone if you decide to do this). Remove (as a community) any and all obstacles that can be moved from this zone: trees, large rocks (if you can), debris and all unoccupied buildings. Make it a strict rule that no solid item larger than a baseball is to be left in this zone… you don't want a stray plow or vehicle providing cover for inbound raiders, after all. Make this zone stretch at least 100 yards out from the community in all directions.

- If terrain or the area won't let you set up a proper open zone, then do what you can to make it treacherous for any attacker.

- Once you get that zone set up, start looking at a little something called *interlocking zones of fire*. What this means is, rig things so that defenders stationed all around the perimeter only needs to fire in one direction, in a narrow 'cone' facing outwards. Make sure that all of those cones cover at least a part of the cones next to it. The narrower that cone, the better you can defend your community (though this will require one person per cone, so be sure that you don't plan for more than you have people). The idea is that each defender only has to shoot in one direction, only has to cover that one narrow bit of land and that his neighboring defenders can also cover it for him should he not be able to.

- Once you have some idea of a defensible zone and where you want to put everyone around it, start setting up more permanent hard points from which each defender can fire. Reinforce these areas with stonework, brick, masonry, or even just piles of rock. If you want to carry it even further, set up small walls or trenches (depending on terrain) to allow defenders to move quickly under cover from one place to the next. You can use rock/stone/concrete-based rubble to accomplish this.

- Finally for defenses, if you do find places which offer raiders cover coming into your community, do whatever you can to remove that cover, or make it so dangerous or impossible for them to use it that they leave it alone.

- Once you can defend your community, start locating common water areas. You've done this before in earlier chapters, but this time you need to formalize it. Be sure at least some of those water sources can be defended and then start cleaning them out. Reinforce the walls of any well with stone or (if you have it) metal. Be sure to have a routine to keep it clean and clear of debris and trash (especially if it is a spring or creek). Keep a more rigid set of rules as to what people can and cannot do at the water site, or at least make sure it is marked as drinking water and to keep people from doing anything stupid (bathing, urinating, dumping, etc) in it (or especially upstream of it!)

- Once you get your water sources formalized, get your trash disposal organized. Set up a place outside of your community to dump the trash - make it close enough to reach easily without exposing individuals to danger, but far enough away to not stink up the town. Make it in an area where large pits can be dug and the trash easily buried (or burned). Then, start getting people to take their trash there. No need for bags - just use trash cans. Once you have a dump, get everyone in your community to start cleaning up, to take their trash (that they don't already burn) out to it.

- Organize a cleanup with the whole community. Clear out any and all debris; take the stuff with potentially usable parts (junked vehicles, etc) to a designated area for stripping later. Take the outright trash to the dump. Set unclaimed wood aside as a communal firewood pile. Encourage recycling building materials from half-wrecked and abandoned buildings (abandoned buildings that are intact should be set aside and put to use later).

- If your little community market is actually doing well, maybe it's time to start improving it. Expand it a bit. Set up good drainage, build sidewalks, improve the outhouses, start building better tables and stalls, maybe giving them a coat of paint if that can be had or made. Set aside an area where merchant sales are not allowed and build small picnic tables for it. This gives people a place to sit and chat, for travelers to eat some lunch, whatever. You may even want to consider building a small stage at one end of the area, which can be used for entertainment and (more importantly) outdoor announcements and civic events.

- If you have a church in your community, take the time to clean it up and bring it back to good shape - even if you don't worship there. If you don't have a church but you have people meeting for prayer services or similar, then see if an abandoned building or other suitable place can be set aside as a church, synagogue, or whatever. No matter your views on religion, or a given religion in particular, know that religion can be a civilizing force and it does help to bring people together. If you have more than one religion meeting on a regular basis (e.g. Catholic, Protestant, Jewish, Muslim, etc), arrange interfaith events, coming up with ways to get people together on a friendly basis - encourage this friendship and sense of community.

- If you have a public (or even impromptu public) library in your community, treasure it. See what can be done to procure more books for it and donate books that are otherwise not being used. Without electronic media and diversions, books offer a wonderful means of getting away from the crap and drudgery that will comprise a lot of post-collapse life. If you can appoint an elderly person or two as librarians (paying them in food and such), then by all means do it. For now, restrict the privilege of checking out books to just community residents - you will be amazed at how popular your library will become and how quickly it becomes that way.

- Is there a teacher or two among your surviving neighbors? See about putting them to work! A good start would be to get the surviving children ages 12 and under and setting up classes in the basic skills of reading, mathematics and basic sciences. Children that age are better served by putting them into a classroom.

Most importantly, take the time to take a look around and to think ahead. See beyond the crisis and look to the days when things become more peaceful. Try to prepare for those days in addition to preparing for the here and now.

About Yourself

While you're cleaning up around your home and around town, take some time to sort things out in your own mind. You've likely spent the past year or more in a constant crisis mode - foraging, fighting, rationing, praying, hoping, crying, with rare but occasional patches of joy, laughter, accomplishment and elation.

Maybe it's time to start planting a few test seeds come the first spring to get some practical experience in growing food now - and if it's peaceful enough, to start planting for keeps. Perhaps it's time to work on some useful projects (time permitting) to make things more comfortable around the home. Examples of this would be to improve the fireplace (if it's a lash-up you had to build just to stay warm), or improve/build a water filtration system. You could turn that quickie outhouse you slapped together into something durable and useful… and comfortable! You can get up skills in turning scrap cloth into blankets, weaving, blacksmithing, or other skills that will be useful as time goes on.

Try to find ways of reaching out more to trusted neighbors. Start coming up with excuses to get you and your neighbors together more. Organize sports and games. Get involved in a community religious group if your views fit theirs.

Most important of all, take a bit of time and look around you. Look at those you love and enjoy them more. Look for ways of making your kids' lives better than yours once they grow up. If you have the resources to do it (paper or blank diary, pencils, etc)? Write down your life story before, during and post-collapse. Write down as much detail as you can pack into it about the way life used to be for you. Describe the conveniences and technology as much as you can do so, right up to the limits of your own personal and technical ability. Add stories and histories of your parents and their parents. Maybe add a bit at the end about why you think things crashed the way they did. Most of all, add a note describing to future generations what it is you hope for out of them.

But as for now, back to work with you. There's a lot to be done and it won't do itself…

Chapter 5

In The Long Term

(2 years and Beyond)

You're Almost There?

Well, okay, that's (almost) a lie, but you could use some encouragement and you are far closer to better times than you were a year ago. Before you celebrate though, know that there is no finish line here. This is your life - more specifically, this is your new life and it's never going to return to the days of technology and convenience, unless you (by some miracle) travel back in time, or happen to make it that way now. Not going to happen either way.

I'll let that sink in a little, but not for too long. Now before you get all depressed and clutch a gaming console to your chest with sobs coming out of your throat? Stop and think about this for a minute. Oh and put the gun down. Certainly, a lot of the benefits and life-enhancing things that we know and love are going to be gone and will likely not return within your lifetime. Definitely, mortality will skyrocket and life expectancy will drop. Without a doubt, you're going to miss a whole lot of things terribly, both large and small.

How-ever! You have actually gained a lot up until now. You have taken literal charge of your own destiny. You can laugh at the banks, insult the IRS in public and, assuming the guy is still alive, you can give your ex-spouse's divorce lawyer the middle finger - all with total impunity. Okay, all humor aside, the fact is that Divine Provenance has pressed the great big reset button on history and if you made it this far? You are by now one of the lucky few who are positioned to take maximum advantage of that. Yes, you may have lost loved ones, but hopefully, you will have found and built relationships with new loved ones. If your spouse and kids have made it this far both intact and alive, you've accomplished something that very few human beings have done and you should be damned well applauded for a job well done. You may have noticed that you take the time to talk to your neighbors now. You've likely fought alongside them and the experience has drawn you into relationships that are tight and unshakeable. You will have learned to love more deeply and to laugh using your very soul - that is, whenever chance presents the increasingly more common opportunity to do so.

Seriously - along with the benefits being gone, a whole lot of the clutter and garbage that came with the pre-collapse world has also been swept away. As a bonus, new benefits have begun to appear. With a little luck and a lot of work, you can help build a better society and together, create a new world that your descendants will actually be grateful for living in.

So how do we stop being a rag-tag gaggle of survivors and become a self-sufficient, prosperous and growing society? Well, let's look at what we have and where things most likely are.

Two Years? Why The Long Wait?

I picked the 2-year mark after complete collapse before even thinking of going full-tilt at rebuilding. I did this for a few reasons:

- Two years after the government came apart, the vast majority of humanity will be, well, dead. There may be some pockets of civilization that have managed to hold out on government-distributed supplies for this long, but odds are very good that all the freely available food as we know it will have been consumed. Most human beings can survive for up to 45 days without food and up to 180 days on a diet averaging 900 calories per day. The generally-available food supplies (silos, granaries, warehouses, large-scale farms, etc) will have run out at the end of the first year, the 180 day near-starvation period will have finished off the bulk of the first wave of survivors and then we added a 6-month pad beyond that to wait for the bulk of violent activity to start dying down.

- Most of the available wildlife will be too scarce to support the population as well and foraging will start to get real tough to do by this point. Given that you (hopefully!) saved up at least 2 to 2-1/2 years' worth of food (and have managed to keep hold of it!) you'll notice that you're running low by now. It's time to start planting, big-time and to do it for keeps.

- Society at large will have become mostly isolated, save for whoever braves the roads between isolated communities.

- Nearly all large cities will have degenerated into a collection of weakened enclaves, with constant in-fighting over rapidly dwindling pre-collapse resources. They will also likely be low on (or even likely out of) most firearm-based weapons and ammunition. Some of them may have wandered out into the countryside (where do you think those raiders came from?) However, the ones who remain will (mostly) be ill-equipped to project any real force beyond what little territory they continue holding on to. There will be an exception or two to this (people being the creative critters that they are), but by and large, most cities will be savage wastelands and as time passes any gangs left in there will weaken further.

- Because most of the population is dead now, there is (relatively) less danger on the roads in most places and determined efforts in clearing the road of criminals and raiders will begin to see success.

- Thanks to the reduction in population, scavenging efforts for non-perishable goods will be more fruitful and there will be more stuff per person (land, water, other resources) around.

- Most efforts at resurrecting the now-dead governments will be universally ignored at this point and few if any would even want to try by now.

- The most important reason of all? This far along, any communities that are left and are still going will start focusing ever more on self-sufficiency and less on domination or conquest. This means they are more likely to reach out to you for trade, mutual support and similar beneficial endeavors.

In Trust We Trust

By this point in time, those in your community who are still alive can be viewed with less suspicion and more with an eye towards friendship and rebuilding. In earlier days, everyone rightfully held back (at least a little bit) from each other, suspecting neighbors of plotting to take each others' food supplies, among other things. Those who are still alive by now have learned to work together in foraging and scavenging food and may even have worked on the beginnings of agriculture, industry and more.

You're finally going to have to start trusting your neighbors beyond what little you've had to in order to stay alive. If you have already been able to safely do this? Congrats! If you haven't yet, you really need to start reaching out. Consider this official notice that hey, it's time to stop hunkering down so much. Start with those neighbors who have done you favors, or who you owe a (small) favor to. Then, work your way up to neighbors you know well, then those you know well enough.

Remember, this trust thing works both ways - if someone isn't willing to trust you, then you have no cause to trust them.

This doesn't mean suddenly revealing everything you have and everything you know. What it means is, you start small and work your way up. If you have kids and they have some spare time, arrange 'play dates' with the neighbors. Spend time over at the neighbor's home. Invite them over to your home. Start talking to each other. Share a bit of your hopes and dreams. Share ideas about how to improve things. Try to learn how they've managed to survive so far and teach them a few of your own tricks and tips. Share small gifts and be charitable if you can do so without revealing everything you have or without putting yourself in a bind. Eventually, you get to the point where you can start making plans together. Do this with a couple of neighbors the first time, alternating whom you visit with and then getting everyone together. Once you have a small group, expand it a bit, one neighbor at a time. Those neighbors you decide that you cannot trust, or that do not trust you, remain friendly with but keep at arm's length.

Cleaning House

At this stage of the game, you and your community will want to get together and do two things: Start rebuilding a workable society and to start cleaning out the trash and crime.

The first part you should already be well on your way towards doing. You've hopefully established a working government and the rest of this book will deal with how to get the rest of that going. The second part is a bit messier and will require a lot of work. This chapter is what will cover that.

Your Mission

You remember all those raiders who consistently ambushed travelers on nearby roads and attacked or stole from your community? Well, it's time to clean them up and make things safe again.

The goal is as simple as it is harsh: you're not going to take them alive. While a jury trial would be the most just, you're going to have a very hard time finding an impartial jury and the raiders aren't going to quietly let themselves be cuffed. The reason you have to do this is straightforward, because it boils down to either you cleaning them out, or them perennially cleaning you out. While things may be getting better and you may be able to survive and even thrive in spite of them, you cannot let them continue. Raiders, if left unchecked, will continue to steal, rob and rape. They could even get lucky and really leave you and your community in a very big and very fatal bind.

As a side benefit of cleaning out the roads and the woods, killing off the raiders will give you and your community more resources - things they have been hoarding all of this time and likely things they have stolen from you in the first place. Cleaning them out also gives you access to the wild areas in between communities.

I know that this sounds amoral and even evil in some aspects, but let me make this clear: The raiders will represent one of the last barriers to your continued survival and either you kill them, or they will eventually kill you - either by luck, or by attrition. To this end, we need to get started.

Know Thine Enemy

Don't just go out and tramp through the woods looking for raiders to kill - this is most definitely not a simple video game. You're also undertaking something that should be approached with logic and intelligence, not just emotion and a lust for blood. Take the time and do this right, because you probably won't get a second chance and mistakes will cost you in both blood and treasure.

To this end, you need to get some information first. Here's how you do it. Get together two sharp shooters and 3-4 soldiers, maximum. There's no real safe way to set up decoys and tromping through the woods in a large group will just tip them off, causing them to just get out of your way and lay low until you go home. Your party will want to stay away from the roads and stick to the hidden places, moving quietly and under concealment as much as possible. Work your way carefully towards the areas where ambushes are most common, then start looking for trails, tracks and other evidence of raiders. Keep a very sharp eye out and make liberal use of binoculars to seek out raider encampments. If/when you come across an encampment, simply sit back (way back, concealed) and watch quietly for awhile - at least 24 hours, from a distant and concealed location. See if the occupants are actually raiders, or just ordinary survivors. See if you recognize any of them from previous raids. Try to figure out how many there are, what weapons they have, if they still have any operable vehicles and what they use for shelter and/or cover. Figure out what routines they have and make good note of them. Be especially observant for any lookouts, sentries, or other guards.

For the ones you can confirm as raiders, get back to town (quietly!) and gather a more substantial force.

The Best Laid Plans

…of mice and men, right? Well, forget the old saw about how plans fall apart in combat. Yes, it'll get chaotic, but the team who has the best plan and can best stick to it (and don't forget - adapt as necessary) has the best chance of winning. Put together a solid plan and do it quietly, just among the hunting party members.

If you choose to attack conventionally, then choose the best time to attack based on observations. If there are a lot of them, maybe the best time to attack is during those times when raiders are out attacking something else, reducing the numbers. If it is a small party, this may mean all but a few token guards will leave, which gives you the opportunity to attack (once the main party has left) and destroy their base of operations. You can then wait for them to return and finish off the rest.

If the raiding party is too big or too well-defended, you may want to opt for flushing them out, or even simply locking them in. You can accomplish this by booby-trapping their bunker doors (or burying them), by simply setting the building they're in on fire, or even the forest they're hiding in on fire (just be sure innocent folks aren't downwind, or in a position to be put in immediate danger by the fire.) Another option is to booby-trap the paths the raiders take to and from their camp or enclave. You can also destroy or cripple their vehicles with traps and even covert sabotage. If you cannot take them out directly, take them out slowly. Taking away their vehicles will take away their ability to conduct attacks with any speed.

Another option? Poisoning their animals or local water supply (as long as they're not sharing the same local watershed/water-table, of course.) How about quietly trailing raiding parties (especially small ones) as they leave the main encampment and pick them off once the party is outside of both view and earshot from the main camp?

Little Versus Big

There are a lot of options available to you when it comes to clearing out raiders. The only real difference is in tactics and time.

Small groups can be attacked and taken down rather quickly. A lot of times, just having a much larger force and literally overwhelming them is sufficient. If they're entrenched (in a house, bunker, etc), it will take a bit of work and planning, but it can still be done with somewhat relative ease, as long as you have superior numbers, tactics and force. Most often, a small raiding party that finds itself under attack will likely pull up stakes and leave, post-haste. The rest of this section needs to focus on the big groups, though read on and take what lessons you feel are appropriate.

Large raiding groups will need to be picked off slowly and treated to a campaign of denial in order to wear them down. This may take a bit of time and will likely result in reprisal attacks. In preparation for this, take out the smaller groups first before going after the big kahuna, to reduce distractions and clean out any potential for having to fight residual battles afterwards.

Your first goal in taking out a large raider group is to deny them the means to travel quickly or easily. Barricade them in with traps along roads, paths and trails. Sabotage or destroy every vehicle they can muster. Unless they happen to be using Humvees and/or armored personnel carriers, this should be fairly easy to do, especially at long distance and with rifles. Aim for the engine area first, as it is the biggest and most easily-hit target. After that, go for the gas tanks and then the tires. Don't try and shoot the driver or occupants, as those are harder to hit and the glass refraction will screw up your aim anyway. Using well-placed sharp objects and other concealed tire-eating obstacles will also help you out greatly. For the more ambitious, a concealed small but deep ditch in the right place will stop most vehicles cold. The beauty of such traps is that you don't really have to tend them too much - just keep a concealed observer in the area to watch over them from a distance.

Once you manage to take out their means of travel, you can then work on them at a more intimate level (just keep a very close guard on any vehicles you may have, because they'll likely try to steal them). Picking off any small parties that wander too far from their encampment is a good start, as is finding ways to cut off their water supply and in stopping any communications or any inbound reinforcements they may get. No need to charge in or attack - just snipe them at the edges for now and deny them whatever you can.

Eventually (or quickly, depending on how smart they are), they'll do one of two things: Hole up, or strike back. If you're the nearest coherent community to them, they may lash out at you while they still can, so always keep that in mind. Use a lot of the same tricks and tactics we've discussed earlier (previous book, "Civil War?" chapter) to keep them at least a little at bay.

If they hole-up, you only need to wait them out, provided that you have a large enough force and enough time to wait around. Use the time to look at what you're up against and quietly take out as much as you can. If they're at the base of a hillside or mountainside, see about starting an avalanche or landslide in that direction. If they're in some sort of underground bunker, it's a bonus - see about blocking the exits with vehicles, or even burying them outright - do the same to any ventilation shafts you see. Otherwise, look at where they get their water. If it comes from any sort of tank that you can see, shoot a hole or four in it. Take the time to shoot out any generators or cut any other sources of power (wind turbines, solar cells, whatever). Divert any streams or creeks away from them.

While waiting them out, always focus on one or two things and always deny them those things as much as possible. So far, we've focused on power and water, because without either, they will eventually get desperate and desperate people (more often than not) do dumber things, moving the odds more towards your favor.

If they break out and attack, you'd better be prepared for it. Odds are perfect they'll try to surprise you and haul out of there with as much stuff as they can carry. Your big priorities are to focus on disabling any vehicles present and then you start taking out the soldiers. Hold your ground as long as you can and take no prisoners during the fighting. If it holds a weapon, it gets shot at. If they flee, make sure they do it on foot and take down as many of them as you can so they don't return to attack later.

If by chance they do manage to drive you back, be sure nothing is left behind for them to use and retreat walking backwards. Don't let up on them if at all possible. Continue working on the vehicles first (if any), as much as you can.

Once the raiders are either vanquished or gone (assuming they fled), your very first order of business is to secure the area. Kill off any seriously wounded or still-fighting raiders on site. Keep guards and soldiers outside, watching for any reinforcements or return. Do a complete and thorough sweep of every possible nook, cranny and potential hiding spot in the whole area. Send large teams in to sweep carefully, with a main force coming in behind them holding each room or area while doing a closer inspection, weapons at the ready and taking out any hidden raiders.

Do this carefully and completely - I cannot stress this enough. One hidden raider can kill up to five people (or more!) just because your team got sloppy.

Only after you have secured the area and have rooted out any hidden raiders, should you proceed to take inventory of what can be taken back to town. If you drove them all out or killed them all fairly early in the fight, you may find a lot of things that the people in your community can use. If things really dragged on, or if the raiders were themselves at the end of their rope (or were smart enough to run away with the bulk of their stuff before you got there), you may not find much of anything. To be honest, it will likely be a crap-shoot in either event, but a bonus if you find useful items.

Divvy Up The Booty

With an eye towards being charitable if possible, divide things up as evenly as possible, with bias and priority towards the families of anyone in your community who were crippled or killed in taking down the raider encampment/stronghold/etc. You'll always want to repay those who gave the ultimate sacrifice and as a close runner-up, repay those who fought alongside in getting the job done.

Take as much as you can carry but still be able to defend yourself on the way back to town. If there is simply too much stuff, organize the community to do it in relays, saving the soldiers as an armed contingent to help make sure the goods (and everyone involved) all get back safely. Once completely emptied, take a good, hard look at the place. If there is some strategic value to it and it is close-by, you can consider re-purposing it for the community's needs. Otherwise, destroy it as thoroughly as you can, rendering it useless (save for maybe salvage parts) to anyone else. The last thing you need is to go through all that trouble to clean out a group of raiders from it, only to have another group of raiders move in (or for the same ones to move back in).

Finally, have scouting parties go back periodically to see if any of the raiders try to return to it. You may even want to consider posting guards to watch over the spot for a period of time.

Prisoners?

Okay, even though I mentioned a few times that you shouldn't bother, human nature is likely going to dictate that you may well end up with a prisoner or two. You just never know, there may have been stray children left behind, kidnap victims and totally innocent parties. There may also be wounded (but not too wounded) raiders and outright cowards who surrendered to the first person that opposed them directly.

For the truly innocent (e.g. any newly-orphaned child under the age of, say, 8-9), you're going to have to take them in and provide for them if you can. There's no moral alternative around that.

Anyone over the age of 8 or 9? It's up to you. If they protest their innocence, you can set up a trial to determine that. Note that while women may be quite valuable post-collapse, they will also be just as duplicitous, devious, thieving and as combative as men, so don't feel the need to take gender into account, no matter how pretty or helpless she looks.

If they're wounded raiders (that is, they were active combatants up until you incapacitated them), or just plain cowards who were fighting you right up until you pointed a gun at their bellies, just take them back to the community and then get it over with in a day or two at most. Set up a place and time of execution and then carry it out. Use the most humane methods available to you to perform the executions (without wasting ammunition) and give them a chance to do any last-minute religious things they may feel the need to do (within reason, obviously).

If you're wondering about the harsh methods and decision (and you really should), the answer is simple. Your community is still too young and fragile to withstand any 'reformed' raiders taking vengeance or reverting back to theft and pillage and in a post-collapse world, even kids as young as 10 will be vicious about it (in fact, kids around that age and up are often far more vicious than the adults, because base survival without remorse or morals is all they're going to really know by this point in time.)

Finally, you simply don't have the facilities to incarcerate anyone for any real length of time and you likely won't for a few more years yet.

As for the rest? Any raider that does survive and manage to escape your grasp is going to think very long and hard before setting up an operation anywhere near your community and any new friends he meets up with are going to hear all about it.

Governance And Leadership

In the more turbulent times between collapse and the two-year-mark (and likely for a little while still), leadership was whatever it took to get the really important things done for the most people in your community. By now, it is hoped that you've assembled and elected at least some form of formal government by now, but what we want to do is to dig a little deeper and help you figure down (and implement!) what true leadership and good governance really is.

First, Do No Harm

Yes, that is the beginning of the Hippocratic Oath for doctors. No, you're probably not a doctor. However, it is still good advice, especially for any aspiring politician. Your job as leader (or whoever got elected to the job) is to make sure that you don't end up tearing down the town just to make a point, or make a royal mess to get some trivial thing done. The entire center of your job description is to make sure the community stays intact, alive and prospering as much as possible. You really shouldn't be paying attention to anything else.

Your big goal in all of this is to make things reasonably safe, reasonably secure and to get the hell out of the way of prosperity and happiness whenever it's reasonably possible to do so.

To this end, these are the main things you should pay attention to:

- Making sure the water supplies are (and stay) clean
- Making sure that any residual raiders are kept at bay.
- Keeping any criminals restrained or detained until such time as they are deemed safe to be among society.
- Organizing and implementing a means to reasonably prevent and put out any fires
- Planning for (and if necessary implement) means to prevent or contain the spread of infectious diseases
- Plan, put in and maintain a decent sewage system
- Make sure that trash is kept to an absolute minimum
- Put together occasional and periodic events to boost morale
- Keep the markets fair, but do not overbear on them
- If your government takes anything, it must also give back of equal or greater value.
- Keep as few restrictive laws on the books as possible

Meanwhile, these are things you should really, really pay attention to:

- Do not try to make things 'too fair'. What I mean is, do not try and use rules and laws to try and make life idiot-proof. If someone is systematically ripping off people by actual fraud (watering down milk, for instance), then you deal with that person for fraud. If someone was too dumb to realize that an item for trade was not what he or she wanted until after the trade, then that's his or her problem - not yours and certainly not the town government's.
- Never use law or rule to implement any kind of compulsion or ban based on purely religious or "moral" principles. For example, if there is a demand to ban dancing, alcohol, or prostitution? Refuse the demand, period. If the public doesn't want something, they can refuse to partake in it without any help from government or law.
- Do not write or implement any law or rule designed to hinder anyone based on religion, gender, race, ethnicity, ideology, or any other aspect that does not (in and of itself) cause

harm to another human being. If they're not hurting anyone, don't hurt them. Live and let live - anything less will cause strife, division and eventually conflict.
- Never appropriate any individual's possessions except in a grave (life-or-death) emergency and in that one particular case, you will always reimburse that person (or persons) with equal or greater-value resources - either of the exact type, or whatever form the person(s) is satisfied with.

The next big thing should go without saying, but it unfortunately needs to be said anyway. Don't use the job to enrich yourself. You're there to serve, not scam.

Make your government as transparent as possible and always keep it open to public scrutiny and review. The citizens should always get in the first and last word on anything that is proposed. Obviously, some things need to be kept secret (e.g. plans to attack a raider camp), but anything having to do with local governance should be wide open. By the way, this also goes for trials and any kind of accusatory actions, in case anyone should forget that.

Mechanisms of Democracy

While you should already have sufficient knowledge (and books) on how democracy works, I feel it prudent to add a couple of thoughts here, to cover those parts that are usually left out.

Fully democratic governments which vote on everything simply do not scale up very well. Having everyone vote on every decision is easy and straightforward when there are only 10-20 people in your community. When that number gets up to around 1,000? Not so easy and it begins to eat into everyone's time. Once you get over 100 voting citizens, you may want to just post minutes and have anyone interested come to the meetings. Once you get over 2000, you may want to seriously get some council members put together as representatives of the population and you can either always keep them at an odd number (to avoid tie votes), or always keep them at an even number (with the Mayor as tie-breaker).

After at most the first election, you will also want to start holding election seasons. Give folks a chance to pick from among themselves. Early on, have an election season that runs for a month or two, to give folks time to nominate candidates and to give everyone a chance to hear each one of them out.

After a couple of years, or if the population is large enough (say, over 500 people), have a full and formal election season (say, nine months.) Also try to host a formal debate among all candidates, with a trusted and intelligent resident (who is also not running for office) moderating it.

When the population is small (less than 500), you can usually get by without any sort of veto mechanism. However, once you get a lot of people, start doing it. A veto is simple… According to the US Constitution, an executive who exercises a veto on a proposed law or rule sends it back to the legislative group (that would be the town council). In order for it to become law, it has to be passed again, but this time with a 66% vote or better (a "supermajority"). If it passes with that 66% or better, it is now law.

Something else to consider: Even with a small population, you'll likely want a judge. At first, this can be the market judge doing double-duty. This person should be familiar with the laws (the ones you passed) and should be someone who is fair and impartial.

Blending the two? If any council member, Mayor, or other official is caught doing something dumb and/or criminal, have something in place to remove that person from power immediately. If it's just dumb, that removal can happen by a simple "no confidence" vote by the council and Mayor (if it's just the mayor, obviously only the council will vote on that.) If it's criminal, that person is arrested and handed over to the judicial system and a replacement is elected as soon as possible. Let the population in on it as well if you want - for instance, if a council member is acting up, then the no-confidence vote can be held by the people that member is supposed to represent.

In my opinion, a nice addition could be made to the whole works: Every year at the same time, everyone votes on whether or not to keep the officials they elected the year before. If that vote turns out to be "no", then someone else should get the job, so maybe have candidates on the ballot as well (it can be as simple as: "do you want to keep your current Mayor/member/etc? If not, who should replace him or her?", with a short list of nominated candidates.)

Finally, get someone to write all of this down, make a copy or two and always keep copies in a place for the public to see and read at any reasonable time they desire to.

Mechanisms Of Crime And Punishment

You should already have the relevant texts that explain how basic criminal law and trials work, but just in case...

All laws are divided up into two types: civil and criminal. Just to be sure, let's run through a very brief description here. Criminal acts are those which deprive or harm others (or a whole community) through violence, the threat of violence, extortion/blackmail, theft, fraud, by perjury (lying under oath), or in contempt of court. Wikipedia has the following definition, which sums it up nicely:

> *"Criminal law [...] is the body of law that relates to crime. It might be defined as the body of rules that defines conduct that is not allowed because it is held to threaten, harm or endanger the safety and welfare of people and that sets out the punishment to be imposed on people who do not obey these laws."*
> - http://en.wikipedia.org/wiki/Criminal_law

A criminal is basically someone who will intentionally satisfy their own wants or desires at the expense of others, in turn either potentially or actually causing loss, injury, or death. These are people who need to be removed from society as soon as possible. It needs to be done either temporarily (until they learn from it, or because the act wasn't bad enough to warrant execution), or permanently (either by execution or by locking the person up permanently.)

Civil laws on the other hand are those which concern otherwise private matters between any two parties. The Missouri State Bar defined it quite nicely in this way:

> *"that part of the law that encompasses business, contracts, estates, domestic (family) relations, accidents, negligence and everything related to legal issues, statutes and lawsuits, that is not criminal law."*
> -http://legal-dictionary.thefreedictionary.com/civil+law

What that means is, civil law is anything which harms another person, usually involves a dispute between any two parties and may or may not also involve a criminal act.

It is very important to keep these two concepts separate, but know that a given incident can often result in both criminal charges and a civil lawsuit. Or it could mean either one - a criminal charge could be dropped but the person may still be found liable in court for damages. It all depends on the incident at hand. Now if it comes around that someone did something criminal, the criminal trial happens first and then any lawsuit the victim wants to bring up can happen later. For instance, if Joe Sixpack got stupid and killed his neighbor during a drunken fight one night, he's going to get arrested and tried for the crime of murder. Whether or not he is found guilty, the victim's widow can still sue Joe for the act, which had deprived her family of the victim and his skills. Even if Joe is found innocent of the crime (maybe the two got in a fight that Joe started and the other guy had a heart attack?), he can still be found liable for damages in the later lawsuit and be forced to give up his possessions (or portions of them) to the widow, or be forced to perform free labor for her for a few years, or some other suitable method to repay her for her loss.

Now why am I going at length to explain all of this? First and foremost, because knowing the differences and how they work will prevent your council from doing something stupid, like making it so that a simple civil dispute over ownership of something doesn't end with one party or the other getting locked-up. Secondly, you want to make sure that everyone knows that there's a difference between a criminal act and a tort (that is, something that can be sued over but isn't necessarily a crime). Finally, we're not going to talk anymore about civil stuff here - we're only focusing on the criminal bits.

Once you've sorted that out, make it perfectly clear that any police force you assemble is only going to go after criminals and that's it.

Before we get too far, here's what you need to have a working justice system:

- Cops - at least one per 50 people and you may want to start with two. The typical cop should be someone who has a good head on his shoulders, is honest, intelligent, relatively fearless and is capable of handling himself well in any kind of fight. The cop should have good working knowledge of the laws of the community, as he will likely also be the prosecutor for awhile. Cops should be hired long-term, but can be fired for misconduct or incompetence by the council or mayor.
- Judge - at least one per 2000 people or so (so basically, you only need one for a good long time). The judge is there to run the courtroom, determine sentence, set the rules and keep order. He or she should not be allowed to determine guilt or innocence for any criminal issue, or for any issue involving a large sum of goods, money, etc. A judge should be elected for a limited period of time and replaced once in awhile, just to keep everyone honest. Judges should also not be allowed to hold any other position than the one he or she has (except as a 'market judge', which we had discussed earlier).
- Jury (only as needed). This should be at least five individuals, but no more than 12. Jury members are drawn at random from among the adults in your community and ideally should be neutral about the case at hand, so as not to influence the outcome on anything but facts and logic.
- A jailer/executioner (one per 2,000 people or so). Someone to help the cop(s) keep the prisoners fed, warm and to prevent them from escaping. Also, this is someone who will have to carry out any executions if/when they occur.
- A jail. You would think this is a no-brainer, but think about it: You have to have a place where you can lock someone up tight enough to where they won't get out. It will have to be pretty strong to withstand a whole lot of abuse from the occupants inside of it. You may want to build something to this effect as soon as you can get your community

organized enough. In a pinch, you can simply handcuff/bind the prisoner to something huge or immovable until the trial.

Once you have the bits in place, the process is pretty simple…

Upon complaint or witnessing of a crime, all citizens should stop the criminal immediately. Restrain him if you can, but kill him if you must. The cop takes legal control of the arrestee (after quickly sorting out what's going on), disarms him, then puts him in confinement until the trial. Witnesses are then noted and contacted. If you can, write down everything and have the witnesses do the same. Inform the witnesses that they shouldn't leave town for a few days (or until after the trial).

You must let at least two days pass at this point! Give it time for things to cool down and to allow everyone involved to prepare. If passions are especially hot, turn that two days into a week or so. During this time, the prisoner must be kept reasonably warm or cool (depending on weather), provided nutritious food at least once a day (if he has family in town, they can provide it) and provided a reasonable amount of clean water for both drinking and washing.

The judge assembles a jury - people who are not related to the defendant or the victim(s) and people with no vested interest either way. If the defendant is unable to speak or normally communicate, the community needs to provide someone to help him do so. The prosecutor, or the community's representative in the trial (until you reach 5,000 people, it'll be the cop) will explain what he knows of the situation and then will present witnesses if he has any. Each witness he presents can be questioned by the defendant. This bit is important - you must allow the accused person the right to confront and question his accusers. After that, the defendant gets the opportunity to explain his side of the story and to present any witnesses he thinks may help his case. Those witnesses in turn can be questioned by the prosecutor. Once that is finished, the prosecutor gets to summarize everything and then the defendant gets to summarize everything.

At this point, the jury goes somewhere quiet and private to talk it over and to come up with a decision. At least 66% (2/3) of them must agree on that decision before returning, though in a fully civilized society the decision should be unanimous. How your town wants to play it is up to you, but make sure there most of them agree on it. However it turns out, the discussion and who said/voted what is kept secret. The jury picks a spokesperson and they all go back to the courtroom. The judge will formally ask them to publicly state their verdict. The spokesperson in the jury states it.

At this point, if the verdict is innocent, the defendant is free to go at that exact moment and the trial is over. If the verdict is guilty and jail/execution is called for, the defendant is restrained on the spot and tied/handcuffed to the chair (or other immovable object.)

In any case which ends with a guilty verdict, the judge can either move directly to sentencing (the judge can deliver any sentence except death without jury involvement), or the judge can let the jury decide on it (separately from the guilty/not-guilty verdict). In cases where execution of the prisoner is called for by law, it is simply called for by law and preparations are made.

If the sentence is a fine, the convict is given 90 days to cough up the fine. Failure to do so should result in jail time, on charges of theft for the amount of the fine that was stated. If the sentence is jail time, then the convict goes to jail at that very moment and the sentence begins on that day. If the sentence is execution, the condemned is taken back to the jail to await execution.

For fines, the cop or someone the judge appoints should receive the fine (in goods or whatever medium of exchange you use) and the fine is either given as part of the usual pay to the council/mayor/cop/etc, or is distributed as provided for by law (e.g. to the victim if there is one, or orphans, widows, the infirm, etc).

For jail time, it gets a bit more complex. First off, if the prisoner has family or resources in town, the family should be responsible for feeding the person, in order to prevent an undue burden on everyone else. Otherwise, insure that (if you can) the prisoner has at least one nutritious meal a day and enough clean water to wash with and to drink. Provide sufficient toilet facilities (even if it's just a chamber pot that gets emptied daily, or in lean times, straw or dried grass spread out on a corner of the room). Keep it warm enough to prevent hypothermia and cool enough to prevent heatstroke. Anything else is up to you and your town.

One alternative you can use between a jail term and outright execution is to expel the offender from the community. Here is how you do it: Strip the offender of all goods except the clothing on his back (and maybe a day or two of food, but no weapons of any kind at all). You then hobble his feet and bind his hands, then take him out blindfolded to a random spot at least 10-15 miles away from the community and drop him off, only removing the blindfold. By the time he unties himself, you will have been long gone and if the person is not from the area, will be completely lost. He is to know beforehand that if he returns, he will be killed on sight. Any attempt on his part to return and he is to be treated as if he were a raider and killed.

For executions, make the process efficient and rapid. Ideally, it should take place the next day and no more than three days after sentencing (anything longer and you're only wasting resources on the condemned). The condemned should have one night's quiet rest, a good last meal (if you have it to provide), access to speak with any family and access to any clergy and/or religious literature that he or she desires. They should also be provided with a pencil and paper to write down any last thoughts, letters, or words. During this time, assemble the place and tools of execution. Make sure that the means of execution are efficient, quick and humane. If it involves any machinery, sharp edges, or rope? Insure that all is in working order, sharp and in good repair. The execution should take place on noon or before noon, to allow time for the person to be well and truly dead and to allow time to dispose of the body afterwards.

During the actual execution, the condemned should be brought to the place under restraint. If it is a public affair, then the prosecutor (or the cop) reads the charge out loud, states that the person was found guilty in a court of law by a jury of his peers and that the sentence is to be carried out promptly. The condemned should then be given the opportunity to speak uninterrupted for up to 10 minutes. If the condemned remains silent the whole time, he remains silent the whole time - his choice. If the condemned requests it, any final religious rituals (e.g. Last Rites) can be performed after that period of silence/speech. He is then led to the device or place of execution, positioned appropriately (by force if necessary) and the execution carried out. The body is to be left in place and under guard for at least an hour or two (to insure death) and any competent medical person can confirm death at that time. If the body is still alive, cut the throat and wait another hour until death is confirmed. Once confirmed, the body is to be promptly buried or cremated.

Okay, So Why The Death Penalty?

I am personally opposed to the death penalty for religious/moral reasons. I find that in a post-collapse world, the idea is to save as many lives as you can, which is why I'm writing this book in the first place. However, I have come to a few realizations that would make the alternatives rather impossible:

- You simply don't have the resources for, say, life without parole. Your community can just barely feed itself. It can barely function. Anything beyond a short jail term becomes a cumulative strain on resources that you simply cannot afford.

- You're not going to have the means to build a proper long-term jail and you likely won't for quite a long time.

- You still have to remove that criminal from society - especially those that kill or rape others - permanently.

- Giving too light of a sentence in this environment is guaranteed to get innocent people killed.

- Expulsion (outlined earlier) is a choice, but some people will not take the hint. They will come back seeking vengeance, requital, purloined supplies, or are basically just too plain stubborn to take the hint and leave your society alone.

Given this, there really isn't much choice in many cases and you're just going to have to kill them. This does not however mean that you have to be a barbarian about it. The methods of execution should be quick and humane, without undue humiliation or pain inflicted on the condemned.

If you intend to hang him (actually a good choice given the simplicity and materials used), your aim should be to break the neck instantly, as opposed to strangulation. To that end, insure that the rope is high enough to allow a good fall (about 4-6 feet), long enough to allow that fall, but short enough so that the condemned doesn't touch ground at the bottom of that fall. You should, if you can, find a way to shield the bottom half of the dropped body from view. This way no one sees the emptied bowels that often accompany a rapid death (just keep that shield a safe distance away from the falling body, so it doesn't break an ankle and so the condemned doesn't try to use it at the last minute to stop the fall.)

Another simple method involves beheading, but this requires a very sharp blade (or axe) and a bit of skill. It also involves having someone swing that blade or axe. The idea, again, is to sever the spinal cord at the neck, causing a quick and humane death.

Whatever you decide to do, do not use firearms or arrows to do the job. Hitting the brain doesn't always cause an instant death (nor does hitting the heart, even directly) and is a waste of ammunition in the case of the firearms.

Always treat the body with respect - discard any urge to "make an example", no matter what the logic may be. Yes, throughout history the executed have had their bodies put on display, their heads severed and placed on pikes and various other means to "warn" potential criminals. However, the practice is barbaric, it has the potential for disease and it *should be* the very opposite of what you want to rebuild.

Checks And Balances

We've covered this in bits and pieces, but let's set it in stone it here and help you set up a few things. You'll need a fair and balanced election process and a means to remove anyone who isn't doing the job. You will also need to know who has what kind of power. We've toyed with ideas and set up some temporary means from which to start things, but now we really need to make it formal.

You start by figuring down the limits of power. Some offices you will definitely want to limit, both in power and in the amount of time any one person holds that office. This keeps things honest and allows new ideas and leadership to come in once in awhile - both of these are good things that we want to foster, because they represent the best chances you have for long-term success.

For each office, you will want to figure out these things: How long is each term of office? How many terms can any one person hold that office? Can any one office override the decisions of another (and if they can, what are the limits for doing that?) How do you remove someone from office who isn't doing a good enough job, or has become corrupt? What does each office actually control? Think these things through very carefully before writing them down. Make sure that each office can override the others in some way and that ultimate power to remove or restrict is held in the hands of the citizens at large. The only exception is whomever winds up being the judge/judicial side of things. That person's word must be final, because otherwise you'll just have a controversial decision go round and round and round… Whomever the first judge is, make sure that person is sharp, honest and looks to the long-term good of the community.

Next up, you will want to figure out how each office-holder gets their respective job. Is it by general vote, by a vote from the council, or by appointment from other offices? Again, choose very carefully. Before we go any further, I strongly suggest that for council members, you set up a staggered system, where you don't have the entire council up for election every time. A good idea is to have half the council up for election in any one election season. For a judge, I suggest that initially you vote him or her in, but from that point on said judge should be appointed by the council.

Once you have all of this in place, then you will want to set up elections. Before we begin, you will want someone to count the votes. This should be someone who is honest to the point of near-sainthood. Then, find a second person to verify all of the counts. Once you have that, then all that is left is to determine how people vote. The best way to do that is one of two ways: If you have enough paper and pencils, just have each person write in their choice and stick the papers in a box. Otherwise, you can have each person pick up one small pebble, marble, or other item and drop it(while being observed) into a box with their candidate's name on it. Or, you can record each person stating their vote in front of the election judges. The idea is that there should be only one vote per person, the vote should be done in a way that is hard to fake and it should be easy to count. Eventually, you will want something that is private and anonymous for each voter, but for now, we work with what we've got.

Finally, figure out what to do in the event of a tie, even if the tie is decided by literally flipping a coin.

Taxation And Levies

In case you don't know it yet, all this leadership and especially the cops and militia work is not going to be cheap, let alone free. Initially, this can all be volunteer work save for the cops and the market/criminal judge, because those two positions are going to take too much time from someone's day. This is time that would otherwise be spent in procuring/growing food, building things and in general surviving. These folks will have to be compensated for their time, obviously. To that end, everyone that is being protected by them will have to give up at least a little of what they produce in order to sustain these employees. By this point, it is hoped that you are able to at least either have some food grown, or are in the process of that. It is also hoped that some sort of industry is going, where people give of their skills in exchange for basic goods or other labor. You may or may not have some form of working money by now, but hopefully you'll have something, if not soon.

Now obviously, not everyone will be at the same stage. Some folks will have what could be best described as plenty, while others will be struggling just getting enough to stay alive. Because of this, initially you will want to not tax people any more than you positively must and when you do, you make it fair. When you levy a tax, mark it as a percentage of the take - for example, 5% of all harvested crops, 5% of a laborer's time and etc. This will keep things fair, since if someone doesn't produce anything, 5% of nothing is still nothing.

Even if you can't successfully get any sort of overall income tax going, there are ways to get enough resources to keep the cops and other individuals paid. You can tax market transactions, property, tax any goods or resources captured during combat (after the widows and/or orphans get their initial shares), put a (light!) toll on anyone with means traveling through your town, or set up scavenging sorties specifically for the benefit of funding the basic government functions. You could even set up specific plots of land just for growing food that can in turn be traded to provide pay. In extreme situations, you could require each resident to provide *something* to help keep the cops/judges fed and able to perform their duties. You could even do a hybrid solution: require a certain amount of labor to assist with certain duties and tasks and tax those who refuse to pitch in.

Initially though, these costs will be very light, so you really shouldn't have to do much of anything in the way of taxes. However, you do want to start thinking about it. You will also want to get agreement from the town residents before instituting anything of the sort. Never, ever just proclaim a tax. In most cases, once you explain what it's for and how it will be used, nearly all residents will likely go along with it, but only under two conditions: one, that it's fair and affordable and two, that there is transparency and accountability in its use. Anything less and you will meet some rather justified resistance.

One final tip: periodically go over any and all taxes at least once a year and see if you, as an ordinary citizen, can justify their continued existence at the level you're taking it in. If you cannot fairly do so, then get rid of or lower it, as needed. It may not be a bad idea to put it up to a vote that must be renewed once every so many years, just to make sure everyone still wants it.

Defense And Use Of Force

As civilization collapsed, this was a no-brainer. You either fought back, or you died. Once things began to stabilize and your community began to coalesce into a unit that looked after itself, it was an easy decision to get out there and clear out any criminals who were attacking travelers to and from town, as well as clearing out the source of those who attacked you.

However, once things become at least somewhat peaceful and all the major sources of attack on your community have either left, died, or were killed? Now you're not so sure. Well, you might be, but residents in your community may start doubting, especially as time goes on.

There is a bit of logic behind the idea of simply disbanding, especially as the weeks and months drag on with no attacks, no raids and no apparent reason to keep people standing guard at all hours of the day and night. But hey, let's look at how things likely are by now before making that decision…

State Of The "Nation"

By now the population will have dropped enough that many of the raiders who managed to live this long may themselves start working on settling down somewhere, perhaps even making their own township or settlement. After all, trying to make a living from stealing can only get you so far. The only targets left are those which defend themselves quite heavily, adding a very high cost to future potential thefts, since all the easy targets have been picked clean by now. The amount of prepared/preserved foods (the kind which are easy to carry) has also dropped to a very low level, with the remainder being defended fiercely by the aforementioned 'targets.' If the surviving raiders want to continue eating, the smart ones will know enough to start making some mechanism by which they can feed themselves - at least not without risking life and limb each time they need food. The news gets even better (or, well, worse, but…) Any surviving gangsters and wannabe dictators will have spent the majority of their ammunition by now and will have run out of fuel awhile back. This means that even the worst criminals will have to start making do with whatever they do have, or can make.

Overall, violence will, in most areas outside of the large cities, decrease greatly by this point in time. Inside the large cities, inter-gang warfare will have decimated the numbers down enough that fighting will be sporadic at best and even there the gangs will have to figure out some way to continue to eat. Since they will likely be unable to travel very far (if at all) by now, this means coming up with some sort of agriculture in-place. To top that off, ammunition will have run low to the point where using it will now take some thought beforehand and will likely only be saved for life-or-death defense.

Now certainly *how* the less moral among us survive from here on out may disturb you, but let's bring it up anyway. Some of them will have established themselves as post-collapse feudal warlords, complete with serfs/slaves and a population which only labor for them and only do it to continue eating, stay warm, etc. Others will have reduced or degraded their lifestyles to the point where the only real difference from being cavemen is the presence of firearms and enough ammunition to still be a threat. Many of these groups will have morphed into societies resembling sheer madness in motion, with nonsensical rules and strict enforcement of them. Finally, some of them may actually evolve, discarding (or at least reducing) the predatory lifestyle and approaching self-sustainability on local agriculture.

Outside of the criminal elements? You will often find individuals or small family groups, doing whatever they can to stay alive and by either luck or refined skill, have managed to hold out fairly nicely, considering the overall circumstances they've found themselves in. You may also find enclaves of people who have devolved into fanaticism, into outright cults, or are surviving only in the hope that the old civilization will arise from the ashes exactly as it was before the collapse, as they cling to the false hope that everything will "soon" be "just like it was".

…and then there's you. Your community should, if you want to survive long-term, be a community that is willing and able to defend itself, that does not fall into the traps of groupthink or cultism and is willing to discard the worst of the old civilization while trying to preserve the best ideals of what once was. In order to do that, you're going to have to defend yourself. By this point in time, those communities which have survived intact to this point, while doing relatively okay, may start getting a bit restless and may want to spread out a bit. Also, those criminal types which are left over, other communities short on goods and even wannabe warlords will periodically still want what you have, so…

Your Militia

Your first lines of defense are going to be diplomacy and deception, but your best and hardest-working one will be your militia. These are your friends and neighbors and have stood with you in helping to keep you safe (and you in turn have been helping them do the same).

Once you get to the point where everyone in your community isn't skipping meals, is sheltered sufficiently and has sufficient access to potable water, the most important task you have is to make sure your militia gets improved. Before, you likely had to get by on a wide variety of weapons, either privately-owned, scrounged, or retrieved from fallen enemies. You likely have no exact idea of what forces your town has at its disposal, nor its weaponry, nor how much in the way of ammunition and supplies you have left. It's time to fix all of that.

Start by assembling your militia and having them accurately report on what they have, how well they can use it, what condition the weapons are in and how much ammunition they have. Find out who doesn't have anything and make it a priority that any future weaponry you come across goes to them. Now note that you're not going to get the whole story, but try to anyway.

Once you have some idea of what you have, it's time to figure out who you have. Figure out who among you are pre-collapse combat veterans. Find out who has the skills to lead men in squads. Learn which men are capable of repairing existing weapons, is capable of reloading skills and who the chemists are among them (to make more explosives, gunpowder,etc).

From this point on (when combat is relatively rare and you go a month or more without seeing any), start conserving your firearm usage. If you have explosives, conserve that too. Save it for defense only, or situations where you have no other choice. Encourage the use of working with only non-firearm weapons for hunting and other non-combat activities. If anyone has reloading supplies and tools, compensate them and press as much of it as you can into working ammunition.

Start looking into procuring and making edged weapons, such as swords, daggers and axes geared for combat. Get started on making bows and arrows, using/copying compound bow and crossbow technology as a means to maximize the power in each shot. Look into making and gaining skill in spears, small catapults and incendiary (flame-producing) devices. Yes, it seems we're traveling back in time here, but these weapons are easy to make, easy to maintain and not that tough to gain skill in using. It also allows each soldier to have a backup means of fighting if he runs out of ammunition.

Another option (in addition to the above) is to look into building (or using) black-powder firearms and making your own gunpowder to supply them. For this, you'll need a few of the higher skills - chemistry, metallurgy and engineering. However, it's fairly simple to make (or re-purpose) a sufficiently thick steel tube that's closed at one end, plus a suitably-shaped piece of wood into a crude musket. Traditional gunpowder can be made from three simple ingredients - sulfur, charcoal and saltpeter (or other high-nitrogen compounds, such as that extracted from guano). Ammunition can be lead balls, or small smooth gravel and some wadding, shotgun-style. The sooner you can get this going, the more powerful you will be (and the more economic influence you will have in the region once things are peaceful).

In addition to the weaponry, make the time to train. Have those soldiers who are skilled in tactics and weapons teach those soldiers who are not. Promote the smarter/more skilled ones to leadership positions. Require that all male citizens over the age of 14 (and under 50-65 or so) participate in the community's defense. They are then to assemble regularly for training in the skills and weapons you have or are building and to answer any alarms at a moment's notice to defend the town. The only exceptions to this would be the men and boys who are mentally deficient, crippled to the point of ineffectiveness, are clergy (for obvious reasons), or who are gravely ill at the moment.

Odds are good that there might be one or two men who refuse to fight, even in defense. It may be on religious grounds, or from ideology. In such cases, try to find a support role for these people if you can, or some other non-combat role in which they can still be useful towards defending your community.

Note that you don't (yet) have the luxury of only using an all-volunteer force and all of your community's residents will know that defense is quite literally up to them. Scavenging and raider-cleaning parties can be all-volunteer at this point, especially if it is made known that those who volunteer get a larger share of the proceeds.

When you make the pitch to assemble and organize the militia, one thing must be apparent to one and all: A militia is still necessary, no matter how peaceful things may seem. The only possible exception to this I can think of would be a post-collapse community living without competition on an isolated island in the middle of the Pacific Ocean. So unless you live on Midway or Pitcairn Island, you've got some work to do. Stress the fact that there are still dangers out beyond the community limits and that a regular patrol and defense is still going to be necessary for a few years to come. You can also use the militia as a backup to the cops if it becomes necessary.

Chains Of Command

The militia is answerable to the people. To that end, you should appoint someone capable and honest to lead the militia and put into law that the leader will be answerable to (and can be fired by) the council upon reasonable cause. No member of your community government should have this job and vice-versa.

Once you have a good militia leader, let that leader organize and set up the rest of the militia, but with some guidance. You should have both leader and council sit together and draft a small but effective set of rules - both in how the militia behaves (e.g. making it lawful to disobey an illegal or immoral order, etc) and in how combat is conducted (whether or not you take prisoners, what to do with non-combatants, etc). These will obviously evolve over time, but I suggest something short, sweet and to-the-point. The council should have final say in it all, but should as a matter of courtesy defer to the militia leader in tactical decisions.

Force And When To Use/Not Use It

Once you have a working militia, you now have something with which to defend and project power. All too often, someone gets the idea that maybe this force should go out and take over other areas. The reasons why even sound logical: Better resource access, access to desperately-needed resources, larger defensive buffer, the 'liberation' of 'oppressed' people nearby, etc.

It's easy to fall for the trap, but honestly, one should avoid using arms to accomplish things that can be done just as easily without them. You should also avoid using arms to accomplish things that would, long-term, just end up being more trouble than its worth.

So when should you use force?

- Defending established territory and established resources
- Providing defense and cover during scavenging sorties
- Clearing out raider/criminal encampments
- Caravan/road security
- As a backup to civil police actions, but only if necessary
- As a means to defeat inbound and/or openly threatening forces before they reach your territory
- As a means to provide temporary labor and rescue during disaster (fire, flood, etc)
- As a means of supporting/defending neighboring communities that may be under attack or are experiencing credible threats thereof (more on that later).

As you can see, things like conquest or expansion of territory (for any reason!) are not listed here and for good reason. You obviously don't and won't have the resources and you don't want or need the headaches. For the suitably tempted, we left out things like theft and plunder for the same reasons - eventually it gets you and yours killed. In short, as you will have likely figured out by now, war is expensive, so there's no sense doing it unless you have no alternative whatsoever.

Hopefully and with a little luck, you never have to find out the hard way.

Money And Economics

You should already have some sort of economic setup going by now. You may well have taken the advice given earlier and have set up a fair, stable and lively market. This is good. Now, let's look a bit further and see how you can improve things a bit and not only have an ongoing economy, but one that eventually builds up and helps everyone in town prosper.

About That "Money" Thing

It should be pretty obvious by now that any paper money which existed pre-collapse may well be worthless. Most modern money in the pre-collapse world was issues by "fiat" - that is, it was based on the full faith and credit of the government that issued it. There is literally nothing else to back it up. If the government fails, then the money becomes worthless, or at most only holds as much worth as the materials it is made of. For instance, whereas a penny pre-collapse may be worth $0.01, post-collapse it is exactly worth the zinc, copper and whatever other trace materials it may be composed of. A $100 bill and a $1 bill hold equal value, post-collapse - they are both worth exactly whatever value is held in the linen-paper and ink that each bill is made of. This means fifty $1 bills are actually worth 50 times a single $100 bill. However, they still have some value left in them. Why? Because paper money folded just right makes for decent fire kindling or oil-lamp wicks, a well-worn bill makes excellent toilet paper and a large enough pile of bills all shredded-up can insulate a jacket or something similar.

So, really, where does this leave you? Obviously one of two things can happen here, since money of some sort is still a necessary component to civilization. You can either re-purpose the existing stuff as money in some way (by alteration and then backing it with some inherently valuable good), or you can (as a single or as multiple communities) issue your own. Either way, you're going to have to come up with something you can call "money."

Now up until this point in time, barter and trade of one valuable good for another has been working just fine and continuing to do so is harmless. However, we're thinking long-term here, so you will want to make something that works. Let's start with defining what money really is and how you can best put it to long-term advantage.

First of all, money needs to be universally recognized, at least within your sphere of influence. That means, everyone in your community knows it's money, knows it's value and is willing to use it in trade or purchase. This is as simple as the community leadership officially stating (and then fully explaining!) What the new monetary system is, how it works and what it does. Just be sure to keep it simple, easily recognized, easily counted and most importantly, easily carried.

Second, unless you have universal and wide-reaching power, money in a post-collapse world is going to need to be backed by something. You could use the earliest form of money, also known as a "commodity" monetary system, where the value is in the actual gold, silver and copper (or other precious metal) coins. You could use "representative" (or "backed") money, which makes it easier to carry around. Or, you could use "fiat" money, which as already mentioned and described, is what most civilizations used before the collapse.

Given that your town's sphere of influence is likely too small to be recognized as a major source of impeccable credit, fiat money is out of the question. Perhaps if you allied yourself with a lot of other towns and communities, you might be able, together, to come up with a fiat money that would stick. However, given that the collapse will have destroyed most folks' faith in such a monetary form, it is doubtful that you could get away with it, at least not for a few generations.

Your other two alternatives is either a commodity system (which means shoveling a lot of precious metals around), or a representative system. The latter is the hardest to set up, but is the easiest to manage and build on - by both town/regional governments and by the people using the money.

So what exactly do you back your money up with? A small note: It can be as simple as backing it with units of labor - the town of Ithaca, New York actually did this. Or, you use some other recognized unit of actual, perceivable, but intangible value. Or, you can use another form of money entirely to back it up, usually at a 1:1 exchange rate. This particular form of money is often referred to as "Scrip" (note that it can also be backed by another monetary unit, such as casinos use, by issuing chips and coins redeemable for an equal value in their national currency). The problem with scrip-based money is that it is only good in places where someone is willing to accept it and usually isn't recognized anywhere else. Given that we're thinking long-term, you'll most likely want to avoid using scrip.

Your best bet is backing it with a tangible but universally valuable good such as a purified precious metal, or some other rare-but-valued commodity that can be stored long-term without corrosion or decay. You can even use multiple items, such as "One Doomsdollar = 1 oz. of pure gold, or 10 oz. of pure silver, 32 oz. of pure copper, or 64 oz. of pure aluminum." (Two things to take note of here: One, yes, pure un-painted aluminum will fetch a premium, as will other corrosion-resistant metals that required modern technology to extract. Second, don't ever call your money unit a "Doomsdollar" - pick at least something that isn't as stupid, okay? I'm only using it as a placeholder name) Also do not go overboard here - just pick a couple of universally-recognized items to back it with (3-4 at absolute most) and stop there. The more items you use, the more complex things get, which is a bad thing given your limited resources.

Setting the actual value will be a very tough guess at first. Too low and you'll be printing too many of them. Too high and you'll need to avoid printing as many as you need. Take your time in deciding the backed value, think long and hard over it and get it right the first time - you won't have a chance to change it (without losing trust) once it's set and announced. To set the value, you'll want two bits of information: How much of that precious backed item (if it's tangible) do you actually have locked away safely? How much you have, limits how much backing you have and if you guess too high on value, you'll have to procure more of those goods before you can print more money. The other thing is that you have to know what those items actually get on the market, which in turn gives you a very good idea as to what the initial unit of money should be set at.

Third, your money will need to be as hard to counterfeit as possible. This should be a no-brainer, but it is amazing how easy it is (and historically, was) to counterfeit money with little-to-no technology involved. While you're at it, make counterfeiting a crime with a real stiff penalty - you don't want people doing it thinking that they'll only get a slap on the wrist here. It's one of the very few crimes that really need to be stomped out as hard as you can without involving a public execution.

Next, it has to be LEGIBLE (I wrote that in capital letters on purpose). It has to be easily recognizable, especially if you're using fractional and multiple units of money on one bit of it. This means printing/minting the value in an easy-to-read format and preferably in a different format than other bits of money that do not have the same value.

Another consideration is that you set a number of units of your money to each unit of tangible good. For example, before The United States abandoned the gold standard in 1971, one troy ounce of gold was equal to $35 US dollars.

Finally, you're going to need a way of controlling how much of it there is out in circulation at any one time. Too much and its value will drop, no matter what you say it may be worth. Too little and it becomes useless (not "worthless", but "useless" - as in, there's none of it really circulating and people are forced to rely on other money forms and on trade/barter to get things done, so what's the point?). The good news is, you can always print/mint/make fractional units of money - as in, "Half Doomsdollar", "Quarter Doomsdollar", etc. You can also print higher values onto a single note, such as "Ten Doomsdollars", etc. Just make sure that the total value of units in circulation don't add up to more than what you have stored, else you get inflation and loss of faith. On the other hand, if you have more backing than you have total value in circulation, you're losing out on how widespread you can make your money.

So let's set up an example for you to show just how ugly this might get, how to fix that and have a stable and useful monetary system.

Setting Up Your Money System

We want to make "Doomsdollars", but need to know how to go about that. Well, let's say you have about 200 residents in your community by now. You've managed to accumulate into the community treasury about 64 oz of gold, 128 oz of silver and maybe 2560 oz (160 lbs) of pure copper. Quite a bit of dough, no? Now before we go any further, note that we're going to use avoirdupois ounces for measurement - that is, ounces that the vast majority of scales out there actually use and the one you actually think of whenever you go near a scale (unless you're a jeweler or precious metals dealer). So why is that important? Well, because precious metals are normally measured in "troy" ounces, which are heavier than the common commercially-used avoirdupois ounces. Therefore, no matter what numbers the metal may have stamped on it, measure it with a precise common commercial scale, because otherwise you may be in for a bit of a surprise.

First, you'd better figure out how much each metal is worth in relation to each other. A 14.5:1 ratio for silver to gold sounds about right (It's the French Livre ratio if you're curious as to how I got those numbers, though I rounded it down harshly because working scales more accurate than ½ oz are going to be quite rare) - that means 14.5 oz of silver = 1 oz of gold. We'll also call 240:1 for copper to gold (a rough approximation that came from the British ratio of Penny to Pound. We may as well use historical examples, since they'd already worked all this stuff out - saves us the time and effort spent in reinventing the wheel).

Now let's say you decided that 100 Doomsdollars equals one avoirdupois ounce of gold. This is called a *Basis Rate*. You would, with your 'Store of Value' (which is what that pile of metal-money is usually called), be able to print out a total of 8349.41 Doomsdollars.

How did we come up with that figure? Well, let's break it down:
- 6400 came from your 64 oz of gold (64 x 100)
- 882.75 came from silver (100 / 14.5 to account for conversion, multiply result by 128 oz, then round it down)
- 1066.66 came from copper (100 / 240 to account for conversion, multiply result by 2560 oz, round it down)
- This leaves you with your grand total of 8349.41 Doomsdollars.

As you can see, you don't need to scrounge for nothing but gold in order to get your money. However, because of the value conversions, you need a lot more of the cheaper metals to equal the value of gold… the good news is, you can always get more, but we'll get into that in a bit. The important part is, we now have money and know its initial value - or at least we now have something to work from.

Now before you release your money, you still have some time to fix a good value for it. If you want to tweak the rate a bit, know that you'll have to adjust everything else to get an idea of how much you have total. So why would you want to tweak it? Because you'll want to take a good, hard look at how much things are going for. If a dozen eggs goes for one ounce of silver, then you may want to tweak things so that, say, 5 Doomsdollars equal that one ounce of silver - make it something people can grasp and think to be fair intuitively. Remember - you only have one shot at it, because you cannot really change this after the money goes into circulation (well, you can, but not without causing a ton of disruptions - better to get it right the first time). To do this, you would have to change your Basis Rate a bit so that you can get your result. In the case of of the eggs (notice that we used silver on purpose here), you will have to do a bit of work...

First, we already know that originally, one ounce of silver equaled around 6.9 Doomsdollars (100/14.5). So, we're pretty close, yes? Now if you want to tweak it down a bit, then you only need to multiply 5 (the cost of the eggs) x 14.5 (the conversion from silver to gold), which gives you your new Basis Rate: 72.5 Doomsdollars per ounce of gold. With the new basis rate, a dozen eggs will cost 5 Doomsdollars upon circulation. As a result, you will also have to figure down how much you can print, which will obviously be less. Just plug it all in again and you get...

- 4640 from your gold (72.5 x 64 oz)
- 640 from your silver (72.5 / 14.5, multiply result by 128 oz)
- 773.33 from your copper (72.5 / 240, multiply result by 2560 oz)
- Grand total for circulation: 6053.33 Doomsdollars

Now 6053.33 is a smaller amount, yes, but the smaller number means less chance of inflation, more value per unit of money and gives folks a far better feeling about using the things.

So - we know what we have and how much we can start with. But what physical form(s) is (are) this thing(s) going to take? Well...

Minting And Printing

It's easy at this point to rig up some sort of electricity to a working copier machine, draw up some "Doomsdollars" with a crayon or two and start cranking out some money, right? Well, maybe not. There's a whole lot more to it than just printing out whole units and expecting it to be used. After all, how will you handle change? In later years, will you expect people to walk around with a massive sheaf of one-unit bills? How on Earth are you going to prevent counterfeiting? How will you insure that the material you use will actually hold up to normal use - in a far dirtier world?

As you can see, for starters, you'll need a few different denominations here. I would suggest starting out with something like this:

- 10-unit note
- five-unit note
- one-unit note
- quarter-unit note
- 1/10th unit note
- 1/20th unit note
- 1/100th unit note

This will look familiar if you look at it a bit. Here, let me help you complete that thought: $10, $5, $1, Quarter, Dime, Nickel, Penny. See it now? This is intentional, so that folks are able to take your money, combine it with what they've grown up knowing about money in general and immediately put it to use without any oddball guessing or excess discomfort at counting, making change, or adding fractions of it up. In effect, it gives you a good head start, since money is, after all, a leap of faith on the part of the people using it.

Next up, we break up the initial money supply into usable chunks. Start by making ½ of the total into "ones" - each one equaling a single unit of your money (we'll stop using the silly term "Doomsdollar" now). Out of the remainder, make ¼ of them into "fives", 1/16th of the total into "tens" and the remaining 3/16th of the total into the fractional units. Of the fractional units, ½ of those should be 'quarters', ¼ should be 'dimes' and 'nickels' and the remainder should be 'pennies' (…see how easy that breakdown translated for you mentally?)

Okay - you've got it all broken down, so now all that's left is to design the stuff. Start with the materials. If you have the facilities to melt metal and make molds, you have the means to make coins. The hard part will be the 'paper' money, so let's tackle that first. Now there is no rule that the basic unit (or larger denominations) actually have to be made of something that folds. You can make those coins too… but let's try to go for folding money anyway, because if it's done right, it's tougher to counterfeit.

You'll need four components for your paper money - a means to print it, a durable material to print it on, ink that you can make locally (and/or enough supply of it to last you a very long time) and a design that will be tough to counterfeit. A printer is relatively easy to build if you do not already have one, but try to go for a printer that embosses - though if you can make one, an *intaglio* printer would be the best way to go (intaglio printing is basically using almost microscopic grooves on the printing surface to hold ink, then using high pressure to force the ink onto a thicker-than-usual paper material - the result is a raised ink printing that you can feel after it is dry. This is common in pre-collapse paper money, incidentally. Try running your finger on some fresh notes and you'll see.) You next needed bit is a durable material. While most folks do in general take some care of their money, it does get dirty, germy, folded repeatedly and wadded-up a lot. Pre-collapse money was usually made of polymers, linen-infused paper, or a combination of the two. You may not have the means or know-how to do either one, so your best bet is to scrounge up some fine or top-quality calligraphy paper of some durability. An alternative is to use a fibrous paper-like plastic (such as that found in certain large business envelopes) that can hold onto ink. The last pre-design bit is going to be your ink. It should ideally be something simple enough to make from locally-sourced ingredients, durable, able to hold up to getting wet and in specific color shades that are hard to easily counterfeit. Before designing your paper money, get these three items together and do some test runs. Beat the crap out of the results and make sure it'll hold up reasonably well (it should make it through, say, 2-3 years of ordinary usage if you can swing it).

Designing your paper money is going to be tough to do, but should have a few components to help you out:

- Serial Numbers: Each bill should be sequentially numbered, so that you can easily spot when there is a counterfeit, or have a quick means to be suspicious. If you have serial #'s 00000 - 09998 in circulation, but someone presents a bill with serial number "14872", it's obviously going to be a fake. Older serial numbers will usually mean money that looks worn-out and used, as opposed to fresh and crisp. If someone wants to fake your money, they'll have to use different serial numbers for each bill, making their job a little harder.

- A Signature and date: Signatures are harder to fake than words. It's also harder to print, but you'll only have to do it each time you make a new printing plate (or you can manually sign each bill, but seriously - engraving the plate is easier on your fingers). As the signatory official changes, the signature and date changes… this means year-old bills with low serial numbers but the signature of the guy who just got the job two days ago are going to be obvious fakes.

- More than one color: It takes more resources to make multi-colored bills that are lined up correctly- this also means it's going to require the counterfeiter to actually put in at least the same amount of work into his creations, making his job harder.

- A different picture or design for each denomination: This prevents counterfeiters from taking a "one" and turning it into a "ten" just by bleaching/erasing the old number and fake-printing a new, bigger one.

Now obviously, this is going to take some work to put together. In a post-collapse situation, you may not have all you need to make this happen. However, there is one way you can come up with instant money and not have to get much more than a lot of pens and writers' cramp: Find some checkbooks.

No, this is not silly. Think about this for a moment: Most people usually have a big box of checkbooks, full of checks (from 100-500 checks in a full box), sitting around somewhere in their house. Each check has a unique routing/transit number and a checking account number in magnetic ink printed on them... magnetic ink that can't be procured post-collapse. Each check is sequentially numbered. Each check has a unique name and address on them. They often have anti-counterfeiting measures on the back (heat-sensitive ink, microprinting, etc). You can sign them. You can use a different check design/color/etc for each denomination, clearly marking each with the denomination in both numbers and words. You can publicly post voided-out checks as samples for merchants and the public to directly compare against. In other words - they're perfect for post-collapse use. Your only concern is finding enough of them and in finding at least three different colors or designs that will suit your purposes (assuming that you want to use coins for change).

Speaking of change... let's focus on the coins for a moment.

Assuming you don't use paper money for everything, you may want to consider coins once you're far enough along, have the skills of a machinist and an artist handy and are capable of doing so. In the meanwhile, using paper to represent coins will do just fine. If you indeed do coins, then you have some options here, but there are a few caveats up front. First off, don't simply re-use existing pre-collapse coins. There are way the hell too many of them out there and they are everywhere, which will wreck your new money system entirely. Don't try and re-purpose any existing small coin-shaped object, either, for the same reasons. Coins will have to get as much attention to detail as the paper money does. Here's what you need to get up some coins...

First off, you're going to need a lot of metal. The metal should be durable, have a high melting point (but not too high obviously and should be just malleable and fine-grained enough to stamp, but still keep fine details. Then, you'll need a means to form and stamp the coins. Getting the press is easy enough to do if you don't mind doing one coin at a time, but it's the dies, a source of heat and getting the metal formed into the correct shape that will take some effort. This is where the machinist will come in handy - preferably someone who has some metallurgy training as well.

Next, you will need a good design. It is suggested that you use a design that is hard to reproduce, that involves some detail work, clearly shows its denomination, shows the year it was minted and shows who minted it.

The initial investment in resources, skill and time will be big, but once you get moving on it and have it going, you end up with a product that is durable, hard to copy and easy to carry.

Spreading The Love

Now that you have your initial money supply about to be printed and/or minted, it's time to introduce it to the town and to the world… before you start printing and minting!

Start with a public meeting. Explain to everyone what you're about to do. First, explain that it is backed by pure precious metals, that it is guaranteed at face value and that most importantly, show them what measures you have taken so far to prevent counterfeiting. Explain in full how each precious metal you're backing it with relates to the other, in what proportions and why. Be certain to explain that barter and gold/silver/whatever means of currency is still perfectly okay, but that your money is going to become the standard for taxation and/or other town government endeavors. Further be sure to state that you are more than willing to buy gold, silver and other precious metals at the exact same rates that you set your standard to, so that residents will have the money from which to pay taxation, carry to your markets, etc.

Take in all questions and comments and be prepared to modify your money plan or composition if any seriously better ideas come along during the meeting. Once everyone (or the vast majority anyway) agree on it all, start printing and/or minting the money.

At a set (and agreed-upon) date, open for business. Keep about 10-20% of the money to yourselves, for government business and the like, but get the rest out there. Buy up gold and your other precious metals at a strict 1:1 ratio, after perhaps a day or two (at most!) of a limited sweetheart deal of maybe (at most!) a 1:1.25 ratio in favor of the metals seller … in our earlier example, a 1:1 ratio means every ounce of gold they bring in means 72.5 units of money. As tempting as it is, you really don't want to 'sweeten' the deal for more than you absolutely have to, because the last thing you need is inflation. Pay for goods and services with it from the market. As a local government, do no business in town with any other form of exchange but your money and unless you're buying precious metals, dealing with another established town, or selling your money to residents in exchange for goods, do not pay anything out in any other medium of exchange.

Keep an eye out as to how much money is circulating out there and keep a tight eye on what your market is doing. If any local seller charges more in your money than he would in gold or other precious metals, consider taxing the offender accordingly and in proportion - payable in your money, or in gold/silver/etc. After all, this is going to be your money and that it should be treated with the same strength and respect as any other commodity good. Sellers who travel into town are folks you should be a bit more lenient with, since they have to spend their medium of exchange elsewhere. Only later, once you get up some alliances, should you consider your money to be usable outside of your immediate sphere of influence.

Speaking of which, a good idea is to invite neighboring towns and communities to use your money and once (or rather, if) they start doing so, actively seek to buy precious backing materials from them.

Making More Money

This one is (or should be) a no-brainer. In order to make more money, you need to take in more precious metals (or whatever you're backing your money with.) You can buy it directly, you can require that all (or perhaps just a percentage of all) scrounged precious metals be sold 1:1 to the town government, or you can do both and more. Just don't get too crazy with it, but always be on the lookout to increase your store of whatever it is you back your money with.

What you definitely do not want to do is to print out more money than you have precious materials to back it up with. That way leads to inflation and to perdition (at least for your monetary system and for any ulcers you may have gotten by now…)

Preventing Troubles

Inflation means that your money is worth less per unit than you need it to be worth and is the biggest danger you face. For example, say that you initially set it up so that a dozen eggs costs one ounce of silver, or 5 whole units of money. However, a month later, you notice that while you can still get that same dozen for the same ounce of silver, it now costs you 10 units of money to buy it. Obviously, your money has lost some value along the way, no? Normally, a healthy system keeps things in check by competitors offering to sell you the same product for less and as long as the majority has faith in your money, there is no reason (outside of scarcity) for the price to go up. After all, your money is backed by something of value and the perceived value shouldn't change without the items backing it up becoming themselves cheaper.

If inflation is the case, then you need to dig down and find out what happened and why. Before you do though, you need to be sure its actual inflation and not some innocuous factor. The problem could just be one of scarcity - but that is fairly easy to figure out, by simply checking to see if prices have gone up across the board, or just for some items but not others. It could be a problem of profiteering due to an upcoming crisis (or rumor of crisis) - again, fairly easy to figure out. If you're certain that it's inflation, then start digging to the root cause. While it is easy to know that inflation in your system is a loss of faith in the currency, you have to dig deeper and find out what is causing that loss of faith. It could be that the population doesn't like or trust the government you have running things. It could be that someone has successfully counterfeited your money, thus diluting the value of it. It could be that the population doesn't think you have the goods to back up the money. Or, it could be that someone in your government got stupid and printed more money than there is backing for it, in an attempt to (even temporarily) increase the buying power of the government.

Each of these have solutions, ranging from the simple (have representatives of the public see, test and count the backing material), to the complex (finding out who is counterfeiting and dragging them to justice, or in cleaning out the government and restoring faith in it).

The big reason that we settled on representative money (that is, money backed by a precious good or goods) is precisely to minimize or remove the uglier forms of economic troubles, such as inflation. The rest is up to you, really. With a strict counting and auditing system, you can prevent most of it and keep the money honest. The downside is that you have to accumulate more precious material to increase wealth overall. I call it a downside because it means that you have to go out of your way to collect and hoard those things such gold, silver, copper, etc. It is also a downside because you have to store it somewhere and you have to guard it very carefully.

There are temptations all over the place when it comes to this. Your government can decide that it needs a lot of money in a hurry and so prints out a lot of it. Maybe the logic even sounds persuasive ("we're attacking that raider encampment and they've got a ton of gold in there!") Resist that temptation at all costs. Never print money that you cannot back as fully as the money you already have in circulation. Only when you obtain more backing can you print and circulate more money, period. There is the temptation by someone to help themselves to the pile of precious metals you have stashed away. Having more than one guard always present (and insure that the two guards are not related or friends with each other) is one way to solve it, as is locking it up in the most secure location you can think of and always doing audits/counts on the material on a very periodic basis.

In cases of emergency, you can tweak money to be more valuable on rare occasion (add more backing but change the value of each unit accordingly) and you can tweak it a little on rare occasion to be worth less individually (by printing more units of money). However, do not do it unless you have no other choice at all. Always seek an alternative first.

Interest Rates, Bonds, Price Controls, Overall Economics

Making treasury bonds and the like is a bit beyond the scope of our present stage of post-collapse development. Keep a mind towards it, but only if you can guarantee growth over the life of the bonds. Otherwise, don't worry too much about that yet. You can do that later, when you want/need to build things like a public well, public sewage, or other public utilities.

Price controls should never, ever be enforced on an ongoing or discriminatory basis and should only be used in dire emergency, if at all. Never try to set the prices on anything by force of law, because doing so will immediately worsen any scarcity that arises of the items being capped. Rule #1 here: Let the market do its job. As an item gets scarce, the price will go up, people will use less (or find alternatives) and eventually the scarcity becomes a surplus, which in turn lowers prices again. There is no need for you to screw with that. As long as more than one person is selling the same (or even a similar) item, prices will remain fair - given a fair accounting of supply and demand.

One thing you may want to consider, however… how about setting up a bank? Now you can only really do this if you can spare the money, manpower and time, but it wouldn't be a bad idea. The reason I bring this up is that, if you can swing it, you can avoid having opportunistic loan sharks or other "lenders" rising up and causing trouble. It seems like a bit of a luxury, but you should never underestimate the power of human greed, or the capacity of one person to screw over someone else. Those folks who are doing better than the rest have a strong temptation to try and exploit those who are in desperate (or near-desperate) conditions and not everyone is able to resist. Having a safe place for your fledgling community to go in order to deposit funds, take out low-interest loans and in general handle money on their behalf is actually a good thing to start with. By doing this, any private entities who found 'banks' and the like will have competition immediately and will be forced to deal fairly and honestly with the public.

Dealing With Fraud And Counterfeiting

One thing that can easily kill your newly-established economic system is some jackass (or jackasses) out there deciding to take advantage of your population by committing fraud, counterfeiting and in general, manipulating your young economic system to their gain and your detriment.

Your best bet is to put into place laws that specifically deal with such activities. It isn't really all that complex, but you do have to identify the activity, establish evidence and to punish the offenders. Most of this can be handled in a civil court, but there are cases when you will have to deal with it.

Counterfeiters should be dealt with swiftly and severely. If one of your citizens commits it, strip them of 75% of their property, leaving them with only some food, some clothing and some possessions. You may even want to consider expulsion from the town and use guidelines we've talked about earlier to make that happen. If the counterfeiting is committed by someone outside your community, use your head, but do not be too lenient on the offender. Just be sure to get the counterfeiter him/herself and not the person who inadvertently passed the bad money. Note that if anyone does receive bad money, they are not entitled to any sort of refund or exchange for good money. Whoever held the fake money last is, well, screwed when it comes to a refund or exchange. This will insure everyone knows to be on the lookout for the stuff and will have a strong incentive in refusing to accept it.

Be quite vigilant in seeking out any and all counterfeiters, especially in the early stages. If another community or town accepts your money, go out of your way to educate them as to what the real stuff looks like and how to spot the fakes.

Odds and Ends

Some odd bits that need tied up a little, which we will include here:

- Always and gladly exchange damaged money for fresh notes, as long as at least 51% of the note is present (or can be pieced together) and the serial number is present and legible. Set up a policy where anyone can bring in old and worn notes in for a direct exchange. If you have a bank set up, use that as a means to cycle out old stuff and to cycle in new stuff.

- Go out of your way to reach out to friendly communities outside your town and either set up some sort of monetary exchange with them, or persuade them to use your money.

- Found money should be held by the town government for up to 90 days and whoever can rightfully claim the money (e.g. knows the serial numbers, can accurately describe the amount, can accurately describe the circumstances in which it was lost, etc) should get it. Any unclaimed money that has sat around for 91 days or more goes either to the finder, or into the town's operating budget.

- You may want to make it a policy that all precious metals found or recovered during military action is to become property of the town government and that all soldiers are paid a share of the loot recovered (as discussed previously), but are paid in your money, at a direct 1:1 basis rate (in our earlier example this means every ounce of pure gold equals 72.5 units of money).

- In case things weren't clear before, all payments the town government makes or takes in will be in your money. Any precious metals paid to the town government will be converted to the money first, then counted on the balance sheets as equivalent income in the form of your money.

- As your money and economy takes root and begins to grow, you really want to start looking for someone who is honest and has solid business accounting skills. When you find that someone, hire him or her immediately if you can. It'll likely become the first non-military/police/leadership role you'll have to fill.

Diplomacy

Once you've reached a point where things are (relatively!) peaceful, you have a (at least somewhat) stable economy and money system going and folks are actually beginning to prosper? It's time to start officially reaching out to folks outside your town and doing so in ways that don't involve pointing weapons at them.

Diplomacy is half art, half science and half bullcrap (but very necessary bullcrap!) This job requires someone who is not afraid to tell you to kiss his ass, but is smooth enough to tell you in a manner that causes your lips to pucker involuntarily. It requires someone who is not only unafraid to stare death in the face, but will either spit in its eye or offer it a drink …and sometimes both. You will need someone who you can trust with your very life, because often, that's exactly what you and your entire community will be doing. This person is going to have to be capable of talking to desperate people and offering them hope. They will have to convince highly-armed people that they would lose in a conflict, even if they have the tactics and weaponry to easily win. This is going to be someone who can use mere words to get something that a show of force or a ton of expensive gifts would get at far higher prices.

Once you have managed to find someone of this caliber, you're good to go and can almost skip the next section. If you haven't, well, time to get a quick education here…

How To Be A Diplomat

First and foremost, you're going to need nerves of steel. It doesn't matter how weak your position is in comparison. It doesn't matter if you have nothing tangible to provide in return for something. It doesn't matter if there is a real knife to your actual throat. You have to get your point across clearly, in a manner easy to understand, but do so in a way that gets you what you need and at the same time doesn't compromise your town's ability to survive in the long-term.

Your next greatest skill is to know who you're talking to. No, this doesn't mean the basics (name, age, education, fighting ability), but to get to know someone on a very personal level. This means that you need to know their personality, to know who their family is, know their current situation completely and to know them well enough to get a very good idea as to how you treat them individually. Know their hopes. Know their fears. Know what it is they really want to happen, in spite of what they tell you.

A very important skill is to know what the real story is behind everything you see and hear. Appearances are one thing, but look at the details, because you have to stare down the devil in this job and among the details is exactly where the devil lives. A community may say they're doing well, but how nourished do the poorest of the people look? Where do they get their water from and what condition is it in? Do the spectators and passerby look about with bright, clear, level eyes, or are those eyes bloodshot (fatigue, drugs), glassy (extreme fatigue, malnutrition), or skittish (deception, fear, or worse)? What does the body language tell you? What does their stance tell you when they stand, or sit? Can you move freely in a strange town, or are you being treated to a 'show'?

Continuing on that vein, there are lots of other questions that you should have solid answers to just by looking. For instance, how is their state of sanitation? How well do they keep things tidy in general (if it's a ragged pit, it may be a sign of deprivation or a hoax. If it's as neat as a pin and they insist that they're doing awesomely well, they may be hiding something…) What condition is their food in? Is it fresh, from storage, foraged/hunted, or what? This tells you a lot about how well they're eating and where they're getting their food. In case you haven't gotten the idea by now, intelligence is a beautiful thing to gather while you're out and about - the more, the merrier. Why? Because doing so tells you a lot about what they're willing to tolerate, how ready and able they are to trade, what kind of mood they're in and much more.

Next up, always keep an eye out for opportunity. Are they running diesel generators and you have (or can get or even make) a lot of diesel fuel that could be used in them? Sounds like a potential opportunity for trade. If, in negotiations, the other party slips up and presents you with valuable information about something they really need from you? You can quietly use that to your advantage. If the community is hostile but eyes something you carry (and can make a lot of) with a bit of obvious lust in their eyes? It's now leverage that you can use if you know how.

Finally, know how to lie with a straight face, while under pressure and stress and do it in a totally convincing manner. Let me give you an example: Pretend you're describing a business trip to your fully menopausal wife during her very deepest bouts of hot-flash-laden PMS, but do it while neglecting to mention the two pretty nubile prostitutes and the gallon of fine whisky that were waiting in the hotel room for you. Oh and don't forget to neglect the fact that her best childhood friend was also there waiting for you, draped across the bed wearing nothing but lingerie and a smile. Oh and she discovered that there was no justifiable business reason for going on the trip in the first place. Did I mention the mountain of home-filmed porn-laden "souvenirs" that you now have sitting in the video camera, the used condom inadvertently tucked into your luggage and the new tattoo on your left buttock? Now remember that as a diplomat, you'll have to do it in a way that sort of tells everything, but tells nothing, all at the same time.

Yes, it's that kind of sleazeball lying skill that you need to muster convincingly, but multiply the stakes up to literal life-or-death. If you can lie through all of that and do it in a way that leaves the other party in a cheerful mood (or whatever mood you want them in)? Well, you're well on your way towards being a fair-to-middling diplomat.

As humorous (*or if you're female, not*) as that previous example was, it's a nearly complete approximation. You have to project an image of one thing when you know the reality is often the opposite. You have to be able to quickly explain any inconsistencies or embarrassing questions that the other party may bring up. You need a nimble mind that keeps track of everything you said and constantly check everything new you say for logical inconsistencies against the old, because lying requires that. You have to do it under some rather fantastic pressure, while at the same time appearing cool and calm and pretending that everything you just said was the absolute truth (unless the charade you're projecting dictates otherwise).

Mind you it's rarely about lying blatantly, because that's just too easy to catch. Most often, it is about the more subtle and sensual forms of lying (misdirection, misuse of semantics, half-truths, misattribution, weasel-wording, amplified impressions, quiet deceptions, etc) and involves those dark areas where science needs to stand aside and let artistic endeavor take over.

Finally, it's all about knowing when the other party is trying to spread a little manure on your fields, so to speak. It takes a great liar to catch a good one after all, so even if your policy is to always shoot straight and to be perfectly honest, an excellent liar is still a great asset to your community's relations with everyone else.

The fun part is, as a diplomat you have to balance all of this in your mind at the simultaneously and do it without breaking a sweat. The ideal candidate for this kind of work has to be intelligent, quick-witted, good with people, outgoing and at the same time able to keep both your secrets quiet and his lies straight. An excellent memory is also helpful here by the way and for obvious reasons.

All else aside, a diplomat is there to help bridge friendships and to make them into long-term ones. The longer the relationship, the less deception and BS you have to apply and the more honesty and trust you can foster between yourself and the other party (be it a person, community, town, region, nation, whatever). Once you have one, we now have to figure out how we're going to relate to our neighboring communities, so…

Policies And Behaviors

Once you have (or become) a good candidate to relate to others, you now have to figure out *how* you intend to relate to your neighboring communities. For this, you will also have to know a few things beforehand:

- Where are you now? In other words, at what point in recovery is your town? Can everyone in it feed themselves? Can you defend yourself well? Are you looking ahead towards building projects that help bring folks back towards a more modern age? How healthy is your population? What skills do you have as a town and individually among its inhabitants? What resources are you chronically short on and what do you have in excess?

- What do you need? What this means is that you have to figure out what it is you need. Do you need certain skills in your town (as opposed to knowledge, though not necessarily exclusive of it)? Do you need certain resources, or more of them (clean water, fertile land, weaponry, medicines)? Does a large percentage of your population who pray a certain way, but have no clergy to help them address it fully (e.g. Catholic, Jewish, Muslim, etc)? Is there knowledge that you need, but do not have on hand (say, you have no books in town on electrical engineering or medicinal plants, etc. Note that this differs from skills)? Find out what you need to reinforce the goal of self-sustenance for your town.

- What can you give? Are there resources you have in such abundance, that neighboring communities lacking them could use with little cost to you? What about skills or useful-but-harmless kinds of knowledge that you can export? For example, if you're a coastal town who does very well with fishing and a neighboring town is inland, maybe you can bring some of that excess protein to them in exchange for something from them. Perhaps you have a force strong enough to patrol the roads nearby to keep them free from raiders and can provide this service for a fee, a tax, or in trade with nearby communities.

- What are you willing to tolerate? While this is usually something you would handle on a case-by-case basis, now would be a good time to set some expectations with your neighboring communities and areas. If their population is hunting or logging like crazy in your woods (but leaving forests in their territory largely alone), you may want to put a stop to that pretty quick, or to let them know in no uncertain terms that doing so is going to bring bad consequences.

- What if they are belligerent? Do you automatically go on a campaign of burning their homes to the ground, killing everyone there and salting the earth, or do you use a more graduated response, showing just enough force to let them know that angering your town is a bad idea?

- What if they trust too much, or are too desperate to care for themselves properly? While few communities indeed are going to cut their own throats (…not this far in, they won't), there may be a group of folks who are desperate enough to give or do anything to find relief from whatever chronic crises they find themselves in.

- What if they are double-dealing, or you catch them in a lie or trap designed to put your town at risk? Figure out what to do then and stick to it.

- What if they request (but not demand) you join them and become a part of their government? If they are larger and/or more powerful and their culture along with their long-term interests align perfectly with yours, it could be a consideration. On the other hand, how much power and autonomy would you retain? Would it be to your ultimate benefit, or your ultimate demise?

- Last but most importantly - what are your own long-term goals? Do you want to grow in unlimited size and power, or are you content with just making sure there is a peaceful and prosperous local region for now? Do you want to ultimately become a force in trade and technology, do you seek conquest, or what?

There are of course many other considerations to work out, but these are the really big ones. Hash them all out and have tentative plans in place. This will allow your diplomatic team to have something to go on. It will give them an idea of how they will react, how they will behave and how they will carry forth the ultimate goals of your town.

Tools And Leverage

Before you get out there and start making yourselves known in the new world, take some stock of how you sit and what options really are available to you. You can impose your will in many, many ways. You have both carrots and sticks to work with. Here are the basic tools that a typical post-collapse town will have going for it - we'll start with the "stick" first.

First, you have the old-fashioned 'stick' called conquest. If you have more than enough weaponry, soldiers and skills to project power, then you have a means of punishment. If a neighboring community is screwing you over, or causing harm to your residents, then you can simply march over there and flatten them, depending on how much of a defense they can put up. Even though this is listed first, you really should reach for it last, when there are no other options or alternatives. Seriously. War is expensive and can often destroy the victor as completely as the victim, so think this one through really, really, *really* hard. There are far more powerful ways of punishment that don't require as much blood and direct confrontation, after all. What would those be? Well, depending on your situation and theirs, you have quite a few. Let's go over some.

Resource denial by force comes up pretty quickly as a more useful 'stick'. In areas where vital resources are scarce, you can control a belligerent population simply by controlling a vital resource or two that they need. In a desert region for example, this will mean water - if you control the water supply to a community that is acting up (say, you're upstream of them along a year-round creek or river) and they can't dislodge you? It won't take long at all until their forces will desert or defect, their population weakens and their leadership becomes a lot more willing to work with you. The only downside is, you will have to defend that resource tightly and as they get desperate, you will have to be fierce and unwavering about it. When they become desperate enough, they will talk, migrate away, or die off.

Denial of trade is another powerful 'stick' type of incentive. This is the little brother of outright resource denial by force - if the people aren't getting the goods they really want or need, they are less likely to support anything which acts against the supplier's best interests. This is a subtler way to deny resources and takes a lot longer to sink in. However, unlike the unifying rallying cry that denial-by-arms would create, this method stands a better chance of creating dissent within the other town or community's ranks.

Another good 'stick' would be to bolster or ally with a similar-sized community that is friendly to you, but neutral or not friendly to the target community. One against one of equal strength is an even playing field, but the potential of one against two suddenly persuades the "one" to be a little more willing to negotiate with the "two", or at least make "one" less likely to do anything stupid.

A solid but subtle 'stick' in your arsenal, especially in a post-collapse era, is to deny protection to a weaker community. This can take many forms. For example, you defend and protect your own caravans of traders, but leave theirs defenseless on the road. Make it known far and wide that you will not come to their aid if they are attacked.

There are literally a ton of far subtler, far more complex ways to punish, but let's get up some 'carrots', or incentives with which another community might rally to your town, or at least remain open and friendly towards it.

First off, yes you can provide gifts, but only a few token ones at most. This is serious business and giving away the whole store is a good way of being portrayed as a sucker, as weak, as desperate, or worse. The only time you should give gifts is to give a taste of what could happen if the target community becomes friendly and open to you. For instance, let's say that enterprising members of your town have come up with means to make life easier and more comfortable, using only locally-scavenged materials (e.g. wind/water generators, biodiesel, useful medicines made from local herbs, soapmaking, things like that). You could assemble a token amount from each of these goods and present them freely to the neighboring community. You then remind them that there is a lot more where that came from, but only if trade were freely and fairly established - of course, only if favorable terms could be agreed on for some challenge or issue, or if you were to begin cooperating in some mutually beneficial endeavor.

Another 'carrot' to present is to offer mutual assistance in matters that involve both you and the target community. These matters can involve defense and protection along roads, large-scale projects that would provide a cleaner (or more abundant) resource to both you and them, mutual defense against an external enemy, a meshing and normalization of monetary systems and things of that nature. Things which both sides can see a good benefit from and will find useful.

Among the 'carrots' you can also present is to render aid and/or assistance if the neighboring community is struck with a disaster that you managed to steer clear of. At this still-fragile stage of post-collapse, one good, hard forest fire, tornado, or earthquake can wipe out an entire town without affecting neighboring ones. A good way to utilize this particular 'carrot' is to pitch in and help out immediately after such an event devastates the neighboring community, only stopping long enough to ask permission and to let them know you're not there to pillage. If they remember you as the helping hand (and not the ones who ignored or kicked them when they were down), they will likely become fiercely loyal to you and yours. Besides, who knows? You may need help someday and their chances of helping you out would increase dramatically if you had similarly helped them in the past.

The best way to start and continue relations is to be friendly at first (but not so friendly that you get suckered or accept a pile of BS) and only resorting to uglier means when/if the other party does something that is willfully stupid. The rest is up to you and the situations you find yourself in.

Go Forth, And …

Once you have a good idea of what you have, what you want and how you will go about getting it, your next item of business is to get out there and do that. Of course, this isn't going to be as easy as walking to the next community over and making demands.

I would suggest starting off with an escorted caravan of traders and willing buyers from your town, to go out and set up a market in the other community (or just ask to use their existing market if they have one and there's enough room). Let trade do the easing-in. While that's going on, ask to see the leadership (if it exists) and see about having a chat with them. Keep it light (work-wise) at first, talking about how you've been doing in general, swapping news, telling jokes and whatever. Once everyone is at ease with each other, bring up the subjects you came to talk about. It could be trade proposals, alliances, an invitation to join, or whatever other things that you've hashed out in advance.

The meeting could be public and open, or it could be closed-door, or it could be a combination of the two, depending on what the other party wants to do. Let them figure out how they want to do things. If they bring food and drink, great! If not, offer them some of what you brought. If they want to talk later, offer a time and place to meet up. If they don't want to talk at all, no problem - just go back out to the market and enjoy the crowds.

If you do get to talking and any firm commitments are made, get out some paper on the spot, write it all down in duplicate (so both parties have a copy) and everyone involved (leadership-wise) signs both copies. If no firm commitments are made, set up a date and time where you can talk about it further. No matter how it turns out, keep the conversation going, if you can.

If they don't want anything to do with any kind of deals or relationships, but are friendly, just let the market continue and go out to have a good time together.

If they're completely intractable, even to the point where they refuse to let you and your caravan come into town, no problem. Just head for home - not every community will be at a point where they want to (or can) ally with anyone. Explore your options and compare notes on what you found there. Try to contact other neighboring communities. Eventually, they may well come to you wanting to talk, but only when they're ready.

Subjects Of Discussion

In a pre-collapse society, subjects that nations and even states discuss with each other are often complex and subtle. They often rely on long-range plans, prevention of warfare (or expansion and conquest) and are often done between peoples of differing cultures, languages and ideologies.

Your job will be a whole lot easier than that. You really won't have a need to discuss immigration issues, but you will have to hash out a few things that will become painfully obvious to you once you get the hang of the whole survival thing. Topics that will come up between your town and neighboring communities are, among others…

- Water Rights. In areas that get a lot of rainfall, this won't even be a mention, but everywhere else, it will be the top subject. The drier an area is, the bigger the subject will be. If you live in an area that gets less than 25" of rainfall per year, or in an area where clean water is a problem (swamps, former urban and heavily-populated areas, former mining areas with a lot of pollution, etc)? You're going to want to bring this up post-haste for any water source that affects you and neighboring communities (especially those upstream of you).

- Hunting/Foraging/Fishing territories. Most wildlife will be dead, but there will still be a few about and there will be a lot of opportunity for fishing (esp. on coastal areas) and foraging (in rural areas). It will be inevitable that these territories will be in dispute.

- Mutual Defense. If you're close enough to support each other, one or both of you will want to entertain the idea of helping each other out in fending off any inbound troops. This could also include such things as going into the wilderness together and clearing out any criminal/raider camps in territory that lies between you and the neighboring community. As another almost separate topic, this can finally mean working together to have functioning and regular armed caravans that travel between the two of you.

- Trade. Yep, we covered this one almost to death, but keep it on your radar nonetheless, because if you don't bring it up, they likely will.

- Communication/News. They want to know what you've heard and odds are perfect you'll want to know the same from them. Working on a regular exchange of news, credible rumor and anything else of mutual interest will be profitable for the both of you.

- Salvage/Scavenging/Mining Territories. As things calm down and people become brave enough to actively salvage what used to be civilized areas outside your town limits, you'll want to bring this up. Perhaps you can mark areas where a given town/community has first right of refusal for anything salvaged from a particular area. As some areas will have richer pickings than others, you'll want to see that things are divided equally. Note that this is equally true for any raw materials (ores, minerals, etc) found in quantity.

- Agricultural Territories. Areas outside the community limits suitable for growing food on a larger scale will become rather prized at this point in time… by everyone. As far as communities and towns go, this is more for taxation reasons than anything else, since a communal effort is likely not going to be possible given limited resources. On the other hand, it also sets limits as to how far out your residents can farm, or how close to you some other community's residents can farm. At this point, setting boundaries entirely would be a good idea, with each party taxing their side of the boundary line. Your job is to set those lines and do it in a way that is either fair, or to your town's distinct advantage.

Keeping The Peace

Once you have established and proven solid relations with other communities, you will want to maintain those relations as often as you can. Here are some tips to help you do that…

- Keep communication open and regular. Pass word along by a courier, via trusted traders who regularly travel back and forth, or by way of any means of communication you can come up with and trust.

- Periodically check in, in person and on an official basis, to look over any agreements together. Do this to make sure they are sufficient and necessary, if they need to be changed or updated and if any new ones need to be put together.

- Always reply promptly to any message or request from a neighboring community, even if you think it is trivial. People get their egos bruised if they feel neglected and political entities are even more susceptible to this than individuals are.

- As long as they're friendly and can trustworthy, plan and carry out a periodic get-together of entertainment and recreation between your town and any neighboring communities. This allows folks to know each other better, allows younger residents to meet prospective mates from other areas and helps bond everyone together far better.

- At the very first sign of any problem, immediately act on it by contacting the leadership at the neighboring community. Tackle the trouble together while it is still small and easy to fix and not wait until after it has blown up and has people taking sides. The sooner you fix it, the less passion and emotion is attached to it.

Long story short, never stop finding ways of keeping communications flowing and never stop finding ways of talking through the tough spots.

Making Treaties And Pacts

Odds are good that if you become friendly with another community, you're going to want to start looking out for each other's mutual interests. These interests usually center around trade, defense and sharing common resources (clean water stands out as a big example here). To that end, you both will likely want to formalize your bonds in some sort of written agreement, pact, or treaty, one which spells out the rights and responsibilities of each party.

Before you start signing things, take a bit of time to think it through - do not rush to any decision until you know enough of the facts and surrounding situation that you don't get screwed. Think it through carefully and perhaps bring it up for debate with the town leadership and even the town public. Get all opinions and listen carefully to all interesting thoughts and ideas which come up. Especially when it comes to shared resources, you will want to make sure that your town gets enough out of the deal to make it beneficial in the long-term and that you don't find yourself boxed into a corner later on. Always make sure that the important contingencies and outcomes are written down and agreed on. Be willing to change something or to refuse parts of a deal if you need to.

Make sure you can, in fact, hold up your end of the deal through thick and thin. You may be able to give away most of the clean water from a shared stream now, but in a drought the whole deal may change dramatically as other water sources dry up, leaving you with only the partial output from that stream to survive on. You might be more than willing to provide troops to help defend your neighbor and snuff out raiders, but if the leadership changes and that community decides to attack other neighbors, you may not want to support them in doing that - especially once someone else retaliates by sending troops to stomp them out of existence. If there are any circumstances that arise which will make you refuse to keep your end of the deal, be sure to state them up-front and to include them in writing. A good rule of thumb is to promise delivery only under certain general circumstances, but leave in escape clauses for everything else.

Make certain that the deal doesn't strain your resources or your people - you may not be able to keep your end of the bargain up if things go from bad to worse. Make the deal only if you can provide it even under the worst of circumstances. Always be sure to think ahead.

Never make a promise based on projections and/or anticipated outcomes. If you don't have it now, you may not have it later. If the other party insists, be sure to write the deal down as being contingent on the projection or anticipation coming to fulfillment.

Note that not everything has to be a formal contract, but that the important things should be, especially if it requires you to provide something and doubly so if that something has to be provided or sworn to on a continuous basis, or done so in perpetuity.

As noted before, once something is agreed on and written down in final form, make at least two copies and make sure that both copies are signed by everyone involved. Each copy goes to each party. Note that these copies are what you will be referring to should any disputes arise, so store them carefully and securely.

Handling a Crisis

First and foremost, get as many facts as you can with as much time as you safely have on hand to get them. There is no sense making policy decisions based on rumor, or worse, lies designed to force you to do someone else's bidding. Reach out immediately with whomever it is that is the source of the 'news' and also reach out to all other parties involved, if you can. If possible, get all of these people together and hash it out.

If you don't have time to act on it, always think 'defense' first. Do not immediately go for the offensive in any aspect unless you are both certain of the facts and the nature of the crisis calling for it. This is just added insurance against doing something stupid.

If the crisis is indeed quite real and it is due to something that the other party did? Give them a chance to explain what happened and why. If they refuse or did it intentionally to antagonize, you can prepare whatever means you feel necessary to meet the threat or to remove the pressure on your town. If it's accidental on their part, it is up to you (and them) as to what steps should be taken to make things right.

If the crisis is real and it's your fault (or rather, your town's fault), be prepared to do what is necessary to make things right and to deal with (and punish) the people responsible. Depending on the crisis, you can either handle it internally, or you can hand over the perpetrator(s) to the other community for disposition. Doing the latter will prevent anything stupid from happening in the future, as the perpetrators will only do it knowing full well that they do so without the town's support. No matter how you resolve this, always do so with an eye for balancing your own town's mood and morale against what needs done.

The good news is, if you keep regular communications in the first place, most crises will be solved long before they become something that affects the whole town. If it keeps growing, start using those carrots and sticks we mentioned earlier. If it gets uglier than that, only then and as last resort, should you ever consider armed conflict.

Even if you're in the midst of armed conflict, if the other party actually does want to talk, it certainly can't hurt to do so - just take the right precautions. Always meet in a neutral area and with equal numbers of forces watching over the proceedings. You can also, if the other party is reaching out, invite that person to come talk, but to not let that person see anything in the way of defenses, preparations, or suchlike. Instead, meet him or her just outside your own defensive perimeter and talk there. Insure that the person arrives unarmed, or leaves all arms a good distance away from the talks. Even if you have to set up some sort of tent to host the person outside of your perimeter, do it. Always treat that person with basic dignity and do not harm, or hold him or her for ransom or as a bargaining chip. Offer refreshments if possible and if the person is staying overnight, provide for that person as best you can without allowing entry into your community.

Conclusion

We've only scratched the barest surface of inter-community communications and relations, but hopefully we've uncovered enough so that you can get a good idea of where to take things from here. Keeping things peaceful and cooperative is your goal. Why? Because with a little luck and a lot of work, the two entities may someday become one and together you can bring others into the fold. Just not yet, though… there are still a few things left to do before we even think about that, though…

Civil Engineering

So… things are peaceful (that is, you're not fending off invasion every other month) and things are even prospering (that is, the town is not in immediate danger of starving wholesale), yes? If not, get on that until you get to a sufficiently stable point, because now we're wandering into territory that will require a lot more resource and focus than you can otherwise afford. It is one thing to get people and society locally stabilized. It is another entirely to start building things - especially things that are going to require upkeep and work to keep running.

What To Leave In, What To Leave Out

Before you start happily pushing plans, you may want to consider a few things here and scale back your enthusiasm…

Electricity: Unless you have the means of manufacturing useful electrical and electronic components from scratch, you may want to put off building a power plant for a little while. Even with renewable energy, large-scale power sources use a horrendous amount of energy, material and work to get up and running (even if you're rehabilitating a broken one), let alone keep running. Even if you manage to get a working local grid, what are people going to use it for right now? Light bulbs burn out. That TV won't have any content on it anyway and video players are fragile. Computers break down in a scarily short amount of time. Even more durable things like refrigerators and heaters are going to eat a ton of power, and eventually need repair or replacement. Most electronic gear is made to be disposable, not serviceable.

It's far better to scale down electrical generation to power things that are useful and/or have some durability without a high consumption, or specific uses that have an immediate and obvious use (in your case, communications gear stands out rather large, with perhaps medical gear close behind). There will of course be other uses, but they don't have to be done at a governmental level... private industry will likely pick it up and run with it. Examples? Enterprising individuals may run a 'movie theater' with one or two televisions and DVD players running from their own solar-and-battery station. Someone could set out a 'charging station' next to a riverside hydro generator, for people to recharge their laptops and smartphones (for a fee, of course. Incidentally, the smartphones aren't for calls, but the apps and "offline" maps –and even GPS– are going to be at least somewhat useful and/or entertaining). They may even rent (with a hefty deposit) fully-charged devices pre-loaded with movies or games, which the customer brings back once the battery is drained. Another person entirely could have a wood/coal-fired generator and a couple of refrigerators as a cold-service, offering to keep items cold (or frozen) for a customer on a per-day or per-week basis... for those times when you stumble across a really big haul of fish or animal and you can't quite preserve it all in time. Only when you can be reasonably certain of two things should you attempt to re-create any kind of local power grid: When you have (or have access to) the means of making electrical devices that are useful and when there is enough demand for those devices.

<u>Water</u>. Water distribution is actually a good thing here. A means of delivering clean water to homes is a huge benefit to your town. A means of improving fire prevention and mitigation is an awesome benefit all by itself. The infrastructure is largely already in place and is fairly simple (umm, in concept) to maintain and build. The hard part is to cleanse the water before you deliver it. This can be done to a large extent without specialized chemicals, but will be somewhat labor-intensive to do. Soft (mineral-free) water will be easier to work with than hard (mineral-laden) water. There is the task of either restoring existing systems, or building new ones. If gravity works in your pre-collapse infrastructure's favor, this will be relatively easy to do. If the pre-collapse infrastructure relies on pumps or electricity to operate, then you've got a lot more work to do and it may prove almost impossible without some real creative solutions. It isn't completely impossible - ancient Romans were able to pull it off with relative engineering ease. It may however become very labor-intensive depending on water sources, especially in construction.

<u>Sewage</u>. This is fairly easy to pull off, since nearly all sewage infrastructures built pre-collapse will rely on gravity. They are easy to build and fairly easy to maintain. You will want this if you set up anything that resembles running water. The hard part will be the sewage treatment before discharge. Given that the total population will be low at this point in time, you may opt to skip this step, but do it knowing full well that wherever you dump it will have consequences downstream and the 'neighbors' may well complain... loudly. You could build a compromise solution that involves settling ponds and huge leach fields (or if sanitation otherwise allows, dump it in a place where you don't want any future raiding parties coming in through).

<u>Roads</u>. Roads are beautiful things, especially post-collapse. They allow for rapid transportation, provide easy passage and make things smoother for wheeled vehicles (whether hand-drawn, horse-powered, or if you're lucky, rated in horsepower). Existing roads can of course be rehabilitated and will be easier to maintain. A means of dealing with potholes and any large cracks that arise will have to be made, even if it means a lot of recycling with used asphalt, used oil, used/old roofing tar mixed with aggregate, etc. Choose which roads outside of town that you want to maintain and which you want to use as a source for material to patch the roads you want to keep. Consider going with gravel as a road material instead of asphalt or concrete, as gravel is easier to get hold of and work with. Gravel also happens to be (a little) easier on horse hooves, which is something to consider. For roads in-town, keep them up as best you can. Maintain at least one lane on your main roads (and any large roads leading to/from the market areas) for vehicular traffic, but the rest can be foot traffic. Also keep in mind that a road that

encircles town limits is a good thing for defense - if you don't have one, you will want to make one.

Bridges. Existing bridges will still require maintenance and inspection. Steel bridges can be maintained with simple tools and either grease or paint, but it will be labor-intensive. If you do not maintain metal bridges, they will eventually rust and quickly become unstable or dangerous. Concrete bridges are somewhat easier to keep up, but you will have to come up with a way to patch and repair them. If a bridge is already too unstable or has collapsed? Instead of trying to repair or improvise with it, consider running a parallel bridge over the obstacle and building/using ferries over waterways. Any really large bridges will be way out of your resource limits, so simply avoid them unless they're new enough and there is enough skill around you (and you have enough material available) to maintain them. Ignore any bridges (no matter how impressive, durable, etc) that do not lead to anything useful from your town. You may even want to consider destroying them in order to shut down any potential invasion/attack paths.

Fuels and Biofuels. It has a lot of benefit, but a lot more in the way of drawbacks. Old-school petroleum will be difficult to extract (even where it is otherwise plentiful) and then there is refinement to the point where the resulting material could be useful in an engine without destroying it. Did I mention that refining fuels is a process that mixes highly flammable liquids and gases with high heat? Modern automotive gas engines are especially picky about what they will burn, to the point where you have to be rather precise in how you refine that petroleum. Diesel engines are a lot less picky and thus a lot more useful. On the other hand, there is still that whole extraction and refinement process. There is the promise of biofuel and it is far easier (and safer!) to do. However, making biofuel relies on a whole lot of biomass (read: plants and fats/greases of the right type) and at this point in time you'd be better off growing plants for food and fats/greases will have other, more immediate uses. However, keep the knowledge around if you can, because there will come a point in later days where it will come in very handy.

Internet. Well, probably not. Yes, the Internet was (originally) designed to withstand thermonuclear warfare. However, what it has become requires a lot of moving and highly delicate parts to stay up and moving: routers, fiber optic lines, switches, servers, repeater stations, etc. The odds of having enough of that still up and running to be useful to the average joe is, well, slim-to-none (I'm favoring "none" quite heavily here). If it still exists and is running in some form, odds are perfect that what's left of the military will have access to and use of it, but no one else will. If there is still some bright spot of civilization in the world after all of this and you think you can reach it with a network, then certainly try… just don't waste too much time and material to do it, though.

How To Make Electricity

In spite of going out of my way to tell you to not really bother with it, it is important to know how exactly one can make electricity in a post-collapse world. Yes, light bulbs will eventually burn out. Most electronics will fry, with no feasible replacement parts. Given the largely disposable nature of electronic devices in the pre-collapse era anyway, fixing what's broke won't be possible for most of the time. However, if you can spare the energy and parts, you can make use of what you do have before it finally dies off once and for all - *as long as you do not rely on it for life or limb!* What I'm saying is, electricity can, until its devices are gone, be very useful to you. With a little luck, maybe it can be useful for long enough until some enterprising soul has (or regains) the knowledge to make new generators and basic electrical/electronic devices once again…

So, let's begin with a (very!) quick tour of the basics…

Electricity is the flow of atomic electrons over a conductor from a point of excess, through a working load, to a point of deficit. In plain English, you have four things that make an electric power source and circuit: One, a point at the power source where there are too many atomic particles called electrons (this point is called the "negative terminal"). Two, a load, or device that uses electricity to do something useful. Three, a point back at the power source where there is a severe lack of electrons (the "positive terminal") - enough to eagerly accept all those extra electrons coming off of the negative terminal. Finally, you need something to connect it all together - and it has to be able to pass those electrons along very efficiently (the "conductor"). Outside of theory and in the real world, there is a fifth item to consider as well - a means to insulate the circuit so that the electrons only flow where you want them to.

There are four basic components to electricity itself that you need to know about: First, we have current - the number of electrons that are flowing past a given point on the conductor at any one instant of time. This is usually measured in Amperes, or more commonly, "Amps". Second, we have the force with which the electricity is flowing, which is usually measured in Volts. Third, we have the amount of work electricity can perform (either in movement or heat), which is measured in Watts. Finally, we have the amount of resistive force that is built against the electron flow. It is commonly measured in Ohms (usually shown as the Greek Omega, or "Ω" symbol). In case you're curious, the speed of the electrons themselves are never measured, because they all flow at one speed - the speed of light to be exact (186,000 miles per second).

The final concept you need to know is the difference between DC (Direct Current) and AC (Alternating Current). DC is a basic type of current where electrons always flow from one point of the power source to the other and always in a single direction (flowing from negative to positive). It has the advantage of being simple to construct, design and build components for, but that's about it. AC is the more advanced type of current where the current flows first in one direction, then the other, switching many times per second. The speed at which it switches over is called Hertz ("Hz"). The typical North American household current (the kind from the wall plugs) switches over 60 times per second, or at 60 Hz. This is important since AC electrical devices made for use in North America require that speed. (Note that that power also needs to be somewhere around 110 - 120 volts)

Making electricity will require some form of power source - a means to generate that electron difference between positive and negative terminals to get a flow going, or a means to store that difference chemically (usually called a battery) for later use. There are four ways to create electricity: One, you can create it by chemicals (for example, using metal plates and a strong acid or acidic paste between them). Two, you can create it by way of forcing an inherent reaction in a given material (for instance, solar power, which uses photovoltaic reactions, or piezoelectric power, which generates electricity by creating a physical shock in certain materials). Three, you can generate it via static electricity with certain dissimilar materials and friction. Four, you can do it by disrupting magnetic lines of force. The last of the four is usually the most powerful and longest-lasting, so we will put most of our focus on it.

To generate electricity by disrupting magnetic lines of force (an electromagnetic generator, you need two things: a spinning magnet and a coil of wire. A conductor (wire) passing through a magnetic field will generate electricity. A lot of wires passing through a magnetic field rapidly will generate a lot of electricity. The way most electromagnetic generators work is by spinning the magnet so that its field is shoved past the surrounding wires, causing electron flow (that is, electricity). The good news is, even in a post-collapse world you will have a lot of devices that are not only capable of this, but are also quite durable. Automobile alternators, truck generators, that odd gas generator laying around, purpose-built generators and even most permanent magnet DC motors will generate electricity if you spin them fast enough. The most commonly found type, the car alternator plus associated circuitry, can reliably generate around 12-14 volts of electricity at around 40-80 amps if we spin the alternator shaft at any variable speed between 500 and 3000 RPM (revolutions per minute). Be certain to not only use the alternator unit itself, but also the "field coil" assembly along with it (most have that included in the unit, but many older cars have it as a separate device). When you strip the alternator out of the vehicle, note that the black wire is the negative ("-") terminal and the red wires are for the positive ("+") terminal. Your best bet if you can find it is to get hold of an actual generator that was specifically built for generating electricity. These can come from actual gas or diesel generators, from gas/diesel-powered portable welding kits, large trucks, or other similar sources. Most of these purpose-built devices also come with an *inverter*, which gives you AC power right off the bat (which in turn allows you to run most electrical and electronic devices).

As you may (or may not) have figured out, the inverter is required if you want or need to power AC devices - and basically anything made for household use requires an AC power source. Fortunately, in spite of being complex devices, inverters are actually fairly common items. You can get them in light-duty (which plug into an automotive cigarette lighter), to medium-duty (which are available at most common stores with an automotive or RV supplies department) and even industrial ones (either scavenged off of a purpose-built generator or from a wide variety of industrial/heavy equipment.) Inverters are rates by the amount of watts they can produce. You simply need to add up the wattage of the equipment you intend to run off of it, adding a 30-50% cushion to the figure just in case.

We have our generator with all the basic bits and we need to make it go, so let's get that spinning so we can actually make some electricity.

When it comes to actually driving the generator, you can do it many ways, usually with gears or pulleys to convert the relatively slow speed of your spinning source power (from wind, water, steam, whatever) into the higher speeds necessary to produce electrical current. Pre-collapse, the most common way to generate electricity was to generate heat in some fashion from fuel (coal, gas, oil, nuclear), which heated water in an enclosed-loop pipe system to high-pressure steam and the steam in turn would push against vanes in a generator, causing rapid rotation. Post-collapse, you could do the same thing if you have a huge and constant supply of fuel - wood, coal, natural gas perhaps. However, the hard part would be to make a reliable steam system and making a reliable steam engine to provide rotation (as it requires devilishly tight precision).

One thing you will want to get hold of if you're building/rolling your own generator solution is an inverter.
All that said, you will be better off using a renewable resource, such as flowing water, a constant wind, or a similar supply of naturally moving fluid. The rest is fairly straightforward, but we'll cover each in turn.

Let's start with wind power. Wind power is actually among the oldest forms of power generation there is, though for most of human history, wind power has been harnessed for mechanical, not electrical means. However, if you live in or near a place where the winds are somewhat constant and average at least 3-5mph, you may have a decent source of power on your hands that you can put to use.

The advantages are many, though different: Wind power runs at almost anytime, day or night. It is a fairly compact source of power when it comes to the amount of land required. Anyone can make a wind-powered generator from scavenged materials and if built with some skill, can also be silent. There are many different designs that you can use to best harness the wind available in your area.

There are a few disadvantages, however. Wind power requires that the weather cooperates - you're obviously not always going to get enough wind to generate the power you want or need. It requires a fair bit of skill to build a balanced set of blades and to build them so that they don't shred apart in a windstorm. You have moving parts that are going to need maintenance. You will definitely need spare parts on hand in case something breaks and need to have have the skills to replace them. The tower sticking up and moving will certainly attract attention for up to miles away. In colder regions, you will have to keep an eye out for certain icing conditions and do something about it quickly if it threatens to un-balance the blades.

But let's talk about siting your wind tower. The best general areas to put a turbine would be in coastal areas (as the wind blows almost constantly either on or off the sea), open plains areas that are free from trees, the tops of rounded hilltops, gaps in mountains (or large hills) and similar. One consideration you may want to keep in mind, though… a wind turbine on a tall tower is going to stick out like a moving sore thumb if you're looking to stay discreet - either as an individual or as a town. It will be a big welcome sign pointing right to you, stating that you have electricity on tap. Just keep that in mind if things are still a bit too hinkey out there to be advertising your tech prowess.

Now, about those blades… you can actually recycle them off of a lot of things fan-like, but make sure that those blades are tough, durable and able to hold up to the weather. They should also be large enough to catch the wind and create useful *torque* (rotational force) for spinning the gearbox/pulleys and generator, but not so big that only a strong storm will be able to move the whole thing. While you're working on rigging the blades, you will also want to pay attention to two other bits: the braking system and the gearing. The braking system is actually fairly simple to build and can use either spring-loaded weights around the shaft to activate a brake, or can use tabs on the blades to automatically change the blade *pitch* (the angle at which the blade faces the wind) in order to slow it down. The gearing can either be a well-oiled gearbox, or a small system of belts and pulleys. The reason you have these is to step up the rotational speed to something that can spin your generator enough for good electrical generation. Note that pre-made wind turbine assemblies will either have that built-in, or won't need them. After you get all of that sorted, you'll want to add a *vane* (a big fixed blade) in the back to keep the blades always facing the wind and to carefully balance the whole thing so that it swivels easily on top of the tower once assembled

Let us pay a bit of attention to the generator for a moment. An electrical generator can be scavenged from car alternators, diesel engines, portable electrical generators and a whole host of other sources. The power output will vary, depending on the generator you manage to get hold of… for most automotive alternators, the output will be 12 VDC and average current output will vary from 45 to 90 amps. The newer ones will generate more (to feed power-hungry car accessories), so shoot for those. A scavenged generator from a gasoline-powered one will have all the bits you need to generate 110V AC, but current will vary greatly with speed. You will want to size your blades so as to create as much torque as you can so that when you step up the gearing, you can maximize the RPMs that you can deliver to the generator at lower speeds. About that gearing? You'll want to step it up enough so that you can get at least 500 RPM for a car alternator, or for most common vehicular alternators/generators.

Next up, if you can get the turbine/blade assembly above the ground clutter (and possibly trees), you're better off than using one close to ground. In fact, the taller, the better - but remember that you have to perform maintenance on the things, so the tower either has to easily come down on occasion, or it has to be strong (and safe!) enough for you to climb up there once in awhile. The best bet here is to have a tower that is easily lowered on a hinge, suing guy wires (supporting wires) to hold it up. The materials are easier to get (or make) and you don't have to risk your neck (or someone else's) every time you need to work on the thing. Unless it is a very small tower, raising and lowering it will require a bit of help from family, friends or neighbors, though with enough rope and pulley action (or a manual winch), one person can do a fair job of working with almost any small tower.

The smaller but almost-as-important things is a means to get that electricity down to the ground, a means to convert it from DC to AC (you'll need an inverter) and perhaps a bank of batteries to store the extra juice for those times when the wind isn't blowing. A couple of copper rings and metal wheels can do it easily enough, plus a way to keep it covered in the rain. After that, it's just a question of having a safety switch, perhaps a battery charger and bank of batteries to store extra power for times when the air is calm and an inverter if you want AC power instead of DC.

Maintenance will have to be regular, but is fairly straightforward: It starts with watching for corrosion (from weather) and insuring that all moving parts are sufficiently lubricated and clean. This will mean occasionally taking the (small) tower down and working on the assembly, or getting up there if it's a larger tower that can hold you up. You will also want to keep an eye and ear out for vibration while the generator is operating (especially in stronger winds), which is a sign of an unbalanced set of blades. Blades will occasionally need repair and you will want to find a way to protect the generator itself from being exposed to wind, rain, snow, ice, etc. If you use any fan belts, you will definitely want spares, but you will be better off using a direct gearing set if you can build or find one. Batteries aside, you should be able to build and maintain a wind generator for years to come and be able to find/scavenge enough parts to keep it going far longer than most other forms of power generation.

How about some hydro-electric power? Hydroelectric power, like wind, has been around for a very, very long time, though again, mostly for mechanical output. Some of the earliest forms of electrical generation relied upon it. The principle is rather easy: Get moving water to push a wheel and gear the wheel output up enough to spin a generator for maximum output. The advantages make this method a near-perfect means of getting electricity. With the right water source, the output is constant and continuous year-round, day or night. Aside from ice or drought, the weather will not affect its output. A well-built and well-maintained hydro power generator will last for as long as you can maintain it. The best advantage of all is that with enough torque, you can not only generate electricity, but provide mechanical energy (for grinding grains, running large tools, etc). This is actually the easiest to do and the most consistent source of power, but it's the hardest to find a good spot for. So let's start with the hardest bit first, shall we?

Finding a good site to run water-power off of is going to take a bit of research and some long-term knowledge of the water source - let me explain why. Small water sources may dry up in summer, especially in arid regions. If you live in an area where snow and ice are common in winter, you'll have to contend with a lot of ice. Huge water sources (e.g. rivers) may be prone to flooding, or at least a rising/falling water level - if your design doesn't account for that, you may end up with a submerged generator, blades that are high and dry, or are frozen solid with ice - and either way not generating electricity.

There are also a few disadvantages to keep in mind: Even a small hydro station is fairly visible and thus exposed to sabotage or tampering. You're generating electricity in a location that is obviously very close to water (two things you definitely do not want to mix together).

Depending on design, corrosion will be something you definitely need to keep an eye out for and prevent wherever possible.

Your best bet is to find a year-round source of strong flowing water and then designing and engineering your way around the rest. Rivers with rising/falling water levels can be accounted for by building a floating platform for the wheel and generator kit, then anchoring that platform to a spot near the riverbank or bridge for easy access. For smaller streams and creeks, you can design a small earthen dam to create a pond, then be sure to build in a outflow pipe or trough that flows downhill to your water-wheel and beyond back to the water source and then an overflow setup for those times when heavy rains send too much water at your pond. Instead of the latter, you can also have the water wheel dip directly into the current of a larger stream or river. Be sure to have some sort of trash grill installed, to keep the uglier stuff out (especially if you're using an outflow pipe instead of a trough or directly using the current).

Once you have a good water flow going, the rest is fairly easy: build a large(-ish) water wheel and use either gearing or pulleys and belts to increase the speed of rotation to something your generator can use. There are far more efficient designs that involve turbine blades and other submersible devices to turn water flow into movement, but you will want something simple and easy to repair or replace as needed. Water wheels are also less prone to blockage by trash and debris, or breakage by rocks. It also means less chance of a sudden flow restriction caused by algae buildup, the odd fish that got too close, clams, or snails.

Your big hazards here will involve all that water. Building something to keep the water off of and away from the generator without getting in the way of the belts or gears isn't too hard to do. It should be done though, if only to keep corrosion down. Insulating the conductors and connections will be a big concern and grease will be your best friend for covering these points.

You can conceal a hydro-electric system somewhat, but for large-scale power generation, that water wheel is going to stick out - just not as much as a wind turbine would and with a bit of planning, you can hide one somewhat decently.

Unlike any other form of renewable electrical generation, this one is, as mentioned, likely to be constant. However, this also means that the maintenance will also be nearly constant. You will have to insure that all bearings are lubricated, in an environment where water will constantly threaten to wash that lubricant away. You will have to insure that the waterway feeding the wheel is free from trash and anything else which would get in the way. You'll have to maintain and repair any wooden or other easily-degradable parts of the system. If you get ice in the winter, so will your system - act accordingly (though the good news is, flowing water freezes at far lower temperatures than still water, which gives you some time to work with). You'll have to scrape off any algae buildup and plan for outage days when you have to stop the wheel and do all that aforementioned maintenance.

Of course, there is Solar Power… but this can take two forms here: The first is purely photovoltaic - that is, it converts light directly into electricity. These are what you have when you think of solar panels. The second type is the conversion of solar radiation into heat, which in turn creates steam to drive a generator. This type usually involves mirrors which focus light onto a surface containing water, which is them heated. For obvious reasons (starting with that fact that post-collapse, perfectly curved mirrors that can automatically track the sun are going to be rare if non-existent), we'll focus on just the photovoltaic type.

The good news with photovoltaic ("PV") solar power is that you can generate it every day and that there are no moving parts. Maintenance is a snap and only involves cleaning off the panels once in awhile. Top-quality PV solar panels can last for decades, usually retaining 80% of rated output by the 25-year mark and still capable of generating between 30-50% of its rated output by year 50. Given post-collapse average life expectancies, that's not bad at all. The downsides, however, are quite a few. First, there's the obvious fact that they won't work at night and their efficiency drops a lot when days are overcast. Then there's the fact that you'd better have them already, because making PV solar panels post-collapse will be nearly impossible. You will have to place and point them well, to insure that they point to the sun and get direct sunlight through as much of the day (side-side, or *azimuth*) and year (up-down, or *inclination*) as possible. If you have all of that, you're usually good to go from there and the rest is maintenance. Just be sure that in a full home installation, that you set the home circuit breakers so that you're not bleeding power off into a now-dead neighborhood grid.

There are of couse the disadvantages. The obvious one is that you won't get any power generation once the sun goes down (rain and clouds will still allow you to get a 50-60% output), so you will certainly need a backup power source, or a bank of batteries to store excess electricity generated during the day. You have to site and orient (that is, point) the panels properly, else the efficiency drops to practically nothing; this also means avoiding any permanent shade (trees, buildings, hills, whatever). Panels are notorious for dropping output to ¼ or less of their output if even one or two cells get shaded by debris (e.g. a leaf), so keeping them clean is a must. Finally, the power output is going to be variable throughout the day - early morning and late afternoon sunlight will produce less power than the noonday sun will. Clouds and rain will cause lower outputs.

In all however, Solar power is not a bad idea at all when combined with other means of power generation, due to its simplicity and longevity. Maintenance is fairly simple - clear off any snow, dirt, or dust from the panels on occasion and that's it.

For the scavengers out there, you can modify and/or re-purpose existing PV panels to your own uses. You will have to (carefully!) un-mount them and (just as carefully) move them to where you intend to generate power. You will want to point them so that they get the best sun, as often as possible throughout a typical day (which also means accounting for shadows cast by trees and the like). In general (and in North America), this means you have to point them somewhat towards the equator at a gentle up-sloping angle. The further north of the Equator you live, the sharper the angle will need to be. A rough calculation is to first make the angle equivalent to your latitude (if you know it) and then tilt it back by 15 degrees. For instance? Portland, Oregon is at roughly 45 degrees latitude north of the equator. Here, we would face it dead south, calculate the initial angle to 45 degrees and then tilt it back so that it is angled 30 degrees. This allows us to account for both winter and summer, as the sun changes its own angle relative to the Earth's surface here.

While obviously not a constant source, the one advantage solar power does have over wind power is some form of predictability. While you may not be able to predict how much power you'll generate (cloudy days generate much less output than sunny ones and summer gives you more daylight than winter), you can at least rig your charging and electrical devices so that they can perform their tasks during the daylight hours and do without at night. It can give you somewhat of a schedule to work off of. For example? If you have a small chest freezer full of meat, you can have it actively keeping food (or whatever) frozen, powering it only during the day. Leaving the freezer closed at night will keep the contents frozen enough for safety reasons, until the next morning can get the power (thus the freezerr) going again. Adding a large block of ice, in an airtight plastic bag (to prevent melting from defrosting cycles) to the freezer will even things out perfectly temperature-wise. Of course you'll want to work out some sort of long-term plan, but this rig would work in the interim.

At time of writing, panel outputs range from tiny hand-held units to large home/commercial panels capable of 255Wp output per 4x8' panel ("Wp" = peak wattage). Four of the top-end panels can produce a continuous kilowatt of electricity under full sun. The needed equipment consists of three parts and five if you want battery backup: The panels, an inverter or power conditioner (depending on whether you want AC or DC) and a main switch. The battery backup adds the battery charging gear and a bank of batteries sufficient to hold enough to keep your post-collapse home going at night. The downside is that all of this may well be rare or hard to get hold of post-collapse (especially the batteries) and as time goes by, the parts will be harder and harder to get (though if you have a setup, replacing parts should be a very rare occurrence outside of the batteries). A final consideration is that photovoltaic cells are made by a very exacting process and will require materials and equipment which you simply will not have post-collapse (unless you're lucky enough to live next door to a solar panel manufacturer, that is). Unless you already have them or can procure them, you likely will not be able to make them from scratch.

Steam Power? It has the benefit of being always-on and will operate no matter what time of day, no matter the weather and no matter how wet or dry things are. The downside is in the care and constant feeding of the power source and inherent danger. It is only renewable if you have an existing geothermal well (a natural source of steam), or a constant supply of wood or charcoal to feed it. You could also feed it pretty much anything flammable, but we're getting ahead of ourselves here.

Your library should have the specifics, but let's give you an overly-simplified version of how this works. What you will need are four main components: a boiler, a reservoir, a steam turbine/engine and a return that links it all. Everything(!) In the system has to be able to withstand some very high pressures and will require constant inspection as well as constant maintenance to run safely. The loop consists of a high-pressure side and a low-pressure side. It all starts with the boiler - heat turns water in the boiler pipes into steam and the steam creates pressure. The pressurized steam is then piped to the turbine, where it spins the turbine blades (or vanes). After it has done its useful work, it leaves the turbine and condenses into water as it travels (and cools) along the low-pressure side of the lines and into the reservoir. The reservoir then collects the water, where it feeds into the boiler and begins the process again.

Before you build such a system, you must be certain that you have a constant supply of fuel to keep it burning. This could be anything flammable, really, but for the least labor, you will want a fuel that carries a lot of potential heat per pound. For this job, your best bets are coal, natural gas, or petroleum (crude or refined). Since petroleum is going to be impossible to get (and would be more useful elsewhere) and natural gas would require an existing well with a huge reserve, coal will be your best bet overall (well, if you can get it). Before you begin, you will need to find a coal seam (or a natural concentrated deposit) that you can defend if necessary and in sufficient quantity to justify building a steam-rigged system. Note that coal is also useful for heating homes and cooking food (in a stove), so you will have to balance those needs out as well. If all else fails, you could use wood, but only if there is enough wood available to use (for example, a forest, a huge tract of abandoned suburban houses, etc). No matter what source you have, you must also be certain that you can afford to spend the energy in constantly getting that fuel out of the ground (or out of abandoned homes) and into the boiler. If you cannot do any of this (or even part of it), then stop and consider another form of power generation.

The next obstacles? Post-collapse, you're really not going to have a readily available machine-shop to make the steam turbines. Because of this, you will have to improvise, or have a ready-made turbine or two handy. Improvisation should only be done if there is enough mechanical and engineering skill to build or forge a turbine that can handle the high pressures of live steam. You will also need someone capable of designing a system that can prevent the pressure from turning your generating rig into a literal bomb. Any weakness in the system will allow steam to escape, often with rather explosive results - especially in the boiler or on the high-pressure side of the loop. So why the high pressure? Because in order to be useful, steam has to be generated (and kept) at temperatures far higher than the boiling point of water (100 C or 212 F) and at pressures far higher than that of the atmosphere around you (usually at 150-200 psi, but can go way higher).

If you have all of this, then your last obstacle is manpower. Do you have enough spare labor to keep the boiler fed on a constant basis and to maintain the system (usually on the lookout for corrosion and stress fracturing)? Do you have someone with the skills to keep an eye on operation, to keep the temperature and pressure from getting too high, yet keep it high enough to run the system efficiently? It takes a long time to get up enough pressure to get the system started and a lot of skill to keep things running safely.

There is one final consideration. If you're hoping to conceal this thing? Well, good luck. The fire alone will generate copious amounts of smoke (especially if it's from coal), with all of it pointing right at your boiler/firebox assembly. Any escaping steam is going to be noisy. The trail from bringing a constant supply of fuel will be easy to spot, even from a moderate distance. Long story short, you're not really going to hide this thing, even on a small scale.

Assuming you have all of that, then you should be good to go. Once you get things started, you get to keep it running, save for scheduled downtime to perform maintenance checks and inspections. It is imperative that you do this, because it is easier to schedule an outage than to deal with one.

Alternate Power Schemata... There are of course many other ways to provide power, ranging from a suitably rigged bicycle for small installations, to something powered by farm animals if you're so inclined (and have the necessary, um, horsepower to do it). The trick is to make it as simple, effective and as maintenance-free as possible. If you can turn an energy source into a rotating motion and give that motion enough speed and torque to turn a generator, you should have electricity.

In summary, there are a lot of ways to make electrical power. At this stage of the game, it can be useful for small comforts and conveniences (or critical community equipment) and in some aspect may well be worth the time, resources and effort put into generating it. Long-term, you will want to look into it as a means of rebuilding industrialization. The rest is going to be up to you.

How Running Water Works, Or Doesn't.

Plumbing is as old as the Roman Empire and provides one very important hallmark of civilization: running water. With running water, you get better overall health, lower incidence of disease, a better means of cleaning clothing (or people!) and a convenience that means less hauling around of water all of the time.

But, since collapse has pretty much left us without mass production, copper solder, plastic glue, virgin pipe, or other conveniences, we're going to have to make do what what's around. Fortunately, this isn't as big of a problem as you think.

Before we begin, we're going to want to do a little planning, starting with the water source. You will want your water source to be physically higher up than any pipe in your home. If you have a means to collect clean water from a place uphill and you have enough spare pipe to pull it off, this isn't too much of a problem. If you can set up a large clean drum onto a platform that is higher up than any point in your home (or close enough to it) and are able to get water into that drum without expending too much effort, you're also good to go. All other means are going to involve the same thing: using gravity to create enough water pressure so that you have running water in your home.

Most times, you shouldn't have too much trouble finding a water source that is at a physically higher elevation. The problem however is three-fold:

- Is the water source clean, constant and readily available?
- Is the water source close enough to be practical?
- Is the elevation difference large enough to create good pressure?

The first is obvious and primary. If the water source is polluted with waste, gone in summer, or controlled by someone who wants it worse than you do? You now have a problem with it and will probably have to look elsewhere.

The second isn't something most folks think of, but should… if it requires you to build a literal quarter-mile of pipeline or more to get, it may not be worth the effort you'd spend finding and assembling all that pipe. Don't get me wrong - you could still use it if you have buckets (assuming the first condition is met) and may well have to do so. However, the resources and effort to put something that large in place may be something that you would want to do as a community and not as a single family. Of course, if you already have the resources handy, then it will be far easier to do - just keep in mind the labor that will be required.

The third and last is something that can be worked around to an extent with a bit of engineering, but in cases where you have multiple sources of water uphill from you, use it as a means to figure out which one to go with. This doesn't mean building three water lines and choosing (though more than one water source would be a good thing), but it does mean that you should get the best bang for your buck, so to speak. If you have no options for water pressure (hey, it happens!), then there's no need to fret. If you have a running river nearby, there is a means to make your own water pressure using water-powered mechanical devices to pump water to a higher elevation. If you have a constant source of wind, you could get together with neighbors and build a large windmill - you use it the same way that the Dutch originally used it: as a big wind-powered pump.

Once you get that water to your home, you'll want to clean it and store a large amount of it. If you have a clean source of water, then filtering it should be relatively easy to do. This filtering can be as easy as using a perforated pan lined with cloth and full of sand. However, if you intend to drink or wash with any of this water, you'll want to do a couple things more. Alternating layers of (top to bottom) gravel, cloth, sand, cloth, charcoal, cloth and sand will clean the water enough to remove almost all of the chemicals and yuck that is commonly found in water. You will however still want to boil any water that you drink, since you want to kill any germs still lurking in there. Of course you could use bleach or a commercial filter, but honestly? Both methods eventually wear out or run out, so unless you can make chlorine from scratch (we show you how later on if you can get the right bits to do it) or build micro-porous filters out of nothing, you may as well get into the habit of boiling the water for now.

You'll want the local water storage for a few good reasons: First, you'll want a constant source of water pressure and second, you will always want the spare water hanging around, just in case. Another great reason is that you can collect multiple water sources into one spot. For instance, you can run your main water source to it and at the same time run the rain gutters on the house into it at the same time. This doesn't require coming up with a massive cistern, but it does mean that you will want to store at least 50-100 gallons of water nearby. As mentioned far too many pages ago, you can do this easily with a clean drum or barrel (or an old hot water tank, suitably modified), elevated above your indoor plumbing with a strong wooden pedestal. The bottom of the drum/barrel should be placed higher than at least one faucet in the home, to allow for complete drainage and to insure good pressure from at least that one faucet.

With a basic water system in place, let's talk about hot water. There are actually quite a few ways to get some of that hot bathing without the need for high tech. In the old days, most folks settled for boiling what they needed and mixing it in a bathtub half-full of cold water, adding the occasional bucket of boiling water to keep things warm (this is why the older bathtubs you see are far deeper than what you find in more modern homes today - because by the time that bath was finished and 2-3 people went through it in one go, there was a whole lot of water in there). You can step up from that (as mentioned many chapters ago) by piping water through a fireplace, usually with a re-purposed all-metal automotive radiator, or a large A/C condenser coil assembly. In hotter climates, a black-painted tank plumbed in-line and set out in a sunny area will give you some nice warm water and with very little effort. You can also re-purpose an existing hot-water heater, if you're not already using it elsewhere… but if it is electric, you will have to modify it a bit. To do that, first strip off the outer shell (the white metal casing) and all the insulation. Leave any electrical attachments in place (minus the wires) and you then have the inner tank which can be used as a boiler of sorts. Speaking of which, do not damage the safety valve or the pipe that runs from it and be sure to make sure that valve and piping points away from anything you don't want steam-blasted. Once you've stripped the tank, plumb it so that it lays sideways above a firepit, safety valve up, safety pipe running along the top if possible. Now all you have to do is light a long small fire under it, keep the fire going for around 45 minutes to an hour and you will have 20-50 gallons of nice hot water. Since you stripped the insulation off, it won't stay hot for too long, but it should last about 20-30 minutes, or just long enough to fill a bathtub. The safety valve at the "top" of the tank will kick out, providing relief if the internal pressure gets too high, so no worries there.

If you have a hot water tank that uses natural gas, you're not in as much luck as you think. You can still use it as it is, though you will have to remove all the gas/burner gear from the bottom and the fire will be smaller (which means a longer heating time). Or, you can strip it and use it the same way you would an electric model. As a bonus though, if you do that, you get a bunch of free exhaust ducting which you can put to good use elsewhere.

From there, you imagination can run wild, but keep a couple of things in mind:

- Remember that heating water in a tank or in closed pipe(s) equal a lot of pressure buildup - always have a means to either vent that pressure, or have a safety valve in place for those times when things get too hot. When indoor plumbing was new, a common hazard involved hot water heater explosions… always plumb in some sort of safety (even an open-topped pipe stretching up to the roof is sufficient). A system involving a radiator of some sort should also be similarly equipped.
- You'll still need pressure. If you're re-purposing a hot water heater in-place (that is, making do with one, but without moving it), keep an eye on gravity. You have to be able to get water into it and get water out of it. For this reason, a solar hot water heater on the

roof is not going to work (unless your water source is higher than the roof or a pump to move water there, naturally).
- Unless you're using a solar rig, you will literally be playing with fire. This means you need to have a fire-proof pit or place to burn and sufficient exhaust for the smoke and gases that any fire will create.
- Since you have no real way to regulate the temperature outside of guesswork (or a lucky find of a thermometer gauge), you will want to be sure that the resulting hot water doesn't scald anybody.

Now that we have all the inbound water sorted out, let's take a good look at the sewage, or outbound stuff. Most suburban and urban plumbing will drain sewage just fine… for awhile. But then, things eventually back-up, as sewage pipes fill up with no treatment plant at the other end to continually empty them. In these cases, you do have somewhat of a solution, though it will require a lot of shovel-work and some open space to work in. If you have a suitable area downhill from your home, you can re-route the existing sewage pipe downhill a bit and out to an open area and dig a leach field - a series of perforated pipes that branch off from the sewer line and drain into the surrounding dirt. Note that this will require a lot of pipe and you may likely not have that just lying around (though with a bit of skill, you can make clay pipe). Another option is to use a nearby wetland or swamp, which will act as a natural filter. Just note that if too many people get the same idea, that area is going to become a hotbed of disease and odor. Until you reach a stage where you can manufacture your own clay (or other) piping, you may want to keep to using an outhouse for your bodily functions (especially anything involving feces) and run the home's sewer pipe into an existing storm drain, or into a natural culvert or ditch.

To explain why, let's break things down a bit. You have two types of sewage - graywater and blackwater. Graywater is the stuff that runs down the drain after washing and blackwater is your urine, feces and other substances that you really won't want to be around when you've finished expelling them. Graywater you can dump out in most places, especially post-collapse, where you're (eventually) not likely to be continually dumping complex non-biodegradable chemical wastes such as solvents, cleaners and the like. You do want to take care to avoid dumping it into (or near) any flowing water, however, as substances like laundry soaps (even if homemade) contain chemicals that will eventually wreck the day of anyone downstream of you. You also want to avoid dumping in waterways because for most folks, that's their source of drinking water and poisoning it with your used bath water is a sure way of causing fights.

Blackwater is a bit different. Until you can properly dig a septic tank or leach field (or already have one of either), you'll want to put that where the surrounding dirt can act as a natural filter (hence the outhouse, where the underlying pit contains the stuff).

For those of you blessed with a septic tank or leach field, you have a different set of chores and rules to follow, but your initial preparation just got a whole lot easier. You can use the existing system, but you have to remember a few simple rules…

- Know that a septic tank relies on keeping bacteria alive which eventually breaks down the solid waste. Don't throw anything down the drain or toilet that can kill those germs… you need them to keep things alive down there. Avoid dumping anything down the drain stronger than laundry water. If you have to clean something with a stronger chemical, do it outside, or at least somewhere that the chemicals you're using don't find their way into a house drain.
- If you're using improvised toilet paper, don't flush it! Pre-collapse toilet paper is made to break down quickly and it dissolves in very short order. Newspapers and tissues break

down much more slowly, which can eventually clog the sewage system. You can always keep that stuff in a closed bucket and burn it later.
- Once in awhile (if you have a septic tank), open up the tank and check to see how much solid waste is in there versus liquid waste. If there is too much solid stuff, you may have to cut back on the solids going into it, add more liquids, or in extreme cases dig it out (be sure to devise some sort of protective clothing and don't reach in there with any cuts or abrasions on your skin!)
- While it is tempting to put bricks in the toilet tank in order to cut back on the amount of water you use, I wouldn't suggest it; septic systems require lots of fluids to keep everything nice and soaked. In drier/semi-arid areas, you can do it, but be gentle about it, cutting back only what you have to. In really dry areas, you may just want to use an outhouse, which uses no water at all.

Some more suggestions, no matter what kind of operable sewage you do or don't have? No problem. Keep the kids and animals away from the stuff at all costs. Post-collapse, almost any disease can kill quickly and you don't want them exposed to the germs unnecessarily. Take the time once every few years (or whenever you notice any slowdown in drainage) to get a drain 'snake' into the pipes and clear them out. Tree roots can slowly but eventually break through into most sewage pipe systems and start clogging things up.

Building a Good Road

There is an old Chinese saying: *"A new road is good for ten years, then bad for a thousand."* However, it does not have to be that way. There are Roman-built roads in Europe and the Middle-East, even today, that were built a couple thousand years ago and they still hold up to light modern pre-collapse traffic just fine. However, a good road takes a lot of planning, a surprisingly large amount of material and a whole lot of work. While previous sections dealt with engineering which can be done on a personal and family level, roads are firmly going to be done on the community/town level. However, it's not as bad as you may think.

First of all, you actually have a lot of advantages that the primitive pioneering types never had. You have established roads already in existence and only two years in, they're still going to be there. Certainly, some roads will have been blocked by natural or human-initiated actions (landslides, barriers/barricades and the like), but by and large the majority of roads will be in perfectly usable condition for quite a long time. So, well, why dedicate space and time to talking about them? Because this far post-collapse, you will have noticed a few things happening to the highways and byways… you will notice plants growing in the cracks and potholes will show up in older roads as winter freeze/thaw cycles take their toll without repair or mitigation by road crews. Bridges will start showing signs of corrosion - even worse than before society collapsed.

While nearly all automotive traffic will have pretty much dropped to a tiny trickle by now, you may notice an uptick in traffic that consists of improvised wagons, foot traffic and perhaps even horses. While you would think that automotive traffic would be hard on a road surface, in reality any kind of wheel that doesn't have a pneumatic tire mounted on it will be far harder. Why? Because the surface area which contacts the road is smaller, thus weight distribution is far less - it means more weight is concentrated into a smaller area. Sure, you can re-purpose existing automotive tires, but tires (and the means to keep them pressurized with air) will become more and more scarce as time goes on.

In the short-term, roads can be left as-is for now, with only the occasional weeding, clearing of debris/obstacles and patching (which we will cover shortly). Roads will, for the most part, hold up to existing traffic for quite awhile before needing anything major.

Bridges on the other hand will need more immediate attention. Corrosion of steel, breakdown of concrete and water erosion around the pilings/supports will cause an untended bridge to collapse in as little as a decade. While strong and massive bridges will hold up for quite a long time (up to 25 years or more) without constant maintenance, over time and without maintenance they become less stable and more unsafe. This is an especially large concern if your town is in an isolated area and needed supplies or trade requires a working bridge or two to keep things going.

Obviously, you're not going to have a lot of resources (if any) or manpower (let alone machinery) to keep things in top condition or to do proper repairs. Because of this, the first thing you need to do as a community is to identify which bridges, if any, that you really want to keep intact. On an individual level, this also means looking to see if there are any bridges going to your home from the nearest stable/friendly community. They're likely going to be smaller, but you will still want them to stay up and working.

If the bridge is too big (maintenance would require too many resources, manpower, is technically impossible due to lack of equipment or skill, etc), then you'd better start working on alternatives. These alternatives can range from simply building a road around it through the shallow depression it spans, to finding a shallow place where the spanned river or stream can be forded (then build a road to that spot), to building ferries that can span a waterway, to starting a whole new bridge with a simpler design and easier-to-obtain materials (e.g. wood). Otherwise, if you have a couple of bridges and think you can keep them up and going, let's look into exactly how you can do that…

The good news is, if aggressively maintained, metal and concrete bridges can be kept going for a very long time. Metal bridges require two things: wire brushes and lots of paint. Both of these can be made from local materials once the pre-manufactured supplies run out. Note that you will also have to occasionally re-make parts of it (depending on climate, weather and condition). You can improvise these repairs using the appropriate tools and existing/suitable steel parts - though it won't be easy to do and will take a bit of skill. Also note that your repairs won't bring the bridge back to full strength (though given the lack of heavy trucks and other cargo loads, this really isn't too much of a problem… yet).

But, we're getting ahead of ourselves here. You're likely still wondering how on Earth you're going to make paint for this thing. Fortunately, you can get that in Chapter 7, "Recipes", if you can get the right ingredients. Just note that you'll have to make a lot of it to get something big painted. Wire brushes, we trust, can be made with far less skill, but just in case, you can take any small steel cabling (bicycle cables, for instance), break it down into very short sections, embed one end of each section into a wooden handle, repeat numerous times, then fray out the threads on the free ends.

If you don't have the time or materials to make paint (and who can blame you?) You can use heavy applications of automotive or other type of waxes you may come across (got an abandoned auto parts store and the pillaging mobs took all the useful stuff? Odds are pretty good that heavy paste-consistency car wax will be still sitting there in large quantities. Help yourself.) Just note that using any kind of wax will require a more frequent maintenance schedule, since waxes tend to melt with heat, erode (slowly) from water and break down under solar UV rays.

If concrete is involved, things get a little more tricky. Most concrete bridges are actually a combination of concrete and reinforcing rods made of steel. To patch it, you will have to make cement, which you can mix with water, sand and gravel to make more concrete with. Making cement requires a bit of work, but it is not impossible (though it will feel that way). Use the concrete to patch the potholes and cracks as they arise, but only after you have tested a piece of the post-collapse concrete for solidity and strength. Forms can be made from wood into almost as many shapes as the skill of the carpenter who makes them.

For new roads, your best bet is to carefully make and level the new roadbed and make a gravel road. You do this by clearing out stumps or other debris, digging down past the topsoil and tamping down the subsoil with heavy, flat weights. You then insure the bed is level, then give it a coat of gravel at least 1-1/2" thick. Maintenance on this or any other gravel road is as simple filling in any holes, ruts, or erosion with packed subsoil, then applying another thin (1") layer of gravel atop the whole road. Note that without machinery, this will take a lot of manpower and a lot of work, so use existing roads as much as you can.

For new bridges, you will need quite a few things, more than what we can provide here. Hopefully you have a book on basic engineering in your library. However, there are a few things to keep in mind: Insure that each side of the gap you're wanting to cross has a solid foundation and that the bottom of the gap (riverbed, ditch bottom, etc) has solid points where you can drive in your pilings or support beams.

Steam Power

It seems that popular fiction has begun to embrace a little something called "Steampunk". The problem with many of these fictional stories is, most of the devices are made for looks and plot, not for actual use. So, if you had any ideas of creating any such fashion or society, you may as well not bother.

Real steam is dangerous, dirty, a high maintenance item and overall it's a whole lot of work. However, if you have access to a huge bounty of readily available fuel, sufficient skills to build steam systems and/or engines and enough material and tools to pull it off, you can create a valuable means to getting useful work done.

You can use steam to power a steam engine, a turbine for an electrical generator, or simply providing heat via radiators. We'll leave the actual mechanics to the library, but we do want to cover everything else surrounding its use. The hard part is figuring out whether or not you want to bother with building one. We've covered the caveats and the pros & cons with regard to generating electricity, but even for other things, you will want to take all of them into consideration and more. There are however instances where you may want to put it to use…

A good idea for steam is to use it the way your great-great grandfather knew - railroads. Now, there's no way you'll be able to credibly gather up enough material and know-how to build a full-blown 1890's steam engine, but with enough effort, you can build a small-scale version that can pull a couple of tons down existing railroad tracks. This could make transportation and freight hauling very profitable to the first survivor who pulls it off. You could even take a small steam engine/boiler set and use it to power boats, which will help coastal and river/lakeside communities regain mobility without being held hostage to currents or wind patterns (which in turn opens up commercial fishing, or at least provides a somewhat steady source of food from the sea/river/lake/etc). Another small but good use would be heating - with safety valves (scavenged off of hot water heaters - tweaked, then tested) and improvised radiators, it is possible. Steam can also be used 'open' or in a pressurized chamber, as a stationary means of disinfection and sterilization for nearly any non-porous object that will fit into the chamber you build and even as a signaling device if you know how. If you're adept (and have enough materials) to make a functioning steam engine, you have a mechanical power source to run equipment from (usually via tensioned belts). Anything beyond small-scale will likely need to wait until your locality (re-)gains enough skill and resources to tool up and actually cast/forge metal for anything large. The good news is, your library (and any existing/surviving books) will have most of what you need to know already figured out - you just have to do the work and get the bits you will need to make it.

While I went through a lot of trouble to caution you against taking this route, long-term you will definitely want to at least consider it. Until large-scale industrialization comes back into operation, steam represents your best bet to get civilization back on its feet again. I say this because the technology is relatively low (compared to pre-collapse standards), fuel can be readily available in most areas and unlike water power, it can provide a source of powerful mechanical energy almost anywhere you want it to be. Just be sure to approach it carefully. In the early days of steam power, explosions were not only a real threat, but were common and often powerful. However, with a bit of planning, an eye for safety and a bit of intelligence in design? You can use it to be the literal engine to reconstruct civilization.

Entertainment And Diversions

This sounds pretty odd, but you will be surprised at two things here: One, that you might actually find time to enjoy them and two, just how badly the human psyche (and a community) needs a bit of entertainment now and then. From collapse to now, you've gone from scrambling to stay alive, to scrambling for critical necessities, to working your butt off in implementing those basic necessities, to working on sustainable sources for the things you need to stay alive. Very little time has popped up in which to kick back and relax. Certainly, there were stolen moments that came up where you were idle (especially in winter), but they were likely fleeting and few.

By now it is likely that you have a going lifestyle of sorts and have even managed to put in a couple of relative luxuries (never thought you would miss a hot bath that much, did you?) So why not get together and formalize some time to goof off? I'm not saying that you need to cast aside all worries and cares from here on in - life is still going to be quite hard and probably still full of nasty surprises. What I am saying is that once things get somewhat stable, set aside a bit of time on a regular basis to sit back and let your body relax. Share a laugh with some friends. Do something stupid (in a good way) once in awhile. Celebrate the good parts of what once was, even if you're still trying to figure out exactly what all got lost when civilization did. Set aside some time to actually grieve and to work through the grief, because you will probably still have good reason to. Take some time to celebrate the religious moments that pre-collapse had become superfluous and over-commercialized. Overall, the human mind really needs to take time to relax a little, to sing, to dance, to respect, to celebrate and to focus on nothing in particular, even if only for an occasional day.

Community Events

Every individual and community has things that pretty much needs everyone coming together in order to accomplish. In more classical/ancient times, this meant dating/courtship, harvest and inventory and remembrance/reminding of why everyone came together in the first place. Pre-collapse society had done away with a lot of it, since harvests were individual, mechanical, or for-hire - holidays were usually rare occasions where one could be with the spouse and kids and dating/courtship usually happened in school, in dance clubs, or on the Internet, depending on age.

Well, collapse has pretty much removed all of that. Young (and older single) adults still have the urges, but now there's no formal outlet or form in which to search for a mate, at least not in a way that is socially polite (and without coming across as really weird). Without machinery, both planting and harvests are going to require manpower and lots of it. Without a literal million pre-collapse diversions, religious and secular holidays alike are going to regain the special meanings they once had - instead of just being a day off or an excuse to buy presents. Most people who survived collapse this far are going to get back in touch with what it means to be alive. Once they have enough time to stop for a minute or two, they will regain a passion for celebrating life and the events in it and do so in any way they can.

So what does this have to do with the town at large? Well, community-based events are also events in which you can establish and keep cohesion, loyalty and even brotherhood between neighbors. They give everyone a sense of identity and of community. They go a very long way towards helping everyone realize that we're all in this together and can create and foster friendships that bond neighbors tightly together.

A word about doing this as an official or formal affair: To avoid leaving someone (or a group) out of things, it will help to keep these events secular (that is, non-religious) in nature, or at least non-denominational. If you want events that have some sort of ethnic angle, then go out of your way to declare everyone an honorary member of whatever ethnic group is being celebrated - even if only for that day. If you're going to pick a subject though, it is best to make it a subject that everyone can relate to (or at least understand) and that provides a reason for everyone to come and celebrate it.

Your best bet for an official community event is to keep it as neutral as possible throughout and center it around an activities that everyone can relate to, enjoy, or at least be inclined to watch. Be certain that it is open to all citizens and/or residents, that it doesn't cost anything (but donations are more than happily accepted when needed) and encourage everyone to participate as much as they can. Oh and be sure everyone knows when it is going to happen well in advance, so that preparations can be made by all interested parties. About 2-3 weeks (at most) should be plenty of notice.

It sounds too simple, doesn't it? It's not, but the results are still worth it. To best illustrate why, let me lay out an example and walk you through it:

Survivor Day!

By way of example, our town is going to establish "Survivor Day". It's purpose is simple, but quite real: to remember those lost and to celebrate those still alive. We want to set aside one day for it each year, just after most crops are harvested so everyone has time to visit. Here's the events we plan to host *(with explanations as to why and how in parentheses)*:

Opening Ceremonies : Speech by mayor/leader welcoming everyone and explaining the purpose of this day. *(Should be held at mid-morning or so - noon at the latest. Church bells work great for calling folks in. Give it about 30-45 minutes for everyone to get there and get settled. The Mayor should give a welcome, a solemn speech of about 20 minutes, tops, followed by a quick explanation of what the rest of the day holds. An opening prayer wouldn't be a bad idea either.)*

New Introductions: For the very first celebration, each family head introduces him/herself briefly to the crowd - their names (of self and family members), where they're originally from, what they can do and a short story of how they got here. Each subsequent year will only have any new arrivals doing this. The crowd is encouraged to shout "Welcome Home!" after each introduction (*This gives everyone a name and is no longer just another anonymous face. It implants a sense of community and the encouraged shouts are a statement of no uncertain terms that each resident is indeed home now - a major psychological boost. It also allows everyone to let everyone else know who they are. Each story should be five minutes or less, but thanks to the fact that most folks are at least somewhat averse to speaking to crowds, this should be easy to do. Note that initially, this will take quite awhile. If there are no new arrivals in a given year, replace this part with something appropriate - say, 20 minutes for each neighbor to greet the folks around them.)*

 Small contests: *(if not yet noon)* Host a series of small but silly contests with a handful of contestants each - the contests should apply to daily life, for small but useful prizes. *(For example? A spouse-calling contest with a small handful of contestants, a 'loudest child' contest, things of that nature.)*

Break for Lunch: (*Do this at noon or so. After all, folks gotta eat, no?*)

Talent Show: A free-for-all talent show, again, for small-but-useful prizes, or perhaps ribbons and token prizes. (*You should hopefully have a stage for this. Have interested parties sign up in advance and judges should not participate or be related in any way to the participants. Prizes should be kept small and/or token and only for bragging rights. That will prevent any funny business from getting in the way of a day set aside for community. Participants can bring their own instruments and/or props.*)

Dunk/Shave The Mayor: Contestants provide food, money, or small usable items in exchange for a chance to knock the Mayor (and town council) into a dunk tank, or shave his head bald. Proceeds are to go to charity. (*This actually serves a lot of good purposes here. One, it teaches the leadership humility, which is always a good quality to have when you're leading the public. Second, a dunk tank is laughably easy to build if you have enough water and materials around and lacking that, shaving a head isn't very skill-intensive. Third, it quietly urges everyone to get involved and provides entertainment. Finally and most importantly, it provides the beginnings of charity, as all donations will be given to those most in need - any perishable food can be put to use later on today to feed everyone. Other alternatives can include auctioning off dances with the Mayor while he's wearing a dress, or other silly bits like that. The idea is to combine a bit of self-efficacy and charity all in one entertaining package.*)

Idle Time: Spend a couple of hours before dinner milling around meeting and greeting. Town leadership is on hand to answer all questions and take all suggestions. (*In fact, the town leadership should actively walk around, seeking out comments and suggestions during this time! It also gives volunteers time to cook and assemble the potluck dinner.*) As an alternative, host workshops by volunteers on how to do certain things that everyone can find useful.

Feast: Everyone gets to eat the potluck dinner, coupled with any food donated earlier. (*It's as old as mankind, really… the group that eats together, sticks together. It also, even if temporarily, reinforces the idea that being together brings prosperity.*)

(optional, time permitting) Open Mic: Have an open stage for anyone to stand on and expound on their dreams, their hopes, or whatever comes to mind. The leadership should start off, then invite everyone else to come up one at a time. (*Sounds silly, doesn't it? It's actually not. It's a chance to put out great ideas and to gather great ideas. It also gives everyone a chance to speak their mind openly and maybe plant a few seeds in others' minds.*)

Talent Show: Hold a small contest - encourage people to sing, play a musical instrument, read poetry, tell jokes, whatever. *Let the audience decide (by applause) who wins. (Again, a chance to show off. It also provides entertainment without all the work and expense of doing it yourself.)*

Dance: Clear a huge space, get a bunch of at least somewhat-talented musical folks onto the stage and open the dance floor. (*This is actually a no-brainer. Dancing gives folks a chance to work off some energy in a friendly way and gives the single folks a perfectly social excuse to get close with someone they may be interested in.*)

Closing Ceremonies: The Mayor gets up, gets everyone calmed down for a minute. He then summarizes the past year, lends a few good words towards those who have passed away and perhaps even raises a toast to them. The closing speech should note what has passed, but should end with a recounting of the progress made and with hopes for what is to come. A (non-denominational) closing prayer is optional, but recommended. If by some miracle you happen to have or have made fireworks, let 'em fly. *(It is always good to end the day with good things and to remind everyone that we're all in it together. A good closing speech will do that and will reinforce the sense of togetherness that you'll want to foster for your town.)*

Now note that this event was just an example and can definitely be modified to taste and theme. However, the important parts were all in the commentary and the fact that you want to lift spirits up and to bring people together. After all, people that have eaten, danced and played with one another …are less likely to strangle one another.

Resurrecting Old Holidays

You would think that all holidays would be eagerly re-created, no? After all, any excuse for a day off is a good one. Thing is, unlike pre-collapse society where taking lots of breaks is no big deal, folks are going to be a bit too busy trying to grow food, defend themselves and provide for their families in general to be taking every other day off.

There is also the fact that when civilization collapses, a lot of the rationales for celebrating collapsed with it. For instance, in just the US calendar, there are plenty of holidays which tend to lose meaning once the government behind them goes under… Labor Day, Presidents' Day, Veterans' Day and even Independence Day may lose all meaning as time rolls on. Even now, how many people do you suppose actually remember and celebrate the reasons behind these days? For most people, it's just a day or two off with no thought to the meaning, much like Christmas morning to a Jewish or Muslim family. There may be a passing thought, but how much really goes into them?

And then we have the purely religious holidays: Easter, Ramadan, Eid, Passover, Hanukkah, Samhain and even very recent ones such as Kwanzaa. Naturally the adherents to various religions will want their particular days celebrated, but how many of them need to be an official event for everyone in town?

The reason I'm bringing this up has nothing at all to do with wanting you to discard all holidays or to change your outlook on keeping them. What I do want is for you to think about the pre-collapse holidays carefully and consider what you wish to keep and what you wish to quietly let folks celebrate on their own. Mind you, this is nothing new, either. The ancient Roman calendar was packed tight with holidays for their various gods and events, as was the calendar for ancient Egypt, ancient Greece, ancient Nearly-Anybody. In spite of that, few people if any celebrate Saturnalia or the annual Nile flood today. Over time, holidays and festivals evolve, disappear with the civilization that celebrated them, are quietly forgotten, or simply get co-opted (at least one Christian holiday, Christmas, was timed to coincide with an existing pagan holiday in order to co-opt it). Even with an existing culture that has existed for millennia, most holidays are quietly forgotten or ignored (example? A read through the Old Testament of the Bible shows that ancient Israel's calendar was packed to the rafters with holidays and mandatory celebrations. After all, how often do you figure the Festival of Booths gets celebrated nation-wide in the US? Nowadays, it's down to a handful of ancient-rooted holidays at best and those are usually celebrated quietly among like-minded friends and family.)

In spite of all of this, holidays are indeed still important. You really do want to mark milestones and remembrances. Even if the original holiday has lost all meaning, you can co-opt it to mean something only slightly different. For example, the US holiday of Thanksgiving is a purely secular celebration of early colonial/settler life - and of one settlement in particular. Why not convert it to give thanks towards surviving the end of civilization and at the same time celebrate a good harvest? But before we get all creative, let's break these things down a bit...

Religious Holidays

As a community, you will want to keep the list of religious holidays short. Let the people be your guide. Many holidays (especially religious ones) will regain their original meanings and in a town full of mainstream Christians, ditching Easter and Christmas would be a quick way of making everyone angry (or at least it would completely kill morale). On the other hand, there's no need to get extreme. An extreme example? All Saints' Day, Ash Wednesday, The Assumption of Mary, Ascension, Corpus Christi, Palm Sunday, Triduum (from Good Friday to Easter Sunday), Advent Sundays, Lenten Fridays and all individual saints' days off (all 500+ of them) and marking them as official events? That would kill your calendar, your budget and would be pointless to those in the population who aren't Roman Catholic.

Seriously, keeping that example? Even Catholics don't do that, because we wouldn't get anything done if we did. We just quietly remember it if it isn't major, we do a short religious celebration on the evenings of the important weekday-timed ones and incorporate the rest into the existing Sunday masses. A good example outside of Christianity? No problem: the month-long Islamic celebration of Ramadan is built so that you still get all your work done during the day (while fasting) and do your celebrating/feasting at night, on your off time.

Another reason to keep the holidays short is for exclusionary reasons. Even if 99% of your town is Christian and you keep the holidays generic, the 1% who aren't are not going to be celebrating them and for obvious reasons (Jews don't recognize Jesus Christ as their savior and Muslims only recognize Jesus as a prophet at best, but only as a secondary/prelude to Mohammed. Atheists and other non-religious types will just think the whole thing is silly.)

All that said, the old system of officially recognizing Christmas and Easter (in typically predominantly Christian populations) isn't a bad one. If you're going to have to choose between the two, go for Easter - it will have the most applicable meaning, as Christmas has become far too commercialized.

You may find to your surprise that other religious groups will want their days included as well. If the religion in particular is the majority, you may well want to do that. Otherwise, go with one or two at the most and otherwise leave it be. Unless your leadership has somehow become a religious/clerical-led government (it had better not be), it's best to keep your government from becoming too entangled with any religion - even your own.

National Holidays

This one is going to be touchy. Not because suddenly all of your US-born survivors are no longer American citizens in the technical sense, but because some folks will want to forget the whole thing, while others will want every scrap of it remembered and perhaps add a few more just in case.

Your job as the leadership is to sit down and figure out what parts you do want to keep. Celebrating, say, the birth of a now-dead government may make no sense, but how exactly do you and your people feel about someday rebuilding the same government (though perhaps somehow better-run than the old)? If you want to resurrect the US government someday (or see it resurrected), you may want to remember the most important points in its history and keep them in the public consciousness. Doing so keeps hope alive, while at the same time remembering the best concepts and ideals of the old. What you may want to do regardless is to keep at least one or two of the most important ones around. A small suggestion would be Independence Day (though this time in the hopes that the ideals and Constitution will be resurrected in a better-operated fashion) and Memorial Day. The rest will be up to you.

Other Secular Holidays

There will of course be all those other holidays that don't tie to a religious or patriotic theme. There may even be some regional holidays that are popular enough to warrant looking into (for example, Utah's *Pioneer Day*, held every July 25th. The whole state practically shuts down, there are fireworks, parades, etc). If you have a large Mexican population, Cinco de Mayo (Mexico's independence day) may come up. Other holidays that may crop up would be Thanksgiving Day, Labor Day and the like. In all honesty, you can examine most of these and simply ignore them on an official basis, or co-opt a few of them to fit your current situation, hops, dreams and visions. Overall, just remember the basics. Time and demands won't let you celebrate too awful much, but at the same time you will want to be sure and at least provide some officially-sanctioned and officially led means for folks to relax, remember, celebrate... and come together as a community.

Education For Young And Old

Just because we've all managed to live this long and have managed to live a far simpler lifestyle does not mean that we can stop bothering with education. I daresay that education will become more important than ever, post-collapse.

What's Happened Until Now

Once things began to crash, each individual's education was stopped, then was immediately re-directed towards the skills and knowledge needed to survive. For the children, this means a whole lot of vital lessons have not yet been taught over the past couple of years, or were taught haphazardly at best. For the older, this means a whole lot of now-useless information is knocking around in your heads doing nothing in particular (e.g. intellectual property law, .NET programming, Human Resources management, Marketing, business management, etc.)

The things you have learned over this time have been taught rather harshly - with little instruction, very little contemplation and even less tolerance for mistakes or ignorance. The price for failing to learn most of these lessons has often been death. By now, there are only three types of adults who are still breathing: The astronomically lucky, the highly intelligent and the very well-prepared - two types will have learned these lessons well enough to remain alive and the third will need to start learning real fast.

The lucky will find their odds (and numbers) decreasing as the law of averages kicks in, so they are the ones who will need to learn quickly. The intelligent will have to gain practical skills in order to put their insights into practice, as scavenging existing technology becomes less fruitful with each passing day. The well-prepared will still have to learn and work towards self-sufficiency into the future. Odds are good that every still-living adult possesses all three of these aspects (luck, intelligence and skill gained through preparation) in varying proportions. The trick now as a community of people is to spread the knowledge and skills and help your neighbors and children make their own luck as time goes on, since luck doesn't self-generate - at least not for very long.

Life Lessons, Post-Collapse

It is going to be quite obvious by now that the lessons you needed to live life pre-collapse aren't exactly going to match up with post-collapse realities. Among the changes? Let's enumerate them:

- Living semi-isolated in some nameless apartment or suburban home is no longer an option. From now on, you are dependent not on impersonal delivery and machinery, but on other people to stay alive and thrive; people you will meet and come to know face-to-face.

- You can no longer afford to let a good opportunity pass you by. The consequences are no longer maybe a career or love-life issue, but may well be life or death itself.

- Your wealth isn't going to be measured in dollars, home size, or make/model of automobile… it will be measured in food, warmth, children, raw materials and skills.

- Speaking of children? Your retirement plan is no longer a number or acronym and won't come with an advisor. Your retirement plan will be your kids and grand-kids - be they adopted or birthed. This means that if you can feed and clothe and care for the little guy (or girl), taking in a young orphan child or two is actually going to be a very smart investment towards your own future.

- Entertainment will be surprisingly easier and less complex than you think. Things that were once ignored as too simplistic, niche or childish will now find you taking delight in them. Examples include watching a sunset, playing made-up games, watching crops grow, dancing to homemade music and similar things.

- If it clearly belongs to a living person, then you don't take it without permission. The consequences are far greater now if you do.

- There is no slacking off or 'working the system' to live leisurely. You either work, or you starve. The quality of your work directly determines the quality of your life.

- Trust is now earned, not granted. The consequences for failing to sufficiently investigate before trusting a stranger can be fatal post-collapse.

- It is paramount to start new relationships with neighbors in a friendly tone and manner and to be polite as long as there is no immediate danger in doing so. Minding your manners is more than just a social skill now.

- No one is a victim, or should be afforded special treatment - at least not without an obvious severe physical or mental handicap.

- Troublemaking merely for the desire or love of drama/excitement is a good way to get beaten, expelled, or killed.

There are of course many more, but you get the idea. What you need to do though is to start teaching them to those who don't quite know them yet - specifically, the children. Surviving adults will either already know by now, or will figure it out quickly (either from a friendly whisper or to the sound of gunfire). Kids however will need to be taught.

Basic Curriculum and Why

This simply means a listing (and rationale) of what will be taught and at what stages of life. It will be radically different from pre-collapse education, simply because of two things: The skillset for success is going to be different and you'll have to accommodate the fact that children are now going to be put to work again.

What hasn't changed (too much) is the ages at which skills can be taught and processed by the child. It will however be a bit more compressed than usual and for a few good reasons. First, the sooner a child acquires a given skill, the sooner the child can put it to use in keeping both family and community alive. Even something as simple as potty training means that the sooner a toddler can crap in the outhouse on his own, the sooner his mom can stop having to wash soiled diapers, leaving her time open to other things. On more complex levels, the sooner a child learns to assist in farming, the faster the plowing/planting/weeding/harvest can get done. As soon as a teenaged boy can use a weapon decently and gain some combat/hunting tactics, the community gains one more defender and the family gets one more person who can put meat on the table.

This also extends beyond basic physical skills. On a cognitive level, children grow quite fast. Most children can read simple words by the age of 4 and write them by 5. They can perform basic mathematics by the age of 6. The only things really holding them back are the availability of instructions, the instructors and the amount of time one can spend on teaching. Believe it or not, a properly-educated 12-year-old child in 1880 could speak and read at least one additional language (usually two - they were almost always Latin and French) and had the relative knowledge and skillset of a modern (pre-collapse) high school graduate. The reason you don't see as much of this in modern pre-collapse schools has more to do with averages than with ability. Mass schooling is required to progress at a level slightly below the average cognitive growth of a child, so as to allow slower children to catch up (thereby keeping overall grades and scores higher as a bonus). Add to this the additional time burdens of teaching children "self actualization" and socialization skills? One wonders how a pre-collapse high school graduate can even solve a simple algebraic problem without assistance. However, the initial reason for the slow-down is because schooling went from being voluntary (where most parents didn't send their kids to school, or at most not beyond the very basics), to being mandatory. This meant that children who were mentally or intellectually deficient went from being fated to menial or agricultural labor, to being forced through the system.

One of the (very!) few benefits of a post-collapse world is this: A lot of the cruft and mental experimentation from (pre-collapse) public school systems can be cast aside and you can focus on what a child needs to *know*, not what he or she needs to *feel*. You can also make education beyond the very basics an option, reserved for those parents and educators who together feel that a given child is intelligent and sharp enough to handle the more advanced subjects.

So, post-collapse, what are the very basics? Believe it or not, they're the same as they were pre-collapse, with one twist:

- Literacy/Grammar (reading and writing skills)
- Basic Math (add/subtract, multiply/divide, percentages/fractions, basic algebra/geometry)
- Basic sciences
- Civics (legal system, citizenship, rights and responsibilities, etc)
- History
- Survival skills (foraging, combat, farming, etc)

The literacy/grammar skills should be enough to (at completion) allow a student to read, write and understand something as complex as, for instance, this book that you're reading. Math skills are pretty much listed and should get as far as figuring out the very basics of algebra and geometry. The basic sciences are to cover what an average adult should know about weather, biology, chemistry, physics and etc… most of the focus will be as it relates to the needs of a more agricultural society than a technological one. However, civics will be a vital component of rebuilding civilization. While not an indoctrination, it is a way to teach children what is expected of them as a citizen, how your government works, what the laws are, how politics works and how to successfully keep things humming and growing. History is taught (in spite of it being reset) for two reasons: First, it binds everyone together, because it is the story of us. Second, it helps prevent repeating the bad bits of it.

Something to keep in mind, however, especially when you're scrounging for textbooks. Oftentimes, the subjects go by different names. Here's what to look for: Literacy and Grammar textbooks often go by subject names such as "Language Arts", "English", or even "phonetics". Math and Science textbooks are pretty self-explanatory, but are often divided by specific subjects. Civics and History are often rolled together in classrooms and called "Social Studies" - you can adjust your curriculum to do the same if needed.

The 'twist', or last item mentioned, is vital due to your current situation. Teaching your children how to handle themselves through combat tactics (individual and group) is vital towards helping your community defend itself, as well as for self-defense. It also involves what is commonly known as 'bushcraft' - that is, self-navigating, improvising shelter, hunting, foraging for food, making weapons and other vital skills. It should be tuned to your particular region.

Pre Schooling

Let's begin with the little ones. As soon as the child can begin to vocalize and move under his/her own power, the parents –or better, grandparents if possible– should begin teaching the child basic dexterity, self-feeding and speaking skills. As soon as the child is able to walk on his/her own, teaching personal skills such as potty training, self-dressing and (some) self-grooming is a must. Be patient with the child, as it may take some children longer than others to get the hang of these basic bits. Gentle encouragement and positive reinforcement will get you faster and better results than anger or frustration will.

Once the child has all of this down (roughly the age of three, sometimes four), begin to introduce chores to the child. Obviously the little guy isn't going to be able to wash dishes or do laundry just yet, but simple things like picking/straightening up, making beds and other minor tasks are quite useful as well as helpful. Useful post-collapse-specific skills involve having the child 'keep lookout' and quietly come up to the nearest parent to announce anyone who comes towards your home. Otherwise, keep them with you as often as possible while they are toddlers, patiently (and repeatedly) explaining everything you do as you do it. Try to keep as much regularity in your daily routine as you can, so that the child can be at once comfortable with the routines and at the same time start learning by watching. Periodically have the child 'help' by performing small tasks and/or steps in the chores that you do. Start infrequently and with small easy-to-do steps and over time involve them more and more.

Be sure to have a regular 'play time' set aside for the child - both time that he or she can spend alone with a few toys and time with other similarly-aged children if possible. Do this as early as you can and do it on a regular basis. The alone-time allows the child to digest everything he or she has learned (and to simply be a kid) and the group play-time allows for early socialization. As a bonus, you can rotate out the 'play date' portions so that one parent (or grandparent, or teenager) watches the group of kids one day and another does the supervision the next day. This in turn gives all the other parents a small break and possibly some alone time.

At least once during the group play-times lead the children in a song, a dance, games, tell a story, or do some activity that requires the children to cooperate together in a playful or quiet manner.

At this stage, keep the children under constant supervision. This is because of the need to gather the kids together in case of trouble and because kids still tend to get into things (or do things) they really shouldn't. In a post-collapse world, this is doubly true and the chances of getting a fatal disease or running into a fatal accident is far higher.

Once the child reaches the age of five, start teaching him or her to read and write. Start with the basics (alphabet, letters, numbers, etc). As an alternative, at this age you can start sending the child to school if he or she is bright enough (let the teacher make that determination, however).

What About The Slow Kids?

Not every child is going to be a genius. Post-collapse? Malnutrition, stress and sub-optimal pregnancy and/or birthing environments will almost guarantee that a disturbing number of children will come out less than perfect. Human beings are amazing in their ability to thrive under some rather unreal environments, but not all will be able to cope, especially in a world where pre-natal care and proper, balanced nutrition is non-existent.

There will certainly be developmental problems in the occasional child, but this doesn't mean the end of the world for that child. If the defect is physical, then focus on developing the child's mind and vice-versa if the defect is mental. It isn't always easy to tell at first, but there are clues that can point you to the right path.

If the problem is only slight, you should get with the teacher as schooling age draws near and discuss the options at hand. The teacher may be able to determine if he/she has the time and skills to help the child along in spite of the slowness and will help determine at what age the education should stop in some subjects, but continue to progress in others. Your best bet is to keep the child's interests in mind, but consider that pushing a slower child will only lead to excess frustration and resentment. The opposite is also true - willfully holding back a child who is capable of keeping up is a disservice to the child. Therefore, always try if you can and if it requires extra tutoring (and there is enough time at home to do it), then by all means do so.

Those children who are obviously unable to learn due to mental retardation or similar should still be given a chance to be a useful and productive member of society, even if it means teaching a limited set of menial but useful skills. In all cases, strive to teach the child enough so that he or she can live independently someday.

The Geniuses Among You

Sometimes you will find an exceptional child - one that excels beyond all expectations and can definitely learn at a far faster pace. In this case, work with the teacher to see about accelerating the child's learning. Early on, you can do this without undue burden on the teacher by tutoring the child at home at a level ahead of the child's peers. At later stages, the teacher will have to assign and monitor progress for the older child.

If you indeed have children of this caliber, challenge them to do more, up to the point where it appears that they cannot learn any faster. However, always be sure to check the child's retention of knowledge by frequently 'testing-back'. To do this, test the child on subjects that he or she had completed mastery of in previous months.

The teacher can ease a bit of his or her own burdens by setting aside some time during the school day for the child to explore subjects that aren't in the standard curriculum. You can also have the child study further within a given subject with other textbooks that are a bit more advanced than the child's age level, to tinker with concepts and physical models (if the child has a knack for engineering), or to come up with problems (then solve them) above and beyond the level of a given subject in class. Finally, in later years, such a prodigy should be given real-world problems to solve, up to and including problems which you as a community may face.

Apprenticeships

Post-collapse, as lost and/or ancient skills are rediscovered, it would be a good thing for students of a non-academic nature to take up apprenticeships with working tradesmen in various fields. For instance? Gunsmiths, blacksmiths, millers, engineers (of various disciplines) and woodworkers/carpenters will gain in demand as the population grows, so training students directly in these fields would insure that skills rediscovered do not become lost again.

This will require a bit of work with the person teaching the trade, otherwise known in old-school parlance as a Master. A curriculum of skills will need to be drawn up, a period of time in which the apprentice works in the program before becoming certified will need to be agreed on and all conditions and terms should be worked out in advance, as much as possible. The certification authority should remain with the school itself (to prevent the forming of guilds), but the acceptance, number of apprentices and final determination of fitness for the trade should be made by each master of the given trade.

College?

This should only be approached if and only if, your town has grown large enough to become self-sustaining and is at peace enough with all neighboring communities in the vicinity that a sufficient population of students can be gathered. There should be about 15-20 students to begin with and should be able to devote time and energy towards studies, without the distractions of having to stop and fend off raiders, plow the fields and such. The students should be supported by the parents and the parents should have enough to not only care for the students, but to afford a stipend of goods or money for the faculty as tuition.

The rest will need to be worked out by the faculty of such an organization, but it should include, at the very least, an undergraduate program of instruction that includes a variety of advanced subjects. Doing so creates a well-rounded person and exposes them to potential fields of further study later on. This curriculum should contain an appreciation for the arts, sciences (chemistry, physics, biology, astronomy, basic electronics), some medicine (further medicine can be taught as a specific degree later on), proficiency in at least one foreign language, history (both pre- and post-collapse), mathematics and advanced skills in the English language.

Of course, if an actual expert can be found in certain fields (medicine, physics, chemistry, etc), then by all means offer those classes for advanced degrees.

Continuing Education, Workshops

As most adults know, learning never stops. This is especially true as new skills are learned post-collapse and as old, long-dead skills are brought back to life. Do not hesitate to offer classes, workshops and lessons (for a fee if you wish) to pass these skills around. This can be done on an ad-hoc basis, or done on a formal schedule.

Boot Camp

As your town grows, you likely won't quite have enough of a population (or tax base) to support full-time soldiers, but you can still set aside time to teach and drill in the art of warfare. Veterans abound and will likely be found in greater numbers post-collapse. Take a few of the most proficient ones and hire them full-time as commanders and officers. In turn, they can teach and drill the male (and female if you wish) citizenry in the art of combat. Courses of instruction can include tactics, strategies, weapons usage and maintenance, general drills, leadership, hand-to-hand combat skills and physical conditioning. It is especially recommended that you do this even in times of peace, since you will never know what may come over the horizon.

Expansion And Confederation

So, things have (finally!) Become peaceful? You and your immediate neighboring communities are friendly with each other? Good! Now let's figure out how to expand our influence, gain new trade and in general become stronger together.

Until now, alliances and treaties were usually born of convenience. You made friends and partnered up with other communities either because they provided combat strength, or because they had needed (or even critical!) things that you could trade for. Now that you have all the basics and maybe a few luxuries, it's time to reach outwards and to make friends even further afield.

So, well, why would you want to do this? The answers are as plentiful as they are simple. You will want a larger buffer of peaceful countryside between you and any potential armies that may come your way. You (and your neighbors) will want a wider variety of opportunities for trade goods and resources. The larger area means more people, which in turn means more shared knowledge that can be passed around. You can join with other communities and towns to share the load of highly complex functions, such as advanced schools. A larger base of friendly towns allows some individual towns to specialize in a given area, such as manufacturing. The grand total of available skills becomes larger. There's more people with whom your kids can grow up, meet and marry. Greater numbers means a greater potential army if the need arises for a common defense. Shortages in one community can be offset by overages in another. Finally, having more friends is a very good thing.

Looking Inward

Before you do anything, look at what you as a community have now, in total.

What resources do you need as a town, but find yourselves consistently low or short on? These resources are going to heavily influence which direction you want to start reaching outward, because those things are going to be very high on your citizens' collective list of things to trade for.

What resources do you all have in quantity (or better, in excess) that others out and beyond would be most willing to trade for? For example, let's say that you're literally sitting on a coal mine, in a plains region where trees just aren't that plentiful. The neighbors to the east live in an area of heavy timber where mountains begin to rise, while your neighbors to the west live in an even more treeless plains area. The folks to the west are going to be more willing to work with you, ally with you and most importantly, trade with you. After all, you have something they need (specifically in this case? Fuel).

How able are you to marshal an effective armed force? The last part isn't for conquest, but just in case you do reach out and find something very bad - or worse, something very rapacious that didn't know you existed until now. If you accidentally open a can of worms (due to diplomatic screw-ups, coming across a powerful community full of gangs, or worse), you will need to be able to fend off any approaching forces.

Next, look around your local area of influence. Who do your neighboring towns ally with and who do they avoid or fight with? Are there any points of mistrust or hostility between any two given communities in your local area? You will need to either neutralize, calm, or solve any conflicts within your local group of communities before you reach outward, lest a chronic-but-minor problem in the local area turns into something larger and uglier later on.

Finally, what kind of long-range transportation do you have on hand, or can build? If you live near a good-sized river, near a large lake or on a coastline, shipping becomes a real possibility. If you have a lot of horses, they can be put to use. If some enterprising soul is brewing ethanol or biofuel (or, say, has converted automobiles to use natural gas from a nearby well), you have a somewhat reliable means to move people and haul goods. Even on the extreme, a refurbished steam locomotive from a museum can be put to use on existing rails. You are only limited by two things in this regard: existing resources and your own creativity. Thing is, you do need to have a reliable means of transportation and in quantity. If you're going to start reaching out and increasing your influence, you will need that transportation for both trade and for potential troop/military movement.

Looking Outward

Next, look out beyond your local sphere of influence. What do long-range scouting parties and trading groups report, in all directions? Is there any particular direction that leads to potential trouble, or potential wealth? This will influence which direction you will want to start expanding.

What does the terrain look like around you? Hard-to-travel terrain (thick forest, mountains, rivers, etc) means longer travel times and more places for ambush parties to hide. Burned-out cities or lawless areas are likely to still be nests of crime and raiders and going through (or by) them will mean trouble.

Take a few exploratory trips out to distant towns in the areas you think may be profitable (both in goodwill and in resources). At first, just go out towards the closest town in the direction you want to go in two small(-ish) parties - do it as if you're out trading, one party roughly a mile or so behind the other, keeping a distant but discreet eye on each other. This helps test out the route to see how safe it is. While in this town, keep out a good eye and take lots of notes, as you did in closer towns, then come back the same way you came out.

This time, weigh the pros and cons of forming an alliance with the town and if possible, take a few exploratory trips outwards in other directions as well (so long as those directions are safe enough to do so).

As always, be cautious and check things out thoroughly before committing. For now, all we're doing is expanding our trade, our influence and the buffer between our town and the untamed masses beyond. However, as time goes by and there are enough towns working well together, the subject of combining governments is certain to come up. So before someone else decides it for you, let's have an idea to bring to the table - an idea that gives maximum equality, maximum freedom at the individual town level and the highest degree of flexibility. To that end, we have Confederation… with a twist.

Confederation (Well, a Hybrid One)

As you grow in size and influence and as towns and communities begin to bond together, the time will come when you will all want to get together and become one large political entity, with individual towns ruling together by consensus. Confederation is the best form of government initially post-collapse for a couple of good reasons. First, it requires the least amount of commitment (which everyone will be loath to do) and minimizes the possibility of any one town, person, or group taking power over the whole area. At the individual citizen level, allegiance can be given to the confederation, or just to their given hometown, or to no one and can do so without reprisal for stating their preferences either way. It is a form of overall government that is voluntary and does not require conquest, military action, or any other form of coercion. It also makes for equality among member towns, which can be in proportion to population, or with each town having an equal voice. Unanimous consensus is not always required, either - so long as a majority or supermajority (2/3) rule occurs (depending on issue) on the big or common issues that arise.

There are a ton of benefits here: A confederation allows towns to work together on large issues, but still decide individually on the smaller ones. It also prevents a central government (or any one town) from growing too powerful and dominating everyone. Security on a military level is multiplied immensely and coordination between towns can all but eliminate raiding parties and criminals from the countryside they share, which in turn allows for expanded farming and mining activities. The confederation structure allows for disputes between towns to be resolved quickly and more peacefully, as opposed to two separate entities which would likely fight each other over a dispute. Towns can pool resources and (finally) afford, together, to put up projects and endeavors that would strain the resources of a single town. These projects can include a working jail for non-capital criminal offenses, irrigation, waste disposal, a working court, mining, manufacturing and much more.

There are however a ton of pitfalls. The monetary system may be poorly-run, or engineered to benefit one town to the detriment of all others. The larger towns may begin to bully and subtly plunder the smaller ones through perfectly legal means. Given that there has to be at least a majority on large issues, an immediate-but-complex problem may find the confederation slow to meet it, as everyone debates the pros and cons of various solutions (or even debating whether or not there is a problem). There is also the fact that because it is a collection of towns voluntarily working together, it can dissolve much faster than any other form of government. Finally, tax collection and law enforcement at the overall level may be weaker than under any other form of government.

There are of course other forms of government and eventually you would want a purely federal form of government, but given the high level of distrust, just barely a couple of years post-collapse? Confederation is likely the best you can hope for. However, we can make a hybrid of sorts, to get a few good benefits from unified governments without the central power of one.

So why do we want to make it a *hybrid* system? Well, we want to unify the money and major economic structures. One solid form of money and policies that everyone can identify and use will make trade and economics much easier to keep a handle on. We will want to unify the military forces, which means that instead of relying only on cooperation between towns, we get the benefits of a multiplied force, but do it under a single chain of command. We should really unify laws for most criminal activities and have a consistent court system. Finally, one set of standards for trade is a definite plus.

Thing is, the previous paragraph lists all the benefits that can only come from a federal or unitary government, which a true confederation would not otherwise allow. You will want to help make those benefits happen, but the question is… how? After all, a unitary or federal government has a central structure and ultimate power over how things run. However, if you can get the member towns to agree, you could, together, set up a council to oversee and set policy up for an approval by member towns, for each of those listed benefits and without the need for an overarching central government that could be easily usurped or dominated.

To get the ball rolling, get the leaderships (or authorized representatives of the leaderships) of all towns and communities together in one spot. Introduce the idea and get everyone's input after doing so. Work together to draw up your initial articles of confederation. For your convenience, we've included a copy of the original US Articles of Confederation (and the US Constitution as it exists at time of printing) in our Recipes section at the end of the book. They are included as an excellent example of what to follow when drafting a document up. Just remember - it is not only important to draft the articles as best you can, but to have plans to put each of them into place (as well as any timetables) so that you don't end up with a bunch of laws that never get put into place or enforced.

Once everyone agrees to it, don't vote on it just yet - instead, be sure to make exact copies of the resulting articles and spread them far and wide. Take the time to have each town explain it all to the population and make sure everyone knows about it. Invite commentary, criticism and suggestions. After a period of time for public commentary (a month should do it - take your time and do it right), bring back all of that commentary, all of the criticisms and all of the ideas. Work together and make any necessary changes. At that time, hold a final vote on making the document as amended and/or changed the official law of the land. In this one case, it should be approved unanimously before being called official. The day a unanimous vote is approved, all parties should sign it - it will take effect either on the date specified on the document, or the second that the ink dries on the last signature. The choice here is yours.

Once official, exact copies should be made, providing at least three fully-signed copies for each town, who can in turn make copies as needed or desired.

Chapter 6

The Distant Future

(10 years and Beyond)

...Now What?

If it's been 10 years past collapse and you're still reading this, I won't need to tell you that it's a whole different world out there. You're in territory that quite frankly, I can only guess at. I doubt that things are along enough that everything is at the same level of luxury and comfort that was common pre-collapse (though some individuals may indeed be living such a life, who knows?) The good news is, you're still alive and with a little hope and luck, you may have a home with a loving spouse and maybe kids playing about the place.

It is hoped that you have some sort of working and benevolent government going by now, though you may have gone through a couple different ones to get there. It is doubly hoped that things have become peaceful enough now that you don't have to have a round-the-clock guard going at each home. Being able to sit back in front of your home as the sun sets while watching the crops grow and breathing a big sigh of satisfaction? Looking into your spouse's smiling face, knowing that together you may have been through hell, but are now living together in something that can someday become heaven? Shouting a joke at a passing neighbor on his way home and hearing a funny riposte coming back at you? These things may not be like bragging on a big-screen television or a brand new car, but I bet you'll enjoy them a lot more.

On the community front? By now everyone in your town has learned to work together, defend each other and in general look out for each other. Well, more or less and more rather than less. Anything less and you likely wouldn't have made it this far. Odds are good that a few enterprising souls are busy in impromptu workshops, feverishly trying to bring old technology back to life, or rediscovering lost techniques to bring back a little of the pre-collapse world. Skilled tradesmen are passing along everything they know and have learned to a new generation, so that vital skills and knowledge are not lost. More than a few of your fellow humans are writing autobiographies... after all, there are potentially millions of gripping stories around the world describing danger, adventure and survival ...with no two alike. No individual human being will come through untouched.

In what were once great cities, most of the die-hard gangs will have killed each other off by now, where only weakened and emaciated 'warriors' remain. Those who haven't settled down will be clinging to their last few bullets - forced to eat rats, each other, or the occasional hapless traveler for sustenance. The rest have either fled, died, absorbed into some community or other as respectable citizens, or have settled down. Occasionally, portions of the cities temporarily become small-but-fertile sources of material reclamation expeditions, but only by heavily-armed trading groups seeking great riches. Perhaps in 10 more years large-scale reclamation can begin in earnest, but for now it's best to leave it all sitting there - a realm lorded over by the crows, the dead and the desperate.

Speaking of bullets, those things will likely be worth 10,000 times their weight in pure gold by now, as there are very few left and few if any facilities on the planet left which could possibly make more. Their use is still occasionally necessary, but creative individuals all over have by now rediscovered the art of making black gunpowder and muzzle-loading guns.

The amount of dead humanity is beyond description - upwards of 60-80% of the world's population has died of violence, starvation, or other sudden ends. 10 years on? You can still trip over a corpse on occasion if you're out walking around beyond town limits. But then, most of them arc skeletons by now and will begin turning to unrecognizable piles of bone and dirty cloth scraps within the next couple of years or so.

In what used to be suburbs and countryside, signs of prosperity are popping up in fits and starts. Individual communities have progressed from questionable subsistence to self-sufficiency and many have progressed to the point where trade is considered to be bustling.

Trucks and autos have been replaced by handcarts, a growing number of horses and horse-drawn wagons and the occasional-but-rare alcohol or biofuel burning car or truck. Over time, worn-out parts will reduce the number of the latter, while equine birthrates and desirability will increase the former as a means to travel without walking. Once in a great, great while, you may stumble across a steam-powered rail engine or car, though most of those will be built from scavenged parts. Most bulk transportation will likely be carried in boats - powered by poles, perhaps steam, or by sail. There may be pockets of continued petroleum production and refinement, but with the lack of new vehicles, increasing lack of parts to maintain existing ones and with corrosion destroying the immobile ones, it is uncertain how much longer that situation will be necessary.

Even this far along, there may well be final bastions of pre-collapse life and technology. In small regions and even just cities, there may be enough materials stockpiled by dying governments (and super-wealthy individuals) to eke out a continued existence at pre-collapse levels of technology and power. How long they will continue to exist in that state is uncertain, as they are simply feeding off of whatever they have stockpiled.

There will likely be a few still-surviving isolated strongholds and bunkers, but for the most part they will have either aligned themselves with a nearby community, or will have abandoned their ideas of permanent isolation forever. Most of the actual bunkers are either completely empty or full of corpses at this point.

Militarily, there may be a civil war or two still raging, as parties still strive to control and claim the governments that once were and the powers they once held. However, given that warfare tends to eat ammunition at a prodigious rate and that vehicles break down or are destroyed quickly? These will most likely simmer down to slow and sporadic border skirmishes, as war materials quickly disappear without replenishment and armies break up wholesale with soldiers deserting for home to help their families survive.

On the international front, there is largely isolation as dead and dying central governments leave populations to their own devices. There may be wars over resources, but few nations will have remained as ongoing (or identifiable) entities by this point. Even China, which has a strong enough cultural identity, massive technological resources and iron determination? They will have been far more than decimated by starvation and deprivation, causing them to spend all of their efforts towards containing internal unrest and revolt. It will likely be another 25-50 years before any nation-sized government arises with the material and strength to claim much of anything - at least beyond the previous government's borders.

Technology may still thrive and even be pushed in some places, but this will be in fits and starts and most research will be forced towards adapting (and preserving) existing technology to a post-petroleum (or at least post-global) paradigm. That said, overall there will likely be a rebirth of many less-technological industries. Textile production, simpler metallurgy, alcohol production, fishing, paper-making, woodworking, milling and many other simpler industries will arise and thrive to accommodate new needs and new demands. With a bit of time and a bit of luck, industrialization will start to make a comeback.

Back on the home front, it's going to be a world unlike any other. People who once kept an eye on balance sheets or network routers now keep an eye on crops or livestock. New routines will have been established and things have come to a point where it is no longer a question of sheer survival, but more a question of what lies ahead.

The most shocking thing you will find (as if your life hasn't been shocked enough up to now) is the fact that there are children now who will be asking their parents what life was like in an age of tablets, satellites and instant global communication. They're asking because they will have been born after the collapse began in earnest and will have never seen it. A new generation will have been born in this more primitive, desperate age and will be completely unaware of the things you had once taken for granted.

On a more personal level, you will be a different person - for good or evil. You will have had your measure taken. You will have, in old British parlance, "seen the elephant". In the decade between the time things began falling part until today, your emotions will have been at their lowest and their highest. Events both horrific and angelic will have passed before your eyes. You and your fellow survivors will be among the last of the golden-age. In a couple of generations, only historians will be able to explain how an elevator works, what a CNC machine looks like, or how to drive a car. In a couple of centuries, most of the world you knew and lived in pre-collapse will have been forgotten entirely.

Some Parting Thoughts

By the time you reach this point, I may or may not be alive, so I'd like to end this tome with a few parting words and a wish for your continued success.

Wait - you seem shocked. I am however quite serious. You see, even though I have prepared exactly as described in this book and then some, I cannot even guarantee my own survival long-term. Violence, disease and just plain bad luck can just as easily remove me from this mortal coil as it can anyone else. I'm only doing my best to improve my odds - just as I hope I'll be doing for you with this book. As I mentioned way back in the beginning, there are no guarantees, not even for myself. Anyone promising otherwise is a liar and had probably died within weeks of collapse. There are only preparations and a substantial increase in the odds that you and I will make it this far - but no guarantees. God willing? We both did okay.

But, there is something I want you all to know. The pre-collapse world was wonderful, exciting and packed with a massive variety of opportunity. The ability to converse instantly with someone on the other side of the planet is a wonder, a business necessity and an endless source of enlightenment and entertainment. One can wander down to the grocery store in late February and within minutes, purchase fresh berries and vegetables. News and information was instant and constant. Transportation enabled one to work in a distant place but sleep every night at home - up to hundreds of miles away from the office. New technologies abounded - many on the brink of immortality, interplanetary travel and the potential for riches beyond counting. Never forget that.

Of course, on the other hand? Yeah there were a lot of pitfalls that we just couldn't shake. Greed, short-sightedness and callousness borne of societal isolation wound up pervading every crevice of our world and in many cases, even our souls. Corporations became predatory organizations, hungrily doing everything they could to achieve their one holy goal - maximize value for their shareholders. In the process, they lost sight of the long-term and whenever they could, they happily manipulated the laws of both man and nature to achieve that goal. Politics became polarized, assisted by network of news organizations who were more interested in reaping drama for advertisement dollars than in presenting events in honest and neutral terms. Self-appointed gurus and experts arose, casting aside ancient and venerable religions in favor of self-invented (and self-serving) reasoning, in favor of self-invented theologies and of pseudo-rites, brought out of antiquity and modified to fit various agendas.

Finally, the fault. It lies with us, individually and as a whole. We became isolated from our neighbors and were self-focused as well as self-absorbed. To our credit, we were overwhelmed with information and propaganda and the highly technological society we lived in forced us to specialize heavily. If I were to sum it in one phrase, we let intellect and desire outrun our wisdom and restraint.

I'd apologize, but it was everyone's fault.

Just do us all a favor: When you rebuild it all, don't screw things up this go 'round, okay?

— — —

One final word of advice for those reading this pre-collapse: Keep one eye on the gathering storm and one hand towards provisioning against it, but try to make some time so that you can enjoy the sunshine while it still warms your face.

Chapter 7

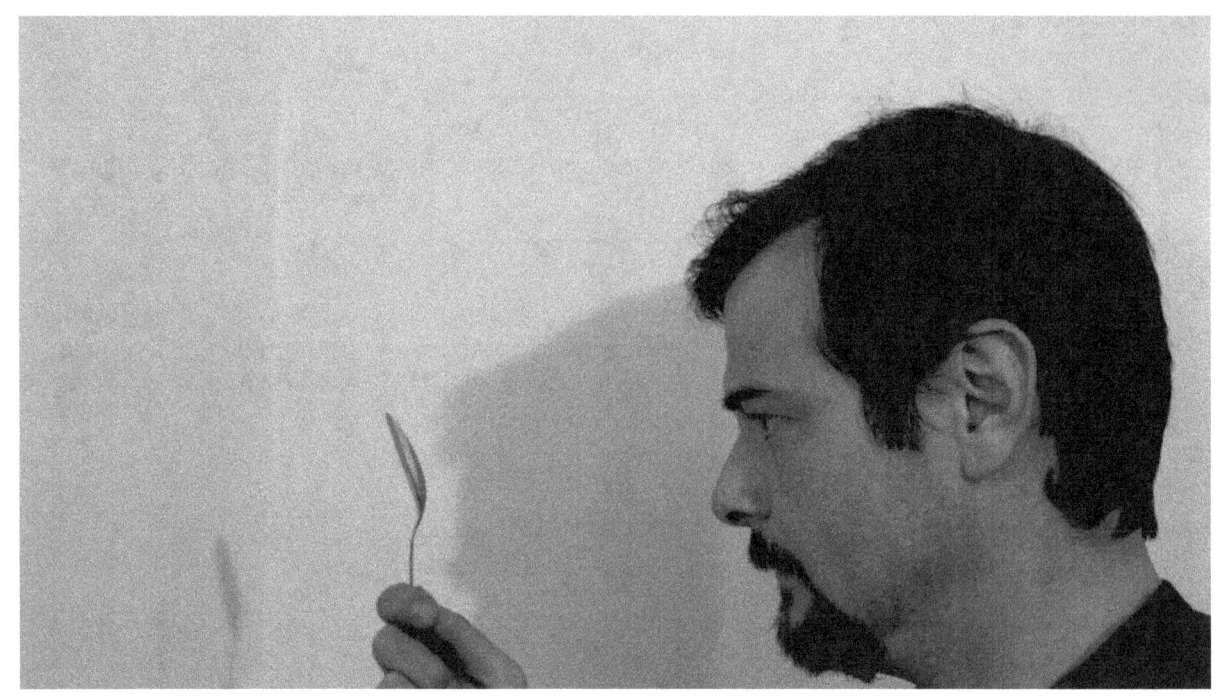

Recipes

Recipes?

It's an odd title, but a useful description of what you will find in the following pages. We've included the vital things you will need, with an eye towards making as much of it from commonly-available or natural materials as possible.

So why the particular choice of recipes, procedures and how-tos? Many of these items are, each useful for multiple reasons: Cleaning/disinfection, as an anti-corrosive agent, building/binding material and most importantly, as critical chemical components in other more complex things that can make civilized life a lot easier when there's no civilization around to otherwise provide them. Also, most of these items come from raw materials that are fairly easy to find or grow and the 'waste' from these materials can be put to other very good uses. As an added bonus, all of these products can easily provide you with items that will be in demand by everyone else, which means you can barter with them for money or needed goods.

While I have done due diligence to insure accuracy, I cannot guarantee that these will work to perfect satisfaction in your location or situation, so you really should test these things out and adjust the amounts to taste, strength and suitability. Note that some of these recipes and/or ingredients may not be legal in your locale, so do a bit of discreet checking on laws and regulations before giving any specific recipe a go.

A word of warning: Note that making most of these things can get very dangerous in a hurry - especially for the inexperienced, the incompetent and the careless. We're talking the possibility of uncontrollable fire, chemical burns, explosions, poisoning and worse that can happen… this isn't like using a kid's chemistry set. Therefore, take all caution and precautions while you test each of these out for the first time and always be careful around it thereafter - especially as you graduate up to larger batches.

A lot more was left out, but only because they are mostly secondary, or were left out in the interest of brevity (we'll be pushing 400 pages as it is… we don't want to make this thing too heavy, now do we?)

So… without further ado, let's get together the bits you need and get cooking.

Household Chemistry

In this chapter, you will find the means to make the chemicals that you will need for everyday life - both at home and around town…

Distillation Equipment And Its Use

We really need to start here, because quite a few items which follow will require a technique known as distillation to procure. So, to make a still, you will need to know a few things about how one works.

Distillation works by using heat and vapor to separate the chemical we want, from whatever it is that is holding that chemical in a matrix - the base material. For instance, we can use heat to extract turpentine from pine tree resin, drinking alcohol from fermented grains, or pure drinking water from wet sludge.

First, in the cooking pot, we heat the base material to a temperature where the liquid we want turns to vapor, but the stuff we don't want stays behind. For instance, the boiling point of ethyl (drinking) alcohol at sea level is 172 degrees F, but water's boiling point is 212, so if we gently heat an alcohol mash (the fermented slop you make to get alcohol) to around 172 degrees, we can extract the alcohol - leaving the water, yeast and soggy grain behind. Turpentine on the other hand boils at around 309-347 degrees (it's a mixture of individual chemical types), so we need a lot more heat to get the job done. To distill or purify water, we simply boil it to 212 degrees, which leaves all the mud, bacteria, rust, or whatever behind. (Note that you always take the temperature in the 'head', or up where the vapors form. Also note that as you rise in altitude above sea level, boiling point temperatures drop. But more on that bit later…)

Second, we have to capture the vapor and condense the stuff back into a liquid, away from the cooking pot. The lower the outside air temperature, the faster this can happen.

Finally, we collect the condensed liquid and put it into containers for further processing or storage. For instance, we would want to store drinking alcohol into wood casks or in a container with a chunk of certain woods in it to let it 'mellow' into a whisky or bourbon and maybe add some water or other liquid ingredients to bring the alcohol content down to something comfortable and tasty to drink. If the alcohol is for medicinal or solvent use, we simply store it in an airtight container at full strength until we need it. For turpentine, we want to store it in an airtight container for later use. For distilled water, we might want to collect it and then perhaps boil it again in an open, clean container, to boil-off/remove any chemicals or solvents that may have collected along with the condensed water.

It seems easy at first, but rest assured that it is a process that takes a lot of time and patience to do correctly. First and foremost, you have to have the condenser and collector scrupulously clean in order to avoid contaminants. Next, the fire has to be built up low and slow and the process takes quite a bit of time for each batch you distill. You have to keep a close eye on the internal temperature of the liquid and keep it constant (especially with alcohol), so you only get the chemical you want and none of the ones you don't. Consistency is the key here, either for taste (alcohol) or potency (everything else), so each batch has to be run the same way.

To build your still, first you need a good cooker. Any clean, metal water-tight drum will do for a start, but you will want to modify it a bit. First, you will want to cut the top off for a condenser hood. You will also want to add a means in the upper-front side to add your cooking material - a door with a heat-proof seal will work wonders here. You'll also want to put in some means towards the bottom to empty out your cooker of old mash/resin/sludge/whatever. A simple but large (2" or larger) tap will suffice for most things, but for hard substances like turpentine resin? You will want to avoid this and just make your upper door large enough to reach in and scrape out all of the leftover goop. A small hole somewhere just above the waterline, where you can stick a thermometer into the cooking mixture, would be a very nice touch as well - just be sure you can seal/close it when not taking a temperature reading.

Your next step is to make a good hood for the cooker. The hood collects all the vapors coming off the cooker and directs them to the coil, so you will need to be handy with metalworking at this point. The hood should most definitely be air-tight in its attachment to the top of the cooker, so you will have to weld, rivet, or solder the hood on. A cone-shaped hood with a fairly small outlet for the coil will work nicely and can be easily fashioned out of a large sheet of metal. Another option is to have a cooker with a dome-shaped top already built in (for example, pneumatic braking tanks from under a semi trailer), which lets you pound out an upside-down funnel of sorts in the center of the dome (coming outwards), leading to the hole where your condenser coil connects.

If you're sufficiently efficient (okay, lazy), a pressure cooker is perfect for this role. Instead of the safety valve weight on the top of the pot, you run your condenser coil off of that little valve (and put the weight to use in your collector pot) and you're all set. On the other hand, you'll have a hard time using that pot for anything else, so keep that in mind.

When it comes to the cooker and hood assembly, size matters - the bigger they are, the bigger a batch that you can run at any one time, which cuts down on the time spent cleaning and preparing it for the next batch, then gently heating it all up again. Just note that too big means you end up burning a lot of fuel to keep it hot.

The condenser coil is a must and your absolute best bet is for it to be made of copper (which conducts heat very nicely - something you'll want at this stage). The condenser coil should be fairly long - at least 10' long in length and preferably 20'. The vapor passes through this, dumping off its heat to the metal walls of the coil and condensing into droplets of the product you want to make. Weld or solder one end of the coil to the condenser hood and loop it into coils, collector end higher than the open end, gradually sloping the whole thing downwards (with no 'valleys' or any upwards changes in elevation). Either rig a means of running cool water over the condenser coil, or put it in a tub with a small hole in the bottom of that tub. When you start the condenser, you will fill the tub with cool/cold water, or start running the cool water over the coil.

Finally, you have the collector. Here, you only need to have it clean, made of metal (to transmit away any residual heat) and it should have a tap on the bottom to drain the liquid. Though you can use an open-topped container, it is likely better if you didn't in order to keep dirt and contaminants away. You might thing that as an alternative, you could simply add 5-10' of condenser coil and put a tap on the end of that, using it to fill airtight containers as you brew a batch. That said, you will really want a proper collector pot at the end to help blend multiple batches together as you brew them, for consistency reasons. You will also want the collector for safety reasons, as pressure does build up in the system as you cook it.

Speaking of pressures, be sure that everything is at least somewhat pressure-tight in the system, all the way until you reach the collector. You then make sure that the collector allows for any extra pressure to escape - the reason why is that the vapor is further encouraged to flow through the condenser coil to the collector. A small hole with a very small (¼" at most) vertical pipe and a small weighted cap over the end of that pipe (you'll have to calibrate it a bit) will act as a good safety mechanism.

One final consideration: To keep your home clear from fumes and prevent potential fires/accidents from reaching anything important, try to keep the still at least 50' away from your home, if possible. Also, whenever you see the word "solder", note that you should always use lead-free solder for any still whose end product is for human consumption.

Linseed Oil

<u>What It Is, And What You Use It For</u>:
Linseed Oil is useful for many different purposes. You can use it as a varnish (or by itself as a wood protectant), as a binder in making paint, resins and putties. You can also use it in soaps. It is completely edible in its pure and natural form and can be used as a mild spicing agent or as a nutritional supplement. Note that you cannot use it as a cooking oil.

Ingredients You Need:
Flax plant seeds - a lot of them. About 30-40% of the weight will become oil when pressed. (The rest makes a high-protein feed for livestock, incidentally.)

How To Make It:
You press the seeds as hard as you can. A workable press can be built from a hydraulic automotive jack (a "bottle jack"), a sturdy metal frame to contain the press, a thick, round metal plate and a thick metal pan with a couple of small holes in one side of the its bottom. You also need something to catch the oil as it comes out of that pan. As an alternative, you can use the round plate and the pan, then have an extremely heavy weight (like, say, a broken-down truck?) do the pressing.

You set up the pan and the weight or frame/jack combo. Fill the pan about 1/2" thick on the bottom with the flax seeds. Then, you assemble your press and start pressing it for all you're worth. Keep the pressure up until you max out and/or until the oil stops dripping out of the holes. If you get no oil, you'll need greater pressure (or more weight). You should see a maximum of about 1 liter of linseed oil (33 ounces) from about 18 lbs of flax seed.

Alternatives:
- Poppyseed Oil
- Walnut Oil (extracted from the nuts)
- Safflower Oil (extracted from the seeds)

Safety Precautions:
Note that Linseed oil is highly flammable. Rags and cloths soaked in the oil have been known to spontaneously combust. Keep the oil and any rags out of direct sunlight when not in use and do not bunch up or pile up (spread them out after rinsing to dry). Note that if you intend to make "boiled" linseed oil it is not as simple as boiling it! The boiling point is far higher than that of water and has to be monitored very closely to prevent a flash fire or even explosion of the liquid.

Other Considerations:
Linseed oil is only edible for up to a year after manufacture. After that, it is still useful for everything else, but do not ingest it, as it will go rancid. Store in an airtight container and out of sunlight. Note that homemade linseed oil will dry slowly - far more slowly than commercial preparations. The flax plant itself is useful as well - the plant fibers from the stem's skin can be prepared and made into cloth (real linen is made from the flax plant's fibers). The flax plant is still grown commercially throughout the northern plains states and in the plains provinces of Canada, but can be found in patches and on farms throughout the United States. It is also found occasionally as an ornamental garden plant. Whole flax seed can be kept stable for years in an oxygen-free container, but ground flax seed can go rancid in as little as a week.

It should be fairly obvious that your sources for linseed oil (or similar) will either have to be grown, or found in/on an abandoned farm. As a short-term alternative to plant extraction, linseed oil can be stocked-up on pre-collapse (though it is somewhat expensive), or flax seed itself can be bought in bulk and stored (at a moderate cost).

Turpentine

What It Is, And What You Use It For:

Turpentine is a solvent distilled from the resin obtained from live trees, usually pines. It is used as a solvent and as a treatment for lice. It can also be used in mixture with other detergents as a cleaning fluid and as an antiseptic. Note that while it was once used for internal medical purposes, it certainly not recommended at all that you actually consume this stuff, for any reason. Try to avoid breathing too much of the vapors, as in concentration it can damage your lungs and central nervous system. A small amount will go a very long way.

Note that this will be old-style or crude turpentine and will contain a mixture of both alpha- and beta pinenes. It isn't as effective as modern ultra-purified preparations, but it will work admirably for most purposes.

Ingredients You Need:
- Access to a lot of medium-to-large mature pine trees
- A small axe and a machete or a very large knife
- "Turpentine Pots", or small metal pots with one side dented in to hug the tree and a 1/4" hole punched into the upper side of the can, with the lip on the 'tree side' hammered thin, cut off, or filed down. One pot per tree.
- A hammer and medium-sized (around 8d-sized) nails (note: be sure to always remove the nails when you're done with the trees - one nail per tree).
- Buckets to lug the sap from the trees to the 'still.
- Distilling equipment (which should be purpose-built for this chemical and used for no others).

How To Make It:
- First, find a large number of pine trees that are at least 20' tall and have trunks at least 24" in diameter. These trees should be very resin-rich (many if not most varieties of pine tree are), which can be tested by making a small gash in the trunk, then returning a few days later to see how much resin has oozed out. This will give you a (very rough) idea of how many trees you will have to 'work' in order to get the final product.
- In the spring, as the weather begins to warm up, shave off a 8" tall vertical portion on the bark, about the 1-¼ times the width of your turpentine pot and with the bottom of the cut about 3-4' from the ground to start with.
- Nail the turpentine pot to the bottom of this shaved area, oriented so that the pot's dent hugs the tree and the lip can catch the resin as it runs down.
- Make a few "v"-shaped cuts up and down in the shaved area, orienting the cuts so that the resin runs down to the turpentine pot.
- Wait a day (or two, depending on air temperature), then return to collect the resin. You'll likely find it easier to collect the resin out of each pot and into a bucket, then run the bucket full of resin to the still for processing.

Next, you will want to distill it. A still should be purpose-built just for turpentine, as the chemicals involved in production are noxious and notoriously hard to clean out sufficiently. The cooking pot will leave a sticky hard-to-remove residue behind, so it makes more sense to only use one particular still for this one particular chemical. Note that you'll be working with smaller batches, so a small still (with a cooking pot capable of holding perhaps 2-3 gallons of water) will work just fine initially.

Load up the cooking pot with your gum/resin. Be sure the stuff is reasonably free from dirt, pine needles, bark, or other debris. Rig up a thermometer through a cork or other stopper into a hole on the side of the cooking pot, with the bulb submerged where the liquefied resin will be (as long as it isn't touching bottom, you're okay). You can use a standard candy thermometer for this, as it will (barely) read the temperature range you'll want. Seal the cooking pot and begin to gently add heat (either with coals, or with a fire directly underneath the pot, but be sure you can add or remove fuel as needed to maintain a consistent temperature). Empty out any liquids from the collecting pot until the thermometer begins to register around 300 degrees or so (depending on altitude - higher altitudes mean it'll happen at slightly lower temperatures, but at sea level the boiling point begins at 309 degrees). At this point, the vapors will be turpentine and you'll want to keep the results. Do not let the temperature get beyond 347 degrees (again, remember altitude differences) for the entirety of the batch. Periodically empty the collecting pot into an airtight jar or container. When the flow slows to a trickle or stops, you're done with that batch. Remove the heat and quickly (if possible, while still hot) dump the cooking pot contents into something you don't particularly care about, for later disposal or for whatever use you want (it'll make a nice tar or ultra-sticky glue if you ever need it). At that point, you can ready your next batch and repeat.

The best turpentine will be a clear color, but it is still useful even if it is honey-colored or anywhere in-between. You may notice some water in the fluid, but that is harmless and will naturally settle out. It can then be removed.

Alternatives:
Few, if any - usually mineral spirits and the like, but the alternatives do require a source of petroleum to extract and refine, which we obviously won't have. Other types of tree oils (camphor, eucalyptus, tea tree, etc) can do similar work, but are rarer and harder to extract in a pure enough form.
Note that you can also distill (or in this case 'crack') turpentine from chipped pine wood, but the conversion rate is horrible, averaging 32 oz of fluid per 1500 lb of wood at 15% moisture.

Safety Precautions:
Turpentine will require airtight and chemically-resistant containers, period. Sealed metal cans or tightly-sealed glass jars are highly recommended for storage. It is highly flammable, has a very low evaporation point and can easily replace kerosene as a fuel. Therefore, keep it out of sunlight, away from excessive heat and be sure to rinse (or just burn) any rags used with the fluid. Like linseed oil, you do not want to have turpentine-soaked rags in a pile, for any reason.

An important thing to remember is that you're using heat to distill a highly flammable liquid. All safety precautions should be taken to avoid bringing the finished turpentine in close proximity to the heated portion of the 'still.

Finally, do not, under any circumstance, use the same distilling equipment for turpentine that you would use for drinking alcohol! Distilling drinking (ethyl) alcohol through a turpentine still will very likely poison anyone who drinks the resulting booze.

Other Considerations:
You will only be able to collect the resin from early spring to late fall during warm days, when the tree resin is actually fluid and running. Spring, or "virgin" resin collected early in the season will be the most potent and highest-yielding, while the autumn resin will be the least potent. The gummy leftover from distillation is often called 'rosin' and can be used as a tar or adhesive if sealed immediately into containers for later use, or set aside to dry and become a hard resin for various other uses.

When you are done with a tree for the season, always take care to remove the nail and fill the nail hole with a small plug of leftover resin. Removing the nail prevents accidents when sawing the log later on and filling the hole helps prevent diseases and pests from entering the hole to the core of the tree.

A tree can be 'worked' for multiple seasons. With each successive season you shave the trunk a few inches higher and to the side of the previous season's shaving, taking care to leave about a 5" strip of vertical untouched bark between the two shavings. After about 5-6 seasons, the tree should then be left alone for an additional 5-10 years to allow for at least some healing-over and regeneration of the tree itself.

Outdoor Paint

What It Is, And What You Use It For:
…and now you know why we were bothering with linseed oil and turpentine. Both are useful towards making paint. Paint is important because there are only two ways post-collapse to prevent corrosion and rot: you either oil/grease it, or you paint it. Oil (especially petroleum-based oil) is going to become harder to come by over time and doesn't work too well on everything, especially outdoors, where rain and sun take their toll on any kind of oil. This is where paint comes in…

Ingredients You Need:
- Linseed Oil, or other *drying oil* (as a binder/finishing agent)
- Turpentine (which acts as a solvent to help paint flow)
- Powdered chalk, gypsum powder, dry clay powder, etc (as a filler and pigment)
- (Optional) some sort of pigment/coloring powder
- A nice ventilated area to mix it all together

How To Make It:
First, make a small pile of the filler material on a surface you don't like. Make a small crater in the top of that pile and pour a little linseed oil into it. With a stick or a small dull knife, work the oil into the filler material, adding more and more linseed oil into the pile until the whole thing becomes a very thick paste.

Place the wad of paste into a container that you can seal airtight. Slowly mix in turpentine thoroughly until the whole thing has the consistency of liquid paint (if you're not sure what that would be, it's about the thickness/goop-factor of used motor oil.)

Alternatives:
Not much in the way of alternatives for paint itself, though you could use waxes or other similar bits. However, you do have a lot of play/leeway with the ingredients. Mineral spirits can replace turpentine and there are a lot of alternatives for the filler (for instance, gypsum powder can come from grinding up drywall). Basically, any find dust-like powder of a consistent color and texture can be used as filler (just don't use iron/steel dust, as that would actually speed up corrosion if painted on a metal surface).

Safety Precautions:
Not too many, just make sure you're in a well-ventilated area and that you keep the paint containers sealed. Also, note that the paint is going to give off flammable vapors until it dries, so keep it well away from flame.

Other Considerations:
Keep it sealed until you're ready to use it - as long as it is sealed airtight, it should be good for up to five years. Keep it stored in a cool, dry place and definitely away from anything where moderate heat or moisture would be generated. Although it will have a far lower freezing point

than water, try to avoid cold temperatures. Apply when temperatures are roughly at least 60 degrees Fahrenheit so that it dries quickly and smoothly.

Charcoal

What It Is, And What You Use It For:
Charcoal isn't exactly what you think it may be at first. While those bags full of briquettes from the grocery store technically counts as being charcoal, that stuff is only good for cooking and little else. What we're after is charcoal that can be used for cooking, water filtration, air filtration, marking and as a usable ingredient in many other concoctions. In our case, it's good for two things: intense heat and water filtering. Charcoal is basically wood or other organic matter that has been pre-heated in a low-oxygen environment to remove any volatile materials and water. What's left behind will burn at far higher temperatures than wood or other ordinary non-petroleum fuels and contains a massive number of pores which you will find useful as a filtration agent.

Ingredients You Need:
- Wood or other dense organic matter (hardwoods are best, but any type will work)
- A charcoal kiln in which to make charcoal
- A supply of fuel

How To Make It:
Let's start with a charcoal kiln… unlike most types of kiln, this one has two parts - one contains the burning fuel and the other is an airtight container with a vent for escaping gases, in which you tightly pack your wood (or other material) to make your charcoal. A typical arrangement involves a metal drum atop an ongoing fire. Before firing, you fill the drum with wood as tightly as possible and seal the lid. The lid or drum (depending on orientation) should have a small hole (1" or so for a 55-gallon drum, smaller for smaller drums) at the top to allow gases to escape. The fire should be very close to the drum at all times.

You keep the fire going under the drum for 6-10 hours, depending on load. You shouldn't see a lot of smoke, even from the fire underneath. As a good way to test for readiness, try to (carefully) light the drum exit hole gases on fire - when you can no longer make that happen, things should be ready. Let the fire die and the drum cool down and then open the lid: the drum should be full of charcoal.

Alternatives:
None, actually. Good news is, this is fairly simple stuff to make.

Safety Precautions:
Note that yes, you're literally playing with fire. The gases that escape from the drum hole are highly flammable.

Other Considerations:
Keep as intact as possible (it will be fairly fragile). Keep in a cool, dry place. If you intend to make charcoal for filtering, keep that particular charcoal in an airtight container until ready to crush and use.

Lime

What It Is, And What You Use It For:

Lime is useful for a lot of things - from making mortar, to fertilizer, glass, paper and even to process raw corn/maize kernels into a *Masa* paste for Latino-style foods. Lime is also useful in making whitewash, which makes a good basic paint for porous surfaces such as wood. There are literally dozens of other uses as well, from plaster to cement and much, much more. The best news comes from mortar made from lime - lime mortar eventually sets into something approaching real limestone, making a stone structure mortared with lime into a something that can last literal centuries.

Lime comes from one of the most common rock types in North America and is fairly easy to make. The most common type of lime is going to be quicklime, which is slightly corrosive. Other types are slaked or hydrated lime (made with the addition of water in the process) and hydraulic lime (which has the ability to set underwater.) We'll concentrate on the first two.

Ingredients You Need:
- Clean limestone (as free of impurities as possible)
- A means to crush the limestone into powder
- A kiln capable of varying temperatures (see below)
- A lot of fuel for that kiln
- A means of using it quickly, or storing it in an airtight container.

How To Make It:
First, you're going to need a kiln. Since a modern kiln and sufficient fuel is going to be impossible, we'll have to go primitive.

To make a primitive but useful Lime kiln, you will need to create a fairly large brick or (non-limestone) above-ground firepit that is fairly tall and fairly thick at the sides. Thick concrete blocks work perfectly here. Insure that there is a trench coming out underneath it and a grate at the bottom above that trench. You want to design this so that air enters at the bottom and rises upwards through an air grate, then up through the layers of fuel and limestone chunks.

For fuel, you will be using either charcoal or actual coal (actual coal if you can swing it) and a set of bellows that will constantly feed the fire to a high temperature if you have charcoal. You will want a proportion of two parts coal to one part limestone, or three parts charcoal per one part limestone.

To load the kiln, alternate a big layer of coal fuel at the bottom, a somewhat large layer of limestone rocks (about 1-2" or so), a layer of fuel, a layer of limestone, a layer of coal, etc until you're all loaded up. Light the fuel and sit back for awhile… a long while.

Once the fuel is burned out, sort out the limestone (while still hot) from the ash and get it into a metal bucket or container to cool. What you have at this moment is quicklime. To test it out: take a small freshly-cooked rock, let it cool, then squirt a bit of water on it. The little test rock should literally 'fizz' as it absorbs ("slakes") the water. At this point you can crush it to powder (it'll take a bit of force to crush the rock) and put it into dry, waterproof containers.

You can also take that powder, add water, wait for the fizzing and steaming to stop, let it set up as a paste or as a moist powder (depending on how much water you added) and you almost have slaked lime. Let it 'mature' under a layer of water into lime putty, in a closed container. You need to 'mature' it before use - it can be used in as little as 24 hours, or you can let it sit for years (for longer periods of time, you usually let it sit as a damp powder in a closed container until you need to make putty). The traditional time for maturing lime putty is around 2 months.

Alternatives:
None of note, given the wide variety of products you can make from it.

Safety Precautions:
We're messing with extremely hot stuff. Finished lime rocks will be literally red-hot as you remove it from the kiln. Take all precautions, definitely wear gloves and protective gear and use metal tools when removing anything hot.

Other Considerations:
Note that you may want to do some tweaking initially, else you may over-cook the whole thing and end up with "dead lime", which is useless. You can certainly and easily scale this up as time goes by and if demand goes up sufficiently.

Chlorine Bleach

What It Is, And What You Use It For:
Chlorine can be used for a lot of beautiful things - cleaning, disinfection, extermination and as a stepping stone to more complex chemical products.

Our main focus post-collapse will be for disinfecting water, because killing germs in your water is a beautiful, critical thing. The added benefit is that you no longer have to burn fuel to boil drinking water so often. This is relatively easy to do as long as you have two vital bits: plenty of salt and a steady source of electricity. Yes, electricity. However, the amount of electricity we're talking about can be generated with a small solar, hydro, or wind-powered station. This means it is still doable to an extent post-collapse, as long as you can keep enough parts, spares and materials on hand.

Ingredients You Need:
- Table salt, or any fine-grained salt
- A source of electricity and wiring that can provide 12 volts DC at a moderate (but not massive) current.
- Sufficient (14-gauge) wiring and a sturdy on/off switch
- Two large electrodes (these will be periodically replaced) made of iron - and don't worry too much about rust on them (a bit of surface rust actually speeds things up).
- A small handful of rust flakes (about a teaspoon or two)
- A large plastic (non-conducting) container - a 3 or 5-gallon bucket will work.
- A 2-3 gallons of clean water (depending on size of bucket - you want a little room in there)
- A *very* well-ventilated area.
- (Optional) a car battery to provide an initial surge of power, to speed things up a bit.

How To Make It:
First, assemble the power source, wiring, switch (wired in-line) and electrodes. It is recommended that you use metal clips to connect the wiring to the electrodes. While leaving the switch off, affix the electrodes close to each other, about 1-2 inches apart from each other.

Insure the bucket is in open air (outdoors), or in a very well ventilated area. Fill the container with a strong brine solution (for you folks living on the coast, add a bit of salt to that ocean water.) The stronger the solution, the faster the results. Turn the switch on and let it run. You will know it is finished by dipping a small sliver of dark-colored cloth on the end of a wooden stick into it. If the cloth bleaches into a lighter color (or even straight to white) in a minute or two after dipping, you're good to go. If not, run it some more until you get that result. Be sure to use a wooden stick and not dip anything with your hands.

The reason we keep the electrodes close is that the negative electrode produces sodium hydroxide, which the freshly liberated chlorine gas from the positive electrode will mix with to produce sodium chlorate. The iron on the electrodes liberates the excess oxygen from the chlorate to make bleach (sodium chloride).

Keep the results in an airtight container and if you suspect it is too strong, mix with water until sufficiently weakened. You can throw out the sludge, obviously - but only throw it in a place where you don't want plants or anything useful to grow.

Alternatives:
There are actually plenty of ways to make bleach, but most involve chemicals and equipment that are going to be hard-to-impossible to come by. It's going to be hard enough to get and keep a supply of electricity going as it is.

Safety Precautions:
Three really big hazards here: electricity, hydrogen gas and chlorine gas. If you wire it correctly, the electricity is not as big of a problem - just keep your hands out of the liquid at all costs, keep the area (and you) dry while you do it and you're good to go.

The dangers of hydrogen gas is that the stuff is extremely flammable. As long as you're not having a fire going or smoking (hah!) around the bucket, you should be just fine.

The chlorine gas on the other hand is dangerous stuff. It is highly toxic and can kill you in high enough concentrations. This is why you should be doing this outdoors. If the area smells way too strongly of a swimming pool and you cannot get upwind of it? Shut off the power and get well away from the area until it dissipates.

A somewhat lesser hazard is that if you make it too strong, it will burn your skin, throat, eyes, lungs and a whole lot of other body parts you don't want chemically burned.

Other Considerations:
Bleach has a whole lot of uses, but you need to remember to keep it in an air-tight container. Use sparingly (not because you can't make more, but because if done right, it will be rather strong - most pre-collapse bleaches are only 3% sodium chloride.) A couple of drops per gallon of clean, filtered water will disinfect it in short order.

Soap

What It Is, And What You Use It For:
I think we all know what soap is and if you're still not sure, then don't bathe for a couple of months. You'll either get the idea soon enough - that, or everyone around you will clue you in with alarming speed and forceful voices. Post-collapse, soap is going to be some pretty awesome stuff and a rare-yet-critical ingredient in staying clean (and therefore healthy). You may have a literal ton of the stuff stocked up in preparation, but either you're going to run out, or your neighbors will. This means with the right amount of skill in making the stuff, you can have an item that you can use as barter and to sell outright.

Ingredients You Need:
- Cold, white hardwood ashes. The stuff can come from oak, maple, apple, or various other hardwood trees. Make sure it is dry and kept in an airtight container.
- The cleanest water you can scrounge - distilled water, rainwater or spring water is preferred if you have it.
- Animal fats and/or grease. Leftover grease from cooking can be used if it is filtered out. No animal oils? You can use Plant oils, or any kind of fragrant vegetable oil
- Salt.
- Some plastic buckets - 1-3 gallon or larger
- A 5-gallon bucket with a tap in the side of it, at the bottom. Be sure you have an airtight lid for it.
- 3 large cast iron or steel pots to boil things in
- Wooden spoons or stirring sticks
- Soap molds - basically small wooden molds shaped like bars of soap.
- Clean cloth and/or rags (to use as filters) - old shirts work if there are no holes in the cloth.

How To Make It:

First, we're going to have to make a chemical known as lye. The best homemade lye will come from the white ashes of oak or apple tree wood. If you don't already have any, then make a clear and clean fire pit, load up the aforementioned wood and get a good, hot fire going. Try to do this in a place with little-to-no wind. Keep feeding it the wood until you have a good-sized pile of white ashes underneath the fire, then let it go out on its own. Collect the ashes once they are cold and put them into the 5-gallon bucket that you can seal airtight. Keep burning wood in batches until you have enough cold white ash to fill the bucket completely. Note: It's okay if you use that wood to heat something useful, such as food, water, your home, etc.

Next, we have the water. The reason we want the water as clean as possible is because hard water (that is, water with a lot of dissolved minerals in it) is tough to use for making soap. You can mitigate this a little by either distilling the water, or by adding baking soda to it little by little until you can easily make soap suds in it with existing soap.

When you are ready, heat half of the water to a hot (but not quite boiling) temperature and gently (and very slowly) pour it into the 5-gallon bucket of ashes, leaving the tap open - if liquid begins dribbling out of the bucket, close the tap. If you don't see water coming out of the tap yet, you gently pour cold water into the bucket until liquid dribbles out of the tap, then close it. Add more ashes and water if you have them, until the bucket is full, but only until you have damp ashes. Let the mixture stand at least six hours, or preferably overnight. In the morning, slowly drain the water out through the cloth into a bucket. Close the tap and gently dump the water back into the ash bucket. Let it drip through the ashes again, filtering back into the bucket.

Seal that first bucket, then break out another. Run fresh water slowly through the ash bucket as before, collecting the drippings on the other end. Continue until the liquid coming out is no longer tinted brown. This batch will be weaker than the first bucket, but it is still useful. You can dispose of the ashes by burying them somewhere, or dumping them in the outhouse.

A word of caution at this point: the liquid is going to be incredibly caustic, so be sure to wear long gloves, an apron and if you have it, eye protection. To test this mixture, you will need either a raw egg or a chicken feather. Since eggs are going to be at a premium, let's use the feather. A feather set on the surface of the liquid will start to dissolve. If it doesn't, boil the liquid down a bit until it does in subsequent tests. If the feather melts instantly, it's a bit too strong - may want to add a bit of soft water to it.

Next up is the fat - slowly cook it down into grease. Add the grease you have been saving for the occasion as well. Bring it to just under a boil and carefully (with tongs, a stick, whatever) remove the solids. Next we will be cleaning the crap out of the grease - bring the grease to a boil and then get it off of the heat. Slowly (and carefully) add cold water - at least 1 part cold water to 4 parts grease, then let harden. The ick will settled towards the bottom with the water. Once it all cools and hardens, scrape off and save the congealed fat - as much as possible and throw out the goop and water at the bottom. You will have to repeat this process (boil, add water, scrape) at least once (and possibly up to 3-4 times) until you get clear, white congealed fat without bits and contaminants. If the fat/grease has a bad odor or is rancid, you can use a potato or two to soak up the smell while boiling the grease. Once done, store the congealed grease in a clean, air-tight container.

Okay - now we have all the ingredients we need. Time to get it all together and make soap. Before we begin, there are two things to keep in mind: proportion and temperature. Proportion is the ratio of fat to lye. The best mixture is around 1 pint of melted fat to 12 pints of lye (at any scale), but leave a little extra of both to tweak to taste (umm, that's a figure of speech - don't actually taste the stuff). In on pot, melt the fat and try to get it to a temperature of around 100 degrees Fahrenheit. In another pot, heat the lye to a temperature of around 75-80 degrees Fahrenheit. Start heating the third pot up, adding ¼ of the grease, then ¼ of the lye to it, then stirring the mixture with a wooden spoon or stick. Continue adding the grease and lye in the same ¼ portions, stirring the pot very well with each mixing. As you're adding this stuff into the pot, let it sit for a minute and inspect it. If there is a thick layer of grease on top, more lye is needed. If the mixture doesn't appear to thicken, you need more grease. The mixture at this point may also look stringy and muddy-looking. Add additional lye slowly until the mixture clears up a bit.

To test things out a bit, put a few drops of the still-heating mixture onto a clean piece of glass or glazed china - if the oil and lye separate, keep stirring the mixture and add a bit of lye a small bit at a time until they don't separate.

Next up, we need to "prove" the soap. The best way to do it is to take out a one-inch ball of the glop out of the mixture and let it cool on a piece of clean glass or glazed china plate. If it cools mostly clear with whitish streaks in it, you're good. If it cools gray or with grayish edges, it needs a bit more lye. However, if it cools with a gray skin over it or in any other manner, a bit more grease is needed.

Once you know you have the right mix, keep boiling until the froth settles down in the pot. You may hear some bubbles audibly pop at this point. Once the froth is gone, you should have liquid soap that is ready to be used. If you're not sure, take a little bit out, let it cool and try it out on something dirty. When ready, let it cool a bit and then pour into airtight containers.

If you want bars of soap, here's what you do with that liquid soap…

This is where the salt comes in. Heat up the liquid soap, then add a small bit of it (perhaps 1/8 cup at most to start for small batches). The soap will float to the top and a brownish mixture will sink to the bottom. Let the results cool, then take the top layer of soap off. You may have to do this a couple of times, but the result is going to be pure, usable soap.

Collect this soap and melt it gently one more time. At this point you can add perfumes, colorings, or even a bit of clay (to help scrub) or even oatmeal (as a skin softener) if you desire. Pour the result into molds and let them harden a bit overnight. Carefully remove them from the molds the next morning (using gloves, please) and letting them 'cure' in a dark, dry place. This curing can take at least 30 days before the resulting soap is ready to sell commercially (yes, a month), but the resulting soap lasts longer under use than it would as a liquid.

Alternatives:
You could simply re-work otherwise unusable bits, odds and ends of existing bars of soap. For example, if you scavenged a hotel and wound up with a ton of those tiny bars of soap? You can melt them together into bigger bars. At most primitive, you could use a mixture of oil and clay as a means to clean oneself, but it uses a whole lot of water and the results are going to be inconsistent at best.

Safety Precautions:
First and foremost, you're messing with an extremely caustic substance. If that feather melting in the stuff wasn't sufficient warning, then the fact that lye splashed in the eyes can blind you should be. It is a very good idea to keep some vinegar or some sort of strong citrus juice on hand in case any lye touches your skin and use it to wash the affected area. Eye protection is strongly recommended and care should always be taken whenever you're messing around with lye.

Other Considerations:
This soap is a general-purpose chemical, which can be used for hot-water laundry, washing yourself, washing/scrubbing things in your home, etc. Note that any use beyond washing your body or clothes are going to require a bit more scrubbing to get good results.

Note that there will be a lot of trial and error at first. If you have a failed batch, or a batch that fails to set hard, you can always shred, re-heat and modify the balance of lye and grease (using the same tests we've discussed earlier) until you have it right. There is going to be a lot more art than science to this and you will eventually come up with a recipe that is better-suited to your local conditions and the types of fat/grease and wood you can get or do have.

Medicines

This chapter is dedicated towards those things that will come in handy during times of disease and distress. Note that these are no substitute at all for a competent doctor and manufactured medicines. The following recipes are only here for one reason - to make medicines when there are no others available. No exceptions. Also note that if you do not stick with the recipes as listed, you stand a great chance of harming or killing somebody, so pay strict attention here.

So, Where's The Penicillin?

Penicillin is one of the earliest and most common antibiotics that we as a species have put into use. The good news is, it kills a wide variety of bad bugs, which makes it perfect for use as an all-around medicine. Most folks will immediately think: *"Oh, it's just bread mold!"* Umm, no, it's a bit more complex than that. It is a specific type of mold, which is then put under stress to produce Penicillium, the chemical that is the actual antibiotic.

So, what to do? Stockpile in advance. You can actually stockpile multiple antibiotics in advance without needing a prescription, or going to the doctor. You can do this in one of two ways: First, you can buy fish-tank pills. Seriously, the following brands work very well and can be scavenged from any competent pet store or aquarium supply store post-collapse (common brand-name and what antibiotic it contains):

- "FISH-MOX" (amoxicillin @ 250mg)
- "FISH-MOX FORTE" (amoxicillin @ 500mg)
- "FISH-CILLIN" (ampicillin @ 250mg)
- "FISH-FLEX" (Cephalexin/Keflex @ 250mg)
- "FISH-FLEX FORTE" (Cephalexin/Keflex @ 500mg)
- "FISH-PEN" (penicillin @ 250mg)
- "FISH-PEN FORTE" (penicillin @ 500mg)
- "FISH-ZOLE" (metronidazole @ 250mg)
- "FISH-FLOX" (Ciprofloxacin/"Cipro" @ 250mg)
- "FISH-FLOX FORTE" (Ciprofloxacin/"Cipro" @ 500mg)

…and many, many more. One word of extreme caution, though - Even though you can also find tetracycline in there, once past the expiration date, tetracycline becomes toxic (whether for fish tanks or for humans).

Another good place to find antibiotics that doesn't involve a pharmacy is at a veterinary office and/or veterinary supply store (the latter if you happen to live in farm country). There, you will find most of the same antibiotics, just that they're concocted for livestock instead of people. Note that the same prohibition on using expired tetracycline exists, however.

One final bit to keep in mind: Anyone who tries to be his own doctor has an idiot for a patient. Always seek out competent advice. That source of advice can be a surviving doctor, nurse, or even pharmacist (at least for drug advice). EMTs, chiropractors, homeopathy experts and dentists? Let's just say that they aren't going to be good sources of advice to help cure strictly pathogenic ills. Seek out and stick with folks who best know what this stuff is and what it's used for - even veterinary (yes, animal) doctors will have more insight into these things.

All that said, there are natural antibiotics that you can put to use if all else fails. Just note that these are not going to have the same strength and efficacy as the manufactured drugs:

- Garlic
- Eucalyptus
- Onion (common and/or wild)
- Oregano
- Echinacea (a flower)
- Cumin

Natural Pain Killers

This is going to be a touchy subject, because of two reasons: First, a big stockpile of painkillers that aren't over-the-counter are going to be a huge liability should someone outside of your family ever discover its existence in your home (pre- or post-collapse). Second, because there is an all-natural painkiller that is a federal felony to possess, grow, transport, or even have seeds for. Given the massive hazards with the first, we're going to discuss the second… Marijuana.

Marijuana will come in differing strengths and strains. Most of them nowadays are bred to maximize the amount of THC (Tetrahydrocannabinol), which give hallucinogenic effects, but also acts as a efficient painkiller and an amazing anti-anxiety drug.

Unlike most illegal drugs, Marijuana has a low chance of physical addiction, which makes it (at least in that aspect) safe to grow and use as a post-collapse medication. In other aspects however, it does carry a bit of liability. Since it is a desired recreational drug among criminals, marijuana plants are, once spotted, going to be targeted heavily. It is usually best to grow this stuff discreetly and within thickets and well-concealed locations. The good news is, these plants usually grow well under most conditions and require very little care.

Obtaining the seeds for this plant is going to be tough to do unless you already have a license to grow it (Oregon, Washington, California), or know of someone who has seeds. Note that keeping seeds around constitutes a crime in most states pre-collapse, so keeping any around pre-collapse is not recommended. However, if you find any in any post-collapse market, look for seeds, keep them and give the rest of the stuff to whoever your community doctor may be.

One final note about Marijuana: Assuming you have and grow the stuff, consuming is best done orally or by steeping the leaves in a tea (or steep it in hot oil as a suspension). Smoking it is the fastest and most powerful means of consumption, but the smoke has a distinct odor that wafts across a neighborhood and tells any interested passerby with criminal intent that you have something they really want.

Other common and useful natural painkillers are as follows:

- Skullcap (Scutellaria lateriflora): an herb native to North America, this plant is also known as hoodwort, grows in and around wetlands and marshes. The leaves can be dried into a tea, steeped at a ratio of one ounce per cup of hot water. It is known to relieve tension headaches and other stress-caused pains (it does so by soothing nerves).
- Eucalyptus: While the tree is native to Australia, it is commonly grown as an ornamental plant along the Western US coastline (it grows especially well in the Pacific Northwest). Dried eucalyptus leaves in a steaming-hot (or even boiling) cup of water with a towel over the head to concentrate the vapors will clear out any sinus headaches and other sinus-related maladies.
- Cayenne, Habanero, or hotter Pepper: Yep - if you're not used to it, your mouth and lips will feel as if they're on fire, so this isn't something you would initially think of. However, for skin, osteoarthritis and muscle pains, you don't eat it - you rub it on your skin. The hotter the pepper, the better the results, so if someone is, say, growing fresh Ghost or Naga peppers? Buy one, put on a pair of gloves, cut off a chunk of pepper (rehydrate it first if needed) and rub that bad boy right on the affected area, covering it afterwards to keep your clothes from getting sticky. Most commercial pain relief creams contain capsaicin as the active ingredient, which is the same chemical which makes peppers hot. Note that jalapeno peppers won't quite do the trick, since the amount of capsaicin is really too low to do much good (unless you apply it to, say, an infant's skin...)
-

Anti-Inflammatory

Believe it or not, if you have or get arthritis (or any other type of chronic inflammation), eating a whole medium-sized red onion each day will help you out a bit. One medium-sized red onion contains 10 mg of quercitin, which is a flavonoid compound that inhibits inflammation. Now clinical strength (e.g. 500 mg) would require eating fifty onions a day, which is obviously not going to work, but even with one? Every little bit helps and if you have it, you may as well use it. Other good sources of this flavonoid that can be grown locally include broccoli, squash, red grapes, cranberries and if you happen to live in Florida, citrus fruits.

The herb Feverfew is a bit more effective - hopefully you'll have a picture of it in your foraging guides. One ounce of it steeped as a tea in two cups of hot water, done four times a day, will reduce a whole lot of inflammation and more importantly, the pain caused by it.

The Chamomile plant is originally a European plant (and has many varieties), but is now commonly found in many parts of North America, growing both wild and as a common ornamental herb. If you happen to find some, dry the leaves and you can make a tea of it. It works as a combination of things - it reduces inflammation overall, especially reduces stomach inflammation and acts as a mild sedative. One word of warning: it also has a habit of inducing premature labor in pregnant women, so any woman who is pregnant should avoid it like the plague.

Arnica (especially the species *Arnica Montana*) is flower that is known as an anti-inflammatory agent and is common throughout the western parts of North America. The dried roots or blossoms (depending on who you ask) can be ground into a powder and mixed into a cream or poultice, then smeared on a bruised or inflamed area of the skin. It is also good on unbroken post-surgery skin (but not in the incision or wound), to reduce inflammation. Some folks report that it works, others report that it does nothing, so take it as you will. Whatever you do, do not take it internally, as the natural toxins in the plant will cause gastroenteritis and internal bleeding of the digestive tract.

Summary

There are many ways of relieving pain and inflammation post-collapse, though it will require a lot more of the natural materials to get the job done. For antibiotics however, it is best to stock up before collapse, or to find opportunities to procure them in places where you wouldn't otherwise expect to. A solid field guide to edibles and medicinal herbs is strongly recommended, as picking the wrong plant to do the job (or using the right plant wrongly) may cause bigger problems than the one you're trying to cure.
e you're trying to cure.

Consumables And Vices

This section isn't exactly what you would call wholesome. However, human nature being what it is, it is better to do it correctly, than to try and haphazardly do it wrongly. Drinking alcohol made wrong can cause blindness or even be fatal. Not curing tobacco properly will cause you to smoke mold and mildew along with the nicotine, causing lung troubles a whole lot worse than what you're already doing to yourself. Note that illicit drugs for recreational use are not going to be mentioned at all in here. That kind of behavior is strictly not to be tolerated or condoned - this includes the misuse of beneficial drugs as well as the bad ones. So, without further ado…

Distilled (Ethyl/Ethanol) Alcohol

Among the world's oldest vices, distilled alcohol actually has a lot of beneficial uses - as a disinfectant, as a source of vinegar, as a useful ingredient in many food recipes and as a preservative.

What It Is, And What You Use It For:

Alcohol is basically the byproduct of yeast and sugars. It is usually distilled from grain, fruits, or even from pure sugar. Usually, the safest means of distilling alcohol is from pure sugar, but post-collapse, that's going to be tough to come by. However, grains and fruits will be easier to distill from and most contain high levels of sugars to easily accomplish this. That said, if you can keep it strictly to grains with moderate sugars (oats, barley, corn, etc), you have the makings of a good whisky. Something else to keep in mind: you can use the booze to preserve fruits, which keeps the fruit edible for far longer and at the same time lends some good flavors to the alcohol, making it a win-win.

Ingredients You Need:
- Grain, fruits, or other crops high in sugar
- Yeast (either wild or stored, or procured from bread starter) - this is semi-optional.
- A very large pot (to hold your initial mash)
- A fine screen with string or metal handles that fits into the bottom of the above-mentioned pot (to act as a means to filter out the spent grain). As an alternative, you can use a fine–screened colander.
- A 'Still (that is, distillery equipment)
- A good working candy thermometer, or a thermometer that allows you to tell the temperature to within 2-3 degrees Fahrenheit.
- Clean, sterilized containers to store the results
- Clean, purified water to make the mash and additional water to blend with the results (for anything less than 180 proof)
- (Optional) Clean hardwood chips or sticks, preferably cured

How To Make It:

The first step is to make your "mash". You first grind the grain or fruits into a coarse meal and set it aside - make enough to fill the large pot. Set the screen (if you have one) into the pot, then add the grain and clean hot water into a large container. The water should be hot, but not quite boiling hot. Let it sit for a couple of days, letting the starches in the grain convert to sugars as it steeps in the water, or letting the sugars leach out of the fruits, dissolving into the water.

At this point, you can either separate the wort (liquid) from the mash and ferment the just the wort, or you can let the whole thing ferment. Either way, try to keep the water temperature between 75-90 degrees Fahrenheit. If you don't have or want to add yeast, then let it sit for a day or two until you see bubbling occurring naturally at the top of the mash (or wort), or you can add yeast immediately.

Once you have fermentation started, let the whole mass sit (semi-covered) for 4-5 days, so that the yeast can do its work.

Before distillation, do a quick safety check. Go over the still carefully and make sure there are no leaks, cracks, pitting, or rust. Be sure you have enough fuel on hand for all batches and be sure your thermometer isn't broken, out of calibration, etc.

When everything is in order strain the liquid into the still - we only want the liquid at this point. The solids can be used as animal feed when mixed in with other grains. Close up the still and start up the heat. Heat the liquid slowly, taking about 30-60 minutes to reach a boil. While it's warming up, check that thermometer you have just before the condenser coil - once it reads 120° F- 140° F, you can start running cool water over the condenser coils, or fill the condenser cooling tub (depending on how you built it) with cool water and then open the tap at the collector pot. Be sure to have two containers handy for what comes out handy for the results. Oh and keep the fuel constant for the cooker from here on out until the batch is complete.

Once you start seeing liquid drip out of the collector pot tap, throw out the first bit of it, which is called the "head". This first bit of liquid contains methanol and other toxic/foul-tasting substances and you don't want it. If you have 5 gallons in the cooking pot, you want to toss out the first 1/4-1/3 cup that comes out of the still - it's the crap that will blind you or worse if it has a chance. For 10 gallons, the first 1/2-2/3 cup will be toxic, etc.

After that first bit is tossed, you can collect and set aside what comes out next - this is the good stuff. The thermometer should read 175° F-185° F around this time and it'll come out fairly rapidly. Set this stuff aside for now.

Once the thermometer climbs up to 205° F, you're starting to distill out something called fusel oil, which you also do not want. It is often called the "tail". Toss that portion out as well and shut down the still.

After the still has cooled, you have an option: You can re-run the still again with what is still in it, pulling off what little bit of drinking alcohol may still be in there. However, odds are good that you won't get much, if any out of that re-run.

Throw out everything that is left behind in the cooking pot after the run and clean it out scrupulously.

Your end product at this point is rough moonshine - if you want to refine it further, you can pour all of the first batch in to the clean cooking pot (plus any other 'first batch' alcohol you may have), gently heat and re-run the whole thing again. It is typical in commercial scotch whisky yo distill it twice and Irish whisky is usually distilled three times through.

Now you get to decide what to do with it. If it is for non-drinking purposes, bottle it immediately and label it as such. If you intend to drink it, the results will have to age a bit, else it will feel extremely rough on the throat. The best way to age it is to put it into a hardwood barrel and let it sit for 5 years - if you're not that patient, you can let it sit in a clean semi-closed (but not sealed!) container with a small bit of clean hardwood (preferably oak) in it - either a small layer of clean chips on the bottom, or a couple of small slabs of it. The best aging container would be a clean 3-5 gallon bucket with a tight-fitting lid, but a moderate number of pin-holes in the lid to allow for an exchange of air. The holes are kept numerous but small to slow down any evaporation. You should have good results in just a couple of weeks, which makes the whisky smoother and adds a tinge of oak (or other hardwood) flavor. Be careful with what hardwood you choose - some will taste rather nasty. Oak (any type) is usually a good bet. Avoid soft and/or resinous woods at all costs.

Once you're ready to sell or drink it, you will definitely want to cut it with some water down to taste. Experiment with tiny batches until you get what you want, but don't cut it down by too much, especially if you intend to sell it (your reputation hinges on your product, after all.)

Alternatives:

You have beer, wine and cider, no sweat - each has their own process, takes a bit more time (sometimes a lot more time) and the results are various. However, the other alternatives do not produce results with such an amazing longevity and shelf-life. Even wine has to be stored under certain conditions to remain viable and long-lived wines will require exacting conditions to last that long.

Safety Concerns:

We have a couple of them here. First, if you don't mind the temperature gauge, you could end up with product that will blind or even kill. If you don't discard the entire 'head' of your batch (or use it strictly for disinfection, cleaning and non-consumption purposes), you get the same bad results. Same story with the 'tail' of the batch.

You're dealing with ethanol - a highly flammable substance. Keeping it away from flame (including the flame you're heating the whole shebang with) is paramount. You're also dealing with with something that can dull your senses if you consume it while you're brewing it, so for your own safety, stay sober while you're running the still and while you're cleaning it. Scrupulously clean out everything when you're done - the still, the mash pot and everything else you used. If you can boil any part of it or use boiling water to clean it, do so, in order to prevent bad bacteria from ruining your batches at best, or killing someone at worst.

Other Considerations:

Booze is going to be a very tempting post-collapse target for a long time. If you sell and/or barter the stuff in quantity, only take what you intend to sell that day. Never allow anyone to follow you home from a booze sale - at any distance, or under any circumstance. You may be better off acting as a supplier to whatever post-collapse bars and the like that will crop up and selling it on the down-low.

Hard Cider

What It Is, And What You Use It For:

Another means of making alcoholic beverages that will keep for awhile is to make fruits into a cider. This is useful in areas where you have a whole lot of fruit that comes into season at once - apples, peaches, cherries, plums, blackberries, you-name-it (note that some citrus fruits aren't really going to work for this, but success has been found with oranges and tangerines). A good cider will keep in a tightly-sealed container for a couple of months easily and can be a very useful means of preserving fruits.

Ingredients You Need:
- A good source of yeast (from bread starter, wild yeast recovered and kept alive from distilling or brewing, etc)
- Lots of fruit (apple works best, but any sweet fruit will work)
- A means of pressing the juice out of the fruit
- A few 3 or 5-gallon buckets with airtight lids, or large jugs with rubber stoppers and a means of running an airlock through them (more on that in a second)
- The brewing buckets need airtight lids with an "airlock" - an airlock is a means to let CO_2 out, but not let air back in. Pre-collapse, they can be had at any brewing supply store. Post-collapse, you can make one out of a moderate length of ½"clear plastic tubing attached air-tight to a hole in the lid, run out of the top then gently bent in a complete loop, with the top of the tube sticking up and covered with clean cloth.
- At least two large stainless steel or plastic spoons (to stir with)

- A means to sanitize the buckets (bleach works well).
- Bottles, jugs, or other containers that you can tightly seal the contents into.
- A moderate length of clean plastic tubing to siphon out the cider once it's done.
- (Optional) brown sugar or honey - up to 2 lb per 5 gallons, to increase the alcohol content
- (Optional) a couple of clean coffee filters and a funnel that the filters can closely fit into.

How To Make It:

First, sanitize your buckets. You can do this by adding a capful (2 tsp) of bleach to a bucketful of water and letting it sit for 45 minutes - or you can use soapy, boiling water, let stand for 15 minutes, then rinse with clean water. Also take the time to sanitize the spoons and anything else the juice/cider will come in contact with. Rig up the airlocks to the lids.

Next - WASH THE FRUITS COMPLETELY BEFORE PRESSING! Seriously - wash the stupid things - scrupulously clean - with clean water. You're spending all of this time cleaning and sanitizing the tools, but you're going to start pressing fruits covered in bits of dirt and animal poop, carried home in containers of questionable cleanliness and maybe rinsed off a little? Let me put it this way: You either wash the fruits, or you drink the stuff and risk spending days on end with torrents of evil-smelling stuff gushing from both mouth and anus simultaneously - your call. (And on a more serious note, that kind of result can kill you, post-collapse.)

Once you have clean receptacles, get the buckets handy. Press out your fruit, collecting the juice. Filter it once though cheesecloth or screen to keep out stray skin or seeds, but don't get too picky about pulp being left behind.

When you have enough juice to fill a bucket to within 2" of the top, add a bit of yeast and the optional sugar or honey if you have it. Stir it a bit with a spoon to aerate it a bit and to blend everything together. Seal the bucket tightly with the lid and fill your airlock tubing with clean water - just enough to fill the 'trap' and no more. Start with the next bucket, then the next, until you're either out of buckets or out of juice. Keep the buckets in a dark and quiet place, where the temperature is a more-or-less constant 60-75 degrees.

Within a couple of days, you will notice bubbling in the airlocks. This is a desired result, as yeast consumes the fruit (and others if you add them) sugars, converting them into alcohol and giving off carbon dioxide. Keep a twice-daily eye on the batches, as they should continue to ferment for up to two weeks or so. Once the bubbling stops, let the buckets sit for another week to let the dying/dead yeast, pulp and other contents settle to the bottom.

At this point, you can bottle the contents, or clarify the cider further. At this point, it is a bit cloudy, but perfectly usable. Whatever you do, move the buckets extremely carefully at this point, to avoid sloshing up the goop that has settled at the bottom.

If you choose to clarify it, then have additional buckets with lids and airlocks handy - sanitize them as you did the others, then carefully siphon the cider from a finished bucket into the clean one. To siphon the cider out, affix a funnel with a coffee filter in it to the side of the clean bucket, then carefully (and slowly) siphon the cider through the filter/funnel into the new bucket. Stop siphoning when you can no longer efficiently filter out the goop in the bottom of the fermentation bucket. Once the new bucket is full, remove the funnel/filter assembly, seal the bucket tightly with the lid and fill the airlock with water as usual. Let the bucket sit a whole month, or longer if you really want to harden it. Any surviving yeast will finish the job and the results will definitely be clearer.

Whether or not you choose to clarify the cider, bottling is the same: Scrupulously clean, rinse and then boil the bottles/jugs/etc and corks/caps/lids/etc - we want the insides to be as clean as humanly possible. Let them cool before bottling. Carefully and slowly siphon out the cider from the bucket through a funnel lines with a coffee filter or two, directly into the bottles (or jugs, etc). Cap/cork/seal each immediately as it becomes full. You will want to consider pasteurizing the cider once it is bottled. To do so, heat a large pot of water (enough water to submerge half of a standing bottle or container) to 160 degrees Fahrenheit. For regular beer-sized bottles, let stand in this water for 10 minutes, then remove and let cool. For larger bottles, adjust up to 15-30 minutes depending on diameter of the container. Do not let the contents reach boiling temperature, as it will really throw off the taste.

You can consume the bottles as soon as they are cool, or let them sit in a dark, cool place for a couple of months to age. The flavor improves over time, much like wine does.

Alternatives:
None outside of what was already discussed previously with distilled alcohol.

Safety Concerns:
Keep everything clean. Clean everything up and re-sanitize it all immediately when you're done using it and also do so between batches. Avoid drinking anything that appears to have become moldy. Cider has an average shelf-life of up to 2 years or more, but anything beyond 3 years should be tested first.

Other Considerations:
If you happened to score some commercial pre-collapse cider or apple juice, don't use it for this. Most of these juices/ciders will contain preservatives which will kill yeast.

In freezing climates, you can make "Applejack" (or "plumjack", or cherryjack", or whatever) out of hard cider. Simply leave a small covered bucket of cider out to freeze overnight. The alcohol won't freeze, but the water and juice will. The next morning, you pour the liquid into another clean bucket and toss the ice (or use the ice as a fruit flavoring, whatever). Repeat this process with the same liquid for about 3-5 nights and you will end up with a far, far higher alcohol content. Think of it as distillation in reverse. Just note that the impurities (microscopic solids) will remain with the booze, but they're harmless and add to the flavor (as long as you keep everything clean, that is). As a bonus, you can save the refined booze as a means to 'pickle' canned fruits of the same species. For instance, you can use Applejack to can apple slices, etc.

Like we mentioned with distilled/grain alcohol, booze is going to be a very tempting post-collapse target for a long time. However, since there is no still involved, you use less fuel and you can more easily conceal it.

Okay, but where's the beer and wine? C'mon!

There are many, many good books on those subjects and since the ingredients for beer and grape-based wines are a bit specialized, we'll leave them out here. We shot for the two recipes we had because you can use a wide variety of ingredients, the processes are fairly straightforward and aside for a few critical bits in distillation (which are useful elsewhere), are pretty simple. Also, the processes aren't as time-consuming as making usable beer or wine will be. Finally, it is real easy to screw up beer or wine, meaning a higher chance of wastage - something we definitely want to minimize.

Weaponry

Now this is a really odd place to be finding weapons recipes. However, we'll concentrate on two very vital bits you will want to know how to make and use: black gunpowder and homemade incendiary fluid (not quite napalm, but good enough to act like it).

Note that we have peaceful uses for this stuff as well. The gunpowder can be used as blasting powder - not as good as real TNT, but good enough to dislodge pesky rocks and stumps. The incendiary fluid can be put to peaceful use as torch fuel, lamp oil, or other good stuff that doesn't involve a molotov cocktail, or burning someone's town to the ground. So, without further ado…

…wait! Note that I cannot and will not guarantee any of what you see in here. Messing with this stuff can be extremely illegal in the pre-collapse region you live in. These are only included for pre-collapse research and post-collapse use. If you make this stuff pre-collapse and get arrested, it's your own damned fault, not mine. If you fail to observe all safety precautions and use your head, you can very easily kill yourself, your family and in sufficient quantities, half of your neighborhood. Take all precautions, don't be stupid and again - note that this is for post-collapse use only!

Gunpowder ("Black Powder")

Black Powder was the most commonly used bullet (and cannonball) propellant until the invention of the more powerful nitrocellulose ("smokeless") gunpowder in the late 1800's. In muzzle-loading guns and rifles, black powder is still useful as a means of moving projectiles at animal- and people-killing velocities. Making the stuff is relatively easy as far as mixing dangerous chemicals go and can be a valuable post-collapse bartering item.

<u>What It Is, And What You Use It For</u>:
Black powder can be used as gunpowder, cannonball propellant, blasting powder and a surprisingly efficient means of clearing obstacles from roads in a hurry.

<u>Ingredients You Need</u>:
- Saltpeter (we'll discuss how to make the stuff shortly)
- Charcoal (from the recipe earlier on - avoid using activated charcoal or briquettes.
- Sulfur (hard to come by, but is actually optional - we only include it here for tradition's sake)
- Three SEPARATE mortar/pestle kits (one for each ingredient)
- A safe place away from fire to make the stuff
- A few large buckets
- A strong screen or sieve - with 0.05" holes (1.5mm).

<u>How To Make It</u>:

First, we need to make the saltpeter if you don't already have it. Making the stuff takes a long time, but comes from a constantly renewable resource: feces and urine. The idea is simple: You take the stuff out of the outhouses and pour it into a big, seal-able tank (I mean really big - 200-300 gallons preferably), mixing a bit of water in and letting everything brew for about 10 months. Note that it really helps to have a pipe coming out of the bottom with a tap and a means to screen it. Once 10-12 months have passed by, you fix up a large watertight trough to the spigot sloping downwards at a gentle angle (or just set up the trough). Pack the trough tight with ashes (hardwood ashes - but don't use ashes that had previously been used in making soap) and fix a screen at the bottom to filter out any solids. You then gently let the liquid run down the trough and into shallow trays (or into buckets which are then poured into shallow trays). Pack in more ashes as you go. Let the trays evaporate in the sun and repeatedly pour more liquid into the trays, letting evaporation happen each time before refilling. What you will eventually have left behind are crystals of potassium nitrate, which is the technical name for saltpeter. Scrape it out of the trays and store it in dry bags for later use. As an alternative, you could always urinate in a shallow tray that is covered to keep rain out, but still well ventilated. Over enough time (about a year or more), you will get the same crystals. Note that there are other and less 'icky' means of finding, mining and extracting saltpeter. However, the aforementioned method of fermenting/leaching/evaporating human (and animal) waste is the most common and by far the most reliable.

Next up is your charcoal. We've discussed making charcoal as one of our recipes, but this time you have to be picky about what kind of wood you use. Always prefer to go for certain soft woods: willow, softer pine woods (with no knots), redwood, cottonwood, or cedar. Once you have charcoal made, keep it whole and set it aside in a dry place until you are ready.

It should be noted at this point that sulfur is strictly optional, but many people insist that it works well as a means to prolong the 'burn time', which in turn generates more gases behind the bullet or shot. The only problem is, unless you live near a dormant volcano, a sulfur mine, or happen to have a whole lot of the stuff stored away, it's going to be extremely hard to make your own sulfur in any quantity post-collapse without a wide supply of chemicals and equipment. Fortunately, sulfur is no big deal and isn't needed. However, we'll include it as an option from here out.

The next steps are actually quite easy, but are also the most fraught with danger. First, find a dry place away from anything you don't want blown up. Bring in your ingredients to this spot, as well as your buckets, mortars/pestles (or grain grinders) and everything else you need. Shoo away the spouse, the kids and dogs and allow no fires anywhere near this area. Note that you should either have a mortar/grinder for each ingredient, or you should scrupulously clean the thing between each step.

First, grind your ingredients. Start with charcoal in one pestle (or mill, or what-have-you) into a fine powder, about the consistency of baking flour. When you have enough, set it aside for a bit. In another pestle/grinder/whatever, grind your saltpeter into the same rough consistency (and if you insist on using sulfur, do the same with it in its own mortar/grinder/etc). Note that the finer you grind everything now, the less fouling you'll find in a blackpowder gun later on.

Once you have your ingredients ground finely, you have some options. If you want a slow-burning powder, you will blend it dry. If you want a fast-burning powder, you'll blend it wet. Keeping that in mind, here are the proportions you will need :

Without Sulfur:
- 100 oz. Saltpeter
- 24 oz. Charcoal

With Sulfur:
- 100 oz. Saltpeter
- 18 oz. Charcoal
- 16 oz. Sulfur

For wet-grinding, you put the ingredients into a mortar in equal parts and grind on it for about 10-15 minutes. Add a little water (around 8-9% of the volume in the mortar) and mix the water in until the mixture has the consistency of pottery clay. Pound and fold the mixture for about 15-20 minutes further, then roll into a ball.

For dry grinding, you do the same thing, but don't bother with water- instead grinding slowly for about 30-45 minutes (to avoid too much heat build-up), then packaging as a slow-burning of fuse powder.

The next step with the wet-grind powder is called corning. You take those balls of damp powder and rub them one at a time against the mesh or sieve, onto a sheet of dry paper or other dry, clean surface in a single layer (for fast drying). Let the powder thoroughly dry at this point, then store it, ready for use.

Alternatives:
Not too many at this point. One thing to keep in mind: do not think that you can scavenge from existing bullets, shells, or suchlike, because those mixtures are going to be far more powerful and the results will be unpredictable.

Safety Concerns:
Plenty! While the powder won't ignite spontaneously when wet, the dry stuff is going to be flammable and should be kept far, far away from excess heat, sparks and any source of flame until you're actually using it. Do not, under any circumstance, mix it with any other form of gunpowder unless you're making a bomb, because the results will be very unpredictable.

Other Considerations:
If you do make dry-grind gunpowder, you can make a fuse with it by wrapping it in a paper tube, by gluing a thick layer to a long strip of paper, or by carefully stripping an electric wire and filling the insulator with the dry powder.

When it comes to using this powder in a muzzle-loading gun, odds are good that you will need 1.5-2.5x as much homemade powder in a charge as you would for modern pre-collapse black powders (mostly because the pre-collapse stuff was ground by enormous millstones to extremely fine granules, causing a more consistent burn). Finally, note that your gun will foul up faster with the homemade stuff, depending on the skill with which the powder was made and depending on the purity of the ingredients going in.

Incendiary Fluid

What It Is, And What You Use It For:
No, we're not going to call it "napalm" and for two reasons: First, the stuff isn't likely to be quite as sticky and second, because there are going to be a surprising number of peaceful uses for it as well, mostly having to do with starting and keeping fires lit. There are going to be multiple variations here, because there are multiple ways of getting stuff that gives similar results.

Ingredients You Need:

Type A (and most commonly known):
- Old Gasoline
- Styrofoam Cups

Type B:
- Soap
- Old gasoline
- A metal bucket
- A large double-boiler that you really do not like

To make Type A:
Procure some old gasoline. Anything older than 2 years after refining counts (because it'll be practically worthless as engine fuel), but it should have at least some vapor action (you'll be able to smell it). Fill a container with it. Start melting styrofoam cups into the gasoline and continue doing so until you end up with a thick gelatinous consistency and no more cups can be dissolved in it. Store it in tightly-sealed glass jars until ready to use.

To make Type B:
Mix soap shavings and the old gasoline into the bucket at a 1:1 ratio. Put the water in the bottom of the double-boiler and get it boiling. After it has boiled for a little while, take it away from the heat (well away!) to a safe place. Dump the bucket of gasoline and soap in to the top part of the double-boiler, stirring it until the soap is completely dissolved. If you find the mixture cooling too early (before it's all dissolved), there is another means of heating the mixture. Build a fire against a largish flat rock, until the rock is insanely hot. Remove it from the fire (with huge tongs or such) to a safe distance and use that rock to heat the pan with the gas/soap mixture.

The result is a viscous fluid that you can use as an incendiary device. Store in tightly-sealed glass jars until ready for use.

Alternatives:
Plenty of 'em and way too many recipes to list. Thing is, most involve gasoline. You can also use kerosene or straight grain alcohol to an extent, but the results will be inconsistent and will take a bit of experimentation to make it work.

Safety Concerns:
One big obvious one: this stuff is highly flammable and making (or storing) it around open flame is monumentally stupid thing to do.

Other Considerations:
Getting the gas for this stuff is going to be rather tough to do and for two reasons: First, because gasoline is going to be extremely rare as time passes. Second, because what you do find as time passes is going to be less potent due to most of its vapors dissipating. However, you can have some luck from abandoned vehicle tanks (if they haven't already been punctured/siphoned), or by 'dipping' it out of underground storage tanks at abandoned gas stations.

Government

One thing up front: What type of government and its operation you settle on is up to you, but you really, really need to know what the United States once had in concept. To that end, I'm including the three major original founding documents: The original Declaration of Independence, Articles of Confederation and US Constitution. Note that to save ink, the font is going to be a little smaller than what you have been reading.

The reason why we include each are as follows:

- We include the Declaration of Independence because it is an excellent example of what an oppressive regime is capable of and what you do not want your government doing.
- The Articles of Confederation are important, as they served as a bridge between a tentative collection of rebellious colonies and a permanent enduring government.
- Finally, the United States Constitution - a permanent foundation of government that allows for modification under strict conditions, which allows government to evolve, but only with consent of the governed.

Declaration Of Independence

IN CONGRESS, July 4, 1776.
The unanimous Declaration of the thirteen united States of America,

When in the Course of human events, it becomes necessary for one people to dissolve the political bands which have connected them with another and to assume among the powers of the earth, the separate and equal station to which the Laws of Nature and of Nature's God entitle them, a decent respect to the opinions of mankind requires that they should declare the causes which impel them to the separation.

We hold these truths to be self-evident, that all men are created equal, that they are endowed by their Creator with certain unalienable Rights, that among these are Life, Liberty and the pursuit of Happiness.--That to secure these rights, Governments are instituted among Men, deriving their just powers from the consent of the governed, --That whenever any Form of Government becomes destructive of these ends, it is the Right of the People to alter or to abolish it and to institute new Government, laying its foundation on such principles and organizing its powers in such form, as to them shall seem most likely to effect their Safety and Happiness. Prudence, indeed, will dictate that Governments long established should not be changed for light and transient causes; and accordingly all experience hath shewn, that mankind are more disposed to suffer, while evils are sufferable, than to right themselves by abolishing the forms to which they are accustomed. But when a long train of abuses and usurpations, pursuing invariably the same Object evinces a design to reduce them under absolute Despotism, it is their right, it is their duty, to throw off such Government and to provide new Guards for their future security.--Such has been the patient sufferance of these Colonies; and such is now the necessity which constrains them to alter their former Systems of Government. The history of the present King of Great Britain is a history of repeated injuries and usurpations, all having in direct object the establishment of an absolute Tyranny over these States. To prove this, let Facts be submitted to a candid world.

He has refused his Assent to Laws, the most wholesome and necessary for the public good.

He has forbidden his Governors to pass Laws of immediate and pressing importance, unless suspended in their operation till his Assent should be obtained; and when so suspended, he has utterly neglected to attend to them.

He has refused to pass other Laws for the accommodation of large districts of people, unless those people would relinquish the right of Representation in the Legislature, a right inestimable to them and formidable to tyrants only.

He has called together legislative bodies at places unusual, uncomfortable and distant from the depository of their public Records, for the sole purpose of fatiguing them into compliance with his measures.

He has dissolved Representative Houses repeatedly, for opposing with manly firmness his invasions on the rights of the people.

He has refused for a long time, after such dissolutions, to cause others to be elected; whereby the Legislative powers, incapable of Annihilation, have returned to the People at large for their exercise; the State remaining in the mean time exposed to all the dangers of invasion from without and convulsions within.

He has endeavoured to prevent the population of these States; for that purpose obstructing the Laws for Naturalization of Foreigners; refusing to pass others to encourage their migrations hither and raising the conditions of new Appropriations of Lands.

He has obstructed the Administration of Justice, by refusing his Assent to Laws for establishing Judiciary powers.

He has made Judges dependent on his Will alone, for the tenure of their offices and the amount and payment of their salaries.

He has erected a multitude of New Offices and sent hither swarms of Officers to harrass our people and eat out their substance.

He has kept among us, in times of peace, Standing Armies without the Consent of our legislatures.

He has affected to render the Military independent of and superior to the Civil power.

He has combined with others to subject us to a jurisdiction foreign to our constitution and unacknowledged by our laws; giving his Assent to their Acts of pretended Legislation:
For Quartering large bodies of armed troops among us:

For protecting them, by a mock Trial, from punishment for any Murders which they should commit on the Inhabitants of these States:

For cutting off our Trade with all parts of the world:

For imposing Taxes on us without our Consent:

For depriving us in many cases, of the benefits of Trial by Jury:

For transporting us beyond Seas to be tried for pretended offences

For abolishing the free System of English Laws in a neighbouring Province, establishing therein an Arbitrary government and enlarging its Boundaries so as to render it at once an example and fit instrument for introducing the same absolute rule into these Colonies:

For taking away our Charters, abolishing our most valuable Laws and altering fundamentally the Forms of our Governments:

For suspending our own Legislatures and declaring themselves invested with power to legislate for us in all cases whatsoever.

He has abdicated Government here, by declaring us out of his Protection and waging War against us.

He has plundered our seas, ravaged our Coasts, burnt our towns and destroyed the lives of our people.

He is at this time transporting large Armies of foreign Mercenaries to compleat the works of death, desolation and tyranny, already begun with circumstances of Cruelty & perfidy scarcely paralleled in the most barbarous ages and totally unworthy the Head of a civilized nation.

He has constrained our fellow Citizens taken Captive on the high Seas to bear Arms against their Country, to become the executioners of their friends and Brethren, or to fall themselves by their Hands.

He has excited domestic insurrections amongst us and has endeavoured to bring on the inhabitants of our frontiers, the merciless Indian Savages, whose known rule of warfare, is an undistinguished destruction of all ages, sexes and conditions.

In every stage of these Oppressions We have Petitioned for Redress in the most humble terms: Our repeated Petitions have been answered only by repeated injury. A Prince whose character is thus marked by every act which may define a Tyrant, is unfit to be the ruler of a free people.

Nor have We been wanting in attentions to our Brittish brethren. We have warned them from time to time of attempts by their legislature to extend an unwarrantable jurisdiction over us. We have reminded them of the circumstances of our emigration and settlement here. We have appealed to their native justice and magnanimity and we have conjured them by the ties of our common kindred to disavow these usurpations, which, would inevitably interrupt our connections and correspondence. They too have been deaf to the voice of justice and of consanguinity. We must, therefore, acquiesce in the necessity, which denounces our Separation and hold them, as we hold the rest of mankind, Enemies in War, in Peace Friends.

We, therefore, the Representatives of the united States of America, in General Congress, Assembled, appealing to the Supreme Judge of the world for the rectitude of our intentions, do, in the Name and by Authority of the good People of these Colonies, solemnly publish and declare, That these United Colonies are and of Right ought to be Free and Independent States; that they are Absolved from all Allegiance to the British Crown and that all political connection between them and the State of Great Britain, is and ought to be totally dissolved; and that as Free and Independent States, they have full Power to levy War, conclude Peace, contract Alliances, establish Commerce and to do all other Acts and Things which Independent States may of right do. And for the support of this Declaration, with a firm reliance on the protection of divine Providence, we mutually pledge to each other our Lives, our Fortunes and our sacred Honor.

Articles of Confederation

To all to whom these Presents shall come, we the undersigned Delegates of the States affixed to our Names send greeting.

Articles of Confederation and perpetual Union between the states of New Hampshire, Massachusetts-bay Rhode Island and Providence Plantations, Connecticut, New York, New Jersey, Pennsylvania, Delaware, Maryland, Virginia, North Carolina, South Carolina and Georgia.

I.

The Stile of this Confederacy shall be

"The United States of America".

II.

Each state retains its sovereignty, freedom and independence and every power, jurisdiction and right, which is not by this Confederation expressly delegated to the United States, in Congress assembled.

III.

The said States hereby severally enter into a firm league of friendship with each other, for their common defense, the security of their liberties and their mutual and general welfare, binding themselves to assist each other, against all force offered to, or attacks made upon them, or any of them, on account of religion, sovereignty, trade, or any other pretense whatever.

IV.

The better to secure and perpetuate mutual friendship and intercourse among the people of the different States in this Union, the free inhabitants of each of these States, paupers, vagabonds and fugitives from justice excepted, shall be entitled to all privileges and immunities of free citizens in the several States; and the people of each State shall free ingress and regress to and from any other State and shall enjoy therein all the privileges of trade and commerce, subject to the same duties, impositions and restrictions as the inhabitants thereof respectively, provided that such restrictions shall not extend so far as to prevent the removal of property imported into any State, to any other State, of which the owner is an inhabitant; provided also that no imposition, duties or restriction shall be laid by any State, on the property of the United States, or either of them.

If any person guilty of, or charged with, treason, felony, or other high misdemeanor in any State, shall flee from justice and be found in any of the United States, he shall, upon demand of the Governor or executive power of the State from which he fled, be delivered up and removed to the State having jurisdiction of his offense.

Full faith and credit shall be given in each of these States to the records, acts and judicial proceedings of the courts and magistrates of every other State.

V.

For the most convenient management of the general interests of the United States, delegates shall be annually appointed in such manner as the legislatures of each State shall direct, to meet in Congress on the first Monday in November, in every year, with a powerreserved to each State to recall its delegates, or any of them, at any time within the year and to send others in their stead for the remainder of the year.

No State shall be represented in Congress by less than two, nor more than seven members; and no person shall be capable of being a delegate for more than three years in any term of six years; nor shall any person, being a delegate, be capable of holding any office under the United States, for which he, or another for his benefit, receives any salary, fees or emolument of any kind.

Each State shall maintain its own delegates in a meeting of the States and while they act as members of the committee of the States.

In determining questions in the United States in Congress assembled, each State shall have one vote.

Freedom of speech and debate in Congress shall not be impeached or questioned in any court or place out of Congress and the members of Congress shall be protected in their persons from arrests or imprisonments, during the time of their going to and from and attendence on Congress, except for treason, felony, or breach of the peace.

VI.
No State, without the consent of the United States in Congress assembled, shall send any embassy to, or receive any embassy from, or enter into any conference, agreement, alliance or treaty with any King, Prince or State; nor shall any person holding any office of profit or trust under the United States, or any of them, accept any present, emolument, office or title of any kind whatever from any King, Prince or foreign State; nor shall the United States in Congress assembled, or any of them, grant any title of nobility.

No two or more States shall enter into any treaty, confederation or alliance whatever between them, without the consent of the United States in Congress assembled, specifying accurately the purposes for which the same is to be entered into and how long it shall continue.

No State shall lay any imposts or duties, which may interfere with any stipulations in treaties, entered into by the United States in Congress assembled, with any King, Prince or State, in pursuance of any treaties already proposed by Congress, to the courts of France and Spain.

No vessel of war shall be kept up in time of peace by any State, except such number only, as shall be deemed necessary by the United States in Congress assembled, for the defense of such State, or its trade; nor shall any body of forces be kept up by any State in time of peace, except such number only, as in the judgement of the United States in Congress assembled, shall be deemed requisite to garrison the forts necessary for the defense of such State; but every State shall always keep up a well-regulated and disciplined militia, sufficiently armed and accoutered and shall provide and constantly have ready for use, in public stores, a due number of filed pieces and tents and a proper quantity of arms, ammunition and camp equipage.

No State shall engage in any war without the consent of the United States in Congress assembled, unless such State be actually invaded by enemies, or shall have received certain advice of a resolution being formed by some nation of Indians to invade such State and the danger is so imminent as not to admit of a delay till the United States in Congress assembled can be consulted; nor shall any State grant commissions to any ships or vessels of war, nor letters of marque or reprisal, except it be after a declaration of war by the United States in Congress assembled and then only against the Kingdom or State and the subjects thereof, against which war has been so declared and under such regulations as shall be established by the United States in Congress assembled, unless such State be infested by pirates, in which case vessels of war may be fitted out for that occasion and kept so long as the danger shall continue, or until the United States in Congress assembled shall determine otherwise.

VII.
When land forces are raised by any State for the common defense, all officers of or under the rank of colonel, shall be appointed by the legislature of each State respectively, by whom such forces shall be raised, or in such manner as such State shall direct and all vacancies shall be filled up by the State which first made the appointment.

VIII.
All charges of war and all other expenses that shall be incurred for the common defense or general welfare and allowed by the United States in Congress assembled, shall be defrayed out of a common treasury, which shall be supplied by the several States in proportion to the value of all land within each State, granted or surveyed for any person, as such land and the buildings and improvements thereon shall be estimated according to such mode as the United States in Congress assembled, shall from time to time direct and appoint.

The taxes for paying that proportion shall be laid and levied by the authority and direction of the legislatures of the several States within the time agreed upon by the United States in Congress assembled.

IX.

The United States in Congress assembled, shall have the sole and exclusive right and power of determining on peace and war, except in the cases mentioned in the sixth article -- of sending and receiving ambassadors -- entering into treaties and alliances, provided that no treaty of commerce shall be made whereby the legislative power of the respective States shall be restrained from imposing such imposts and duties on foreigners, as their own people are subjected to, or from prohibiting the exportation or importation of any species of goods or commodities whatsoever -- of establishing rules for deciding in all cases, what captures on land or water shall be legal and in what manner prizes taken by land or naval forces in the service of the United States shall be divided or appropriated -- of granting letters of marque and reprisal in times of peace -- appointing courts for the trial of piracies and felonies commited on the high seas and establishing courts for receiving and determining finally appeals in all cases of captures, provided that no member of Congress shall be appointed a judge of any of the said courts.

The United States in Congress assembled shall also be the last resort on appeal in all disputes and differences now subsisting or that hereafter may arise between two or more States concerning boundary, jurisdiction or any other causes whatever; which authority shall always be exercised in the manner following. Whenever the legislative or executive authority or lawful agent of any State in controversy with another shall present a petition to Congress stating the matter in question and praying for a hearing, notice thereof shall be given by order of Congress to the legislative or executive authority of the other State in controversy and a day assigned for the appearance of the parties by their lawful agents, who shall then be directed to appoint by joint consent, commissioners or judges to constitute a court for hearing and determining the matter in question: but if they cannot agree, Congress shall name three persons out of each of the United States and from the list of such persons each party shall alternately strike out one, the petitioners beginning, until the number shall be reduced to thirteen; and from that number not less than seven, nor more than nine names as Congress shall direct, shall in the presence of Congress be drawn out by lot and the persons whose names shall be so drawn or any five of them, shall be commissioners or judges, to hear and finally determine the controversy, so always as a major part of the judges who shall hear the cause shall agree in the determination: and if either party shall neglect to attend at the day appointed, without showing reasons, which Congress shall judge sufficient, or being present shall refuse to strike, the Congress shall proceed to nominate three persons out of each State and the secretary of Congress shall strike in behalf of such party absent or refusing; and the judgement and sentence of the court to be appointed, in the manner before prescribed, shall be final and conclusive; and if any of the parties shall refuse to submit to the authority of such court, or to appear or defend their claim or cause, the court shall nevertheless proceed to pronounce sentence, or judgement, which shall in like manner be final and decisive, the judgement or sentence and other proceedings being in either case transmitted to Congress and lodged among the acts of Congress for the security of the parties concerned: provided that every commissioner, before he sits in judgement, shall take an oath to be administered by one of the judges of the supreme or superior court of the State, where the cause shall be tried, 'well and truly to hear and determine the matter in question, according to the best of his judgement, without favor, affection or hope of reward': provided also, that no State shall be deprived of territory for the benefit of the United States.

All controversies concerning the private right of soil claimed under different grants of two or more States, whose jurisdictions as they may respect such lands and the States which passed such grants are adjusted, the said grants or either of them being at the same time claimed to have originated antecedent to such settlement of jurisdiction, shall on the petition of either party to the Congress of the United States, be finally determined as near as may be in the same manner as is before presecribed for deciding disputes respecting territorial jurisdiction between different States.

The United States in Congress assembled shall also have the sole and exclusive right and power of regulating the alloy and value of coin struck by their own authority, or by that of the respective States -- fixing the standards of weights and measures throughout the United States -- regulating the trade and managing all affairs with the Indians, not members of any of the States, provided that the legislative right of any State within its own limits be not infringed or violated -- establishing or regulating post offices from one State to another, throughout all the United States and exacting such postage on the papers passing through the same as may be requisite to defray the expenses of the said office -- appointing all officers of the land forces, in the service of the United States, excepting regimental officers -- appointing all the officers of the naval forces and commissioning all officers whatever in the service of the United States -- making rules for the government and regulation of the said land and naval forces and directing their operations.

T he United States in Congress assembled shall have authority to appoint a committee, to sit in the recess of Congress, to be denominated 'A Committee of the States' and to consist of one delegate from each State; and to appoint such other committees and civil officers as may be necessary for managing the general affairs of the United States under their direction -- to appoint one of their members to preside, provided that no person be allowed to serve in the office of president more than one year in any term of three years; to ascertain the necessary sums of money to be raised for the service of the United States and to appropriate and apply the same for defraying the public expenses -- to borrow money, or emit bills on the credit of the United States, transmitting every half-year to the respective States an account of the sums of money so borrowed or emitted -- to build and equip a navy -- to agree upon the number of land forces and to make requisitions from each State for its quota, in proportion to the number of white inhabitants in such State; which requisition shall be binding and thereupon the legislature of each State shall appoint the regimental officers, raise the men and cloath, arm and equip them in a solid-like manner, at the expense of the United States; and the officers and men so cloathed, armed and equipped shall march to the place appointed and within the time agreed on by the United States in Congress assembled. But if the United States in Congress assembled shall, on consideration of circumstances judge proper that any State should not raise men, or should raise a smaller number of men than the quota thereof, such extra number shall be raised, officered, cloathed, armed and equipped in the same manner as the quota of each State, unless the legislature of such State shall judge that such extra number cannot be safely spread out in the same, in which case they shall raise, officer, cloath, arm and equip as many of such extra number as they judge can be safely spared. And the officers and men so cloathed, armed and equipped, shall march to the place appointed and within the time agreed on by the United States in Congress assembled.

The United States in Congress assembled shall never engage in a war, nor grant letters of marque or reprisal in time of peace, nor enter into any treaties or alliances, nor coin money, nor regulate the value thereof, nor ascertain the sums and expenses necessary for the defense and welfare of the United States, or any of them, nor emit bills, nor borrow money on the credit of the United States, nor appropriate money, nor agree upon the number of vessels of war, to be built or purchased, or the number of land or sea forces to be raised, nor appoint a commander in chief of the army or navy, unless nine States assent to the same: nor shall a question on any other point, except for adjourning from day to day be determined, unless by the votes of the majority of the United States in Congress assembled.

The Congress of the United States shall have power to adjourn to any time within the year and to any place within the United States, so that no period of adjournment be for a longer duration than the space of six months and shall publish the journal of their proceedings monthly, except such parts thereof relating to treaties, alliances or military operations, as in their judgement require secrecy; and the yeas and nays of the delegates of each State on any question shall be entered on the journal, when it is desired by any delegates of a State, or any of them, at his or their request shall be furnished with a transcript of the said journal, except such parts as are above excepted, to lay before the legislatures of the several States.

X.
The Committee of the States, or any nine of them, shall be authorized to execute, in the recess of Congress, such of the powers of Congress as the United States in Congress assembled, by the consent of the nine States, shall from time to time think expedient to vest them with; provided that no power be delegated to the said Committee, for the exercise of which, by the Articles of Confederation, the voice of nine States in the Congress of the United States assembled be requisite.

XI.
Canada acceding to this confederation and adjoining in the measures of the United States, shall be admitted into and entitled to all the advantages of this Union; but no other colony shall be admitted into the same, unless such admission be agreed to by nine States.

XII.
All bills of credit emitted, monies borrowed and debts contracted by, or under the authority of Congress, before the assembling of the United States, in pursuance of the present confederation, shall be deemed and considered as a charge against the United States, for payment and satisfaction whereof the said United States and the public faith are hereby solemnly pleged.

XIII.
Every State shall abide by the determination of the United States in Congress assembled, on all questions which by this confederation are submitted to them. And the Articles of this Confederation shall be inviolably observed by every State and the Union shall be perpetual; nor shall any alteration at any time hereafter be made in any of them; unless such alteration be agreed to in a Congress of the United States and be afterwards confirmed by the legislatures of every State.

And Whereas it hath pleased the Great Governor of the World to incline the hearts of the legislatures we respectively represent in Congress, to approve of and to authorize us to ratify the said Articles of Confederation and perpetual Union. Know Ye that we the undersigned delegates, by virtue of the power and authority to us given for that purpose, do by these presents, in the name and in behalf of our respective constituents, fully and entirely ratify and confirm each and every of the said Articles of Confederation and perpetual Union and all and singular the matters and things therein contained: And we do further solemnly plight and engage the faith of our respective constituents, that they shall abide by the determinations of the United States in Congress assembled, on all questions, which by the said Confederation are submitted to them. And that the Articles thereof shall be inviolably observed by the States we respectively represent and that the Union shall be perpetual.

In Witness whereof we have hereunto set our hands in Congress. Done at Philadelphia in the State of Pennsylvania the ninth day of July in the Year of our Lord One Thousand Seven Hundred and Seventy-Eight and in the Third Year of the independence of America.

Agreed to by Congress 15 November 1777 In force after ratification by Maryland, 1 March 1781

United States Constitution (with amendments)

The Constitution of the United States

Preamble

We the People of the United States, in Order to form a more perfect Union, establish Justice, insure domestic Tranquility, provide for the common defence, promote the general Welfare and secure the Blessings of Liberty to ourselves and our Posterity, do ordain and establish this Constitution for the United States of America.
Article I - The Legislative Branch Note

Section 1 - The Legislature

All legislative Powers herein granted shall be vested in a Congress of the United States, which shall consist of a Senate and House of Representatives.

Section 2 - The House

The House of Representatives shall be composed of Members chosen every second Year by the People of the several States and the Electors in each State shall have the Qualifications requisite for Electors of the most numerous Branch of the State Legislature.

No Person shall be a Representative who shall not have attained to the Age of twenty five Years and been seven Years a Citizen of the United States and who shall not, when elected, be an Inhabitant of that State in which he shall be chosen.

(Representatives and direct Taxes shall be apportioned among the several States which may be included within this Union, according to their respective Numbers, which shall be determined by adding to the whole Number of free Persons, including those bound to Service for a Term of Years and excluding Indians not taxed, three fifths of all other Persons.) (The previous sentence in parentheses was modified by the 14th Amendment, section 2.) The actual Enumeration shall be made within three Years after the first Meeting of the Congress of the United States and within every subsequent Term of ten Years, in such Manner as they shall by Law direct. The Number of Representatives shall not exceed one for every thirty Thousand, but each State shall have at Least one Representative; and until such enumeration shall be made, the State of New Hampshire shall be entitled to chuse three, Massachusetts eight, Rhode Island and Providence Plantations one, Connecticut five, New York six, New Jersey four, Pennsylvania eight, Delaware one, Maryland six, Virginia ten, North Carolina five, South Carolina five and Georgia three.

When vacancies happen in the Representation from any State, the Executive Authority thereof shall issue Writs of Election to fill such Vacancies.

The House of Representatives shall chuse their Speaker and other Officers; and shall have the sole Power of Impeachment.

Section 3 - The Senate

The Senate of the United States shall be composed of two Senators from each State, (chosen by the Legislature thereof,) (The preceding words in parentheses superseded by 17th Amendment, section 1.) for six Years; and each Senator shall have one Vote.

Immediately after they shall be assembled in Consequence of the first Election, they shall be divided as equally as may be into three Classes. The Seats of the Senators of the first Class shall be vacated at the Expiration of the second Year, of the second Class at the Expiration of the fourth Year and of the third Class at the Expiration of the sixth Year, so that one third may be chosen every second Year; (and if Vacancies happen by Resignation, or otherwise, during the Recess of the Legislature of any State, the Executive thereof may make temporary Appointments until the next Meeting of the Legislature, which shall then fill such Vacancies.) (The preceding words in parentheses were superseded by the 17th Amendment, section 2.)

No person shall be a Senator who shall not have attained to the Age of thirty Years and been nine Years a Citizen of the United States and who shall not, when elected, be an Inhabitant of that State for which he shall be chosen.

The Vice President of the United States shall be President of the Senate, but shall have no Vote, unless they be equally divided.

The Senate shall chuse their other Officers and also a President pro tempore, in the absence of the Vice President, or when he shall exercise the Office of President of the United States.

The Senate shall have the sole Power to try all Impeachments. When sitting for that Purpose, they shall be on Oath or Affirmation. When the President of the United States is tried, the Chief Justice shall preside: And no Person shall be convicted without the Concurrence of two thirds of the Members present.

Judgment in Cases of Impeachment shall not extend further than to removal from Office and disqualification to hold and enjoy any Office of honor, Trust or Profit under the United States: but the Party convicted shall nevertheless be liable and subject to Indictment, Trial, Judgment and Punishment, according to Law.

Section 4 - Elections, Meetings

The Times, Places and Manner of holding Elections for Senators and Representatives, shall be prescribed in each State by the Legislature thereof; but the Congress may at any time by Law make or alter such Regulations, except as to the Place of Chusing Senators.

The Congress shall assemble at least once in every Year and such Meeting shall (be on the first Monday in December,) (The preceding words in parentheses were superseded by the 20th Amendment, section 2.) unless they shall by Law appoint a different Day.

Section 5 - Membership, Rules, Journals, Adjournment

Each House shall be the Judge of the Elections, Returns and Qualifications of its own Members and a Majority of each shall constitute a Quorum to do Business; but a smaller number may adjourn from day to day and may be authorized to compel the Attendance of absent Members, in such Manner and under such Penalties as each House may provide.

Each House may determine the Rules of its Proceedings, punish its Members for disorderly Behavior and, with the Concurrence of two-thirds, expel a Member.

Each House shall keep a Journal of its Proceedings and from time to time publish the same, excepting such Parts as may in their Judgment require Secrecy; and the Yeas and Nays of the Members of either House on any question shall, at the Desire of one fifth of those Present, be entered on the Journal.

Neither House, during the Session of Congress, shall, without the Consent of the other, adjourn for more than three days, nor to any other Place than that in which the two Houses shall be sitting.

Section 6 - Compensation

(The Senators and Representatives shall receive a Compensation for their Services, to be ascertained by Law and paid out of the Treasury of the United States.) (The preceding words in parentheses were modified by the 27th Amendment.) They shall in all Cases, except Treason, Felony and Breach of the Peace, be privileged from Arrest during their Attendance at the Session of their respective Houses and in going to and returning from the same; and for any Speech or Debate in either House, they shall not be questioned in any other Place.

No Senator or Representative shall, during the Time for which he was elected, be appointed to any civil Office under the Authority of the United States which shall have been created, or the Emoluments whereof shall have been increased during such time; and no Person holding any Office under the United States, shall be a Member of either House during his Continuance in Office.

Section 7 - Revenue Bills, Legislative Process, Presidential Veto

All bills for raising Revenue shall originate in the House of Representatives; but the Senate may propose or concur with Amendments as on other Bills.

Every Bill which shall have passed the House of Representatives and the Senate, shall, before it become a Law, be presented to the President of the United States; If he approve he shall sign it, but if not he shall return it, with his Objections to that House in which it shall have originated, who shall enter the Objections at large on their Journal and proceed to reconsider it. If after such Reconsideration two thirds of that House shall agree to pass the Bill, it shall be sent, together with the Objections, to the other House, by which it shall likewise be reconsidered and if approved by two thirds of that House, it shall become a Law. But in all such Cases the Votes of both Houses shall be determined by Yeas and Nays and the Names of the Persons voting for and against the Bill shall be entered on the Journal of each House respectively. If any Bill shall not be returned by the President within ten Days (Sundays excepted) after it shall have been presented to him, the Same shall be a Law, in like Manner as if he had signed it, unless the Congress by their Adjournment prevent its Return, in which Case it shall not be a Law.

Every Order, Resolution, or Vote to which the Concurrence of the Senate and House of Representatives may be necessary (except on a question of Adjournment) shall be presented to the President of the United States; and before the Same shall take Effect, shall be approved by him, or being disapproved by him, shall be repassed by two thirds of the Senate and House of Representatives, according to the Rules and Limitations prescribed in the Case of a Bill.

Section 8 - Powers of Congress

The Congress shall have Power To lay and collect Taxes, Duties, Imposts and Excises, to pay the Debts and provide for the common Defence and general Welfare of the United States; but all Duties, Imposts and Excises shall be uniform throughout the United States;

To borrow money on the credit of the United States;

To regulate Commerce with foreign Nations and among the several States and with the Indian Tribes;

To establish an uniform Rule of Naturalization and uniform Laws on the subject of Bankruptcies throughout the United States;

To coin Money, regulate the Value thereof and of foreign Coin and fix the Standard of Weights and Measures;

To provide for the Punishment of counterfeiting the Securities and current Coin of the United States;

To establish Post Offices and Post Roads;

To promote the Progress of Science and useful Arts, by securing for limited Times to Authors and Inventors the exclusive Right to their respective Writings and Discoveries;

To constitute Tribunals inferior to the supreme Court;

To define and punish Piracies and Felonies committed on the high Seas and Offenses against the Law of Nations;

To declare War, grant Letters of Marque and Reprisal and make Rules concerning Captures on Land and Water;

To raise and support Armies, but no Appropriation of Money to that Use shall be for a longer Term than two Years;

To provide and maintain a Navy;

To make Rules for the Government and Regulation of the land and naval Forces;

To provide for calling forth the Militia to execute the Laws of the Union, suppress Insurrections and repel Invasions;

To provide for organizing, arming and disciplining, the Militia and for governing such Part of them as may be employed in the Service of the United States, reserving to the States respectively, the Appointment of the Officers and the Authority of training the Militia according to the discipline prescribed by Congress;

To exercise exclusive Legislation in all Cases whatsoever, over such District (not exceeding ten Miles square) as may, by Cession of particular States and the acceptance of Congress, become the Seat of the Government of the United States and to exercise like Authority over all Places purchased by the Consent of the Legislature of the State in which the Same shall be, for the Erection of Forts, Magazines, Arsenals, dock-Yards and other needful Buildings; And

To make all Laws which shall be necessary and proper for carrying into Execution the foregoing Powers and all other Powers vested by this Constitution in the Government of the United States, or in any Department or Officer thereof.

Section 9 - Limits on Congress

The Migration or Importation of such Persons as any of the States now existing shall think proper to admit, shall not be prohibited by the Congress prior to the Year one thousand eight hundred and eight, but a tax or duty may be imposed on such Importation, not exceeding ten dollars for each Person.

The privilege of the Writ of Habeas Corpus shall not be suspended, unless when in Cases of Rebellion or Invasion the public Safety may require it.

No Bill of Attainder or ex post facto Law shall be passed.

(No capitation, or other direct, Tax shall be laid, unless in Proportion to the Census or Enumeration herein before directed to be taken.) (Section in parentheses clarified by the 16th Amendment.)

No Tax or Duty shall be laid on Articles exported from any State.

No Preference shall be given by any Regulation of Commerce or Revenue to the Ports of one State over those of another: nor shall Vessels bound to, or from, one State, be obliged to enter, clear, or pay Duties in another.

No Money shall be drawn from the Treasury, but in Consequence of Appropriations made by Law; and a regular Statement and Account of the Receipts and Expenditures of all public Money shall be published from time to time.

No Title of Nobility shall be granted by the United States: And no Person holding any Office of Profit or Trust under them, shall, without the Consent of the Congress, accept of any present, Emolument, Office, or Title, of any kind whatever, from any King, Prince or foreign State.

Section 10 - Powers prohibited of States

No State shall enter into any Treaty, Alliance, or Confederation; grant Letters of Marque and Reprisal; coin Money; emit Bills of Credit; make any Thing but gold and silver Coin a Tender in Payment of Debts; pass any Bill of Attainder, ex post facto Law, or Law impairing the Obligation of Contracts, or grant any Title of Nobility.

No State shall, without the Consent of the Congress, lay any Imposts or Duties on Imports or Exports, except what may be absolutely necessary for executing it's inspection Laws: and the net Produce of all Duties and Imposts, laid by any State on Imports or Exports, shall be for the Use of the Treasury of the United States; and all such Laws shall be subject to the Revision and Controul of the Congress.

No State shall, without the Consent of Congress, lay any duty of Tonnage, keep Troops, or Ships of War in time of Peace, enter into any Agreement or Compact with another State, or with a foreign Power, or engage in War, unless actually invaded, or in such imminent Danger as will not admit of delay.

Article II - The Executive Branch Note

Section 1 - The President

The executive Power shall be vested in a President of the United States of America. He shall hold his Office during the Term of four Years and, together with the Vice-President chosen for the same Term, be elected, as follows:

Each State shall appoint, in such Manner as the Legislature thereof may direct, a Number of Electors, equal to the whole Number of Senators and Representatives to which the State may be entitled in the Congress: but no Senator or Representative, or Person holding an Office of Trust or Profit under the United States, shall be appointed an Elector.

(The Electors shall meet in their respective States and vote by Ballot for two persons, of whom one at least shall not lie an Inhabitant of the same State with themselves. And they shall make a List of all the Persons voted for and of the Number of Votes for each; which List they shall sign and certify and transmit sealed to the Seat of the Government of the United States, directed to the President of the Senate. The President of the Senate shall, in the Presence of the Senate and House of Representatives, open all the Certificates and the Votes shall then be counted. The Person having the greatest Number of Votes shall be the President, if such Number be a Majority of the whole Number of Electors appointed; and if there be more than one who have such Majority and have an equal Number of Votes, then the House of Representatives shall immediately chuse by Ballot one of them for President; and if no Person have a Majority, then from the five highest on the List the said House shall in like Manner chuse the President. But in chusing the President, the Votes shall be taken by States, the Representation from each State having one Vote; a quorum for this Purpose shall consist of a Member or Members from two-thirds of the States and a Majority of all the States shall be necessary to a Choice. In every Case, after the Choice of the President, the Person having the greatest Number of Votes of the Electors shall be the Vice President. But if there should remain two or more who have equal Votes, the Senate shall chuse from them by Ballot the Vice-President.) (This clause in parentheses was superseded by the 12th Amendment.)

The Congress may determine the Time of chusing the Electors and the Day on which they shall give their Votes; which Day shall be the same throughout the United States.

No person except a natural born Citizen, or a Citizen of the United States, at the time of the Adoption of this Constitution, shall be eligible to the Office of President; neither shall any Person be eligible to that Office who shall not have attained to the Age of thirty-five Years and been fourteen Years a Resident within the United States.

(In Case of the Removal of the President from Office, or of his Death, Resignation, or Inability to discharge the Powers and Duties of the said Office, the same shall devolve on the Vice President and the Congress may by Law provide for the Case of Removal, Death, Resignation or Inability, both of the President and Vice President, declaring what Officer shall then act as President and such Officer shall act accordingly, until the Disability be removed, or a President shall be elected.) (This clause in parentheses has been modified by the 20th and 25th Amendments.)

The President shall, at stated Times, receive for his Services, a Compensation, which shall neither be increased nor diminished during the Period for which he shall have been elected and he shall not receive within that Period any other Emolument from the United States, or any of them.

Before he enter on the Execution of his Office, he shall take the following Oath or Affirmation:

"I do solemnly swear (or affirm) that I will faithfully execute the Office of President of the United States and will to the best of my Ability, preserve, protect and defend the Constitution of the United States."

Section 2 - Civilian Power over Military, Cabinet, Pardon Power, Appointments

The President shall be Commander in Chief of the Army and Navy of the United States and of the Militia of the several States, when called into the actual Service of the United States; he may require the Opinion, in writing, of the principal Officer in each of the executive Departments, upon any subject relating to the Duties of their respective Offices and he shall have Power to Grant Reprieves and Pardons for Offenses against the United States, except in Cases of Impeachment.

He shall have Power, by and with the Advice and Consent of the Senate, to make Treaties, provided two thirds of the Senators present concur; and he shall nominate and by and with the Advice and Consent of the Senate, shall appoint Ambassadors, other public Ministers and Consuls, Judges of the supreme Court and all other Officers of the United States, whose Appointments are not herein otherwise provided for and which shall be established by Law: but the Congress may by Law vest the Appointment of such inferior Officers, as they think proper, in the President alone, in the Courts of Law, or in the Heads of Departments.

The President shall have Power to fill up all Vacancies that may happen during the Recess of the Senate, by granting Commissions which shall expire at the End of their next Session.

Section 3 - State of the Union, Convening Congress

He shall from time to time give to the Congress Information of the State of the Union and recommend to their Consideration such Measures as he shall judge necessary and expedient; he may, on extraordinary Occasions, convene both Houses, or either of them and in Case of Disagreement between them, with Respect to the Time of Adjournment, he may adjourn them to such Time as he shall think proper; he shall receive Ambassadors and other public Ministers; he shall take Care that the Laws be faithfully executed and shall Commission all the Officers of the United States.

Section 4 - Disqualification

The President, Vice President and all civil Officers of the United States, shall be removed from Office on Impeachment for and Conviction of, Treason, Bribery, or other high Crimes and Misdemeanors.
Article III - The Judicial Branch Note

Section 1 - Judicial powers

The judicial Power of the United States, shall be vested in one supreme Court and in such inferior Courts as the Congress may from time to time ordain and establish. The Judges, both of the supreme and inferior Courts, shall hold their Offices during good Behavior and shall, at stated Times, receive for their Services a Compensation which shall not be diminished during their Continuance in Office.

Section 2 - Trial by Jury, Original Jurisdiction, Jury Trials

(The judicial Power shall extend to all Cases, in Law and Equity, arising under this Constitution, the Laws of the United States and Treaties made, or which shall be made, under their Authority; to all Cases affecting Ambassadors, other public Ministers and Consuls; to all Cases of admiralty and maritime Jurisdiction; to Controversies to which the United States shall be a Party; to Controversies between two or more States; between a State and Citizens of another State; between Citizens of different States; between Citizens of the same State claiming Lands under Grants of different States and between a State, or the Citizens thereof and foreign States, Citizens or Subjects.) (This section in parentheses is modified by the 11th Amendment.)

In all Cases affecting Ambassadors, other public Ministers and Consuls and those in which a State shall be Party, the supreme Court shall have original Jurisdiction. In all the other Cases before mentioned, the supreme Court shall have appellate Jurisdiction, both as to Law and Fact, with such Exceptions and under such Regulations as the Congress shall make.

The Trial of all Crimes, except in Cases of Impeachment, shall be by Jury; and such Trial shall be held in the State where the said Crimes shall have been committed; but when not committed within any State, the Trial shall be at such Place or Places as the Congress may by Law have directed.

Section 3 - Treason

Treason against the United States, shall consist only in levying War against them, or in adhering to their Enemies, giving them Aid and Comfort. No Person shall be convicted of Treason unless on the Testimony of two Witnesses to the same overt Act, or on Confession in open Court.

The Congress shall have power to declare the Punishment of Treason, but no Attainder of Treason shall work Corruption of Blood, or Forfeiture except during the Life of the Person attainted.

Article IV - The States

Section 1 - Each State to Honor all others

Full Faith and Credit shall be given in each State to the public Acts, Records and judicial Proceedings of every other State. And the Congress may by general Laws prescribe the Manner in which such Acts, Records and Proceedings shall be proved and the Effect thereof.

Section 2 - State citizens, Extradition

The Citizens of each State shall be entitled to all Privileges and Immunities of Citizens in the several States.

A Person charged in any State with Treason, Felony, or other Crime, who shall flee from Justice and be found in another State, shall on demand of the executive Authority of the State from which he fled, be delivered up, to be removed to the State having Jurisdiction of the Crime.

(No Person held to Service or Labour in one State, under the Laws thereof, escaping into another, shall, in Consequence of any Law or Regulation therein, be discharged from such Service or Labour, But shall be delivered up on Claim of the Party to whom such Service or Labour may be due.) (This clause in parentheses is superseded by the 13th Amendment.)

Section 3 - New States

New States may be admitted by the Congress into this Union; but no new States shall be formed or erected within the Jurisdiction of any other State; nor any State be formed by the Junction of two or more States, or parts of States, without the Consent of the Legislatures of the States concerned as well as of the Congress.

The Congress shall have Power to dispose of and make all needful Rules and Regulations respecting the Territory or other Property belonging to the United States; and nothing in this Constitution shall be so construed as to Prejudice any Claims of the United States, or of any particular State.

Section 4 - Republican government

The United States shall guarantee to every State in this Union a Republican Form of Government and shall protect each of them against Invasion; and on Application of the Legislature, or of the Executive (when the Legislature cannot be convened) against domestic Violence.

Article V - Amendment

The Congress, whenever two thirds of both Houses shall deem it necessary, shall propose Amendments to this Constitution, or, on the Application of the Legislatures of two thirds of the several States, shall call a Convention for proposing Amendments, which, in either Case, shall be valid to all Intents and Purposes, as part of this Constitution, when ratified by the Legislatures of three fourths of the several States, or by Conventions in three fourths thereof, as the one or the other Mode of Ratification may be proposed by the Congress; Provided that no Amendment which may be made prior to the Year One thousand eight hundred and eight shall in any Manner affect the first and fourth Clauses in the Ninth Section of the first Article; and that no State, without its Consent, shall be deprived of its equal Suffrage in the Senate.

Article VI - Debts, Supremacy, Oaths

All Debts contracted and Engagements entered into, before the Adoption of this Constitution, shall be as valid against the United States under this Constitution, as under the Confederation.

This Constitution and the Laws of the United States which shall be made in Pursuance thereof; and all Treaties made, or which shall be made, under the Authority of the United States, shall be the supreme Law of the Land; and the Judges in every State shall be bound thereby, any Thing in the Constitution or Laws of any State to the Contrary notwithstanding.

The Senators and Representatives before mentioned and the Members of the several State Legislatures and all executive and judicial Officers, both of the United States and of the several States, shall be bound by Oath or Affirmation, to support this Constitution; but no religious Test shall ever be required as a Qualification to any Office or public Trust under the United States.

Article VII - Ratification Documents

The Ratification of the Conventions of nine States, shall be sufficient for the Establishment of this Constitution between the States so ratifying the Same.

Done in Convention by the Unanimous Consent of the States present the Seventeenth Day of September in the Year of our Lord one thousand seven hundred and Eighty seven and of the Independence of the United States of America the Twelfth. In Witness whereof We have hereunto subscribed our Names.

{Signatories removed for brevity}

The Amendments

The following are the Amendments to the Constitution. The first ten Amendments collectively are commonly known as the Bill of Rights.

Amendment 1 - Freedom of Religion, Press, Expression. Ratified 12/15/1791.

Congress shall make no law respecting an establishment of religion, or prohibiting the free exercise thereof; or abridging the freedom of speech, or of the press; or the right of the people peaceably to assemble and to petition the Government for a redress of grievances.

Amendment 2 - Right to Bear Arms. Ratified 12/15/1791.

A well regulated Militia, being necessary to the security of a free State, the right of the people to keep and bear Arms, shall not be infringed.

Amendment 3 - Quartering of Soldiers. Ratified 12/15/1791.

No Soldier shall, in time of peace be quartered in any house, without the consent of the Owner, nor in time of war, but in a manner to be prescribed by law.

Amendment 4 - Search and Seizure. Ratified 12/15/1791.

The right of the people to be secure in their persons, houses, papers and effects, against unreasonable searches and seizures, shall not be violated and no Warrants shall issue, but upon probable cause, supported by Oath or affirmation and particularly describing the place to be searched and the persons or things to be seized.

Amendment 5 - Trial and Punishment, Compensation for Takings. Ratified 12/15/1791.

No person shall be held to answer for a capital, or otherwise infamous crime, unless on a presentment or indictment of a Grand Jury, except in cases arising in the land or naval forces, or in the Militia, when in actual service in time of War or public danger; nor shall any person be subject for the same offense to be twice put in jeopardy of life or limb; nor shall be compelled in any criminal case to be a witness against himself, nor be deprived of life, liberty, or property, without due process of law; nor shall private property be taken for public use, without just compensation.

Amendment 6 - Right to Speedy Trial, Confrontation of Witnesses. Ratified 12/15/1791.

In all criminal prosecutions, the accused shall enjoy the right to a speedy and public trial, by an impartial jury of the State and district wherein the crime shall have been committed, which district shall have been previously ascertained by law and to be informed of the nature and cause of the accusation; to be confronted with the witnesses against him; to have compulsory process for obtaining witnesses in his favor and to have the Assistance of Counsel for his defence.

Amendment 7 - Trial by Jury in Civil Cases. Ratified 12/15/1791.

In Suits at common law, where the value in controversy shall exceed twenty dollars, the right of trial by jury shall be preserved and no fact tried by a jury, shall be otherwise re-examined in any Court of the United States, than according to the rules of the common law.

Amendment 8 - Cruel and Unusual Punishment. Ratified 12/15/1791.

Excessive bail shall not be required, nor excessive fines imposed, nor cruel and unusual punishments inflicted.

Amendment 9 - Construction of Constitution. Ratified 12/15/1791.

The enumeration in the Constitution, of certain rights, shall not be construed to deny or disparage others retained by the people.

Amendment 10 - Powers of the States and People. Ratified 12/15/1791.

The powers not delegated to the United States by the Constitution, nor prohibited by it to the States, are reserved to the States respectively, or to the people.

Amendment 11 - Judicial Limits. Ratified 2/7/1795.

The Judicial power of the United States shall not be construed to extend to any suit in law or equity, commenced or prosecuted against one of the United States by Citizens of another State, or by Citizens or Subjects of any Foreign State.

Amendment 12 - Choosing the President, Vice-President. Ratified 6/15/1804.

The Electors shall meet in their respective states and vote by ballot for President and Vice-President, one of whom, at least, shall not be an inhabitant of the same state with themselves; they shall name in their ballots the person voted for as President and in distinct ballots the person voted for as Vice-President and they shall make distinct lists of all persons voted for as President and of all persons voted for as Vice-President and of the number of votes for each, which lists they shall sign and certify and transmit sealed to the seat of the government of the United States, directed to the President of the Senate;

The President of the Senate shall, in the presence of the Senate and House of Representatives, open all the certificates and the votes shall then be counted;

The person having the greatest Number of votes for President, shall be the President, if such number be a majority of the whole number of Electors appointed; and if no person have such majority, then from the persons having the highest numbers not exceeding three on the list of those voted for as President, the House of Representatives shall choose immediately, by ballot, the President. But in choosing the President, the votes shall be taken by states, the representation from each state having one vote; a quorum for this purpose shall consist of a member or members from two-thirds of the states and a majority of all the states shall be necessary to a choice. And if the House of Representatives shall not choose a President whenever the right of choice shall devolve upon them, before the fourth day of March next following, then the Vice-President shall act as President, as in the case of the death or other constitutional disability of the President.

The person having the greatest number of votes as Vice-President, shall be the Vice-President, if such number be a majority of the whole number of Electors appointed and if no person have a majority, then from the two highest numbers on the list, the Senate shall choose the Vice-President; a quorum for the purpose shall consist of two-thirds of the whole number of Senators and a majority of the whole number shall be necessary to a choice. But no person constitutionally ineligible to the office of President shall be eligible to that of Vice-President of the United States.

Amendment 13 - Slavery Abolished. Ratified 12/6/1865.

1. Neither slavery nor involuntary servitude, except as a punishment for crime whereof the party shall have been duly convicted, shall exist within the United States, or any place subject to their jurisdiction.

2. Congress shall have power to enforce this article by appropriate legislation.

Amendment 14 - Citizenship Rights. Ratified 7/9/1868.

1. All persons born or naturalized in the United States and subject to the jurisdiction thereof, are citizens of the United States and of the State wherein they reside. No State shall make or enforce any law which shall abridge the privileges or immunities of citizens of the United States; nor shall any State deprive any person of life, liberty, or property, without due process of law; nor deny to any person within its jurisdiction the equal protection of the laws.

2. Representatives shall be apportioned among the several States according to their respective numbers, counting the whole number of persons in each State, excluding Indians not taxed. But when the right to vote at any election for the choice of electors for President and Vice-President of the United States, Representatives in Congress, the Executive and Judicial officers of a State, or the members of the Legislature thereof, is denied to any of the male inhabitants of such State, being twenty-one years of age and citizens of the United States, or in any way abridged, except for participation in rebellion, or other crime, the basis of representation therein shall be reduced in the proportion which the number of such male citizens shall bear to the whole number of male citizens twenty-one years of age in such State.

3. No person shall be a Senator or Representative in Congress, or elector of President and Vice-President, or hold any office, civil or military, under the United States, or under any State, who, having previously taken an oath, as a member of Congress, or as an officer of the United States, or as a member of any State legislature, or as an executive or judicial officer of any State, to support the Constitution of the United States, shall have engaged in insurrection or rebellion against the same, or given aid or comfort to the enemies thereof. But Congress may by a vote of two-thirds of each House, remove such disability.

4. The validity of the public debt of the United States, authorized by law, including debts incurred for payment of pensions and bounties for services in suppressing insurrection or rebellion, shall not be questioned. But neither the United States nor any State shall assume or pay any debt or obligation incurred in aid of insurrection or rebellion against the United States, or any claim for the loss or emancipation of any slave; but all such debts, obligations and claims shall be held illegal and void.

5. The Congress shall have power to enforce, by appropriate legislation, the provisions of this article.

Amendment 15 - Race No Bar to Vote. Ratified 2/3/1870.

1. The right of citizens of the United States to vote shall not be denied or abridged by the United States or by any State on account of race, color, or previous condition of servitude.

2. The Congress shall have power to enforce this article by appropriate legislation.

Amendment 16 - Status of Income Tax Clarified. Ratified 2/3/1913.

The Congress shall have power to lay and collect taxes on incomes, from whatever source derived, without apportionment among the several States and without regard to any census or enumeration.

Amendment 17 - Senators Elected by Popular Vote. Ratified 4/8/1913.

The Senate of the United States shall be composed of two Senators from each State, elected by the people thereof, for six years; and each Senator shall have one vote. The electors in each State shall have the qualifications requisite for electors of the most numerous branch of the State legislatures.

When vacancies happen in the representation of any State in the Senate, the executive authority of such State shall issue writs of election to fill such vacancies: Provided, That the legislature of any State may empower the executive thereof to make temporary appointments until the people fill the vacancies by election as the legislature may direct.

This amendment shall not be so construed as to affect the election or term of any Senator chosen before it becomes valid as part of the Constitution.

Amendment 18 - Liquor Abolished. Ratified 1/16/1919. Repealed by Amendment 21, 12/5/1933.

1. After one year from the ratification of this article the manufacture, sale, or transportation of intoxicating liquors within, the importation thereof into, or the exportation thereof from the United States and all territory subject to the jurisdiction thereof for beverage purposes is hereby prohibited.

2. The Congress and the several States shall have concurrent power to enforce this article by appropriate legislation.

3. This article shall be inoperative unless it shall have been ratified as an amendment to the Constitution by the legislatures of the several States, as provided in the Constitution, within seven years from the date of the submission hereof to the States by the Congress.

Amendment 19 - Women's Suffrage. Ratified 8/18/1920.

The right of citizens of the United States to vote shall not be denied or abridged by the United States or by any State on account of sex.

Congress shall have power to enforce this article by appropriate legislation.

Amendment 20 - Presidential, Congressional Terms. Ratified 1/23/1933.

1. The terms of the President and Vice President shall end at noon on the 20th day of January and the terms of Senators and Representatives at noon on the 3d day of January, of the years in which such terms would have ended if this article had not been ratified; and the terms of their successors shall then begin.

2. The Congress shall assemble at least once in every year and such meeting shall begin at noon on the 3d day of January, unless they shall by law appoint a different day.

3. If, at the time fixed for the beginning of the term of the President, the President elect shall have died, the Vice President elect shall become President. If a President shall not have been chosen before the time fixed for the beginning of his term, or if the President elect shall have failed to qualify, then the Vice President elect shall act as President until a President shall have qualified; and the Congress may by law provide for the case wherein neither a President elect nor a Vice President elect shall have qualified, declaring who shall then act as President, or the manner in which one who is to act shall be selected and such person shall act accordingly until a President or Vice President shall have qualified.

4. The Congress may by law provide for the case of the death of any of the persons from whom the House of Representatives may choose a President whenever the right of choice shall have devolved upon them and for the case of the death of any of the persons from whom the Senate may choose a Vice President whenever the right of choice shall have devolved upon them.

5. Sections 1 and 2 shall take effect on the 15th day of October following the ratification of this article.

6. This article shall be inoperative unless it shall have been ratified as an amendment to the Constitution by the legislatures of three-fourths of the several States within seven years from the date of its submission.

Amendment 21 - Amendment 18 Repealed. Ratified 12/5/1933.

1. The eighteenth article of amendment to the Constitution of the United States is hereby repealed.

2. The transportation or importation into any State, Territory, or possession of the United States for delivery or use therein of intoxicating liquors, in violation of the laws thereof, is hereby prohibited.

3. The article shall be inoperative unless it shall have been ratified as an amendment to the Constitution by conventions in the several States, as provided in the Constitution, within seven years from the date of the submission hereof to the States by the Congress.

Amendment 22 - Presidential Term Limits. Ratified 2/27/1951.

1. No person shall be elected to the office of the President more than twice and no person who has held the office of President, or acted as President, for more than two years of a term to which some other person was elected President shall be elected to the office of the President more than once. But this Article shall not apply to any person holding the office of President, when this Article was proposed by the Congress and shall not prevent any person who may be holding the office of President, or acting as President, during the term within which this Article becomes operative from holding the office of President or acting as President during the remainder of such term.

2. This article shall be inoperative unless it shall have been ratified as an amendment to the Constitution by the legislatures of three-fourths of the several States within seven years from the date of its submission to the States by the Congress.

Amendment 23 - Presidential Vote for District of Columbia. Ratified 3/29/1961.

1. The District constituting the seat of Government of the United States shall appoint in such manner as the Congress may direct: A number of electors of President and Vice President equal to the whole number of Senators and Representatives in Congress to which the District would be entitled if it were a State, but in no event more than the least populous State; they shall be in addition to those appointed by the States, but they shall be considered, for the purposes of the election of President and Vice President, to be electors appointed by a State; and they shall meet in the District and perform such duties as provided by the twelfth article of amendment.

2. The Congress shall have power to enforce this article by appropriate legislation.

Amendment 24 - Poll Tax Barred. Ratified 1/23/1964.

1. The right of citizens of the United States to vote in any primary or other election for President or Vice President, for electors for President or Vice President, or for Senator or Representative in Congress, shall not be denied or abridged by the United States or any State by reason of failure to pay any poll tax or other tax.

2. The Congress shall have power to enforce this article by appropriate legislation.

Amendment 25 - Presidential Disability and Succession. Ratified 2/10/1967.

1. In case of the removal of the President from office or of his death or resignation, the Vice President shall become President.

2. Whenever there is a vacancy in the office of the Vice President, the President shall nominate a Vice President who shall take office upon confirmation by a majority vote of both Houses of Congress.

3. Whenever the President transmits to the President pro tempore of the Senate and the Speaker of the House of Representatives his written declaration that he is unable to discharge the powers and duties of his office and until he transmits to them a written declaration to the contrary, such powers and duties shall be discharged by the Vice President as Acting President.

4. Whenever the Vice President and a majority of either the principal officers of the executive departments or of such other body as Congress may by law provide, transmit to the President pro tempore of the Senate and the Speaker of the House of Representatives their written declaration that the President is unable to discharge the powers and duties of his office, the Vice President shall immediately assume the powers and duties of the office as Acting President.

Thereafter, when the President transmits to the President pro tempore of the Senate and the Speaker of the House of Representatives his written declaration that no inability exists, he shall resume the powers and duties of his office unless the Vice President and a majority of either the principal officers of the executive department or of such other body as Congress may by law provide, transmit within four days to the President pro tempore of the Senate and the Speaker of the House of Representatives their written declaration that the President is unable to discharge the powers and duties of his office. Thereupon Congress shall decide the issue, assembling within forty eight hours for that purpose if not in session. If the Congress, within twenty one days after receipt of the latter written declaration, or, if Congress is not in session, within twenty one days after Congress is required to assemble, determines by two thirds vote of both Houses that the President is unable to discharge the powers and duties of his office, the Vice President shall continue to discharge the same as Acting President; otherwise, the President shall resume the powers and duties of his office.

Amendment 26 - Voting Age Set to 18 Years. Ratified 7/1/1971.

1. The right of citizens of the United States, who are eighteen years of age or older, to vote shall not be denied or abridged by the United States or by any State on account of age.

2. The Congress shall have power to enforce this article by appropriate legislation.

Amendment 27 - Limiting Changes to Congressional Pay. Ratified 5/7/1992.

No law, varying the compensation for the services of the Senators and Representatives, shall take effect, until an election of Representatives shall have intervened.

Appendix A | Shopping Lists

The following lists are broken up into a few categories,

Note that the following items are specific for stockpiling and are not all-inclusive. You will find things that you need which are not on the list and you should certainly add them. A good general guideline is to figure out how much of something you use in 2-3 years and stock up appropriately.

We left out the obvious stuff such as food, because it has been covered at length and will vary by person, family, dietary restrictions and etc. As long as you have 2-½ years' worth of food for everyone in your household, plus 10% more for charity, you should be good in that department. Otherwise, without further ado…

Home Goods

Item	Min. Qty.	Reasons / Notes
Pencils and paper	100 pencils, 10 reams of paper	You will be taking a lot of notes, will want paper for schooling, etc
Wind-up watches	2	You will want some form of rough timekeeping.
Wind-up Alarm Clock	2	Because you will still want to get up early… a lot.
Perpetual Calendar	1	As long as you keep up with it, you can keep your dates straight and to plan for planting, harvest, etc.
Pressure Cooker	1	For canning, sterilizing and sundry other uses
Cast Iron pans, pots	various	You should have at least two frying pans (one huge), two dutch ovens and four pots of various sizes
Large (10-20 qt) Cooking pots	5	Heating water for laundry, bathing, cooking, canning, etc
3 or 5-gallon clean food-grade plastic buckets	10	There will be a wide variety of uses for these things, but keep them clean for food purposes. You can also use the buckets you've been storing food in.
A competent set of cooking knives	At least two sets	Do not scrimp or go cheap here. Each set should have at least two large butcher knives, two large bread knives, eight steak knives and four general-purpose cutting knives.
Skinning knives	2	At least one pair of solid skinning knives (one large, one small) will come in handy for turning animals into dinner.
Food saws	2	One large, one small, to cut through bone. The large one should be able to cut a 6" log and the small one could be a clean hacksaw or small wood saw.
Mortar and Pestle	1	For grinding spices and various other dry ingredient grinding
Steel cooking utensils	various	Basically you will want these utensils instead of your usual plastic ones, since metal will be easier to keep clean.
Manual (hand-cranked) grain mill	1	For making flour out of nearly any grain
Can openers (manual)	3	You'll have a lot of cans to deal with…have a good can opener and lots of spares handy (in case of breakage, loss, whatever).
Plastic zip-baggies (sandwich, quart, gallon)	20 large boxes of each size	The idea here is that the bags can be used for nearly everything and with care can be re-used.
Kitchen trash bags (16 gal)	5 boxes of 50	Useful for a myriad of reasons and purposes
Yard/Lawn (33-gal) trash bags	5 boxes of 50	Also useful for a ton of purposes
Rolls of duct tape	20 rolls	This stuff is awesome for a ton of reasons and is useful everywhere
Zip-ties	2000 small, 4000 med, 2000 large	Don't let the numbers scare you - they're cheap and will be useful forever.
Brooms, dustpans	4 of each	Only use one at a time; save the rest to replace each in turn as it wears out
Carpet sweeper	1	Good for rugs and soft upholstered furniture (and yes, if you still have carpet post-collapse, it's good for that too)

Item	Min. Qty.	Reasons / Notes
Laundry soap	2000 loads	The longer you can do laundry without making your own soap, the easier life is for you (you'll be rather busy as it is)
Bleach	20 gallons	Bleach can be used in lots of places and not just laundry. In fact, post-collapse you will want to minimize the laundry usage of bleach
Steel Wool	20 large boxes	Steel wool will be a primo means of scrubbing out tough stains and for various other uses. Be sure to get the kind without detergent
Glass Cleaner	5 gallons	Made primarily of ammonia, it is an excellent means of cleaning a whole lot more than glass, but is cheaper than most other cleaners
Liquid dish soap	10 gallons	Can be used for dishes and also as an all-around cleanser
Powdered Cleanser	20 large cans	While normally used for porcelain bathtubs, it can be used to scrub everything from floors to concrete if needed
Kerosene	20 gallons	Primarily useful as lamp oil; the more you have, the happier you will be
Oil lamps	5	Amazingly cheap (especially at garage sales), they put out better light than candles, burn longer, have a semi-enclosed flame (which means it's safer) and can burn any kind of oil fairly efficiently.
Oil lamp wicks	50	While the wicks will last a very long time, they are also cheap and the quantity insures a lifetime of use
Strike-anywhere matches	20000	Don't let the number scare you - matches are cheap and you will go through a lot of them, especially in colder climates. Set aside a couple of boxes as bartering material.
Cheap cigarette lighters	100	A nice portable means of quickly making fire, you can always keep one on your person wherever you go.
Bar soap	100	We're talking the cheap but pure stuff (e.g. Ivory). The more you have, the less you have to make later. As a bonus, you can use it in place of shampoo if you have short hair.
shampoo	20	A typical bottle of cheap shampoo will last a typical person about 2 months with daily showering. Post-collapse, 20 bottles will keep two adults for at least 3 years. Avoid heavily-scented shampoos.
Stick deodorant	10 per adult	Unscented or lightly-scented works best. Best for those times when you don't have time to bathe for a day or two.
Straight Razor	2 per adult	Gents (and ladies?) Learn to use one - disposable razors will quickly wear out. Also learn to use regular soap as shaving cream. Also have at least two strops.
Blade sharpening stones	2 sets	Each set will have a coarse, medium and fine stone. Also have at least four quarts of clean, light motor or machining oil to use while sharpening.
Toothbrushes	30 / adult	Even cheap ones will last for up to 6 months or more.
Baking soda	10 lbs per adult	Set aside specifically to make toothpaste and other minor non-cleaning household bits
Toilet Paper	50-100 rolls	You can use them in rotating stock; this will be all too useful and for obvious reasons.

Home/Construction Materials

As mentioned before, this is not a comprehensive list, but a small list of critical items that you should have on hand. Please feel free to add to, subtract from, or otherwise change to fit your specific circumstances. Extra lines included in case you think of something you need.

Also note that if you're in an apartment or condo, you may not have enough room to store all of this stuff, so pick and choose as circumstance dictates.

Item	Min. Qty.	Reasons / Notes
Electrical tape	20 rolls	A wide variety of mechanical uses, not just electrical
Duct tape	15 rolls	This is on top of what you have from other lists
Cheap baling wire	200'	An infinite variety of uses
16d nails	30 lbs	
8d nails	20 lbs	
1" philips wood screws	5 lbs	
2" philips wood screws	10 lbs	
3" philips wood screws	15 lbs	
Nylon twine	2000'	
1/4" rope - 20' long	10	Nylon or Propylene preferred
1/4" rope - 50' long	5	Nylon or Propylene preferred
1/2" rope - 50' long	5	Nylon or Propylene preferred
1/2" rope - 100' long	5	Nylon or Propylene preferred
3/4" rope - 20' long	5	Nylon or Propylene preferred
3/4" rope - 50' long	5	Nylon or Propylene preferred
Sandbags (empty)	150	
Large Concrete Blocks	50	Try to use solid concrete blocks, not hollow cinder blocks
Bricks	100	Note that you can use these and concrete blocks as garden walls, sandbox borders and the like.
Hydrated Lime	5	50 lb bags -keep dry
Ready-mix cement	10	50 lb bags -keep dry
Metal Flue Ducting	Approx. 25'	Straight piping, cut into 5' and 10' sections as needed
Metal Flue Elbows	varies	Enough 90 and 45 degree elbows to complete a fireplace flue
1/2" Copper Tubing	50'	Useful as last-second plumbing, distillery condenser and various.
Copper Solder Flux	2 large cans	
Plumbing Solder (95/5)	3 lbs	1/8" diameter and w/o a flux core
4'x8'x½" Plywood	10 sheets	Avoid particleboard, but no need to spring for the pricey veneered stuff, either.
4'x8'x½" MDF	10 sheets	MDF = "Medium Density Fiberboard"
4'x8'x½" cementboard	10 sheets	Also known as "backer board", it can provide a fire-resistant surface in a pinch. Made of Portland Cement w/ fiberglass fibers
20' wide 6 mil plastic sheeting	200'	Usually sold in 100' lengths; also commonly called "visqueen"
14 gauge automotive wiring	50'	Useful for numerous electrical projects and also as baling wire in a pinch
12 gauge THHN home wiring (3-strand)	50'	Same as above, but note that you can always scavenge this out of abandoned/damaged homes.

Medical Supplies

This is a list of critical and useful medical supplies to keep handy around the home. Note that this does *not* include what you should keep in your car(s) or bags.

Item	Min. Qty.	Reasons / Notes
Industrial Burn Kit	4	The incidence of burns are liable to go way up…
Surgical and Suture Kit	3	
Emer. Dental Repair Kit	3	Enough to create temporary fillings, repair fillings, etc.
Large maxi pads	100	Trauma bandages - can soak up a ton of blood.
Sterile 8"x10" surgical pad	100	
Sterile 5"x9" surgical pad	300	
Sterile 4"x4" gauze pad	300	
1" wide surgical tape	50-100 yds	
2" "Ace" style bandages	10	Note that this is washable and reusable up to 10 times
3" "Ace" style bandages	10	Note that this is washable and reusable up to 10 times
Strong nylon thread	10 spools	For impromptu sutures. Just be sure to sterilize before use.
Sewing needles, curved	30	Should be stainless steel, so you can sterilize them
Safety pins, medium	300	Useful in multiple situations - not just first aid
Enema bag, hose and acc.	3	Keep the unused ones oiled to prevent cracking.
Eye patches, medium	5	(You can always improvise more if needed)
Adhesive bandages (std.)	300	While more would be desired, it'll take awhile to go through them and you don't want them going bad on you.
Butterfly bandages	200	Useful externally, when stitches are not quite required.
Snakebite kit	1 *	* If you live in an area where poisonous snakes are common, have at least 4 or 5.
Single-edged razor blades	300	Can be handy for many medical purposes and are easy to sterilize in a pinch.
Disposable latex gloves	200 pair	While not ideal for surgical use, they can be pressed into service as such if necessary. Just be sure to rinse them in alcohol or other safe sterilizing agent before using them as such. They're also surprisingly cheap.
Cotton swabs (bag of 500)	4 large boxes	External use only, but still quite useful. As a bonus, they don't go bad.
Rubbing Alcohol	4 gallons	Can be replaced by grain alcohol if needed
Hydrogen Peroxide (3%)	5 gallons	An excellent general antiseptic, but use sparingly and keep dark.
Pain-relieving cream - menthol-based	5 tubes	If you use it often pre-collapse, have 15 tubes handy. Provides cooling relief
Pain-relieving cream - Capsicum-based	5 tubes	Same as above, but provides warming relief.
Antibiotic ointment	10 tubes	Ideally you'd have many more, but expiration dates will put a natural limit on how many to stock up on.
Ibuprofen - 200 mg	1000 pills	
Acetaminophen - 250 mg	1000 pills	
Aspirin - 200 mg	2000 pills	
Benadryl ® (Diphenhydramine)	1000 pills	A great all-purpose anti-allergen

Item	Min. Qty.	Reasons / Notes
Anti-diarrhetic (e.g. Imodium® AD)	200 per adult	As explained, you're going to need these - badly.
Antibiotic pills	500 / each type	See page 358-359 for a good list
Athlete's Foot Powder	4 large bottles	Can also be used for 'Jock itch'
Nail Fungus Treatment	2 bottles	Just in case.
Petroleum Jelly	4 large jars	Various uses
Senna-based Laxative	200 pills	Under high stress, many folks get easily constipated
Other type of Laxative	200 pills	Same reason as above, but alternate between the two to avoid a digestive dependency on Senna
Milk of Magnesia or Bismuth Subsalicyclate	3 large bottles	Great for generic tummy troubles - Bismuth Subsalicyclate is the pink stuff, also known as Pepto Bismol®
Toothache Pain Gel	4 large tubes	It's fairly obvious as to why you would want this…
Calamine® lotion or similar	4 large bottles	Poison plant rashes, bee stings, jellyfish stings…
Ipecac Syrup	2 large bottles	For certain types of poison
Activated Charcoal	10 doses	For all the other types of poison
Eye drops	3 large bottles	Various eye-related irritations
Buffered Contact Lens Saline Solution	10 large bottles	In addition to eye wash, the contents are sterile until you open the bottle.
Talcum Powder	3 large bottles	For various reasons
Diaper Rash Cream	3 large bottles	Even if you don't have a baby now, you'll be glad you have this stuff when if do have one.
Zinc Oxide	5 large tubes	Various. Not to be confused with the diaper rash cream.
Suntan lotion, 30+ SPF	2 large bottles	For those not accustomed to working outdoors…
Glass thermometers	4 of each type	Each type being "oral" and "rectal". Be sure to learn how to read them correctly. 4 of each type because they're fragile
Sphygmomanometer	2: one small cuff, one large cuff	Blood pressure cuffs - the manual type with bulb and dial.
Stethoscope	2	Get the good kind - used to measure heartbeat, checking breathing for signs of fluids or damage, etc.

Feminine Care

This one is broken out for you ladies out there, because there are certain things that guys aren't really going to be too thoughtful of (most of us sadly assume that the ladies just 'take care of that'). However, given the stark facts of gynecology, well…

Item	Min. Qty.	Reasons / Notes
Feminine napkins (washable)	50	This should also include any belts or similar gear
Pregnancy test kit	At least 5	…because knowing long before the 'baby bump' shows is a good thing. Mind the expiration dates.
Intimate lubricant (water soluble)	4 large tubes	…because guys are too stupid to think that far ahead on such matters.
Menstrual Pain relief pills	200 pills	Most pain meds aren't really engineered to deal with bloating, headaches and cramps.
Menopausal supplements	6 months' worth	This is pretty much only for ladies who are approaching menopause, though you can always give them to someone in need.

Feminine Care (continued)

Item	Min. Qty.	Reasons / Notes
Vinegar	3 gallons	Mixed with water, it's a time-tested douching formula
Enema/Douche bag	2	Be sure it has the correct nozzles. Keep lightly oiled.
Birth control pills	6-8 months	Not strictly for birth control here - they can be used for hormonal therapy if under medic supervision.
Tincture of iodine	3 large bottles	Mixed with water to solve certain types of vaginal infections
Cranberry pills	1000	As UTI's (Urinary Tract Infections) are more common in women, this is a must-have for any woman.

Tools

We're talking normal hand tools here - critical things that will be useful around the house in a post-collapse world. The good news is, you can use them pre-collapse as well. Note that for specific post-collapse trades you will have to get and keep additional tools specific to the pre-collapse hobby/post-collapse trade.

Item	Min. Qty.	Reasons / Notes
Spade shovels	6	All general-sized - you will go through them a lot
Crosscut saws	4	Various sizes
Hammers	6	Two std, two ball-peen, two roofing
Pliers	6 pair	2 small, 2 medium, 2 large
Clamping pliers	6 pair	2 small, 2 medium, 2 large
Slip-joint pliers	6 pair	2 small, 2 medium, 2 large
Needle-Nose pliers	6 pair	2 small, 2 medium, 2 large
Screwdrivers	10 of each type	5 phillips of various sizes, 5 straight-slots of various sizes
Pipe wrench	3	One of each size - small medium and large
Adjustable wrench	6	Two small, two medium, two large
Wire Cutters	3 pair	Get the toughest ones you can
Bolt Cutters	3 pair	Large-sized
Std ¼" drive sockets	2 sets	Socket sizes should at least range from 3/32" - 1/2"
Metric " "	2 sets	Socket sizes should range from around 3mm - 13mm
Std ½" drive sockets	2 sets	Socket sizes should at least range from 3/8" - 13/16"
Metric " "	2 sets	Socket sizes should range from around 6mm - 18mm
Ratchets for above	1 per set	
Std. combo wrenches	2 sets	Range should be from 3/32" - 1"
Metric " "	2 sets	Range should be from 3mm - 25mm
Allen Wrench sets	4 sets	2 std., 2 metric
Crowbar	4 large	You'll need them.
Straight Chisels	6	The type used for metalworking, not woodworking - various sizes
Axes	4	Large enough for felling trees
Large bucksaws	2, with 20 spare blades	As opposed to hand saws, these are used for cutting logs.

Item	Min. Qty.	Reasons / Notes
Chain	Various sizes	Make them substantial, in 20' and 50' lengths
Towing strap	At least 4	20' and longer lengths if possible
hacksaws	4, with 50 spare blades	Don't get the little ones and don't scrimp.
Hand drills	3 small, 3 large	Should fit most drill bits
Drill bit set	2 complete sets	Each set should be for wood and should contain all sizes your drills will fit.
Wire brushes	10, large	
Sledgehammer	4, large	
Snowshovel	2	Believe it or not, it's useful for a lot more than snow.
Metal files	6, large	Two fine, two medium, two rough
levels	2 large, 2 small	
Shop vise	1 medium	Large would be preferable, but medium will work.
Utility knives	various	At least 10 good ones - 5 small, 5 large
Sharpening stones	3 sets	2 rough, 2 medium, two fine

Survival-Specific Goods

Up until now, everything has perfectly good uses in the pre-collapse world, up to and including minor crises. The only difference is, you just have a lot of them. However, from here on out the items may be specifically engineered for use in major disasters, severe crises and especially in the post-collapse world. These items (and all items in every list from here on out) should be kept out-of-the-way until and unless they are needed, but should be checked at least once a year.

Item	Min. Qty.	Reasons / Notes
Water filters	4	These are backpacker-style filters, *not* the home 'pitcher' kind, or the in-line pipe filters. Get the best you can find.
Hand-cranked radio	2	Strong preference for those which have shortwave
'Shake' flashlights	6, large	These flashlights are charged by merely shaking them.
Paracord (550)	500'	This stuff is useful nearly everywhere for small/medium jobs
12'x12' outdoor tarps	10	Useful for covering things, an impromptu tent, or etc
25'x12' tarps	5	Useful for covering a damaged roof
Earth-tone dyes	Various, 6 of each	Can be used to convert bright-colored clothing to something less likely to attract attention.
55-gallon trash bags	50 (large box)	
Road flares	10	
walkie-talkies	1 pair	Useful for communications
"AAA" batteries	50	Note that most top-end batteries have a 5-7 year shelf life
"AA" batteries	100	
Tall rubber boots	2 pair per person	...because walking through muck isn't fun

Item	Min. Qty.	Reasons / Notes
Sleeping bags	2 per person	Over-size them for the kids. Also, be sure they can handle the lowest temperatures that you can expect in your area.
55-gallon plastic barrels	3	Can be cut in half as wash tubs, be used for rain barrels, etc. Be sure they are food-grade, clean, come with lids and are sealed. Fill with clean water and some bleach while storing, changing the water out once a year.
55-gallon metal barrels	3	Useful for trash-burning, can be cut in half and made into a cooking grill, or can be converted into a wood stove. Keep filled with sand, kerosene, or with gasoline (mark appropriately) during storage - or one of each if you desire. If gasoline is used, add stabilizer and rotate the liquid out at least once a year.
Manual (treadle-powered) sewing machine	1	Keep it indoors and in good working condition. You can find them (and parts) in most antique stores and flea markets. Will be extremely useful post-collapse.
Sewing needles	500	Both manual and the type that fits your particular machine
Upholstery needles	50	These are the curved needles useful for sewing tough, thick fabrics.
Sewing thread	50 spools	Mostly darker and earth-colored, unless you live where it snows, then be sure to have a lot of white spools as well. Get the strongest thread you can find.
Sewing pins	500	Are useful in situations well beyond sewing.
Backpacker survival kits	2 per person	Useful for almost anything else

Food Preparation and Storage

This isn't food per se, but extra bits and bobs which will come in handy when it comes to preparing and storing foods long-term…

Item	Min. Qty.	Reasons / Notes
1 qt. Canning jars	100	Note that you can buy these second-hand (as long as they're not cracked or scratched) and boil them.
1 pt. Canning jars	100	Same as above
Rings and lids	200	Be sure they fit the above-mentioned jars
Pressure cooker	1, large	…if you don't already have one.
Jar tongs	3	Because it's easier to pull jars out of boiling water, that's why
cheesecloth	A lot!	Perfect for straining food
Large pot	2	If you don't already have them; should be large enough to boil 12 jars
Large metal bowls	4	The reason we specify metal (stainless steel) bowls is for sterilization
Dehydrator racks	5	For food dehydration - you can also use small framed window screens with clean, new metal screen mesh for this purpose.
Table salt	200 lbs	For salting and brining food. For those who live on coastal or saline lake areas (or within reach of a salt mineor flats), you can cut the minimum by half

Hunting, Foraging, Fishing

Primarily, you will be using these for food, but they can have other pre-collapse uses as well (mostly hobby-related). Note that firearms and other weapons are left off this list, as it is assumed that you have them anyway and they are covered elsewhere.

Item	Min. Qty.	Reasons / Notes
Air rifle (.177 caliber)	2 per adult	An easy and cheap way to hunt small (> 25 lb) animals and birds for food, without wasting firearm ammunition. Also far quieter than a firearm.
.177 caliber pellets	20,000 per adult	These things are laughably cheap and should be stocked up on as much as possible. Look for the pointed variety.
Fishing rods and reels	3 per person	Use gear appropriately-sized for your region.
Spare fishing line	10 100' spools	For obvious reasons
Fishing tackle	various	Per person: Have at least 4-20 of each lure type, depending. Also don't forget at least two of each type of tool and four of each type of knife.
Fish netting, fine mesh	various	Have enough for at least two of each type of net per person if applicable to your area.
Crab/Clamming gear	4 sets / person	If applicable to your region. Include extra shovels, rope and rakes, as salt water and sunlight will degrade them rapidly.
Fur/coyote traps	4	If you have a lot of medium-sized animals in your neighborhood, these will be well worthwhile. If you live in the country, up the minimum to 10. Be sure to practice before using and be very careful around them.
Large shoulder bags with compartments	3 per person	Handy for carrying foraged foods back home.
Garden trowels	3 per person	Useful for digging up edible tubers and roots
Crayfish traps	2 per person	If applicable to your area. Crawdads can also be caught one at a time with tongs or gloved fingers.
Gigging spears	2 per person	If applicable to your area and if frogs are plentiful. Odds are good that in most areas they will be.
3/8" nylon rope	6 - 20' lengths	Useful for dragging game, building traps, bundling and other uses.
Large shoulder bags	2 per person	Useful for holding small game, foraged plants/animals, larger game entrails and similar items as you gather them.
Large trash bags	200	Useful for numerous purposes while getting food; be sure they are the strong/sturdy type.
10' x 12' tarps	4	Useful for dragging larger items, or large bundles of them. Can also be used to cover game or similar.
Tent pegs/stakes	20	Good for numerous uses, especially when trapping or setting nets.
Small folding shovel	2	Excellent for burial of entrails or leftover inedible bits, digging for roots, etc.
Cheesecloth	As much as you can get.	Excellent for curing meats, straining plants and various.

Weapons

This will vary by person, ability, pre-collapse laws and various other factors, so we'll be fairly generic here.

Item	Min. Qty.	Reasons / Notes
Pistol (revolver or automatic)	2 per adult and teen	Excellent close-range weapon. If you choose to use a revolver, be sure to include at least two 'speedloaders' per pistol
Shotgun, Pump-action, Modified Choke	2 per teen and adult	Good for quick-firing where aim isn't going to be perfect. Should be 12 gauge, but 10 gauge or 14 gauge works too.
Rifle	1 per teen and adult	This should be at least a .30-06 caliber or larger; scope is optional, but it is recommended to have a combination scope/open-sight arrangement.
Sword/Machete	2 per person	You want a strong blade at least 16" long as a melee weapon.
Compound Bow	1 per adult	These are useful for quieter attacks/counter-attacks.
Longbow	1 per person	Compound bows cannot shoot wood arrows, but a longbow can. Longbows can also be repaired/made far easier.
Dagger/Knife	2 per person	You will need something personal in the blade department.
Pistol ammunition	2000 rds / person	If you can only get around 500 rds, you'll do okay, but otherwise stock up as much as possible.
Shotgun ammunition	3000 rds/ person	No more than 500 rds should be birdshot and then only after you have the rest in sizes 0, 00 and rifled slug.
Rifle ammunition	3000 rds/ person	The rifle will be your primary defensive weapon as a community, so it pays to stock up. Also, avoid exotic calibers.
Aluminum arrows	At least 20 doz	Get more if you can if you have a compound bow; use hunting points. Substitute crossbow bolts if you chose one.
Wood arrows	30 doz.	See above.
Gun cleaning kit	2 per weapon	Get the good ones and be sure to stockpile solvents, oils and rags/patches.
Spare parts	various	Have enough spare parts to rebuild most of each weapon you have.

Item	Min. Qty.	Reasons / Notes
Clubs	1 per adult	Preferably made of metal; aluminum baseball bats will suffice
Long (6') poles	1 per adult	Preferably 2" diameter; can be used as is or made into spears
"Buckler" type shield	1 per adult	A small shield for the forearm; useful for blunting non-firearm attacks
Throwing knives	Set of 6	Finely-balanced and in experienced hands, they can be extremely effective and very silent.
Mace	1 per adult	A Mace is a sturdy short pole (4' long) with a heavy metal ball (can be spiked) securely fastened to the end. Takes a bit of practice.
Axes	1 per adult	A modified double-bladed wood axe can be put to use here by shortening the handle. However, only do so if you can spare it.

Note that with any weapons you choose, it is highly important to train in them to the point where you are comfortable, accurate and confident in using them. Also note that any weapon is excellent in some situations, but useless in others - know which ones work well in which environments.

Special - Refugee/Vehicle Packing List

For those among you who have decided ahead of time that you will be evacuating at first sign of full collapse, it pays to pre-position the stuff on the previous lists at your destination and concentrate just on this one and the ones immediately after it (bug-out and get-home bags).

This will be easier to do if you take a bit of time in advance and think of one thing - storage. How much room do you have in that car, truck, or SUV? Think in terms of cubic feet and in all three dimensions. Assemble a bunch of storage containers that are both tough and simultaneously easy to open. Avoid the gimmicky stuff like ammo cans and think more like those plastic tubs you store Christmas ornaments in. Don't just stick with one size, either - try to fill every nook and cranny of your vehicle's storage, with containers or dedicated space for bigger stuff.

For additional non-critical items, think of unusual spaces where you can stash things - the engine bay (mind the fan belt), fender wheel wells (mind the steering and vertical wheel bounce), under the seats, under the dashboard, under the driver/passenger's legs, etc. For the outside, you can buy a little shelf that fits onto a trailer hitch and give yourself a couple hundred pounds of additional carrying capacity. Just note that any places outside will need containers that are lashed down firmly and are impervious to outdoor weather. One caveat - try not to rely too much on external racks and pods, as they may get stolen during evacuation if you're ever forced to stop (or even slow down). Use those for non-critical items or spare items instead.

With all of this in mind, figure out how to pack the absolute most into the vehicle. While you do this, make the critical stuff the easiest to get out, so that you can 'grab and dash' if it comes to that. Speaking of which, be sure to leave room for all passengers *and their bug-out bags*, plus any spare gasoline and etc that you may need to make the journey with.

Once you have it all figured down, draw a diagram of it and label the boxes covertly. What I mean is, do not scrawl "food" in big letters on the side. Instead, label the boxes just under the lids where they cannot be seen without a very close inspection; label the bags just inside and just under the rim of the bag. I think you can figure it out further from here.

So what are we putting into them? Let's start with the most critical - things that can fit into a typical five-passenger car (anything smaller and you'll be lucky to fit the bug-out bags in the things)

Item	Min. Qty.	Reasons / Notes
Change of clothing	6 per person	Ordinary but durable clothing. Keep to earth tones and neutral colors. Be sure to include cold-weather clothing.
Socks	5 pr/person	You will always want dry socks
Sleeping bag	1 / person	This can be a light-duty bag if you sleep in your clothing (you will).
Blankets	2 / person	Make sure they're warm and easy to launder.
Boots/shoes	1 pr/person	One pair of good hiking boots, the other sturdy walking shoes.
Spare eyeglasses	2 pr	Replace with sunglasses if you don't wear prescription lenses.
Personal care stuff	At least 2 sets/person	All the stuff you normally care for yourself with - toothpaste, toothbrushes (compact), neutral deodorant, feminine products, etc.
Cheap lighters	20	Useful for starting fires.
Toilet paper	4 rolls/pers.	Compress hard, then pack into a plastic zip-style bags
Rain poncho	2/person	One for you, one as an impromptu tent if needed
Dish soap	2 gal.	Useful as an all-around cleaning soap (dishes, scrubbing, etc).
Bath Soap	30 bars	The large number is due to use for bath and laundry.
Washrags	10	Small ones will work just fine here
Bath towels	6	For drying off after bathing, handling hot utensils, etc
Multifunction tool	1/person	Each person older than 8 years old should have one.
Mechanic tool kit	2 sets	Should be enough to do most work on most cars.
Bolt cutters	2 large	Useful for getting past gates, chained-off areas and etc.
Wire cutters	2 pr.	Useful for getting past barbed wire or wire fences.
Oil lamps	2	Try to get the 'hurricane' style
Lamp oil	2 gal.	To feed the oil lamps
Long-term or canned food	8 wks/pers.	If you have any space left in your car, pack even more of this in.
Portable water filters	3 large	Shoot for the best ones you can find.
20' x 20' tarps	3	Should be waterproof and not cheap.
Backpacking tents	2 min.	Should have enough to house everyone plus a spare.
Folding Shovels	1/person	
Large shovels	3	
Hunting knife	1/person	Make it a very good one and don't forget the sheath
Hand Axes	3	
Large bow saw	2	
Folding saw	2	
Money	About $300	Be sure to keep it in denominations no larger than $10.
First aid kits	3 large	
Maps	2 sets	Should be sufficient to cover county roads to planned dest.
Binoculars, small	1pr/person	
Pistol	1/adult	Personal defense
Rifle	1/person	Each person older than 15 years old. Younger can have air rifles.
Ammunition for above	100 rds	At least 100 rds for each type of rifle.
Small fishing kit	1 kit/person	Should contain monofilament line, hooks, lures usable in your region, etc.

…from there, you start combing through the other lists (except home goods) and pack in what you can. For instance, if you have a large SUV or van, you can start piling in more food, hand tools, etc. If you have an RV, you can pile in all of that, more medical supplies and even small appliances (e.g. grain mill and etc.)

Be creative. Instead of a plastic 'travel pod' full of stuff lashed on the roof of your car, lash a 55-gallon steel drum full of non-perishable stuff onto it, with the lid crimped on. The drum can then become a wood-burning stove at your eventual homestead.

By the way, this isn't just restricted to land vehicles. If you have a large boat (either trailered with your home near a boat ramp, or docked at a nearby marina), you can load up similarly. However, keep in mind that the longer the voyage, the more fresh water and fuel that you will need on board. If your voyage is all on fresh water, you can replace a lot of your water stores with water filters and water sanitizing tablets. Just note that when you reach your destination, you will have to haul all that stuff by hand from shore to your new home; it would pay to stow a couple of furniture dollies or folding handcarts with balloon tires.

For aircraft, motorcycles and other limited-capacity vehicles, your options are going to be very restricted, so cut back accordingly and pre-position most of this stuff at your destination if possible (and seriously re-think your plan, unless you have no other choice). Just know that if your destination has been looted or is occupied by a better-armed/stronger force, you will likely be stuck with only whatever it is that you're carrying.

One final bit of advice: Keep in mind that too much stuff will reduce your fuel mileage, handling ability and top speed, so don't go overboard, no matter the temptation.

Community Supplies

The following list is a collection of things that will come in handy at a community-wide scale. Odds are pretty good that much of this will already be available in smaller/rural towns, but just in case (and if you have the storage for it), these supplies can get very useful to a whole community.

Item	Min. Qty.	Reasons / Notes
Paper	30 reams	Because every government in history ran on the stuff.
Pencils	1000	...gotta write with something.
Pens	500	Most modern moderate-priced pens will store for years, though the real cheap ones may become worthless after about 3 years.
Manual typewriter	3	They're amazingly cheap pre-collapse and easy to maintain. Also keep as many ink ribbons/cartridges on hand as you can afford.
Mimeograph	1	The pre-Xerox way of making copies. Keep plenty of fluid and paper on hand. A great way to make copies until you can build a printing press.
Large corkboards	3	A means of posting notices and passing around information
Pop-up canopies	4	These have become rather cheap and serve as meeting areas, a market judge's tent, or your own bartering stall cover

Much of the rest of what you would expect for gatherings and governmental business (folding tables, benches/chairs, etc) can be scavenged easily from abandoned churches, halls, theaters and the like.

Bug-Out Bags

Note that your very life will depend on these items, so test them completely and tweak as necessary. If all else fails, these items will be all that you have, so treat them well and do not buy cheaply here.

Item	Min. Qty.	Reasons / Notes
Change of clothing	3	Ordinary but durable clothing. Keep to earth tones and neutral colors
Socks	5 pr.	You will always want dry socks
Sleeping bag	1	This can be a light-duty bag if you sleep in your clothing (you will).
Hiking boots	1 pr.	It's an extra pair if you left on your schedule and your only pair if you left in a hurry (put them on once safe and carry your ordinary shoes).
Spare eyeglasses	2 pr	Replace with sunglasses if you don't wear prescription lenses.
Personal care stuff	At least 2 sets	All the stuff you normally care for yourself with - toothpaste, toothbrushes (compact), neutral deodorant, feminine products, etc.
Toilet paper	½ roll	Compress as hard as you can, then pack into a plastic zip-style bag.
Rain poncho	2	One for you, one as an impromptu tent if needed
Washrags	2	Small ones will work just fine here
Small hand towels	2	For drying off after bathing, handling hot utensils, etc
Multifunction tool	2	
Backpacking food	2 weeks	If you have any space left in your bag, pack even more of this in.
Backpacker water filter	2	
1 qt. canteen or similar	1	Make this two if it's not a rigid one, or carry an extra liner if it uses them
Camping mess kit	1	Should be a multifunction cooking/eating kit. Include metal utensils
Small folding shovel	1	Useful practically everywhere
Fire starter kit	2	
Disposable lighter	5	A faster means of fire-making
Pens and small notebook	2 ea.	For taking notes, leaving notes, etc.
Commercial maps	See notes	You should have high-detail (trail-resolution) maps covering your intended route (leave unmarked) and general maps for all other directions (just in case you were forced to take some nasty detours)
Backpack first-aid kit	2	They're fairly small and can come in handy when you need them.
Hunting knife	1	Make it a very good one and don't forget the sheath
Folding saw	1	Useful for firewood-making
Money	About $100 in cash and coins	Be sure to keep it in denominations no larger than $10. Useful in case anyone actually still takes the stuff. Otherwise it's a great firestarter. Don't forget some coins as well.
Ibuprofen	100 pills	Can also be Acetaminophen if that works better for you.
Binoculars, small	1 pair	
Pistol	1	Personal defense
Rifle	1	Personal defense; hunting only if opportunity is perfect.
Ammunition for above	100 rds ea.	
Small fishing kit	1 kit	Should contain monofilament line, hooks, a couple of lures usable in your region, etc.

Get-Home Bag

Unlike your bug-out bag, this is only going to carry enough to get from work, school, or wherever back to your home. It should be counted as useful for about 3 days, maximum. Note that the whole shebang should fit into a small laptop backpack and that the bag should be muted colors (and preferably look a bit ratty to avoid attracting attention).

Item	Min. Qty.	Reasons / Notes
Backpacker Food	4 days	Even if you're only traveling 5 miles to get home, you just never know…
Small canteen	1	
Small blanket	1	Something to keep warm with - should be a muted color
Rain Poncho	1	…just in case.
Backpack water filter	1	
Change of clothes	1	Complete change of clothing
Sturdy hiking shoes	1 pr	
Maps	various	Maps that cover streets and trails between typical locations (work, school, etc) and home, plus a more generic one just in case. Like the bug-out bag, *do not mark routes on them!*
Multifunction tool	1	…should include a small pair of wire cutters
Pocket knife	1	
Hunting knife	1	
Pistol	1	A small concealable handgun is all that is necessary here - enough for close-quarter fighting if threatened directly.
Ammunition for above	50 rds.	Preferably pre-packed into spare magazines or speedloaders

Appendix B | Scrounge Lists

In this section, we're going to show you how to build your own lists. The reasoning is simple - listing all the different types of stores and what can be had in them would be page-eating work and we would still likely not cover all contingencies. Sure, you can safely assume that grocery stores will hold food, sporting good stores will hold weapons and everything will be sitting at the local suburban shopping-club warehouse store. On the other hand, there are a whole lot of not-so-obvious retail outlets that you can scrounge a lot of useful things from, ranging from automotive parts stores to porn shops.

So, you need to do a bit of sniffing around pre-collapse and see what's out there, then make a few lists. A little work now saves a lot of looking and guesswork later.

Priorities

Your priorities are going to be very simple, but are a combination of two things: distance and contents. Something close to you that has top-priority items and are less than 1 mile away will be the first thing you want to go to, items at bottom of the list and are 15 miles off you can get whenever and the rest falls in-between. The following should be your tool when determining priorities:

Item	Priority	Distance
Food	1	Less than one mile
Weaponry	2	1-3 miles
Medical	3	3-5 miles
Toiletries/Household	4	5-7 miles
Clothing and shoes	5	7-10 miles
Building materials	6	Over 10 miles
Tools	7	10-15 miles
Power generation	8	15-25 miles (Well, don't bother at this point)
Communication gear	9	25-50 miles (Seriously? I said don't bother!)
Fuel (in bulk)	10	50-100 miles (No, really... this would be suicide)

If you're not sure, add the priority # of the item to the priority of its distance: lowest number wins. For instance, if a gas station is less than a mile away (*10+1=**11***) and an untouched source of toiletries sit 16 miles away (*4+8=**12***)? You get the gasoline first. If you have two options that match in priority, then the closest one to you wins out. Note that all joking aside, anything further off than 15 miles should be left alone, unless you have no other options available to you.

Finally, note that if you're in a rural area or the countryside, you can fudge the distances outward to compensate.

Your role

As someone who already has preparations in place, your role will best be served by 'directing traffic' during the initial rush - coordinate the truckloads of neighbors as they go out and come back and work on distribution of goods.

Their Role

Have each person in the party pick what they're going to get as they travel. Have them study their bit of the list and perhaps one other. If they finish early, they can throw it in the truck and go back to help someone else.

Hey, Wait… This is Looting!

Actually, it isn't. There is a distinction here, though admittedly it makes for a very fine line. Let me explain that line and just how sharp it is. Then we'll lay down a few firm and unmovable ground rules to insure that you stay on the good side of the line and don't become a looter.

In the final throes of civilization itself, there is going to be a period of intense upheaval, as the masses realize that things are taking the final plunge. It will be much like the passengers of RMS Titanic during the tragedy. A very large percentage of the passengers held out at least some hope for rescue during most of the sinking. When the stern rose sharply, the ship broke in half and she began to take the final plunge, all doubt was removed as to where she was going. It is during that final realization when everyone left on board resigned themselves to fate, screamed prayerful cries, jumped ship, or scrambled like mad for something - anything that might float.

When civilization finally begins to go under, the same thing will happen, albeit on a much larger scale. The masses will resign themselves to fate (suicides will be way up), scream in prayer, get out of town to some imagined (or real) destination, or scramble like mad for whatever they can to help keep themselves alive. Unlike the Titanic, the lifeboats are a bit too metaphorical - suffice it to say that if you are sufficiently prepared, you've made it to one, but it's very small and you can't really row this thing by yourself. So, it's your job to help pull together as many fellow survivors as you can trust without swamping your own boat. Given this, your fledgling community's job is going to be to scramble like mad for whatever can be scavenged. At the inflection point between civilization and collapse's final plunge, there is a small window of opportunity - enough order left to procure the things you need without gunfire, but quickly deteriorating. You have one shot to collect as much as you can.

The rationale is simple enough - either your community gets it, or someone else's does, or the raiders get it. All too soon, there will be no more corporation to own it, police to defend it (though it will be defended, rest assured) and no more of it anywhere to go around.

In a perfect world, you and your neighbors will be all prepared equally and when it comes crashing down, you can come together, set up a defensive perimeter and start modifying your neighborhood to post-collapse living, with no scrounging necessary. In the real world, most of your neighbors will be all too short on the things they need to survive and you won't have enough charity stores to keep them fed. This leaves the hybrid option of scavenging quickly to mitigate the suffering your neighbors will face. Once collapse has hit, you scavenge under different rules, but for this time period, this is what it is.

I do want to put in a couple of ground rules here. These are firm and non-negotiable. If you violate them, you do so at your own risk and you will deserve whatever consequences you get from them, period. These are the rules:

1) <u>Do not scavenge from stores owned by a sole/local proprietor.</u> True independent mom-and-pop operations need to survive too and antagonizing them is bad mojo on all too many levels. Plus, an independent store own will likely shoot at you, where a corporate chain store likely won't. So, logic dictates that you aim your scavenging efforts for the corporate-owned and large franchise operations.

2) <u>If there is money available, pool it and pay for it</u>, even if it means throwing a wad of money at the clerk (or to the guards at the door) on your way out. If the credit card readers still work (and they are equipped to take the things), hand them a known good card and leave.

3) <u>If law enforcement still exists and is enforcing the law, obey it as far as is just</u>. If the rule of law is still in effect and there is otherwise no crime occurring, don't commit any yourself. The good news is, most of our lists are for the less obvious spots that aren't as guarded - we figure that you're smart enough to know what grocery, hardware and outdoor equipment stores have in them.

4) <u>Keep a quick, low and civil profile</u>. Don't barge in with weapons at the ready. Keep any weapons concealed. Quietly get in, get your stuff, pay for it if they still accept money, then get out. Waste no time. If someone else takes it off the shelf before you do, it's theirs - arguments will only waste time, so move on. If another 'shopper' gets too pushy, get backup and ready your weapon for drawing - let him see you do it. The interloper will quickly leave in nearly all cases (and you should immediately leave with your stuff at that point.)

5) <u>Avoid the luxury crap</u>. You're looking to survive, not build a man-cave. Leave the electronics and luxury goods alone. You have more important things to work towards, like keeping the kids fed and clothed.

How to Make a Scrounge List

This will disturbingly appear as if you're 'casing' a joint, but what we're doing is getting an idea of what stores have things to scrounge when everything comes crashing down and how best to do it. The good news is, you're also categorizing what all these stores have, which can come in handy before you yourself go shopping - even when civilization is humming along just fine. Why? Because you'll know beforehand what to expect based on your own previous observations.

Start by printing off a blank list. Fill out what you know of it beforehand. Keep it somewhat generic - don't list individual items, but general groupings (tools, medical bits, food, toiletries…) Then write up a brief-but-descriptive note on each. See the example for details on how best to do it. Next, take the list to the store in question with you. Take your time and walk down the aisles a bit, with an eye on what products would best help your neighborhood long-term in a post-collapse situation. Jot down notes on groups or clusters of items that would be especially useful.

Take a hard look at these items and note their quality (or lack thereof), quantity (also package sizes) and durability if that's applicable.

Buy a few things while you are there - especially small/sample sizes and individual items you think will be useful, so you can test them out later. As you walk around, discreetly look for narrow areas and 'choke points', where mobs can make things tough to get through. While you're at it, look for back doors and other potential alternate exits. Look at the checkout counters and how the layout would (or would not) handle massive, semi-panicked crowds (and mentally look for ways to avoid or get around that). Make notes from all of this. If you don't feel comfortable writing this stuff down on the fly, then make mental notes of as much as you can, then write it down immediately after you leave the store.

Over time, products will be moved around, many will be discontinued (which is why you don't list specific ones) and often the store will get remodeled. Check in once in awhile and update your lists accordingly.

Example: The Local Mega-Drugstore

By way of example, let's analyze the local mega-chain drugstore. It's actually an excellent candidate, since in a collapse, most folks would ignore them (unless it's the only close-by source of anything in a dense urban area, but…)

Mind you, in this particular example, the local mega-drugstore has big caveats - for example: you avoid the pharmacy portion because that's where the local criminal element will run for. Why? Because that's where all the painkillers and other high-octane drugs live. On the other hand, that same store will have a lot of things away from the pharmacy that you will definitely want or need - for example: all that food, camping gear, clothing , blankets, tools, and non-drug medical items that often lurk in many large drugstores.

As you can see on the next page, everything is in quickly-readable form so that your neighbors can get in, get what they need and get out. Be certain to stress that the best way everyone will remain alive long-term is to focus on these items only, in order and to pay attention to all notes. Your best bet is to get a notebook and use each page in a format that is similar to the one on the next page…

Store Name: MegaDrug, Inc

Priority: Moderate Priority. Get the MegaMart store next door if you can, but otherwise go here.

Location: 964 Magnolia Road, Titanicville. (Map on the next page)

Top Resources: Medical, Medicine, Supplements, Food, Clothing, Household Stuff, Toiletries.

Item Type	Aisle # / Location	Notes
Medical/Medicine	Aisle 1,2 and 6 - right-side of store	Avoid The Pharmacy! You don't want to be stuck in that crowd. What you want are the bandages, ointments, supplements, pain relief pills and tubes, gloves and any larger stuff (crutches, etc).
Food	Aisles 12-13, 15	Avoid the snacks and empty sugary stuff and focus on more solid staples - especially dry goods.
Supplements	Aisles 3-5	Definitely worth going after, but warning, it'll be close to the pharmacy so watch the crowds
Household Goods	Aisle 10, 11	Cheaply made, but better than nothing. Try for kitchen utensils, plastic kitchen bowls and containers and things like that. Also zipper bags!
Clothing	Aisle 4	Skip the flouncy crap and go for sturdier clothing. Stock up on socks, belts, sturdy shoes, t-shirts and all coats/jackets
Toiletries	Aisle 16-20 - left-hand side of store	Wide variety here, so don't get bogged down on details. Large soap bar packages, toothpaste, tooth brushes, large shampoo bottles, stuff like that.
Household Chemicals	Back of store	Biggest packages of bleach laundry soap, dish soap, steel wool, pine cleaner, ammonia, cleaners
Camping Gear	Aisle 14 towards back	Cheap but very useful. Focus on water filters, sleeping bags, small tents, canopies and all fuels and matches you find.
Wheelchair	Aisle 4 has a demo model that looks usable	Mrs. Johnson will need a spare when her scooter batteries are gone. Plastic strap can be cut with pocketknife. Idea: Use as shopping cart.
Tools	Aisle 14 towards front	Not worth the time unless there is nothing else to go for, or you are desperately short. Small assortment of tools, but they will come apart quickly under hard use, so avoid them if you can.
Patio Stuff	In front of checkout	Not really worth it. Skip
Matches	Aisle 10 left side towards back	Get all of them!
Lighters	Checkout counter	Get all of them if you can (watch the crowd)

Notes on checkout and front exits:

Normal checkout lines, but open enough to go around if we need to.

Alternate exits:

Exit in the back, but send someone out to have the driver go around.

Appendix C | Your Library

Why This is Important

Unless you are a part of that precious few who are complete savants, your ability to remember everything and pass it along to your children is going to be pretty crappy. You cannot count on the local town library to have everything your town needs to carry forth post-collapse and they certainly not enough copies to go around. Schools aren't going to be much better in this regard. To top all that off, neither school or local library has the funding to seek out and collect specific books that you may find handy in a world without civilization.

When all else fails, it's up to you to keep knowledge and culture alive. Even if you're not intellectual and even if you're not considered a smart person, the books will store what you cannot. Even if you had to evacuate your home long ago and the books got left behind, someone else may stumble across them and find them useful (even if that someone is born years after you died of old age).

Some of these books are going to be priceless to you. Some are around just for entertainment. Some will contain tomes that will lift your spirit, while others will make you think. Either way, these are the books you need to collect at a minimum. Feel more than free to add to them as well, because there's no such thing as having too much information.

Care And Feeding

This part is pretty simple and requires little work. Keep the room they're stored in clean and dry. Do not allow the air to get too moist in there. Handle these books only with clean and dry hands and have a clean table set aside to place the book on as you read it. Do not write in them unless what you are writing is both relevant to the text and useful to future generations. If you can keep the shelves out of direct sunlight, then do so. Treat them as gently as you can - they will have to last at least your lifetime and perhaps a few lifetimes beyond before the printing press makes a sufficient comeback.

Finally, never pass up an opportunity to add to the collection in your shopping and scroungings - even post-collapse.

Where and how to get them pre-collapse

This part is easy - bookstores, online, thrift shops, garage and rummage sales, you-name-it. Don't be afraid to buy used, but shoot for new when you can, especially for books that have been in print for a long time. If there are multiple editions, find both the latest edition and the first edition, so you have something for comparison.

Where and how to get them post-collapse

This part is not-so-easy, but still doable. Books can be scavenged from abandoned libraries first and foremost: public city or town, school, university/college and even hospitals. Abandoned bookstores are excellent sources, though you will have to wade through all the irrelevant stuff to get at what you want or need. Used/rare bookstores are quite excellent, as there are better odds of finding books on subjects that are more relevant to post-collapse living.

I would finally suggest poking through abandoned homes - sometimes you would be surprised what may be lurking on shelves, or in a box stashed somewhere in the odd attic, garage, or basement.

Books To Get

At a minimum, this is what you will want, sorted by type. Author's last name is in parentheses and check footnotes for regional recommendations

General Post-Collapse Skills

Back To Basics, 3rd Edition (Gehring) *	**The Trapper's Bible** (Martin)
Just In Case (Harrison)	**Fishing Basics: Complete Ill. Guide** (Kugach)
The Ultimate Guide To Homesteading (Faires)	*Edible plant guides specific to your region* **
The Survival Chemist (Howard, Flores)	**The Open Hearth Cookbook** (Goldenson, Simpson)
Simply Primitive: Rug Hooking... (Cross)	**The Back-Country Kitchen** (Marrone)
The Book of Looms (Broudy)	**Hand Loom Weaving: A Manual...** (Todd)
Rustic Retreats: A Build-It... (Stiles)	Beer Brewing at Home †
Big Book of Self-Reliant Living (Szykitka)	**Hunt Gather Grow Eat** (Akers)
Hunt, Gather, Cook (Shaw)	**Boy Scout Handbook** (Baden-Powell)††
How To Survive The End Of The World As We Know It (Rawles)	

* Formerly published by Reader's Digest (1981, 1997) and is now independent and on its 3rd edition. Any version will work.
** For my region, I recommend these two -
  Northwest Foraging: The Classic Guide to Edible Plants of the Pacific Northwest (Benoliel)
  Wild Harvest: Edible Plants of the Pacific Northwest (Domico)
† This will vary by availability of grains and other ingredients for brewing.
†† Be sure you get a pre-1980 edition, as the early ones contain a ton of useful information geared for younger audiences.

Medical And Medicine

Backcountry First-Aid and Extended Care (Tilton)	**Let's Talk About S-E-X [...] for Kids** (Gitchel)
Tactical Medicine Essentials (Wipfler, others)	**Dr. Spock's Baby And Child Care** (Spock)
Emergency War Surgery (NATO) *	*Medicinal Plant guides specific to your region* **
Heart and Hands: A Midwife's Guide (Harrison)	**Where There Is No Doctor** (Werner)
	Where There Is No Dentist (Dickson)

* Note that this one is made for medical pros, but you should have it nonetheless. Remember not to use it unless you have no other choice, as in life-or-death. Always seek a doctor or medical professional first.
** For my region, I recommend -
  A Field Guide to Western Medicinal Plants and Herbs (Hobbs)
  Edible and Medicinal Plants of the West (Tilford)

*   An introduction to Astronomy
**   Can be any competent encyclopedia written within the last 30 years. What we're looking for is a general catch-
-all of reference.

Philosophies And Classics*

The Prince (with commentary) (Machiavelli)	**The Art of War** (Tzu)
Complete Works Of Shakespeare (Shakespeare)	**Aesop's Fables**
Plutarch's Lives (Complete) (Plutarch)	**The Complete Sherlock Holmes** (Doyle)
1984 (Orwell)	**Animal Farm** (Orwell)
The Wealth Of Nations (Smith)	**The Histories** (Herodotus)
How To Win Friends And Influence… (Carnegie)	**Swiss Family Robinson** (Wyss)
Illiad and **Odyssey** (Homer)	**Catch-22** (Heller)
Walden (Thoreau)	**Atlas Shrugged** (Rand)
Lord of The Flies (Golding)	**The Book Of Deeds Of Arms…** (Pizan)
The Young Man's Guide (Alcott)	**Leviathan** (Hobbes)
The Republic (Plato)	**The Federalist Papers** (Jefferson et al)
Essential Manners For Men (Post)	**The Maltese Falcon** (Hammett)
War Of The Worlds (Wells)	**Autobiography of Benjamin Franklin**
The Great Railway Bazaar (Theroux)	**To Kill A Mockingbird** (Lee)
Self Reliance (Emerson)	**Brave New World** (Huxley)
South: The Endurance Expedition (Shackleton)	**The Journals Of Lewis And Clark** (Lewis, Clark)
The Divine Comedy (Aligheri)	**The Four Voyages** (Christopher Columbus)
The Travels Of Marco Polo	**A Man On The Moon** (Chaikin)
The Innocents Abroad (Twain)	**Anthem** (Rand)
The Moon Is A Harsh Mistress (Heinlein)	**Lolita** (Nabokov)
Hitchhiker's Guide To The Galaxy (Adams)	**Fahrenheit 451** (Bradbury)
Paradise Lost (Milton)	**Starship Troopers** (Heinlein)
Foundation (Series) (Asimov)	

* Please note that most of this particular list is subjective and reflects my own personal tastes.

Government And Civics

Robert's Rules Of Order (Robert)	**Democracy In America** (de Tocqueville)
Litigation And Trial Practice (Hart, Blanchard)	**Campaign Rules: A 50-state Guide** (Kasniunas)
Hands-On Elections (Tobi)	**Property Law: Rules…** (Singer)
The Road To Serfdom (Hayek)	**On Common Laws** (Glenn)
Studies In Civics (McCleary)	

Religious Texts

Generic Christian

Holy Bible (King James Version)   **Ultimate Bible Study Suite** (James, et al)

Catholic Christian

Holy Bible, Rev. Std. Version (Catholic Edition)   **The Roman Missal, 3rd Chapel Ed.** (USCCB) *

US Catholic Catechism For Adults (USCCB)   **Catholic Book Of Prayers** (Fitzgerald)

Jewish

JPS Hebrew-English Tanakh (JPS)**   **Siddur** (your preference)

Schottenstein Edition of The Talmud (Cohen)   (Optional) **The Classic Midrash** (Hammer)

Islam †

Qur'an (English and Arabic) (Haleem)   **Tawrat (English and Arabic)**
(Optional) **Injil (English and Arabic)**   (Optional) **Zabur (English and Arabic)**

LDS ("Mormon")

Book Of Mormon   **Doctrine And Covenants**
Pearl of Great Price   **King James Bible**

(<u>For Other Religions</u>: Add the appropriate books as your particular religion demands and/or requires.)

---

*   Should include entire 3-year cycle. Note that this is going to be a very large book (1,500 pages)
** Contains the Torah, Nevi'im and Ketuvim in one volume; Hebrew and English printed side-by-side
†   Note that contrary to popular belief, you simply cannot substitute the Torah for Tawrat, New Testament for Injil, or Book of Psalms for Zabur, as doing so will be rather offensive due to subtle differences and beliefs in the veracity/divinity of text and sourcing for each.

Useful Online Resources

These are sites and resources that you really want to scour through a bit and print off what you find there, as you will find them quite useful, thought-provoking and enough to inspire you to explore even more… Especially seek out the huge pile of useful tips and tricks to help you better prepare well beyond the scope of this book:

http://shtfblog.com
http://beprepared.com/article.asp_Q_ai_E_270
http://skepticalsurvivalist.tumblr.com/
http://shtfplan.com
http://www.backdoorsurvival.com/
http://www.survivalweek.com/
http://www.thesurvivalistblog.net/
http://thesurvivalmom.com/
http://www.doomandbloom.net/
http://apartmentprepper.com/
http://www.survivalblog.com/
http://saltnprepper.com/

PS: I'm not going to take responsibility for what folks post on said websites… that's up to the site owners and operators, but you can generally glean a ton of good information from them.

About The Author

I am an odd questioner. I admit it.

However, this is largely due to an intensive education by both parents from a very early age and an exposure to a stupendous variety of cultures and environments - both low and high. Oh and then there was that rigorous parochial school system with a particular instructor that (blessedly) required that I inquire, question, deconstruct and demand answers and seek them out myself if the initial answers didn't satisfy me. Blended in is a crazy-quilt of vocations ranging from ironworker to professor to an engineer on a quiet (and rather overly-guarded) US Air Force test range - just to pick from a few. The product is a man with a mind that is inquisitive, a childlike sense of wonder, but a hard skeptic's eye.

I am the progeny of a strong-minded psychologist with a sense of logic and order and an engineer who is given to dreams and wanderlust. I have, as a result, wound up finding a home and a spouse located far, far from my place of birth. My hometown? I'll let you know when I find it, but my current location is looking really good.

I am a semi-devout Catholic man (courtesy of my wife, who pulled me out of 'lapsed' status quite a few years ago) who has learned the hard way to find value and love in ordinary people, no matter who they are, or what they know, have, or worship (or, well, sometimes don't). That said, I am also the penultimate skeptic when it comes to, well, anything not involving my personal relationship with God.

Long story short? I'm just a regular guy who has lived a highly irregular life. I fully expect it to get even more irregular as time goes by as well - but then, a regular, ordinary life would be rather boring, now wouldn't it?